Methods in Sustainability Science
Assessment, Prioritization, Improvement, Design and Optimization

Methods in Sustainability Science
Assessment, Prioritization, Improvement, Design and Optimization

Edited by

Jingzheng Ren

Department of Industrial and Systems Engineering, The Hong Kong Polytechnic University, Hong Kong, China

Elsevier
Radarweg 29, PO Box 211, 1000 AE Amsterdam, Netherlands
The Boulevard, Langford Lane, Kidlington, Oxford OX5 1GB, United Kingdom
50 Hampshire Street, 5th Floor, Cambridge, MA 02139, United States

Copyright © 2021 Elsevier Inc. All rights reserved.

No part of this publication may be reproduced or transmitted in any form or by any means, electronic or mechanical, including photocopying, recording, or any information storage and retrieval system, without permission in writing from the publisher. Details on how to seek permission, further information about the Publisher's permissions policies and our arrangements with organizations such as the Copyright Clearance Center and the Copyright Licensing Agency, can be found at our website: www.elsevier.com/permissions.

This book and the individual contributions contained in it are protected under copyright by the Publisher (other than as may be noted herein).

Notices

Knowledge and best practice in this field are constantly changing. As new research and experience broaden our understanding, changes in research methods, professional practices, or medical treatment may become necessary.

Practitioners and researchers must always rely on their own experience and knowledge in evaluating and using any information, methods, compounds, or experiments described herein. In using such information or methods they should be mindful of their own safety and the safety of others, including parties for whom they have a professional responsibility.

To the fullest extent of the law, neither the Publisher nor the authors, contributors, or editors, assume any liability for any injury and/or damage to persons or property as a matter of products liability, negligence or otherwise, or from any use or operation of any methods, products, instructions, or ideas contained in the material herein.

British Library Cataloguing-in-Publication Data
A catalogue record for this book is available from the British Library

Library of Congress Cataloging-in-Publication Data
A catalog record for this book is available from the Library of Congress

ISBN: 978-0-12-823987-2

For Information on all Elsevier publications visit our website at
https://www.elsevier.com/books-and-journals

Publisher: Joe Hayton
Acquisitions Editor: Marisa LaFleur
Editorial Project Manager: Aleksandra Packowska
Production Project Manager: Joy Christel Neumarin Honest Thangiah
Cover Designer: Greg Harris

Typeset by Aptara, New Delhi, India

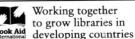

Contents

Contributors xv

CHAPTER 1 Methods in sustainability science
Ao Yang, Ruojue Lin, Tao Shi, Huijuan Xiao, Weifeng Shen, Jingzheng Ren
- **1.1** Introduction 1
- **1.2** Sustainability assessment and analysis 2
  - 1.2.1 Sustainability metrics/indicators 2
  - 1.2.2 Sustainability analysis tools 3
  - 1.2.3 Material flow analysis 4
- **1.3** Sustainability ranking and prioritization 4
- **1.4** Sustainability enhancement and improvement 6
- **1.5** Sustainability design and optimization 7
- **1.6** Conclusion 8
- Acknowledgments 8
- References 8

CHAPTER 2 Business contributions to sustainable development goals 13
Juniati Gunawan
- **2.1** Introduction 13
- **2.2** Literature review 14
  - 2.2.1 Sustainable development goals (SDGs) 14
  - 2.2.2 Sustainability reports 15
- **2.3** Materials and methods 16
- **2.4** Discussion 17
  - 2.4.1 SDGs disclosures based on industrial sector 17
  - 2.4.2 SDGs disclosures based on goals 18
- **2.5** Conclusion 23
- References 24

CHAPTER 3 Sustainability assessment: Metrics and methods 27
Himanshu Nautiyal, Varun Goel
- **3.1** Introduction 27
- **3.2** Need of sustainability assessment 29
  - 3.2.1 Steady-state economy 30
  - 3.2.2 Circular economy 31
  - 3.2.3 Ecological footprints 31

- **3.3** Various methods of sustainability assessment ... 32
  - 3.3.1 Life-cycle assessment ... 32
  - 3.3.2 Socioeconomic impact assessment ... 34
  - 3.3.3 Strategic environmental assessment ... 35
  - 3.3.4 Cost-benefit analysis ... 36
  - 3.3.5 Travel cost analysis ... 37
  - 3.3.6 Social impact assessment ... 37
  - 3.3.7 Contingent valuation method ... 37
  - 3.3.8 Hedonic pricing method ... 38
  - 3.3.9 Multicriteria analysis ... 38
  - 3.3.10 Material intensity per service unit ... 39
  - 3.3.11 Analytic network process ... 39
  - 3.3.12 Environmental and sustainability rating systems ... 39
- **3.4** Comparison of sustainability assessment methods ... 39
- **3.5** Conclusion ... 41
- References ... 43

CHAPTER 4 Sustainability assessment of energy systems: Indicators, methods, and applications ... 47
Imran Khan
- **4.1** Introduction ... 47
  - 4.1.1 Principle of sustainability ... 48
  - 4.1.2 Energy system and sustainability ... 48
- **4.2** Sustainability indicators ... 50
- **4.3** Sustainability assessment methods ... 53
  - 4.3.1 Multiattribute Value Theory (MAVT) ... 55
  - 4.3.2 Weighted sum method (WSM) ... 55
  - 4.3.3 Analytic hierarchy process (AHP) ... 56
  - 4.3.4 Weighted product method (WPM) ... 56
  - 4.3.5 Technique for Order Preference by Similarity to Ideal Solution (TOPSIS) ... 57
  - 4.3.6 Preference Ranking Organization METHod for Enrichment of Evaluations (PROMETHEE) ... 59
  - 4.3.7 ELimination Et Coix Traduisant la REalite (ELECTRE) ... 59
  - 4.3.8 VlseKriterijumska Optimizacija I Kompromisno Resenje (VIKOR) ... 60
  - 4.3.9 COmplex PRoportional ASsessment (COPRAS) ... 61
  - 4.3.10 Other methods ... 62
- **4.4** Sustainability assessment: an application of COPRAS method ... 62
  - 4.4.1 Results and discussion ... 64
- **4.5** Conclusion ... 67
- References ... 67

CHAPTER 5 Sustainability measurement: Evolution and methods 71
Mariolina Longo, Matteo Mura, Chiara Vagnini, Sara Zanni

- 5.1 Why measuring sustainability matters in the current business landscape 71
- 5.2 The evolution of sustainability measurement research 72
  - 5.2.1 Literature intellectual structure 72
  - 5.2.2 Sustainability measurement: a broken compass 74
  - 5.2.3 Contribution to performance measurement and the management literature 76
- 5.3 Methods and tools: the path toward sustainability measurement 77
  - 5.3.1 Sustainability core issues and stakeholder mapping 77
  - 5.3.2 Sustainability performance measurement system 78
  - 5.3.3 Sustainability reporting 81
- 5.4 The future of sustainability measurement 82
- References 83

CHAPTER 6 Industrial sustainability performance measurement system—challenges for the development 87
Alessandra Neri

- 6.1 Industrial sustainability 87
- 6.2 Industrial sustainability performance measurement 87
  - 6.2.1 Why do firms measure industrial sustainability–related performance? 88
  - 6.2.2 How do firms measure industrial sustainability–related performance? 88
  - 6.2.3 Focus of the present chapter 88
- 6.3 Industrial sustainability PMS—toward an effective development 89
  - 6.3.1 Usefulness to internal and external stakeholders 90
  - 6.3.2 Completeness and balance according to a holistic perspective on industrial sustainability 90
  - 6.3.3 Usability and manageability 91
  - 6.3.4 Selection of indicators 92
  - 6.3.5 Context of application 92
- 6.4 A scalable framework for measuring industrial sustainability performance 93
- 6.5 Concluding remarks and future perspectives 97
- References 99

CHAPTER 7 Life cycle assessment: methods, limitations, and illustrations 105
Sara Toniolo, Lorenzo Borsoi, Daniela Camana

- 7.1 Introduction to the life cycle assessment (LCA) methodology 105
  - 7.1.1 First phase 107
  - 7.1.2 Second phase 108
  - 7.1.3 Third phase 109
  - 7.1.4 Fourth phase 110
- 7.2 International standards 111

	7.3	Applications	113
	7.4	Limitations	115
		References	116

CHAPTER 8 Life cycle assessment for better sustainability: methodological framework and application 119
Aman Kumar, Ekta Singh, Rahul Mishra, Sunil Kumar

- 8.1 Introduction 119
- 8.2 LCA methodology 120
- 8.3 Important aspects of LCA methodology 121
  - 8.3.1 Goal setting and functional unit 121
  - 8.3.2 Assigning environmental burdens 121
  - 8.3.3 Credit for avoided burden 121
  - 8.3.4 Consequential LCA 122
  - 8.3.5 Inventory data availability and transparency 122
  - 8.3.6 Identifying data uncertainty 122
  - 8.3.7 Distinguishing risk assessment 123
  - 8.3.8 Reporting quantitative and qualitative information 123
  - 8.3.9 LCA does not always state a "winner" 123
  - 8.3.10 LCA is an iterative process 124
- 8.4 Sustainability approach 124
- 8.5 Application of LCA 125
  - 8.5.1 Sustainable cities 125
  - 8.5.2 Municipal solid waste management 126
  - 8.5.3 Wastewater treatment 127
  - 8.5.4 Solar power 128
  - 8.5.5 Agricultural strategic development planning 129
  - 8.5.6 Biofuels 130
- 8.6 LCA limitations and their probable solutions 131
- 8.7 Conclusion 132
- References 132

CHAPTER 9 Life cycle sustainability dashboard and communication strategies of scientific data for sustainable development 135
Daniela Camana, Alessandro Manzardo, Andrea Fedele, Sara Toniolo

- 9.1 Introduction 135
- 9.2 Ethical definition of sustainable development and communication strategies 136
- 9.3 Life Cycle Sustainability 138
  - 9.3.1 Data report and illustration of results 138
- 9.4 The dashboard of sustainability, a tool for sharing results 141

9.5 The life cycle sustainability dashboard...143
9.6 Other sustainability tools and communication strategies 145
9.7 Conclusions..146
References...147

CHAPTER 10 Multicriteria decision-making methods for results interpretation of life cycle assessment 153
Ana Carolina Maia Angelo

10.1 Introduction..153
10.2 An overview of the multicriteria approach...154
    10.2.1 The MCDM basic process...154
    10.2.2 MCDM methods classification..155
    10.2.3 A brief description of the main MCDM methods 156
10.3 LCA and multicriteria methods integration..157
    10.3.1 Selection of MSWM option..159
    10.3.2 Selection of sewer pipe materials...161
    10.3.3 Selection of poultry production systems..161
    10.3.4 Urban transport systems comparison 163
10.4 Discussion 165
10.5 Concluding remarks 165
References...165

CHAPTER 11 Composite sustainability indices (CSI); a robust tool for the sustainability measurement of chemical processes from "early design" to "production" stages 169
Mohammad Hossein Ordouei

11.1 Introduction..169
11.2 The CSI methodology and applications 173
    11.2.1 WAste Reduction algorithm and potential environment impact balance ...174
    11.2.2 Risk assessment index..175
    11.2.3 Energy impact index...181
11.3 Discussion 185
11.4 Conclusions...192
Reference 192

CHAPTER 12 Sustainability assessment using the ELECTRE TRI multicriteria sorting method 197
Luis C. Dias

12.1 Introduction..197
12.2 ELECTRE TRI in the MCDA panorama 198

12.3 ELECTRE TRI in detail 199
    12.3.1 Origins and purpose 199
    12.3.2 Classification rules 200
    12.3.3 Valued outranking relations 201
    12.3.4 Other variants 203
12.4 An illustrative example 203
12.5 Setting the parameter values 207
12.6 Conclusion 211
Acknowledgments 212
References 213

CHAPTER 13 Sustainability improvement opportunities for an industrial complex 215

Rahul Singh Yadav, Dilawar Husain, Ravi Prakash

13.1 Introduction 215
13.2 Methodology 216
    13.2.1 Design of the various systems and utilities 216
    13.2.2 Ecological footprint (EF) 217
13.3 Case study 218
    13.3.1 Survey of MNNIT Industrial Complex 218
    13.3.2 Data collection 218
13.4 Results and discussion 218
    13.4.1 Rooftop solar PV system 218
    13.4.2 Rainwater harvesting system 220
    13.4.3 Solar day-lighting System 220
    13.4.4 Turbo ventilators 222
    13.4.5 Chilled water air conditioning system 222
13.5 Scope of future work 223
13.6 Conclusions 225
Short Biography of the Authors 225
References 226

CHAPTER 14 Coupled life cycle assessment and data envelopment analysis to optimize energy consumption and mitigate environmental impacts in agricultural production 227

Ashkan Nabavi-Pelesaraei, Zahra Saber, Fatemeh Mostashari-Rad, Hassan Ghasemi-Mobtaker, Kwok-wing Chau

14.1 Introduction 227
14.2 Data collection 228
14.3 Energy in agriculture 232
    14.3.1 Energy analysis 232
    14.3.2 Energy indices and forms 234

14.4 Life cycle assessment...235
    14.4.1 Scope and goal definition...235
    14.4.2 Life cycle inventory...235
    14.4.3 Life cycle impact assessment..240
14.5 Data envelopment analysis..243
14.6 Integration of LCA and DEA..249
14.7 Result analysis...250
    14.7.1 Energy use pattern...250
    14.7.2 Environmental life cycle analysis..252
    14.7.3 Energy optimization by DEA...255
    14.7.4 Mitigation of environmental impacts by DEA + LCA...........................257
14.8 Conclusions...258
References..259

CHAPTER 15 Lean integrated management system for sustainability improvement: An integrated system of tools and metrics for sustainability management...265
João Paulo Estevam de Souza

15.1 Introduction...266
15.2 Literature overview and background..266
    15.2.1 Sustainability..266
    15.2.2 Management systems...267
    15.2.3 Lean manufacturing system...268
15.3 The Lean Integrated Management System for Sustainability Improvement (LIMSSI) model..272
    15.3.1 How to implement the LIMSSI model..273
15.4 Conclusions...279
References..289

CHAPTER 16 Coupled life cycle thinking and data envelopment analysis for quantitative sustainability improvement..............................295
Mario Martín-Gamboa, Diego Iribarren

16.1 Introduction...295
    16.1.1 Life cycle approaches..297
    16.1.2 Data envelopment analysis..298
16.2 Methodological framework..300
    16.2.1 Sustainability-oriented LCA + DEA approaches..................................300
    16.2.2 From LCA + DEA to LCSA + DEA...303
16.3 Progress in sustainability-oriented LCA + DEA..304
    16.3.1 Indicators and sustainability benchmarking..304
    16.3.2 Sustainability-oriented prioritization...308
    16.3.3 Other advancements..308

16.4 Delving into needs in sustainability-oriented LCA + DEA 313
16.5 Conclusions and perspectives..314
Acknowledgments..315
References...315

CHAPTER 17 How can sensors be used for sustainability improvement? 321
Patryk Kot, Khalid S. Hashim, Magomed Muradov, Rafid Al-Khaddar

17.1 Introduction...321
17.2 Sustainability in Civil Engineering 322
17.3 Working principle of sensing technologies for sustainability improvement..........325
  17.3.1 Acoustics sensing methods...326
  17.3.2 Magnetic sensing methods (Hall-effect sensor).................................327
  17.3.3 Electromagnetic sensing methods..328
17.4 Installation methods of sensing technologies in structural and environmental applications 332
17.5 Applications of sensing technology in civil and environmental engineering 333
  17.5.1 Civil engineering applications..333
  17.5.2 Environmental engineering applications..335
17.6 Chapter summary 337
References...338

CHAPTER 18 Sustainable design based on LCA and operations management methods: SWOT, PESTEL, and 7S 345
Manel Sansa, Ahmed Badreddine, Taieb Ben Romdhane

18.1 Introduction...345
18.2 Methodology 346
  18.2.1 SWOT analysis...347
  18.2.2 The 7S analysis 350
  18.2.3 The PESTEL analysis 351
  18.2.4 Integration of the SWOT, 7S, and PESTEL 353
18.3 Creation of strategic scenarios 353
  18.3.1 Developed algorithms 353
  18.3.2 Web application 354
18.4 Illustrative example 354
18.5 Results and discussion..360
18.6 Conclusion...361
Acknowledgments..362
References...362

CHAPTER 19 The importance of integrating lean thinking with digital solutions adoption for value-oriented high productivity of sustainable building delivery 365
Moshood Olawale Fadeyi

- 19.1 Introduction 365
- 19.2 Digital solutions and lean thinking adoption 368
  - 19.2.1 Digital solutions adoption 368
  - 19.2.2 Lean thinking adoption 369
- 19.3 Typical wastes in the management of the construction process ... 371
  - 19.3.1 Defective production 372
  - 19.3.2 Over-processing 372
  - 19.3.3 Waiting 373
  - 19.3.4 Unused talent 373
  - 19.3.5 Transportation 374
  - 19.3.6 Inventory 374
  - 19.3.7 Motion 375
  - 19.3.8 Excessive production 375
- 19.4 Case study 376
  - 19.4.1 Digital solutions for sustainable building delivery 377
  - 19.4.2 Lean thinking for sustainable building delivery 379
- 19.5 Challenges of adopting digital solutions grounded in lean thinking in the industry 381
  - 19.5.1 Challenges facing adoption of digital solutions 381
  - 19.5.2 Challenges facing adoption of lean thinking 382
- 19.6 Conclusion and future directions 383
- Biography 384
- References 385

CHAPTER 20 Robust optimization and control for sustainable processes 391
Alessandro Di Pretoro, Ludovic Montastruc, Xavier Joulia, Flavio Manenti

- 20.1 Robust design challenges in a renewables-based landscape 391
  - 20.1.1 The renewables challenge 391
  - 20.1.2 Recasting the petrochemical industry process design procedure 392
  - 20.1.3 Sustainability-oriented design 393
- 20.2 Feasibility assessment 394
  - 20.2.1 Feasibility limits in biomass processing 394
  - 20.2.2 The biorefinery distillation case study 395
  - 20.2.3 Feasibility assessment via residue curve maps 396

20.3 Flexible, sustainable, and economic optimal process design procedure 399
    20.3.1 Premises on the uncertainty characterization 399
    20.3.2 Introduction to the flexibility indices 399
    20.3.3 Application to the biorefinery case study 402
20.4 Process intensification 406
    20.4.1 General features 406
    20.4.2 Applications to distillation 407
20.5 Process dynamics and control 411
    20.5.1 The concept of switchability 411
    20.5.2 Application of the switchability assessment 411
20.6 Conclusions 413
Greek letters 417
List of acronyms and symbols 417
References 417

Index 421

Contributors

Rafid Al-Khaddar
Built Environment and Sustainable Technologies (BEST), Research Institute, Liverpool John Moores University, United Kingdom

Ana Carolina Maia Angelo
Fluminense Federal University, Volta Redonda, RJ, Brazil

Ahmed Badreddine
LARODEC, Institut Supérieur de Gestion de Tunis, Avenue de la liberté, LeBardo, Tunisie

Lorenzo Borsoi
Department of Industrial Engineering, University of Padova, Padova, Italy

Daniela Camana
Department of Industrial Engineering, University of Padova, Padova, Italy

Kwok-wing Chau
Department of Civil and Environmental Engineering, Hong Kong Polytechnic University, Hung Hom, Kowloon, Hong Kong

João Paulo Estevam de Souza
National Institute of Space Research / Instituto Nacional de Pesquisas Espaciais (INPE), São José dos Campos, SP, Brazil

Luis C. Dias
Univ Coimbra, CeBER, Faculty of Economics, Coimbra, Portugal

Moshood Olawale Fadeyi
Sustainable Infrastructure Engineering (Building Services) Programme, Engineering Cluster, Singapore Institute of Technology, Singapore, Singapore

Andrea Fedele
Department of Industrial Engineering, University of Padova, Padova, Italy

Hassan Ghasemi-Mobtaker
Department of Agricultural Machinery Engineering, Faculty of Agricultural Engineering and Technology, University of Tehran, Karaj, Iran

Varun Goelb
Department of Mechanical Engineering, National Institute of Technology Hamirpur, Himachal Pradesh, India

Juniati Gunawan
Universitas Trisakti, Jakarta, Indonesia

Khalid S. Hashim
Built Environment and Sustainable Technologies (BEST), Research Institute, Liverpool John Moores University, United Kingdom; Faculty of Engineering, University of Babylon, Hilla, Iraq

Dilawar Husain
Department of Mechanical Engineering, Maulana Mukhtar Ahmad Nadvi Technical Campus, Malegaon Nashik, Maharashtra, India

Diego Iribarren
Systems Analysis Unit, IMDEA Energy, Móstoles, Spain

Xavier Joulia
Laboratoire de Génie Chimique, Université de Toulouse, CNRS/INP/UPS, Toulouse, France

Imran Khan
Department of Electrical and Electronic Engineering, Jashore University of Science and Technology, Jashore, Bangladesh

Patryk Kot
Built Environment and Sustainable Technologies (BEST), Research Institute, Liverpool John Moores University, United Kingdom

Aman Kumar
CSIR-National Environmental Engineering Research Institute, Nagpur, Maharashtra, India

Sunil Kumar
CSIR-National Environmental Engineering Research Institute, Nagpur, Maharashtra, India

Ruojue Lin
Department of Industrial and Systems Engineering, The Hong Kong Polytechnic University, Hong Kong, China

Mariolina Longo
Department of Management, University of Bologna, Via Terracini, Bologna, Italy

Flavio Manenti
Politecnico di Milano, Dipartimento di Chimica, Materiali e Ingegneria Chimica "Giulio Natta, Milano, Italy

Alessandro Manzardo
Department of Industrial Engineering, University of Padova, Padova, Italy

Mario Martín-Gamboa
Chemical and Environmental Engineering Group, Rey Juan Carlos University, Móstoles, Spain

Rahul Mishra
CSIR-National Environmental Engineering Research Institute, Nagpur, Maharashtra, India

Ludovic Montastruc
Laboratoire de Génie Chimique, Université de Toulouse, CNRS/INP/UPS, Toulouse, France

Fatemeh Mostashari-Rad
Department of Agricultural Biotechnology, Faculty of Agricultural Sciences, University of Guilan, Rasht, Iran

Magomed Muradov
Built Environment and Sustainable Technologies (BEST), Research Institute, Liverpool John Moores University, United Kingdom

Matteo Mura
Department of Management, University of Bologna, Via Terracini, Bologna, Italy

Ashkan Nabavi-Pelesaraei
Department of Mechanical Engineering of Biosystems, Faculty of Agriculture, Razi University, Kermanshah, Iran

Himanshu Nautiyal
Department of Mechanical Engineering, THDC Institute of Hydropower Engineering and Technology, Tehri, Uttarakhand, India

Alessandra Neri
Department of Management, Economics and Industrial Engineering, Politecnico di Milano, Milan, Italy

Mohammad Hossein Ordouei
Energy Research Center, University of Waterloo, Ontario, Canada

Ravi Prakash
Department of Mechanical Engineering, Motilal Nehru National Institute of Technology, Allahabad, Uttar Pradesh, India

Alessandro Di Pretoro
Politecnico di Milano, Dipartimento di Chimica, Materiali e Ingegneria Chimica "Giulio Natta, Milano, Italy; Laboratoire de Génie Chimique, Université de Toulouse, CNRS/INP/UPS, Toulouse, France

Jingzheng Ren
Department of Industrial and Systems Engineering, The Hong Kong Polytechnic University, Hong Kong, China

Taieb Ben Romdhane
LISI, Institut National des sciences Appliquées et de Technologie, Université de Carthage, Centre Urbain Nord BP, Tunisie

Zahra Saber
Department of Agronomy and Plant Breeding, Sari Agricultural Sciences and Natural Resources University, Sari, Iran

Manel Sansa
LISI, Institut National des sciences Appliquées et de Technologie, Université de Carthage, Centre Urbain Nord BP, Tunisie

Weifeng Shen
School of Chemistry and Chemical Engineering, Chongqing University, Chongqing, PR China

Tao Shi
Department of Industrial and Systems Engineering, The Hong Kong Polytechnic University, Hong Kong, China

Ekta Singh
CSIR-National Environmental Engineering Research Institute, Nagpur, Maharashtra, India

Sara Toniolo
Department of Industrial Engineering, University of Padova, Padova, Italy

Chiara Vagnini
Department of Management, University of Bologna, Via Terracini, Bologna, Italy

Huijuan Xiao
Department of Industrial and Systems Engineering, The Hong Kong Polytechnic University, Hong Kong, China

Rahul Singh Yadav
Department of Mechanical Engineering, Motilal Nehru National Institute of Technology, Allahabad, Uttar Pradesh, India

Ao Yang
Department of Industrial and Systems Engineering, The Hong Kong Polytechnic University, Hong Kong, China; School of Chemistry and Chemical Engineering, Chongqing University, Chongqing, PR China

Sara Zanni
Department of Management, University of Bologna, Via, Terracini, Bologna, Italy

CHAPTER 1

Methods in sustainability science

Ao Yang[a,b], Ruojue Lin[a], Tao Shi[a], Huijuan Xiao[a], Weifeng Shen[b], Jingzheng Ren[a]

[a]*Department of Industrial and Systems Engineering, The Hong Kong Polytechnic University, Hong Kong, China*
[b]*School of Chemistry and Chemical Engineering, Chongqing University, Chongqing, PR China*

1.1 Introduction

Sustainability science, which aims to find a harmonious coexistence method of nature and society, is a new emerging field (Kates et al., 2001, Kates, 2011). The studies on sustainability or sustainable development usually focused on investigating multidimensional factors/criteria, especially in economic, environmental, and social dimensions (Ren, 2020). To achieve the sustainable development (i.e., meeting the needs of present and future generations while substantially reducing poverty and conserving the planet's life support systems), the interactions between natural and social systems should be explored (WCED 1987).

There are three authoritative sustainability indexes, including the American Institute of Chemical Engineers Sustainability Index (AIChE SI), Sustainable Development Goals (SDGs) and United States Environmental Protection Agency (USEPA), that could be used to evaluate the extent of the sustainable development. The AIChE SI reports seven key metrics to assess the sustainability performance of the chemical industry (Cobb and Tanzil, 2009). The SDGs involve 17 goals and 169 targets (World Health Organization 2015), which aim to integrate three dimensions (economic, social, and environmental) of sustainable development corresponding to people, planet, prosperity, peace, and partnership. USEPA commits to protect human health and the environment in economic, social, and environmental aspects. The above-mentioned evaluations could be completed via the triple bottom line of sustainability (i.e., triple P, including people, planet, and profits). The stakeholders/decision makers are encouraged to simultaneously consider the individual's profits, people's live, and the planet achieving sustainable development by using the triple-bottom-line framework.

Three fundamental aspects represented by economic prosperity, environmental cleanness, and social responsibility are always used to assess the sustainable development at the early-stage design. Additionally, some other dimensional indicators in technological, policy and political aspects, etc. should also be integrated to evaluate the sustainable development of alternative systems/processes, because these indicators can affect three fundamental aspects of sustainability. Take the technical aspect as an example, the intensified extractive distillation process with less exergy loss and energy consumption can achieve more net profits and less gas emissions (e.g., CO_2) (Yang et al., 2019).

To sum up, sustainable development is a multidimensional concept. Therefore, multiple factors/criteria are usually prerequisite to be incorporated in the decisions related to sustainability, namely, the so-called "multicriteria decision making (MCDM)." MCDM, also called "multicriteria decision analysis," refers to the evaluation of multiple conflicting criteria in decision making, and it usually involves multiple decision criteria and multiple alternatives. MCDM has become a powerful tool for sustainability issues, including (1) sustainability assessment and analysis, (2) sustainability ranking and prioritization, (3) sustainability enhancement and improvement, and (4) sustainability design and optimization. However, there are still various challenges in these four fields.

1.2 Sustainability assessment and analysis

Sustainability assessment and analysis are usually used to evaluate the economic, environmental, and social sustainability or people, planet, and profits of industrial systems/processes, as illustrated in Fig. 1.1 (Elkington, 1994, Elkington, 1998). Sustainability requires a tradeoff between environmental, economic, and social benefits, instead of considering merely one of them.

1.2.1 Sustainability metrics/indicators

At first, sustainability-related indicators have to be determined in comparing alternative process options to better understand the tradeoffs among multiple objectives. Sustainability metrics/indicators have received much attention because many individual and teams require to measure, track, and compare

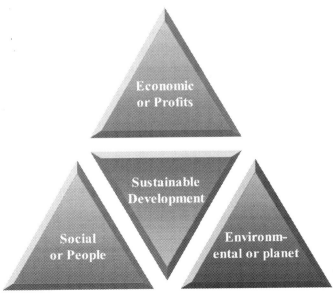

FIG. 1.1 Triple bottom line of sustainability, or "triple P": people, planet, and profits (Elkington, 1994, Elkington, 1998).

the efforts of sustainability indicators in the sustainable development. So far, there has been no uniform consensus on metrics and indices due to the complex definition of sustainable development. Sustainability metrics can be accessed from three authoritative sustainability databases such as AIChE SI (Cobb et al., 2007), USEPA (Martins et al., 2007), and SDGs (Pizzi et al., 2020).

The AIChE SI can help understand the sustainability contribution and evaluate the sustainability performance with seven key metrics (i.e., strategic commitment, sustainability innovation, environment performance, safety performance, product stewardship, social responsibility, and value-chain management) of proposed processes (Cobb and Tanzil, 2009). USEPA could provide available indicators such as poverty, population stability, human health, living conditions, coastal protection, agricultural conditions, ecosystem stability, atmospheric impacts, generation, consumption, economic growth, and accessibility for the evaluation of environment and human health (Fiksel et al., 2012). According to the World Health Organization, SDGs contain 17 goals and 169 targets (World Health Organization 2015). Not only to eliminate poverty and hunger is the goal of SDGs, but also to protect population health, to create inclusive economic growth, to preserve the planet, etc.

Many studies have assessed the current situation and progress of SDGs of a great many countries using various indicators. The choice of indicators is generally based on the specific situation of each country within the framework of the 2030 Agenda proposed by the United Nations (UN). For example, Schmidt-Traub et al. (2017) evaluated the baselines and future progress of 149 of 193 UN member states. Based on data quality and availability, 63 global indicators were chosen for non-OECD countries to assess the SDG baselines, while 77 indicators were selected for OECD countries (Schmidt-Traub et al., 2017). The progress of SDGs of the Arab region over 20 years was evaluated using 56 indicators (Allen et al., 2017). The study revealed that despite promising trends over the past two decades, the Arab region was still falling short of the global benchmarks. The SDGs level of Chinese provinces from 2000 to 2015 was investigated based on 119 indicators (Lu et al., 2019). The results showed that the sustainability of both China and each province improved during 2000–2015. The SDGs of Australia were evaluated based on a proposed scenario modeling approach using 97 indicators (Allen et al., 2019). The results suggested that Australia was off-track to achieve the SDGs by 2030, while significant progress can be possible by changing the development path of Australia (Allen et al., 2019). The UN Sustainable Development Solutions Network (SDSN), which was set up in 2012 under the auspices of the UN Secretary-General, has evaluated the SDGs of different regions, such as Africa (T.U.S.D.S.N. (SDSN) 2020), Europe (T.U.S.D.S.N. (SDSN) 2019), and Arab region (T.U.S.D.S.N. (SDSN) 2019), using appropriate indicators. These indicators, findings, and thinkings can be regarded as references for the future study of the indicator-based SDGs assessment at not only global, regional, national but subnational levels.

1.2.2 Sustainability analysis tools

After the determination of various key sustainability indicators from the above-mentioned database, the sustainable development could be further quantified and compared via the sustainability analysis tools. Sustainability is usually evaluated by life cycle sustainability assessment (LCSA). The LCSA consists of life cycle assessment (LCA), life cycle costing (LCC), and social life cycle assessment (SLCA) for evaluating environmental, economic, and social aspects of sustainability.

The LCA is a typical sustainability assessment for environmental impacts, which is well defined by the international standard ISO14044. In this international standard, the LCA contains four main steps, which are goal and scope definition, life cycle inventory analysis (LCIA), life cycle impact analysis, and interpretation. To improve the repeatability of the assessment and to unify the impact categories, several methods for inventory analysis and impact analysis were proposed and applied in LCA based on the general framework provided in ISO14044. The most commonly used methods include ReCiPe, EPS, CML, and LIME. Among them, some methods can be separated into two types: Endpoint and Midpoint methods. As for the Midpoint method, the indicators refer to the exact substances released to the environment and the results can be obtained after the inventory analysis. As for the Endpoint method, the indicators correspond to the environmental impacts, and the results are generated from the impact analysis. For example, the ReCiPe and LIME have both the Midpoint and Endpoint methods. To assist the life cycle inventory (LCI) and LCA assessment process, there are different softwares and toolboxes being developed including TRACI (tool for reduction and assessment of chemical and other environmental impacts), Boustead Model (computer model and database for LCI), CMLCA (chain management by LCA), SimaPro (LCA software), Spine (LCI database), Gabi4 (LCA software), etc. The use of softwares and toolboxes can help to accelerate the completion of LCA process and improve the efficiency of assessment.

LCC is the latest approach to economically evaluate long-term projects. It helps to evaluate economic performance of a project in the whole life cycle including purchase, installation, operating, maintenance, financing, depreciation, and disposal. The analysis results of economic performances contain life cycle cost, total net revenue, internal rate of return, etc. (Preuß and Schöne, 2016, Woodward, 1997). Some softwares of LCA can be used to solve LCC problem, for instance, SimaPro (Ciroth et al., 2009).

SLCA is an assessment evaluating social sustainability. It could be employed to evaluate social issues (i.e., social acceptability, work environment and impacts on local culture, etc.). The same assessment framework of LCA can be applied in SLCA (Jørgensen et al., 2008). The processes include goal and scope definition, inventory analysis, impact analysis, and interpretation. The difference between LCA and SLCA is that the impact categories of LCA correspond to environmental impacts, while the impact categories of SLCA are related to the social impacts.

1.2.3 Material flow analysis

Material flow analysis (MFA), as illustrated in Fig. 1.2, is a systematic approach to quantify flows and stocks of materials within an arbitrarily complex system defined in space and time, which is considered as a core method of industrial ecology or anthropogenic, urban, social, and industrial metabolism (Islam and Huda, 2019). To analyze the metabolism of social systems, (Sendra et al. 2007) applied the MFA to an industrial park and the companies located with it.

1.3 Sustainability ranking and prioritization

Economic, environmental, and social performances of different alternatives could be obtained via the sustainability assessment and analysis. However, it is still difficult for the decision makers/stakeholders to select the most sustainable alternative among multiple choices, because there are usually various conflict criteria, and one alterative performs better with respect to several criteria, but may perform worse with respect to some other criteria. For example, alternative A has fewer life cycles cost and B has higher social acceptability and less global warming potential, as shown in Fig. 1.3. "Which alternative is more or the most sustainable?" always confuses the decision makers/stakeholders.

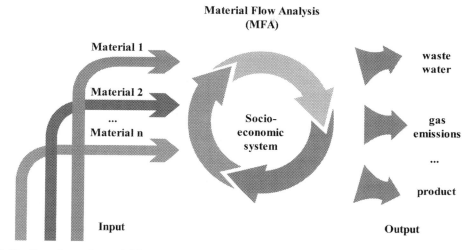

FIG. 1.2 The flow sheet of material flow analysis.

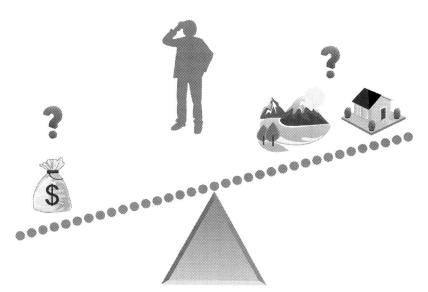

FIG. 1.3 The choice made by decision makers/stakeholders.

As for solving the optimal strategy selection in the specified design projects, developing effective multicriteria decision-making (MCDM) models can help to select the optimal processes and aggregate them as the optimal pathway. In this MCDM model, the LCA tools need to be included, while various criteria and the different roles each criterion plays in each decision should be considered (Balasbaneh and Marsono, 2020, Paramesh et al., 2018). For instance, a novel framework for the prioritization of bioethanol production pathways by combination of social LCA and MCDM is proposed by (Ren et al. 2015) and then the uncertainty condition is also considered to rank the alternative energy systems (Ren, 2018). A systematic method is explored based on the MCDM and LCA for the solid waste collection (Ulukan and Kop, 2009) and then the proposed approach is extended to the planning, designing, and commissioning of green buildings (Kiran and Rao, 2013).

One of the core problems to solve urgently is the determination of weights of different evaluation indicators. Considering the ubiquitous uncertainties and vagueness from the view of multistakeholders, the fuzzy theory combining best-worst method or analytic hierarchy process (AHP) was usually applied to solve such problems (Liu et al., 2020, Lin et al., 2020). Data envelopment analysis (DEA) (Martín-Gamboa et al., 2017) can provide a new pathway to obtain the sustainability efficiency of different alternatives. In this way, the indispensable subject factors introduced into weight calculation can be avoided. For example, *Laso et al.* (2018) explored the evaluation of energy and environmental efficiency for the Spanish agri-food system by using the DEA-LCA approach. The developed method could also be applied to improving the environmental impact efficiency in mussel cultivation (Lozano et al., 2010) and managing the fishing fleets (Laso et al., 2018). Vázquez-Rowe and Diego Iribarren (2015) reviewed the application of DEA-LCA method for energy policy making. In summary, the prioritization and ranking of the alternatives systems/processes could be obtained via the hybrid MCDM or DEA model based on life cycle approaches.

1.4 Sustainability enhancement and improvement

Various factors/criteria influencing the sustainability performances of alternatives can be applied to improve sustainability performance of the alternatives. It is important to note that these factors/criteria are usually interacted. Thereby, it is necessary to distinguish the complex cause-effect relationships for determining the critical "causes" leading to bad sustainability performances.

There are several conventional models (e.g., qualitative, semiqualitative, and semiquantitative methods) and emerging quantitative models (e.g., hierarchical control approach) that can be used to identify the above-mentioned critical "causes" and the interactions among factors/criteria of sustainability. The qualitative models involve fishbone diagram (Lin et al., 2019) and driving forces-pressures-state-impacts-responses (DPSIR) (Scriban et al., 2019) methods, which could be used in different research fields. For example, (Dharma et al. 2019) explored the reduction of nonconformance quality of yarn via the integration of Pareto principles and fishbone diagram and the results indicated that the reason of yarn inequality lies in the technical part (i.e., top roll surface unevenness). The evaluation and drive mechanism of the tourism ecological security has been investigated based on the DPSIR-DEA model (Ruan et al., 2019). The results illustrated that the greening degree and the economic status of tourism were the critical factors affecting state. Semiqualitative and semiquantitative methods mainly include decision-making trial and evaluation laboratory (DEMATEL) (Yang et al., 2008) and analytic network process (ANP) (Chen et al., 2018, Feng et al., 2018, Büyüközkan and Çifçi, 2012) as well as their corresponding improved methods (i.e., fuzzy DEMATEL (Zhang and Su, 2019), fuzzy ANP (Hatefi and Tamošaitienė, 2019), gray DEMATEL (Xia and Ruan, 2020), gray ANP (Rajesh, 2020), intuitionistic fuzzy DEMATEL (Abdullah et al., 2019), and intuitionistic fuzzy ANP (Büyüközkan et al., 2017), etc.). Recently, some emerging quantitative models are developed to improve sustainability performance of the alternatives, such as mathematical and decision support frameworks based on the two-layered hierarchical control scheme (Moradi-Aliabadi and Huang, 2018), multistage optimization (Moradi-Aliabadi and Huang, 2016), model predictive control (Moradi-Aliabadi and Huang, 2018), and integration of vector-based multiattribute decision-making and weighted multiobjective optimization (Xu et al., 2019).

However, there are still various challenges in life cycle sustainability enhancement and improvement: (1) how to incorporate the requirements (sustainability objectives) of the stakeholders in the model for sustainability enhancement and improvement? (2) how to develop a systematic model for making informed decisions on sustainability enhancement and improvement from life cycle sustainability perspective? and (3) how to develop the generic models for sustainability enhancement and improvement that can be used in different scales (process, plant, enterprise, municipal, provincial, and national scales)?

1.5 Sustainability design and optimization

Mathematical models, which could be used to design and optimize the multiscale industrial systems (e.g., unit operation, plant and region) at the early-design stage, are developed to achieve sustainable design and optimization. The above-mentioned processes have discrete and continuous decision variables that are expressed as mixed integer nonlinear programing (MINLP) problem.

To solve the MINLP problem of the chemical process, evolutionary optimization model and multiobjective optimization model are developed and employed. For example, (Zhang et al. 2020) used the self-adapting dynamic differential evolution algorithm to obtain the sustainable extractive distillation processes. (Su et al. 2020) developed a stakeholder-oriented multiobjective process optimization approach based on an improved genetic algorithm for two chemical processes (i.e., extractive distillation and methanol synthesis).

In addition, there are various models such as goal programming model, multiobjective evolutionary optimization model, robust multiobjective optimization model, and MINLP model developed for the sustainability-oriented design and optimization of supply chain in process industries (Govindan and Cheng, 2018). Economic, environmental, social objectives, and various uncertainties are considered in the above-mentioned optimization models to obtain more sustainable designs. For example, (Vivas et al. 2020) developed a hybrid model based on the goal programming model and AHP for evaluating the sustainability of an oil and gas supply chain. (Shankar et al. 2013) and (Azadeh et al. 2017) applied the multiobjective evolutionary approach for the location and allocation decisions of multiechelon supply chain network and environmental indicators of integrated crude oil supply chain under uncertainty, and the results illustrated that the large-scale and MINLP issues could be efficiently solved via the proposed optimization model. Robust multiobjective optimization model is employed to optimize the supply chain of the energy system (Majewski et al., 2017, Wang et al., 2017) and the computation shows that the sustainable/promising design/solution could be easily obtained via the applying the proposed approach. Additionally, the MINLP model could also be adopted to design food supply chain (Mogale et al., 2018) and biofuels supply (Wheeler et al., 2017), which could help the decision maker to obtain a sustainable solution.

In summary, the sustainability design and optimization could be efficiently solved via the different optimization models, while there are still various challenges: (1) these optimization models cannot incorporate as many sustainability objectives/criteria as that used in sustainability ranking and prioritization. Most of the models consider three or less than three objectives. How to incorporate a complete list of sustainability objectives/criteria in the optimization model? (2) how to handle the various uncertainties in the models for sustainability design and optimization? and (3) how to determine subject-oriented model that can incorporate the requirements of the decision makers/stakeholders for sustainability design and optimization.

1.6 Conclusion

To promote the sustainable development, it is necessary to introduce the methods in sustainability science involving sustainability assessment/analysis, sustainability ranking/prioritization, sustainability enhancement/improvement, and sustainability design/optimization. In addition, some sustainability oriented decision-making methods such as multidimensional, multistakeholder, multiobjective, multiscenario, multiscale, and multilevel methods should also be incorporated to obtain a better sustainable solution in a different industry.

Acknowledgments

This chapter was prepared based on the thinking and the structure presented in the editorial (Specialty Grand Challenge) written by Dr. Jingzheng Ren—Ren, J. (2020). Specialty Grand Challenge: Multi-Criteria Decision Making for Better Sustainability. Front. Sustain. 1: 2. (with the Copyright © 2020 Ren). This work was supported by Joint Supervision Scheme with the Chinese Mainland, Taiwan and Macao Universities—Other Chinese Mainland, Taiwan, and Macao Universities (Grant No. SB2S to A. Yang) and T. Shi would like to express their sincere thanks to the Research Committee of The Hong Kong Polytechnic University for the financial support of the project through a PhD studentship (project account code: RK3P).

References

Abdullah, L., Zulkifli, N., Liao, H., Herrera-Viedma, E., Al-Barakati, A., 2019. An interval-valued intuitionistic fuzzy DEMATEL method combined with Choquet integral for sustainable solid waste management. Eng. Appl. Artif. Intell. 82, 207–215.

Allen, C., Metternicht, G., Wiedmann, T., Pedercini, M., 2019. Greater gains for Australia by tackling all SDGs but the last steps will be the most challenging. Nat. Sustain. 2, 1041–1050.

Allen, C., Nejdawi, R., El-Baba, J., Hamati, K., Metternicht, G., Wiedmann, T., 2017. Indicator-based assessments of progress towards the sustainable development goals (SDGs): a case study from the Arab region. Sustain. Sci. 12, 975–989.

Azadeh, A., Shafiee, F., Yazdanparast, R., Heydari, J., Fathabad, A.M., 2017. Evolutionary multi-objective optimization of environmental indicators of integrated crude oil supply chain under uncertainty. J. Clean. Prod. 152, 295–311.

Balasbaneh, A.T., Marsono, A.K.B., 2020. Applying multi-criteria decision-making on alternatives for earth-retaining walls: LCA, LCC, and S-LCA. Int. J. Life Cycle Assess. 25, 2140–2153.

Büyüközkan, G., Çifçi, G., 2012. A novel hybrid MCDM approach based on fuzzy DEMATEL, fuzzy ANP and fuzzy TOPSIS to evaluate green suppliers. Exp. Syst. Appl. 39, 3000–3011.

Büyüközkan, G., Güleryüz, S., Karpak, B., 2017. A new combined IF-DEMATEL and IF-ANP approach for CRM partner evaluation. Int. J.Prod. Econ. 191, 194–206.

Chen, Y.-S., Chuang, H.-M., Sangaiah, A.K., Lin, C.-K., Huang, W.-B., 2018. A study for project risk management using an advanced MCDM-based DEMATEL-ANP approach. J. Ambient Intell. Humaniz. Comput. 10, 2669–2681.

Ciroth, A., Franze, J., Berlin, G., 2009. Life cycle costing in SimaPro. J. GreenDelta TC, August 1–10.

Cobb, C., Schuster, D., Beloff, B., Tanzi, D., 2007. Benchmarking sustainability. Chem. Eng. Prog. 106, 38–42.

Cobb, C.D., Tanzil, D., 2009. The AIChE sustainability index the factors in detail. Chem. Eng. Prog. 60–63.

Dharma, F.P., Ikatrinasari, Z.F., Purba, H.H., Ayu, W., 2019. Reducing non conformance quality of yarn using pareto principles and fishbone diagram in textile industry. IOP Conf. Ser. Mater. Sci. Eng. 508, 1–7.

Elkington, J., 1994. Towards the sustainable corporation: win-win-win business strategies for sustainable development. Calif. Manage. Rev. 36, 90–100.

Elkington, J., 1998. Partnerships from cannibals with forks: the triple bottom line of 21st-century business. Environ. Qual. Manage. 8, 37–51.

Feng, Y., Hong, Z., Tian, G., Li, Z., Tan, J., Hu, H., 2018. Environmentally friendly MCDM of reliability-based product optimisation combining DEMATEL-based ANP, interval uncertainty and VIse Kriterijumska Optimizacija Kompromisno Resenje (VIKOR). Inform. Sci. 442-443, 128–144.

Fiksel, J., Eason, T., Frederickson, H., 2012. A Framework for Sustainability Indicators at EPA.

Govindan, K., Cheng, T.C.E., 2018. Advances in stochastic programming and robust optimization for supply chain planning. Comput. Oper. Res. 100, 262–269.

Hatefi, S.M., Tamošaitienė, J., 2019. An integrated fuzzy DEMATEL-fuzzy anp model for evaluating construction projects by considering interrelationships among risk factors. J. Civil Eng. Manage. 25, 114–131.

Islam, M.T., Huda, N., 2019. Material flow analysis (MFA) as a strategic tool in E-waste management: applications, trends and future directions. J. Environ. Manage. 244, 344–361.

Jørgensen, A., Bocq, A.Le, Nazarkina, L., Hauschild, M., 2008. Methodologies for social life cycle assessment. Int. J. Life Cycle Assess. 13, 96–103.

Kates, R.W., 2011. What kind of a science is sustainability science? Proc. Natl. Acad. Sci. USA 108, 19449–19450.

Kates, R.W., Clark, W.C., Corell, R., Hall, J.M., Jaeger, C.C., Lowe, I., McCarthy, J.J., Schellnhuber, H.J., Bolin, B., Dickson, N.M., Faucheux, S., Gallopin, G.C., Grübler, A., Huntley, B., Jäger, J., Jodha, N.S., Kasperson, R.E., Mabogunje, A., Matson, P., Mooney, H., Moore III, B., O'Riordan, T., Svedin, U., 2001. Sustainability science. Science 292, 641–642.

Kiran, B.A., Rao, P.N., 2013. Life Cycle Assessment (LCA) and Multi-Criteria Decision Making (MCDM) for planning, designing and commissioning of green buildings. Int. J. Adv. Trends Comput. Sci. Eng. 2, 476–479.

Laso, J., Hoehn, D., Margallo, M., García-Herrero, I., Batlle-Bayer, L., Bala, A., Fullana-i-Palmer, P., Vázquez-Rowe, I., Irabien, A., Aldaco, R., 2018. Assessing energy and environmental efficiency of the Spanish agri-food system using the LCA/DEA methodology. Energies, 11, 3395.

Laso, J., Vázquez-Rowe, I., Margallo, M., Irabien, Á., Aldaco, R., 2018. Revisiting the LCA+DEA method in fishing fleets. How should we be measuring efficiency? Mar. Policy 91, 34–40.

Latha Shankar, B., Basavarajappa, S., Chen, J.C.H., Kadadevaramath, R.S., 2013. Location and allocation decisions for multi-echelon supply chain network—a multi-objective evolutionary approach. Expert Syst. Appl. 40, 551–562.

Lin, R., Liu, Y., Man, Y., Ren, J., 2019. Towards a sustainable distributed energy system in China: decision-making for strategies and policy implications. Energy Sustain. Soc. 9, 51.

Lin, R., Man, Y., Lee, C.K.M., Ji, P., Ren, J., 2020. Sustainability prioritization framework of biorefinery: a novel multicriteria decision-making model under uncertainty based on an improved interval goal programming method. J. Clean. Prod. 251, 119729.

Liu, Y., Ren, J., Man, Y., Lin, R., Lee, C.K.M., Ji, P., 2020. Prioritization of sludge-to-energy technologies under multi-data condition based on multi-criteria decision-making analysis. J. Clean. Prod. 273, 123082.

Lozano, S., Iribarren, D., Moreira, M.T., Feijoo, G., 2010. Environmental impact efficiency in mussel cultivation. Resour. Conserv. Recycl. 54, 1269–1277.

Lu, Y., Zhang, Y., Cao, X., Wang, C., Wang, Y., Zhang, M., Ferrier, R.C., Jenkins, A., Yuan, J., Bailey, M.J., Chen, D., Tian, H., Li, H., von Weizsäcker, E.U., Zhang, Z., 2019. Forty years of reform and opening up: China's progress toward a sustainable path. Sci. Adv. 5, eaau9413.

Majewski, D.E., Wirtz, M., Lampe, M., Bardow, A., 2017. Robust multi-objective optimization for sustainable design of distributed energy supply systems. Comput. Chem. Eng. 102, 26–39.

Martín-Gamboa, M., Iribarren, D., García-Gusano, D., Dufour, J., 2017. A review of life-cycle approaches coupled with data envelopment analysis within multi-criteria decision analysis for sustainability assessment of energy systems. J. Clean. Prod. 150, 164–174.

Martins, A.A., Mata, T.M., Costa, C.A.V., Sikdar, S.K., 2007. A framework for sustainability metrics. Ind. Eng. Chem. Res. 46, 5468.

Mogale, D.G., Kumar, S.K., Tiwari, M.K., 2018. An MINLP model to support the movement and storage decisions of the Indian food grain supply chain. Control Eng. Pract. 70, 98–113.

Moradi-Aliabadi, M., Huang, Y., 2016. Multistage optimization for chemical process sustainability enhancement under uncertainty. ACS Sustain. Chem. Eng. 4, 6133–6143.

Moradi-Aliabadi, M., Huang, Y., 2018. Manufacturing sustainability enhancement: a model predictive control based approach, in: Proc. 13th International Symposium on Process Systems Engineering (PSE 2018), 2059–2064.

Moradi-Aliabadi, M., Huang, Y., 2018. Decision support for enhancement of manufacturing sustainability: a hierarchical control approach. ACS Sustain. Chem. Eng. 6, 4809–4820.

Paramesh, V., Arunachalam, V., Nikkhah, A., Das, B., Ghnimi, S., 2018. Optimization of energy consumption and environmental impacts of arecanut production through coupled data envelopment analysis and life cycle assessment. J. Clean. Prod. 203, 674–684.

Pizzi, S., Caputo, A., Corvino, A., Venturelli, A., 2020. Management research and the UN sustainable development goals (SDGs): a bibliometric investigation and systematic review. J. Clean. Prod. 276.

Preuß, N., Schöne, L.B., 2016. Real Estate und Facility Management. In: Preuß, N., Schöne, L.B. (Eds.), Real Estate und Facility Management. Springer, Berlin, pp. 93–130.

Rajesh, R., 2020. A grey-layered ANP based decision support model for analyzing strategies of resilience in electronic supply chains. Eng. Appl. Artif. Intell. 87, 103338.

Ren, J., 2018. Multi-criteria decision making for the prioritization of energy systems under uncertainties after life cycle sustainability assessment. Sustain. Prod. Consum. 16, 45–57.

Ren, J., 2020. Specialty grand challenge: multi-criteria decision making for better sustainability. Front. Sustain. 1, 00002.

Ren, J., Manzardo, A., Mazzi, A., Zuliani, F., Scipioni, A., 2015. Prioritization of bioethanol production pathways in China based on life cycle sustainability assessment and multicriteria decision-making. Int. J. Life Cycle Assess. 20, 842–853.

Ruan, W., Li, Y., Zhang, S., Liu, C.-H., 2019. Evaluation and drive mechanism of tourism ecological security based on the DPSIR-DEA model. Tour. Manage. 75, 609–625.

Schmidt-Traub, G., Kroll, C., Teksoz, K., Durand-Delacre, D., Sachs, J.D., 2017. National baselines for the Sustainable Development Goals assessed in the SDG index and dashboards. Nat. Geosci. 10, 547–555.

Scriban, R.E., Nichiforel, L., Bouriaud, L.G., Barnoaiea, I., Cosofret, V.C., Barbu, C.O., 2019. Governance of the forest restitution process in Romania: an application of the DPSIR model. For. Policy Econ. 99, 59–67.

Sendra, C., Gabarrell, X., Vicent, T., 2007. Material flow analysis adapted to an industrial area. J. Clean. Prod. 15, 1706–1715.

Su, Y., Jin, S., Zhang, X., Shen, W., Eden, M.R., Ren, J., 2020. Stakeholder-oriented multi-objective process optimization based on an improved genetic algorithm. Comput. Chem. Eng. 12, 106618.

T.U.S.D.S.N. (SDSN), Arab Region SDG Index and Dashboards Report, in, 2019.

T.U.S.D.S.N. (SDSN), Europe Sustainable Development Report, in, 2019.

T.U.S.D.S.N. (SDSN), 2020. Africa SDG Index and Dashboards Report, in, 2020.

Ulukan, H.Z., Kop, Y., 2009. Multi-criteria Decision Making (MCDM) of Solid Waste Collection Methods Using Life Cycle Assessment (LCA) Outputs, in: Proc. 2009 International Conference on Computers & Industrial Engineering. Troyes, France, 584–589.

Vazquez-Rowe, I., Iribarren, D., 2015. Review of life-cycle approaches coupled with data envelopment analysis: launching the CFP + DEA method for energy policy making. Sci.World J. 2015, 813921.

Vivas, R.d.C., Sant'Anna, A.M.O., Esquerre, K.P.S.O., Freires, F.G.M., 2020. Integrated method combining analytical and mathematical models for the evaluation and optimization of sustainable supply chains: a Brazilian case study. Comput. Ind. Eng. 139.

W.H. Organization, 2015. Health in 2015: From MDGs, Millennium Development Goals to SDGs. Sustainable Development Goals.

Wang, L., Li, Q., Ding, R., Sun, M., Wang, G., 2017. Integrated scheduling of energy supply and demand in microgrids under uncertainty: a robust multi-objective optimization approach. Energy 130, 1–14.

WCED, 1987. World Commission on Environment and Development. Our common future 17, 1–91.

Wheeler, J., Caballero, J.A., Ruiz-Femenia, R., Guillén-Gosálbez, G., Mele, F.D., 2017. MINLP-based Analytic Hierarchy Process to simplify multi-objective problems: application to the design of biofuels supply chains using on field surveys. Comput. Chem. Eng. 102, 64–80.

Woodward, D.G., 1997. Life cycle costing—Theory, information acquisition and application. Int. J. Proj. Manag. 15, 335–344.

Xia, X., Ruan, J., 2020. Analyzing barriers for developing a sustainable circular economy in agriculture in China using grey-DEMATEL approach. Sustainability, 12, 6358.

Xu, D., Li, W., Shen, W., Dong, L., 2019. Decision-making for sustainability enhancement of chemical systems under uncertainties: combining the vector-based multiattribute decision-making method with weighted multiobjective optimization technique. Ind. Eng. Chem. Res. 58, 12066–12079.

Yang, A., Su, Y., Chien, I.L., Jin, S., Yan, C., Wei, S.a., Shen, W., 2019. Investigation of an energy-saving double-thermally coupled extractive distillation for separating ternary system benzene/toluene/cyclohexane. Energy 186, 115756.

Yang, Y.-P.Ou, Shieh, H.-M., Leu, J.-D., Tzeng, G.-H., 2008. A Novel hybrid MCDM model combined with DEMATEL and ANP with applications. Int. J. Oper. Res. 5, 160–168.

Zhang, X., He, J., Cui, C., Sun, J., 2020. A systematic process synthesis method towards sustainable extractive distillation processes with pre-concentration for separating the binary minimum azeotropes. Chem. Eng. Sci. 227, 115932.

Zhang, X., Su, J., 2019. A combined fuzzy DEMATEL and TOPSIS approach for estimating participants in knowledge-intensive crowdsourcing. Comput.Ind. Eng. 137.

CHAPTER 2

Business contributions to sustainable development goals

Juniati Gunawan
Universitas Trisakti, Jakarta, Indonesia

2.1 Introduction

The Millennium Development Goals (MDGs), which were succeeded by the Sustainable development Goals (SDGs) at the end of 2015, left many valuable lessons on how to improve the world's condition, and were Influenced by developing needs and a changing world. With the promise of "leave no one behind," the United Nations urged all nations to commit to reaching the 17 SDGs based on each nation's needs. The principle of the 17 goals is to preserve the environment and narrow welfare gaps in society to protect our planet and future generations.

Indonesia is one of the 193 Nations that are officially committed to supporting the SDGs. This commitment was realized through Presidential Decree no. 59/2017 and the appointment of a Ministry of National Development Planning that would be responsible for the achievement of the SDGs. All achievements need to be evaluated and monitored periodically and the Ministry has closely worked together with other Ministries since its establishment. One of the evaluation tools used is through published sustainability reports that are commonly used to disclose the companies' practices in supporting the SDGs.

Prior to the Presidential Decree, the Financial Service Authority (FSA) also published a new regulation mandating sustainability reports for all Indonesian publically listed companies. Under their regulation POJK 51/POJK.03/2017, the sustainability report practices have been implemented gradually based on the type of industry. The FSA has clearly stipulated that one of the purposes of mandating sustainability reports is to support the SDGs. However, even though sustainability reports have become mandatory for Indonesian listed companies, this study has identified that the understanding of both SDGs and sustainability reports is still at an early stage (Gunawan, Permatasari and Tilt, 2020; Gunawan and SeTin, 2019).

Having understood that SDGs and sustainability reporting needs to be explored by businesses, their adoption also needs time. Businesses or companies need to look at each SDG target and evaluate each against their business strategies. Rosati and Faria (2019a) said that businesses should report on SDGs that can be aligned with organizational planning, implementing, measuring, and communicating their efforts. This process can be another challenge for many companies, as they need to adjust their operations and strategies to the requirements of the SDGs (Tsalis, et al., 2020).

In this context, the purpose of this chapter is to provide evidence derived from Indonesian companies on how they have responded to the SDGs and disclosed their support in their sustainability reports from 2016 to 2018. Initially, all samples were taken from the whole population of available sustainability reports, and these were explored to see which sectors provided most SDG information. Then, which goals the majority of the Indonesian publically listed companies had disclosed. The discussions provided deeper understanding and ideas on what should be further planned to support the achievement of targeted SDGs in accordance with the National Development Plan.

Indonesia was chosen as a sample as this nation is considered a large developing country that plays an important role in business in Southeast Asia (Gunawan, 2015). Further, Kusharsanto and Pradita (2016) stated that Indonesia is a G20 member with an emerging economy. In September 2020, according to the Asian Development Bank, Indonesia's gross domestic product (GDP) reached around 5% in 2018 and 2019, and it is still expected to grow, despite the current COVID-19 pandemic.

In addition, the World Bank reported that the economic growth in Indonesia is supported by its commodity markets, a large, young population, and a solid macroeconomic policy framework. However, social conditions still need to be improved progressively considering that around 30% of Indonesians are at risk of poverty. Another condition to consider is the large gap between lower, middle, and upper-class society, and the welfare gap.

As well as the social conditions, the environmental issues in Indonesia have drawn global attention due to the large-scale deforestation with majority of it being illegal. Tacconi et al. (2019) noted that the deforestation occurs despite the law enforcement from the government, and is driven by social conditions that influence the surrounding communities to carry out illegal logging without thinking about the consequences. Hence, the environment and social conditions cannot be separated and these will also influence the economic performance (Erbaugh, 2019).

The results provided by this study will not only benefit Indonesia, but also other countries in taking further immediate action, and shaping the SDGs into business strategies. Further, support for SDGs does not only benefit the Government or society, but also gives advantages to companies. Martinez-Ferrero and Garcia-Meca (2020) and Rosati and Faria (2019a) stated that aligning the relevant SDG target with the business strategy will strengthen internal corporate governance and in the long run create business sustainability. Hence, the results of this study may be taken as ideas to create new programs, to educate, and to communicate to help companies and other related institutions gain the most benefit from targeting relevant SDGs.

2.2 Literature review

2.2.1 Sustainable development goals (SDGs)

The 2030 agenda for Sustainable Development Goals, known as SDGs, refers to a new development agreement that encourages changes to shift toward sustainable development based on human rights and equality to promote social welfare, economic stability, and environmental protection. SDGs are enforced with universal, integrated, and inclusive principles to ensure that no one will be left behind. The SDGs consist of 17 Goals and 169 targets to continue the efforts and achievements of the MDGs that ended at the end of 2015.

The SDG program accommodates solutions for wider problems compared to the MDGs. SDGs provide more comprehensive targeted goals, both qualitatively and quantitatively, with resolutions for the goals and objectives. The SDG program defines sustainable development as a means of development that meets the needs of the present without compromising the ability of future generations to meet their own needs. The SDGs call for efforts to build an inclusive, sustainable, and resilient future for humanity and the planet. Over a 15-year period (2015–2030), new targets will be applied universally covering all countries, whereby each country should prioritize their efforts to end all forms of poverty, fight inequality, and tackle climate change, as well as to ensure that no country is left behind. This is the essence of SDGs and the commitment to reach the targets will be the responsibility of each country.

The SDGs program contains 17 goals, 169 targets, and 241 indicators, with 5 main foundations, namely, people (humans), planet, prosperity (welfare), peace, and partnership. Based on these foundations, each country can develop their own initiatives to be incorporated into the National strategies. To ensure alignment with the SDG programs, Indonesia has applied the goals into the Indonesian National Development Plan.

An initial agenda was formulated in 2015 and included increasing awareness and meeting with stakeholders; formulating a Presidential Decree and developing technical notes on SDGs National and regional action Plan (2016); creating SDGs metadata (2017); and developing a 2018–2020 road map, a national action plan that was approved by all stakeholders, and included implementing, monitoring, evaluating, as well as reporting. The essence of the SDGs has also been categorized into Four Pillars of SDGs in Indonesia (sdgindonesia.or.id) (Fig. 2.1).

Through intense communication on SDGs, the Indonesian government has successfully become one of the six best voluntary national reviews countries that is focusing on boosting investment plans, reducing investment gaps, and developing priority projects and sectors.

2.2.2 Sustainability reports

Sustainability reports have been defined using different criteria with no single conclusion, as there is no right and wrong. According to Elkington (1997), a sustainability report refers to a report that contains not only financial performance information, but also nonfinancial information consisting of

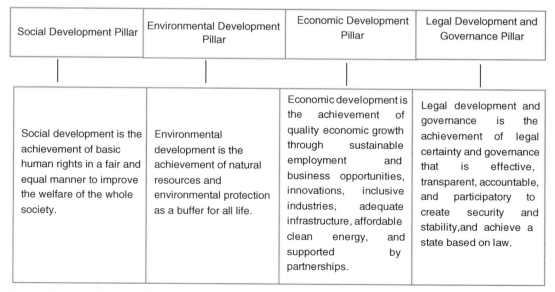

FIG. 2.1 The Four Pillars of SDGs in Indonesia.

social and environmental information that enables companies to grow in a sustainable manner. The Global Reporting Initiative, a well-known sustainability reporting standard setter, defines a sustainability report as "a report published by a company or organization about the economic, environmental, and social impacts caused by its everyday activities." Sustainability reporting represents a mechanism to generate data and measure progress, as well as the corporations' contribution to support the SDGs. By publishing such a report, companies can measure their performance against all aspects of SDGs, including green economy, circular economy, emissions, water, energy, resource efficient, biodiversity, and social impacts.

Sustainability reports have been widely used by companies to disclose their sustainability activities, including activities in supporting the SDGs (Fonseca and Carvalho, 2019; Rosati and Faria, 2019b; Siew, 2015). Fonseca and Carvalho evaluated the quality, environmental, and occupational health and safety disclosures as the three dimensions for sustainable development. They mapped the level of engagement of Portuguese companies in contributing and reporting the 17 SDGs in their sustainability reports, while Rosati and Faria (2019b) examined the institutional theory by investigating disclosures in sustainability reports. In addition, Siew (2015) highlighted the demand for publishing sustainability reports with greater transparency in both environmental and social information. Hence, sustainability reports have become one form of media for corporations to communicate to their stakeholders its social and environmental information, and the activities it has undertaken to support SDGs.

2.3 Materials and methods

This study applies a qualitative approach to explore the SDGs support by examining disclosures in sustainability reports by Indonesian publically listed companies. Attempts were made to acquire all sustainability reports for 2016, 2017, and 2018 to be used as samples, regardless of the type of industries. The 2016 reports were used to ascertain the immediate responses from companies once the SDGs had been announced. However, this study does not intend to provide evidence year by year, but covers the period 2016 until 2018 as the early stage when both the Government and companies were still preparing their readiness to support SDGs (Fig. 2.2).

The content analysis was conducted by awarding a dummy score of 0 (not disclosed) and 1 (disclosure available) according to SDG disclosures accompanied by a logo. This study disregarded any

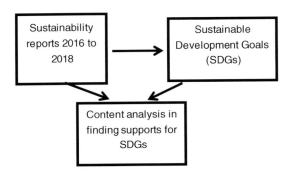

FIG. 2.2 Study framework.

information that may be relevant to SDGs, but was missing any SDG information or logo as this may indicate that the companies were not really aware of supporting SDGs. Content analysis uses a simple method as the only measurement to quantify the qualitative information by giving scores in various ways (Krippendorff, 2018). As the purpose of this study is to identify which goals are the most selected goals supported by companies, a score was awarded for all 17 goals.

Several steps were taken to conduct the content analysis by examining the sustainability reports. First, each SDG was recorded from the list of disclosures, and then data were taken from all of the sustainability reports from the Indonesia Stock Exchange (IDX). Second, to increase the sample size, sustainability reports were taken from the companies' websites. Third, to ease the evaluation and for a better understanding, the samples were categorized by sector. There are eight IDX sectors: (1) infrastructure, utilities, and transportation; (2) trade, services, and investment; (3) agriculture; (4) finance; (5) mining; (6) basic industry and chemicals; (7) miscellaneous industry; and (8) consumer good industry. Finally, after awarding scores, analyses were undertaken to see which goals were selected the most and the least to understand the goals supported by the companies. Thereby the Government can develop the relevant programs and create strategies to ensure that SDGs can be achieved.

2.4 Discussion

2.4.1 SDGs disclosures based on industrial sector

This study attempts to look at 34 companies as a sample over three sequent years, resulting in a total sample of 102 sustainability reports. After awarding scores, all scores were totaled and averaged to see which sectors disclosed the most SDGs in their sustainability reports and which goals were the most targeted.

Fig. 2.3 shows that trade, services, and investment (32%) is the highest sector for SDGs disclosures, followed by basic industries and chemicals (29.8%) and miscellaneous industries (28%). In contrast, the consumer good industry (20.1%), infrastructure, utilities and transportation (23%), and finance (25.05%) were the sectors with the lowest SDGs disclosures. The overall average for SDGs disclosure is 26.32%. This low number can be explained as during the period 2016 until 2018, SDGs disclosures in the sustainability reports was low, with less than 30% of reports disclosing the SDGs information. In addition, the spread from the highest (32%) and the lowest (20.1%) shows a significant difference of 59.2%.

The trade, service, and investment industries were the most responsive sector in supporting SDGs. According to Hoekman (2016), the trade industry is an important process whereby many of the agreed targets can be achieved. To make trade more effective, ways are sought to reduce costs, and this allows companies in developing countries to source their goods in a more competitive way so they can give households better access and thus help improve society's welfare. Many of the SDGs involve trade and service, as well as responsible investment.

On the other hand, the consumer goods industries disclosed little information for supporting SDGs. Having analyzed their business processes, in fact, this type of industry should take more responsibility in using efficient resources so as to become more relevant to "sustainable and responsible production and consumption" (SDG no. 12). Sala and Castellani (2019) stated that sustainable and responsible production and consumption is very important and is becoming the heart of sustainable development. They explained by providing examples of the life-cycle assessment processes that should be implemented in consumer goods industries. The life-cycle process is an integrated approach that utilizes waste and resources, and therefore increases efficiencies and reduces any negative impact on the environment from the production process.

In addition, Oliveira, Tomar, and Tam (2020) highlight that the consumer goods industries can play a significant role in supporting the environment by creating green packaging using technology, or by improving collaborative consumption. Through collaborative consumption, companies can share resources and individuals to access resources at a lower cost. Consequently, when consumer goods industries carry out these activities, they support SDG no.12 (sustainable and responsible production and consumption) as well as SDG no.11 (sustainable cities and communities), SDG no. 9 (industry, innovation, and infrastructure), and SDG no.13 (combat climate change and reduce emission). This support should be disclosed in the sustainability reports to inspire other companies and provide information to stakeholders. Further, the government can also consolidate this information to be reported as the country's contribution to SDGs.

Generally, Fig. 2.3 shows that the SDGs disclosures in the sustainability reports are still limited despite many companies already conducting good and relevant sustainable development. This result can be explained three ways. First, perhaps, the companies do not understand that the activities they have undertaken are supporting SDGs and they should start establishing targets and strategies. Second, the companies may not be aware that disclosing SDGs support could bring benefits for their stakeholders, especially the government, and finally, the companies are not well informed about what SDGs are and what the company has to do. Accordingly, the results indicate that there is an urgent need and demand for more education and socialization about SDGs, including how to map the goals and the benefits for sustainable business.

2.4.2 SDGs disclosures based on goals

The next analysis covers the SDGs disclosures results presented in Table 2.1. It shows that SDG no.1 (no poverty) received the most attention from companies as the average disclosure reached 61.11%, followed by SDG no.12 (responsible consumption and production) with 57.84%, and SDG no.15 (life on

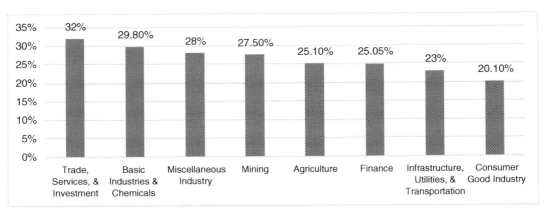

FIG. 2.3 Average scores for SDGs disclosures in sustainability reports 2016–2018.

Table 2.1 Average scores for SDGs disclosures in sustainability reports 2016–2018.

Goal No.	Description	Average disclosure scores
1	No poverty	61.11%
2	Zero hunger	43.73%
3	Good health and well-being	6.27%
4	Quality education	3.92%
5	Gender equality	43.79%
6	Clean water and sanitation	25.49%
7	Affordable and clean energy	13.89%
8	Decent work and economic growth	21.67%
9	Industry, innovation, and infrastructure	21.64%
10	Reduced inequalities	24.65%
11	Sustainable cities and communities	37.45%
12	Responsible consumption and production	57.84%
13	Climate action	21.76%
14	Life below water	33.09%
15	Life on land	49.02%
16	Peace, justice, and strong institutions	6.21%
17	Partnership for the goals	17.38%

land) with 49.02%. On the other hand, quality education (3.92%), peace, justice, and strong Institutions (6.21%), and good health and well-being (6.27%) were the three least disclosed. These findings seem to indicate that for the most disclosed SDGs, companies are well informed on the initiatives needed to support these SDGs, whereas they are ill-informed on the least disclosed SDGs.

Supporting less fortunate communities (which is linked to "no poverty"), using efficient resources (related to "responsible consumption and production"), and managing waste (relevant to life on land) are the initiatives activities the companies consider as worth reporting. On the other hand, training and education (which is linked to quality education), practicing good corporate governance (related to peace, justice, and strong Institutions), and health, safety, working environment (relevant to health and well-being) cover the information most commonly stated as this information has become mandatory, and must be disclosed in any of published report. This may explain why the companies are unaware that the activities they routinely conduct, and the information they are familiar with, are actually considered as supporting SDGs.

Companies in Indonesia have taken many initiatives to help the government reduce the rate of poverty. In line with World Bank concerns, it seems that the issue of poverty in Indonesia is still highly prevalent, even though the economic conditions have improved over a period and millions of its citizens have moved out of poverty. According to the Asian Development Bank midterm 2020 report, Indonesia is considered as a country with the lowest minus GDP rate (predicted minus 1) compared to other Asian countries (predicted minus 5%). However, the issue of "No Poverty" is still significant.

Poverty remains a common problem for many developing countries (Asongu and Odhiambo, 2019; Masron and Subramaniam, 2018; Ngo, 2018). Hanandita and Tampubolon (2015) and McKague, Wheeler and Karnani (2014) stated that poverty is one issue with many multidimensional causes and it needs the support of various institutions, including the private sector, government, and civil society organizations.

Continuing its effort to reduce poverty is high on the agenda of the Indonesian government. It has its challenges as it involves improving living conditions in rural areas where many of the poor are located.

The Government cannot reduce poverty by itself, but needs the support and collaboration from the private sector and other organizations (Larantika et al., 2017; McKague, Wheeler and Karnani, 2014). Realizing that support from companies is needed to reduce poverty, many corporate social responsibilities (CSR) activities have been conducted by companies in Indonesia. This may be the reason why the SDGs disclosures related to combatting poverty have become the most reported information in the companies' sustainability reports.

The major activities, to reduce poverty, conducted by companies involve providing jobs for low-income individuals and helping small, medium-sized enterprises (SMEs). In addition, making donations and philanthropy activities are also considered as helping improve society's welfare. Ngo (2018) stated that empowering society through its businesses is one of these methods to help them become entrepreneurs and independent. Education is also seen as a most essential factor for reducing poverty.

Further, with the COVID-19 pandemic during 2020, the rate of poverty will be much higher in developing countries, including Indonesia. This is the important call for all governments globally to fight the pandemic and to look after the economic condition. Consequently, the issue of poverty will be higher compared to previous years and it is another challenge that the SDG No.1 (No Poverty) target may not be achieved.

The second most disclosed information after "No Poverty" is "Responsible Consumption and Production" (SDG no.12), which was disclosed in 57.84% of the samples. This finding is interesting as the consumer goods industries usually use a lot of resources, and the responsible consumption and production is very relevant for this type of industry. However, this industry disclosed the least SDG information compared to other samples, even though the relevant information of responsible consumption and production disclosed was quite high.

Further analysis found that the most disclosures for responsible consumption and production occurred in the banking sector, not the customer goods industries. The dominant information disclosure related to reducing paper consumption by using digital banking applications. Other information disclosed in support of SDG no.12 related to using less energy by saving electricity.

Other than reducing paper consumption, many of the sustainability reports disclosed information on waste management (reduce, reuse, recycle), especially for effluent from the production activities, and plastic reduction, especially their activities to reduce the use of plastic bottles. The activities supporting this goal are relatively easy for the companies to understand as they are closely related to efficiencies that benefit them. Internal initiatives for reducing waste and natural consumption have been discussed widely even though data showing the reductions are still unavailable. It may take time for the companies to develop data management systems for these activities, in particular calculating the amount of waste reduction. As such, data efficiency is considered as the visible efforts taken by these companies in disclosing their activities and reporting their targets to support SDG no.12.

Following a closer analysis, the process of consumption in the companies is mainly associated with the raw materials needed by the manufacturing industries to produce their products. These processes should be based on green manufacturing processing to be relevant to support SDG no.12. While for the service industries, where they do not use raw materials, the consumption is much more in using paper and energy when they deliver services to customers. Saving energy can be calculated when the companies apply efficiencies, usually using innovation technology, and this mainly occurs in the banking industry through their use of digital banking solutions. In addition, reducing transportation would decrease fuel usage and will lead to less energy consumption. These activities are well reported in the sustainability reports and may increase in the near future, as technology applications cannot be separated from the sustainability context (Cavaleri and Shabana, 2018; Kenneth et al., 2017; Bernal-Conesa, Nieto, Briones-Penalver, 2017). Furthermore, during the COVID-19 pandemic, Obrenovic et al. (2020) revealed that enterprises' operations and productivity could be sustained when they apply the effectiveness of business models by using effective technology.

"Life on Land" (SDG no.15) was the third most disclosed information in the sustainability reports. The information reported refers to major disclosures on biodiversity and managing waste, and was provided by the basic industries and chemicals, miscellaneous, and mining industries. These three industries addressed the issue to show their responsibilities in managing the negative impact of their business operations. For mining companies, land protection involves the maintenance or improvement of the land conditions during mine-closures by applying sustainable land management methodologies (Asmeri, Alvionita, and Gunardi, 2017; Trireksani and Djajadikerta, 2016; Barkemeyer et al., 2015).

'Life on land' is actually very close to the agriculture processes. Having analyzed the findings in this study, it shows that the agriculture industries disclosed 25.1% information, and a further examination was conducted. The results showed that the majority of disclosures by agriculture industries relating to Life on land issues involved the efforts made for land protection through promoting sustainable forestry, management certifications, and preventing soil contamination from all sources by reducing chemical fertilizer usage. Other industries, apart from agriculture, disclosed some information on biodiversity protection and waste management to support Life on land. Besides agriculture being responsible for protecting life on land, Arana et al. (2020) and Nhemachena et al. (2018) stated that any industry that uses a lot of water will significantly impact life on land, with one being the agricultural industries.

Having analyzed the samples used in this study, it was noted that the companies publishing sustainability reports are those categorized as large-sized companies. Similarly, agriculture and mining companies that disclose their compliance in fulfilling the environmental regulations, do so as information on sustainable land management is required by the law. Nasih et al. (2019) also stated that agriculture and mining industries are the two sectors that need to disclose the most environmental information to show their compliance to regulations as well as for transparency and accountability.

Among the most disclosed information, unfortunately, information related to 'quality education' (SDG no.4) was low compared to the other disclosures. Having examined the sustainability reports, in fact, there is a lot of training and education information disclosed by companies to support their employees' capabilities. In addition, social responsibility activities also included many educational programs addressed to the communities. As the disclosures for supporting quality education (SDG no.4) were low, it was observed that this happens as the companies do not put the SDG logo indicating their support. They also did not mention any SDGs connection when reporting their training activities, even though the education was disclosed.

This finding confirms the assumption that the low number of SDGs disclosures is not due to there being no information disclosed, but because the companies did not state that they support the SDGs. A lack of understanding of SDGs could be the main reason, as companies do not understand why they need to refer to and report on SDGs in their sustainability reports. This assumption is also strengthened by the finding of Gunawan, Permatasari and Tilt (2020) who stated that Indonesian companies tend to focus their business and CSR activities in support of SDG no. 4, which is quality education. Further, this explains why poor quality education is one of the major factors leading to community problems, and therefore corporations should help the Government improve education. Previously, Soobaroyen and Mahadeo (2016) also noticed that education is one of the most common areas in need of improvement in the communities. Hence, filling the gaps in the knowledge and benefits of SDGs is essential so as to inform companies that conduct educational programs that they are supporting SDG no. 4. This understanding is important so they can develop strategies to help reach the 2030 educational targets (Allen, Metternicht, Wiedmann, 2019).

Similar assumptions were also proven in the minimum disclosures in "Peace, Justice, and Strong Institutions" (SDG no.16) and "Good Health and Well-being'" (SDG no.3). Having analyzed the disclosures in the sustainability reports, there is much information reported related to these two SDGs. Information, such as stakeholders' engagement process, compliance to regulations, and zero-tolerance policy toward corruption and bribery are some of the major disclosure examples that support SDG no.16. Additionally, information on health, safety, and security in the working place, and helping communities by providing good medical services and equipment are the two major information disclosures in the sustainability reports. Unfortunately, all of this information disclosed did not include any information related to the SDGs they support.

As previously discussed, it seems that a lack of understanding of SDGs is the core problem for the companies in providing disclosures as they may think this kind of information is just "common" mandatory information, and does not need to be associated with SDGs. As suggested by Cernev and Fenner (2020), understanding SDGs is important as the impact of nonachievement of SDG targets may expose humanity to risk. They expressed their concern as SDGs receive limited attention. Further, as some of sustainable development targets are fundamental, including SDG no.3 and SDG no.16, Cernev and Fenner stated the importance of synergy among multistakeholders, especially governments and corporations.

Generally, the results of this study suggest that Indonesian companies tend to disclose comprehensive information when it is mandatory and if not, they like to inform activities dealing with community's needs. Health and education, as a basic need of communities, has become a concern for companies to show that they care for their stakeholders (Cernev and Fenner, 2020). As the communities are commonly cited as one of the main stakeholders for most industries, stakeholder theory is considered to be relevant in explaining this situation (Aguiñag, et al., 2018).

Another discussion is needed to compare a prior study that applied a different sample of reports. A similar study conducted by Gunawan, Permatasari and Tilt (2020) provided quite different findings as they used both annual and sustainability reports for 2014–2016, while this study only uses sustainability reports for 2016–2018. Gunawan, Permatasari and Tilt found that disclosures in annual reports were higher compared to sustainability reports as companies are required to publish annual reports based on the Financial Service Authority Regulations, while sustainability reports are still voluntarily, except for the banking industry were it became mandatory in 2019. Hence, the SDG disclosures in this study may also be affected by the limited extent of SDG information.

The study conducted by Gunawan, Permatasari and Tilt (2020) looked at the years before (2014 and 2015) and early adoption of SDG (2016), while this study looks at all early years of SDGs adoption, from 2016 to 2018. These different periods will undoubtedly influence the extent and type of disclosures, as the longer the SDGs are adopted will create a higher understanding in disclosing SDGs-related activities (Allen, Metternicht, Wiedmann, 2019).

Furthermore, the findings for "Good Health and Well Being" was the second most disclosed in Gunawan, Permatasari and Tilt (2020), while in this study, this information was found to be minor. This difference may be explained by the fact that information on Good Health and Well Being is disclosed more extensively in annual reports rather than in sustainability reports, or there may be a change in SDGs prioritization in supporting the goals year by year. If one company discloses different priority goals in different years, it may lead to another assumption that can be analyzed further. One premise could be that companies may not understand how to prioritize SDGs according to their business strategies. They also may not be familiar with the SDG Compass, which explains the step-by-step approach in supporting SDGs and that is why the disclosure of Goals priorities is different from year to year.

Besides these differences, information related to "Responsible Consumption and Production" has been disclosed dominantly both in the annual and sustainability reports, and hence the findings of this study supports the Gunawan, Permatasari and Tilt's (2020) finding. Similar findings may also explain why the disclosure for Responsible Consumption and Production is consistent each year. Arana et al. (2020) and Franco and Newey (2020) support this discussion by articulating that responsible consumption and production is one of the approaches for wellbeing, including the communities. As companies in Indonesia are likely to disclose information related to communities (Gunawan, 2015), so it may be relevant to explain that responsible consumption and production information should be disclosed consistently both in annual and sustainability reports to show companies are concerned with helping community wellbeing.

2.5 Conclusion

This chapter provides a preliminary review of the efforts being made by Indonesian companies in supporting SDGs by examining the disclosures in their sustainability reports. The total samples gathered came from 102 reports from 34 companies for the 2016, 2017, and 2018 reporting period. The results showed that the company sectors that provided the most SDGs disclosures were (1) Trade, services, and investment sector (32%); (2) basic industries and chemicals (29.8%); and (3) miscellaneous industries (28%). On the other hand, the sectors that provided the least SDGs disclosures were (1) The consumer goods industry (20.1%); (2) Infrastructure, utilities, and transportation (23%); and (3) finance industries (25.05%). The overall average of SDGs disclosure was 26.32%. This finding confirms the early assumption that SDGs information disclosures in the 2016–2018 sustainability reports was still low, as this period was the early stage of SDGs adoption.

The next finding demonstrates that "No poverty" information was the highest disclosure (61.11%), followed by "Responsible consumption and production" information (57.84%) and "Life on land" information (49.02%). This finding is in line with the situation in developing countries, including Indonesia, where they are still combatting poverty as one of the main social issues. In contrast, information related to "quality education" (SDG no.4) was the lowest disclosure. This finding showed that although information about education including training, scholarships, and workshops was discussed widely in the sustainability reports, it was not clearly explained that these activities support SDG no.4. As assumed, the absence of SDGs information references may be due to minimum understanding about SDGs and the benefits of linking the SDGs to the companies' business strategies, including corporate social responsibility activities targets.

This study provides contributions to governments not only in Indonesia but also other countries as the findings may describe the same situation in other developing countries. The government should plan the right programs to socialize the SDGs, especially for promoting the benefits to companies, and on how to create sustainable business by protecting the environment, and increasing local community welfare. Further, the government should also develop collaborations with other governmental institutions, regulators, corporations (private, state-owned, and multinational corporations), educational institutions, and other countries to work together to achieve the SDGs.

To be more in line with the SDGs and priority issues in each country, the Government may have to consider providing guidelines to assist every corporation in selecting the relevant goals. Simple guidelines can be developed to help assist companies with incentives as a driver for companies to practice SDGs as a form of appreciation. These ideas may benefit both governments and corporations, and importantly, the achievement of sustainable development targets.

Another lesson that should be addressed is the importance of SDGs being acknowledged by the companies, where their support for the goals also benefits them in shaping their businesses. It would be too complicated for companies to take all the Goals as one Goal as they contain many indicators. Thus, selecting the relevant goals is important and can be achieved by mapping their priorities in line with the businesses' strategies. Taking only a number of goals based on the companies' efforts should be considered enough as long as the target and initiatives can be implemented and measured. The most important issue is that the results from supporting the Goals can bring a positive impact and support "leave no one behind."

As a preliminarily finding of SDGs disclosures in the early stages of implementation, this study provides a platform for further studies to examine the SDGs supports in the period after 2018. The prediction is that SDGs activities and disclosures should be higher in the future, as over time understanding of SDGs will be deeper, and the Government will have intensely promoted the SDGs. While the benefits of adopting SDGs can be defined, a better adoption will make it easier for companies to link their strategies, not only for the current goals, but also for future targets.

This study has some limitations. First, the limited number of sustainability reports provided may result in some generalization regarding all Indonesian companies. As annual reports are not included in this evaluation, it may also result in a gap in presenting the extent of disclosures only in sustainability reports, even though sustainability reports are considered as the most relevant for disclosing SDGs.

It is suggested that future research should analyze the disclosures more comprehensively, using both annual and sustainability reports. This should be conducted regularly to provide a trend and analyze the improvement of sustainable activities. The types of industry will also be another interesting area to be discussed to see whether there is a pattern of industrial types related to each goal, in line with their business' strategies.

References

Aguiñaga, E., Henriques, I., Scheel, C., Scheel, A. (2018). Building resilience: A self-sustainable community approach to the triple bottom line. J. Clean. Prod., 173, 186–196.

Allen, C., Metternicht, G., Wiedmann, T., 2019. Prioritising SDG targets: assessing baselines, gaps, and interlinkages. Sustain. Sci. 14, 421–438.

Arana C., Franco I.B., Joshi A., Sedhai J., 2020. SDG 15 Life on Land. In: Franco I., Chatterji T., Derbyshire E., Tracey J. (Eds.) Actioning the Global Goals for Local Impact. *Science for Sustainable Societies*. Springer, Singapore.

Asmeri, R., Alvionita, T., Gunardi, A., 2017. CSR disclosures in the mining industry: empirical evidence from listed mining firms in Indonesia. Indones. J. Sustain. Account. Manage. 1 (1), 16–22.

Asongu, A.A., Odhiambo, N.M., 2019. Mobile banking usage, quality of growth, inequality, and poverty in developing countries. Inform. Dev. 35 (2), 303–318.

Barkemeyer, R., Stringer, L.C., Hollins, J., Josephi, F., 2015. Corporate reporting on solutions to wicked problems: sustainable land management in the mining sector. Environ. Sci. Policy 48, 196–209.

References

Bernal-Conesa, J.A., Nieto, C.N., Briones-Penalver, A.J., 2017. CSR Strategy in technology companies: its influence on performance, competitiveness, and sustainability. Corp. Soc. Responsib. Environ. Manage. 24 (2), 96–107.

Cavaleri, S., Shabana, K., 2018. Rethinking sustainability strategies. J. Strategy Manag. 11 (1), 2–17.

Cernev, T., Fenner, R., 2020. The importance of achieving foundational Sustainable Development Goals in reducing global risk. Futures. 115.

Elkington, J., 1997. Cannibals with Forks: The Triple Bottom Line in 21st Century Business. Oxford, Capstone.

Erbaugh, J., 2019. Responsibilization and social forestry in Indonesia. For. Policy Econ. 109.

Fonseca, L., Carvalho, F., 2019. The reporting of SDGs by quality, environmental, and occupational health and safety-certified organizations. Sustainability 11 (20), 57–97.

Franco, I.B., Newey, L., 2020. SDG 12 Responsible Consumption and Production. In: Franco, I., Chatterji, T., Derbyshire, E., Tracey, J. (Eds.) Actioning the Global Goals for Local Impact. Science for Sustainable Societies. Springer, Singapore.

Gunawan, J., 2015. Corporate social disclosures in Indonesia: stakeholders' influence and motivation. Soc. Responsib. J. 11 (3), 535–552.

Gunawan, J., Permatasari, P., Tilt, C., 2020. Sustainable development goal disclosures: do they support responsible consumption and production? J. Clean. Prod. 246.

Gunawan, J., SeTin, 2019. The development of corporate social responsibility in accounting research: evidence from Indonesia. Soc. Responsib. J. 15 (5), 671–688.

Hanandita, W., Tampubolon, G., 2015. Multidimensional poverty in Indonesia: trend over the last decade (2003–2013). Soc. Indic. Res. 128, 559–587.

Hoekman, B., 2016. Trade hot topics. The COMMONWEALTH. Iss 128.

Kenneth W. Green Jr., Pamela J. Zelbst, Victor E., Sower, Jeremy C., Bellah, 2017. Impact of radio frequency identification technology on environmental sustainability. J. Comput. Inform. Syst. 57 (3), 269–277.

Krippendorff, K., 2018. Content Analysis: An Introduction to Its Methodology, fourth ed. Sage Publishing, New York.

Kusharsanto, Z., Pradita, L., 2016. The important role of science and technology park towards Indonesia as a highly competitive and innovative nation. Societal Behav. Sci. 227, 545–552.

Larantika, A.A.A.D., Zauhar, S., Makmur, M., Setyowati, E., 2017. Collaboration as a strategy for poverty alleviation. Int. J. Soc. Sci. Humanit. 1 (3), 40–48.

Martinez-Ferrero, J., García-Meca, M., 2020. Internal corporate governance strength as a mechanism for achieving sustainable development goals. Sustain. Develop. 28 (5), 1.189–1.198.

Masron, T.A., Subramaniam, Y., 2018. Remittance and poverty in developing countries. Int. J. Dev. Issues 17 (3), 305–325.

McKague, K., Wheeler, D., Karnani, A., 2014. An integrated approach to poverty alleviation: roles of the private sector, government, and civil society. The Business of Social and Environmental Innovation 129–145.

Nasih, M., Harymawan, I., Paramitasari, Y.I., Handayani, A., 2019. Carbon emissions, firm size, and corporate governance structure: evidence from the mining and agricultural industries in Indonesia. Sustainability 11 (9), 24083.

Ngo, D.K.L., 2018. A theory-based living standard index for measuring poverty in developing countries. J. Dev. Econ. 130, 190–202.

Nhemachena, C., Matchaya, G., Nhemachena, C.R., Karuaihe, S., Muchara, B., Nhlengethwa, S., 2018. Measuring baseline agriculture-related sustainable development goals index for Southern Africa. Sustainability 10, 849.

Obrenovic, B., Du, J., Godinic, D., Tsoy, D., Khan, M.A.S., Jakhongirov, I., 2020. Sustaining enterprise operations and productivity during the COVID-19 pandemic: enterprise effectiveness and sustainability model. Sustainability 12, 5981.

Oliveira, T., Tomar, S., Tam, C., 2020. Evaluating collaborative consumption platforms from a consumer perspective. J. Clean. Prod. 273.

Rosati, F., Faria, L.G.D., 2019a. Business contribution to the Sustainable Development Agenda: organizational factors related to early adoption of SDG reporting. Corp. Soc. Responsib. Environ. Manage. 26 (3), 588–597.

Rosati, F., Faria, L.G.D., 2019b. Addressing the SDGs in sustainability reports: the relationship with institutional factors. J. Clean. Prod. 215, 1312–1326.

Sala, S., Castellani, V., 2019. The consumer footprint: monitoring sustainable development goal 12 with process-based life cycle assessment. J. Clean. Prod. 240, 118050.

Siew, R.Y.J., 2015. A review of corporate sustainability reporting tools. J. Environ. Manage. 164, 180–195.

Soobaroyen, T., Mahadeo, J., 2016. Community disclosures in a developing country: insights from a neo-pluralist perspective. Account. Audit. Account. J. 29 (3), 452–482.

Tacconi, L., Rodrigues, R.J., Maryudi, A., 2019. Law enforcement and deforestation: lessons for Indonesia from Brazil. For. Policy Econ. 108.

Trireksani, T., Djajadikerta, H.G., 2016. Corporate governance and environmental disclosure in the Indonesian mining industry. Australas. Account. Busi. Finan. J. 10 (1), 18.

Tsalis, T, A., Malamateniou, K., Koulouriotis, D., Nikolaou, I, E., 2020. New challenges for corporate sustainability reporting: United Nations' 2030 Agenda for sustainable development and the sustainable development goals. Corp. Soc. Responsib. Environ. Manage. 27 (4), 1.617–1.629.

http://www.sdgsindonesia.or.id (Accessed on 22 December 2020).

http://elibrary.worldbank.org/doi/abs/10.1596/34163 (Accessed on 22 December 2020).

https://www.un.org/development/desa/socialperspectiveondevelopment/issues.html (Accessed on 22 December 2020).

https://www.worldbank.org/en/country/indonesia/overview (Accessed on 22 December 2020).

https://www.adb.org/countries/indonesia/economy (Accessed on 23 December 2020).

https://www.unescap.org/sites/default/files/Session_1_Indonesia_WS_National_SDG_10-13Sep2019.pdf (Accessed on 23 December 2020).

https://www.globalreporting.org (Accessed on 15 December 2020).

https://www.unenvironment.org/explore-topics/resource-efficiency/what-we-do/responsible-industry/corporate-sustainability (Accessed on 15 December 2020).

https://www.adb.org/countries/indonesia/economy (Accessed on 15 December 2020).

CHAPTER 3

Sustainability assessment: Metrics and methods

Himanshu Nautiyal[a], Varun Goel[b]

[a]*Department of Mechanical Engineering, THDC Institute of Hydropower Engineering and Technology, Tehri, Uttarakhand, India*
[b]*Department of Mechanical Engineering, National Institute of Technology Hamirpur, Himachal Pradesh, India*

3.1 Introduction

Today the world is contending with so many challenges that cannot be overcome without proper strategies and collective effort of all nations. There are so many global issues such as climate change, child education, health, decolonization, poverty, population, economic development, and many more; which restrain the growth of standard of living of human beings and hinder peace in the world. It becomes essential to alleviate all these impediments and proceed toward sustainable future. The catastrophic effects of human activities and exercises on the biophysical environment cause various environmental concerns that affect livability of earth and; devastate natural environment and essential amenities of human beings. Nowadays almost all activities of human beings are associated with discharge of high amount greenhouse gas (GHG) emissions into the air, which staggers balance in atmospheric concentrations (Nautiyal et al., 2015). This unbalance gives rise to various climate disasters such as global warming, acidification, climate change, lessened biodiversity, and so on.

Besides environmental problems, there are so many social issues faced by the people of the society. These social issues increase due to conflict of ideas and notions among the people. Social problems are also responsible for stunt in growth and development of people in society (Kamali et al., 2018). So many social problems can be seen in today's world such as poverty, drug addiction, racism, human rights, and so forth. To diminish all these social issues, so many organizations are working and movements are running throughout the world as it is quite indispensable to strive to get rid of these issues and achieve sustainable development. After environmental and social problems, economic issues are equally influential when dealing with objective to attain sustainable future. The main cause of economic problems is the augmentation of resource scarcity and inconstancy of prices. In the condition of scarcity, allotment of resources to one group may create inadequacy for another group and it further leads to economic unbalance. In fact, the availability of resources in any country or area governs the standard of living of people and that creates inevitability to recognize and fix the scarcity of resources and other pressing economics issues (Anand and Sen, 2000). Issues cropped up from the need of choice to avail resources; are also considered the cause of economic problems (Martin and Petersen, 2019). Therefore, it becomes a vital need to focus on the proper allocation of resources to scale down the situation of scarcity.

Economic issues depending on their intensity can be bifurcated and handled on macroeconomics and microeconomics level. For example, problems such as inflation, recession, living standard, etc. are dealt at macroscopic level, while externalities, pricing policies, availability food, etc. are treated at microscopic level. People all over the world are striving to be independent along with having approach to avail resources to meet their demands and it needs financial independence. Economic sustainability ensures an unscathed economic system in which exercises such as livelihood sources, etc. are accessible to all people (Chiappori and Ekeland, 2006). In general, social and economic issues are addressed separately; however there exists several issues having both social and economic impacts and termed as socioeconomic issues. Some common examples of socioeconomic issues are overpopulation, corruption, unemployment, wealth, etc. All these issues discussed so far have been a great hindrance to a sustainable future. In this regard, it becomes quite essential to consider the subject of "sustainability" for achieving a favorable and good quality of life along with positive wellbeing (Sala et al., 2015).

The world "sustainability" was introduced due to indiscriminate growth of resource consumption by population along with disrespect of environment. Sustainability is not only pertaining to environment rather it also related with social concerns and economy. Apart from natural resources, economic and social resources are also vital for affluent and amicable life in future. Thus sustainability comprises social, economic, and environmental development for attaining an overall wellbeing in present as well as in future (Malvestio et al., 2018). It can be clearly understood by the definition that adopting the concept of sustainability in activities pertaining to the businesses and interaction of human beings with society and environment demands a change from traditional methods to new eco-friendly techniques by which these activities are carried out. Moreover, the definition of sustainability is also significant to realize the importance of economic and social factors along with environment to attain the real development beneficial for human race. It also changes the perspective to recognize the correct essence of development and growth.

Sustainability is a wide approach which gives cognizance about the relation between social science, environmental science, and future technology. It allows to think and explore renewable energy sources, GHG emissions reduction, environmental protection, and methods to make a balanced ecosystem to get good human health, clean ecology, and environment with propulsive technical and economic innovations for better life (Nautiyal and Goel, 2020). The objective of sustainability is to concentrate on fulfilling the present needs of population by being competent to generate and avail resources to fulfill the needs in future. The idea of sustainability comprises three towers namely environmental, economic, and social (Fig. 3.1). This can also be expressed in terms of three P's: Planet, Profit, and People (sustainabilitydefinition.org). Sustainability seeks a balance among these three which is generally based on

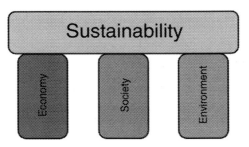

FIG. 3.1 Three pillars of sustainability.

the characteristics of being independent, interconnected, and equal (Poveda, 2017). The concept of sustainability promotes trades and enterprises to adopt policies and practices to accomplish sustainability goals in terms of zero waste, recycling, reductions in energy consumption and emissions, etc.

The objective of environmental metrics is to analyze environmental impacts associated with a given project, process, or technology. This analysis is used to achieve sustainability goals by reduction in consumption of conventional energy sources, release of emissions, and wastes. However, it is not possible to exclude all these factors; therefore, it is worthwhile to assess the rate of absorption of impacts by the environment. Emergy, Transformity, and Embodied energy are the most common concepts in assessing the environmental sustainability. Emergy refers to the total energy associated with the work processes required to produce a good or service. Transformity is expressed as the ratio of total emergy input to energy output. Embodied energy refers to the total energy associated directly or indirectly with a service or production of a product. Economic metrics are used to get computable information and data required for effective decision making. A potent economic analysis requires complete details of investments, previous economic data, all costs, taxes, returns, externalities, annuities, etc. Social metrics are associated with welfare and prosperity of society in sustainable development, which includes life standard, Governance, equity, diverseness, etc.

Nowadays sustainable development is one of the primary objectives of almost all nations in the world and it is being incorporated as a key part in all policy framework related to socioeconomic and environmental development (Roseland, 2000). In past few decades, the concept of sustainability was also seen in eight Millennium Development Goals issued by United Nations for which all members nations agreed to strive to achieve these goals by the year 2015. It was then followed by 17 Sustainable Development Goals or Global Goals adopted by member countries of United Nations in 2015 to balance the development in economic, social, and environmental aspects. All these goals were focused on making the planet more prosperous and peaceful. In fact, various issues such as increasing global population, global warming, climate change, etc. thrust to adopt these goals to overcome the impediments in the way of sustainability (Reinhardt et al., 2019).

The conservation of environment is considered as the most essential thing to achieve sustainable development (Nautiyal et al., 2011; Kariyawasam et al., 2020). This requires collective efforts of all stakeholders, appropriate policies, and decision accompanied with strict regulations. It means that protection of environment is not liability of a specific group of people or authorities rather it is a responsibility of each and every individual to come forward and adopt all essential measures to save environment at all levels (Varun et al., 2016). In environmental sustainability, ecological systems must be balanced with consumption of natural resources by human beings. Moreover, the rate of consumption of natural resources by human beings must be balanced with the rate at which these resources restore in the environment. In the present chapter, discussions commence with a need of sustainability assessment. Further various sustainability assessment methods are discussed and compared; which are then followed by discussion and conclusion.

3.2 Need of sustainability assessment

The need of sustainability assessment begins with awareness to solve social, economic, and environmental issues globally. This can be seen in the response of global policies introduced to mitigate these issues. The environmental concern on the global level was first seen in a conference in 1972 at

Stockholm (Sweden) titled United Nations Conference on Human Environment. The conference was focused on to create attention on the issues of environmental deterioration throughout the world. In respect of more environmental concern and awareness, United Nations constitute a framework named United Nations Framework Convention on Climate Change (UNFCCC) in 1992 in an eminent conference titled United Nations Conference on Environment and Development (UNCED) in Brazil. The object of this exercise was to increase environmental consciousness in nations and focus them toward sustainable development. In the same track, after 5 years of UNCED summit an extremely notable treaty namely Kyoto Protocol was introduced in 1997. The Kyoto Protocol is considered more noteworthy as for first time some specific rules were formed for all nations to set their emission reduction targets and attain them in a definite period of time. Another important consequence of the Kyoto Protocol was to introduce some important mechanisms such as Clean Development Mechanism (CDM), joint implementation, and carbon trading to move steps toward sustainable development (Nautiyal and Varun, 2015). Another important consequence of the Kyoto Protocol was to promote sustainable development in developing countries along with emission reduction in developed countries (Nautiyal and Varun, 2012). Therefore, in this regard sustainability assessment has become a quite important practice to make a balance among economy, society, and environment. In fact, with the help of sustainability assessment, one can come to know where we stand and where we have to go to attain the goals of sustainable development? Nowadays almost all human activities use a huge amount of energy and natural resources to generate products and services. Sustainability assessment is a very influential exercise in making all human activities more conscientious from an environmental point of view (Lopez et al., 2019).

In respect of a sustainability assessment structure, the advancement and effectiveness of sustainable development revolves around consolidation of criteria, standard methodologies, and definitions; and proper application of approaches to produce effective results. Along with the growth of sustainable development, different methodologies for sustainable development have also been matured to set the appropriate sustainability goals from the beginning of a project. There are so many methodologies and processes available for sustainability assessment; however, selection of an appropriate method for a given case is still a complex task. The appropriateness of any method used for sustainable assessment depends on its ability to produce clear results that can be beneficial to promote sustainability with all environmental, social, and economic aspects. Sustainability assessment has always been an important area of research and various approaches have been developed by the researchers to make assessment more realistic and useful. Some important methodologies/framework for sustainability assessment are summarized as follows.

3.2.1 Steady-state economy

The concept of steady-state economy pertains to a balance between growth and environmental protection. The concept of steady-state economy is introduced with an objective of effective consumption of resources along with proper allocation of worth produced by growing these resources. Steady-state economy may be simply understood as an economy of a state or union but the concept can be employed to the economy of a small region to the whole world. As the name implies, a steady-state economy is basically an economy with balance and stability however some moderate fluctuations may be present (O'Neill, 2015). An economy can attain the condition of steadiness after growing or falling in a specific period of time. While considering a steady-state economy sustainable, environmental limits must be taken into account to confine all the processes and activities within.

Balanced population, per capita consumption of resources, gross domestic product (GDP) are the important characteristics of steady-state economy. In addition, the economy also demands increase in reduction of waste during all production and consumption processes. High population growth is largely responsible for low wages and reduction in availability of resources. Due to this, the condition of steady-state economy cannot be thought without control of population. The population growth along with falling per capita consumption of resources may occur in a steady-state economy and reverse of this is possible too. Nevertheless, this situation is not necessarily last for a long period of time. That is why a steady-state economy signifies a balanced population and capital stocks along with stable consumption of energy and resources to provide goods and services along with minimum waste flows to the environment (Kerschner, 2010). Here recycling of materials is also desirable to achieve waste reduction. Overall it can be said that a steady-state economy is the manifestation of stability or moderate variations in the economy. One of the most important characteristics of a steady-state economy is that conventional policies can still be used for the overall growth of economy. To achieve a long-term sustainability, the concept of a steady-state economy should be adopted by the economies, which connotes a constant GDP. However, the expected growth in a steady-state economy has always been a point of conflict. This is due to the fact that here emphasis is always given on the stability, which means that fluctuations in the economy are not expected (O'Neill, 2012). Therefore, a steady-state economy is not considered competent to handle major ups and downs as a consequence of which it always tends to repel extensive risk factors that may either push the growth of economy or decline it.

3.2.2 Circular economy

Circular economy represents an economic system that is structured with the objectives of reduction in waste and consumption of resources (Prieto-Sandoval et al., 2018). It forms a closed-loop cycle by focusing on recycling and reusing of resources and tends to lessen the input resources along with minimization of waste flows and GHG emissions into the atmosphere. In this manner, outputs of any process do not go waste rather becomes inputs for other processes. The concept of circular economy introduces a closed economic model in which production of goods and services is done in a sustainable manner by reduction in consumption of resources and waste (Jorgensen and Remmen, 2018). A circular economy is far distinct from linear economy by aiming at environmental protection along with improvement of efficiency at all phases of a product life cycle. Thus the incorporation of the concept of circular economy is highly favorable to promote sustainable development (Kirchherr et al., 2018). The design of circular economy model is largely appropriate for the promotion of sustainable development. International Organization for Standardization (ISO) has also formed a standard namely ISO/TC 323 for circular economy to establish its procedure, guidelines, tools, etc. (iso.org).

3.2.3 Ecological footprints

Ecological footprints are the measure of impacts of an individual or community on the environment and can be expressed as the amount of resources consumed. The concept of ecological footprints was introduced in 1990 for the evaluation of environmental impacts associated with human activities (Wackernagel and Rees, 1996). Ecological footprints measure the rate at which resources are consumed and wastes are generated. The assessment of ecological footprints is also associated with biologically productive area or biocapacity. Ecological footprints measure the regular demand, whereas biocapacity refers to unceasing supply. The figures of ecological footprints and biocapacity vary year by year with respect to population, per capita consumption, production efficiency, and productivity. In the same track, the components such as carbon footprints and water footprints are originated from ecological footprints; and have become popular throughout the globe as climate change issues become a serious matter of concern (Nautiyal et al., 2019; Shree et al., 2015; Varun and Chauhan, 2014).

3.3 Various methods of sustainability assessment

Sustainability assessment can be considered as a multidisciplinary activity employed to handle variety of objectives. It is a technique to monitor sustainable characteristics associated with a project, process, or policy and; it serves as integrated evaluation of social, environmental, and economic aspects in terms of sustainability. Various methods have been used by the researchers for sustainability assessment. Each method has its own peculiarity, constraints, and complexity. A sustainable assessment method must comprise scope and objectives of assessment, appropriate sustainability indicators, assessment technique, and finally the interpretation and application of assessment (Fig. 3.2). Sustainability assessment uses various qualitative approaches or data collection along with quantitative techniques to produce useful outcomes (Hasna, 2008). One of the important confrontations in sustainability assessment is the tradeoff among society, economy, and environment. There are so many sustainability assessment methods available and it is quite complex to explain all of them at one place. However, some important assessment methods are described briefly in this section.

3.3.1 Life-cycle assessment

Life-cycle assessment (LCA) is a popular approach for sustainability assessment and is widely used for estimating environmental aspects and potential impacts associated with a product or process

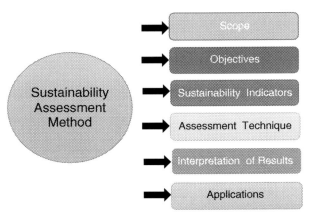

FIG. 3.2 Sustainability assessment method.

3.3 Various methods of sustainability assessment

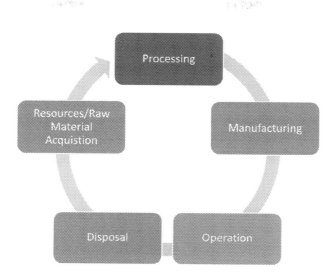

FIG. 3.3 Social impact assessment.

(Fig. 3.3). The idea of LCA was introduced from the concept of comprehensive environmental evaluation of various products and was adopted in United States and Europe in the beginning of 1970s (Boustead, 1996). LCA methodology is an effective tool to study environmental aspects and potential impacts all through a product's life cycle from acquisition of raw materials to its operation and final disposal (ISO 14040, 1997). The full procedure to carryout LCA studies is described in ISO (14040–14043). However, the extent and depth of an LCA study depends on goal and scope definition. It is expected to have clear and transparent scope, assumptions, data methodologies while performing an LCA study.

LCA study is capable to explore big industrial systems, data collection, and its analysis and deliver various environmental information (Varun et al., 2012). Product development and improvement, planning, policies are some important applications of LCA. The procedure of LCA methodology includes goal and scope definition, life cycle inventory analysis, life cycle impact assessment, and life cycle interpretation. Moreover, there are some other items such as functional unit, product system, boundaries, allocation, type of impact methodology, data requirements, assumptions, limitations, etc. which must be considered and illustrated in an LCA study. A well-defined scope with appropriate and adequate depth and details of the study ensures influential and useful results. Moreover, scope of an LCA study is also used to specify the functions of the system to be studied (Nautiyal et al., 2018).

The next step of an LCA study is Life cycle inventory analysis that comprises activities such as preparation of flow charts, collection of data, procedures to perform calculations, etc. to evaluate inputs and outputs associated with a product or system (Varun et al., 2008). It also involves the preparation of life cycle models and quantification of emissions discharged into the atmosphere along with consumption of resources during complete life cycle of a project. The models are prepared during the inventory analysis with respect to the prescribed goals and scope definition. The process of data collection in inventory analysis is quite time consuming and becomes somehow complex when dealing with many items and materials or data location for various remote activities (Sharma et al., 2011).

Life cycle impact assessment is the next step that involves assessment of product systems in terms of environmental consequences. The phase models of life cycle impact assessment selected to address these environmental consequences are known as impact categories. Life cycle impact assessment phase involves identification of opportunities to improve product system, characterization, or bench marking of a product system; relative comparison of product systems on the basis of selected category indicators; and indication of environmental outcomes useful for decision and policy makers (Chauhan et al., 2011). After life cycle impact assessment next comes the final phase that is, life cycle interpretation phase in which the outputs and results obtained in life cycle inventory assessment and life cycle impact assessment (LCIA) phases are summarized and explained. The results are summarized on the basis of results, conclusion, recommendations, and decision making with respect to goal and scope definition. In fact, the goal of life cycle interpretation phase is to study and analyze results, draw conclusions, describe limitations, and give recommendations on the basis of results obtained in life cycle impact assessment (Baumann and Tillman, 2004).

There may be different variants of LCA depending on the type of approach selected in an LCA study. Cradle to grave type LCA performs analysis from extraction of raw materials (or resources) to its operational and final disposal phase. In cradle to gate LCA, analysis is carried out for a part of a product's life cycle that is from the raw materials extraction to the doorway (gate) of manufacturing unit. Apart from this, another partial product's life cycle analysis is gate to gate approach (Jimenez-Gonzalez et al., 2000). The final phase of this analysis is considered before it is transferred to the customer. The next approach is cradle to cradle LCA in which recycling of the product is incorporated. The analysis deals with extraction of resources, manufacturing, operation, final disposal, and recycling to produce some new products. The approach is quite significant to mitigate the environmental impacts associated with products. This is also known as a closed loop of production approach.

In addition to the above, LCA can be classified as process LCA, input-output LCA, and hybrid LCA. Process LCA is commenced with the identification of a specific product that may be either good or a service; as the object of the study (Varun et al., 2009). Then different resources required directly and indirectly to produce the product are examined. After preparing the list of these inputs, evaluation of total energy consumption and associated emissions is carried out for the particular product. This LCA is also known as process chain LCA and it requires extensive data. Input-output LCA gives relatively better comprehensiveness. In this LCA, the supply chains of the product system are modeled with the help of economic flow databases or input-output tables. These databases can be obtained from various government statistical agencies. These databases economically represent the amount/share that each industrial sector expends on the goods and services produced by other sectors. Finally, emissions and other associated impacts are assigned to the different sectors. Another type of LCA is hybrid LCA in which both process and input-output LCA approaches are used to the perform the LCA study (Varun et al., 2010).

3.3.2 Socioeconomic impact assessment

Socioeconomic impact assessment is a methodical procedure in which pros and cons for a whole community or various processes are shown and studied. The objective is to explore and evaluate the objective of a given plan/program along with associated eventual impacts (Ramanathan and Geetha, 2012). Basically in this technique socioeconomic cost is evaluated against the socioeconomic benefit. The method is used to evaluate the economic and social impacts associated with product and processes.

Moreover, it tries to consider all types of social, economic, and environmental impacts and consequences for all users in a community or society. It considers the view of stakeholders and policy makers before making the decisions. In this method both items, that is, social and economic are incorporated in an environmental impact. The procedure allows one to recognize and include various impacts to make decisions; however, there are some constraints associated with it. If the potential impacts are found serious and adverse, then the assessment can aid the planner or developer in environmental impact assessment to find alternatives to mitigate or eliminate these impacts (Berkhout and Hertin, 2000). The important socioeconomic components are health and wellbeing, sustainable wildlife harvesting, land access and use, protection of heritage and cultural resources, business and employment opportunities, sustainability of population, services and infrastructure, ample sustainable income and lifestyle, etc. (Socio-Economic Impact Assessment Guidelines, 2007).

The main aspect of socioeconomic impact assessment is that all identified impacts are expressed in economic terms. Therefore, this method may differ from other methods in the essence of scientific and technical prospect. Depending on the scope of the problem to be handled and data availability, socioeconomic impact assessment may produce different levels of outcomes that may somehow affect the decisions- and policies making processes. However, some constraints pertaining to prediction of impacts, their definition and evaluation, monitoring, application of specific methods, etc. still exist in socioeconomic impact assessment (Brandon and Lombardi, 2011). In addition, socioeconomic impact assessment tries to avoid and mitigate adverse impacts and gives a platform for planning to increase the favorable impacts associated with the proposed plan.

3.3.3 Strategic environmental assessment

Strategic environmental assessment (SEA) is an organized procedure to assess environmental impact associated with a program, plan, or policy. It also helps in providing a medium to explore overall impacts and address them properly along with economic and social aspects at the phase of decision making. Using SEA, environmental-related issues and challenges are appropriately addressed. Moreover, it also provides various alternatives and opportunities to mitigate environmental impediments and support low carbon sustainable development (Unalan and Cowell, 2019; Ludovico and Fabietti, 2018). It is also found that SEA helps effectively at strategic level and gives an effective management on interactions and other cumulative impacts than environmental impact assessment. SEA can be effectively applied in preparation and execution of programs and plans for energy sector, industries, transportation, water and waste management, tourism, forestry, land, fisheries, agriculture, etc. (Stinchcombe and Gibson, 2001).

The framework of SEA is based on the following phases namely screening, scoping, and analysis (Fig. 3.4). Screening is a kind of an examination phase where a program or project is scanned under the legislation of SEA. In the next phase scoping is related to describe the scope, assumptions, and boundaries requisite for assessment. This phase also allows recognition of various issues and their clear descriptions that are to be necessarily discussed in SEA. To perform an effective SEA study, all views and valuable opinions of all stakeholders can be considered in this phase. The next phase is pertaining to analysis where comprehensive analysis is performed, which includes documentation of environmental baseline, evaluation; and assessment of potential environmental impacts, determination of restraints, and opportunities related to environmental issues. Moreover, a study of various performance indicators along with identification of ability of institutions to address environmental issues is

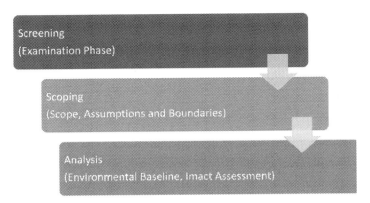

FIG. 3.4 Steps in contingent valuation method (www.ecosystemvaluation.org).

also the part of analysis phase. Finally, it is proceeded to drawing conclusions and recommendation; and making decisions based on the assessment and further monitoring of programs after the implementation of SEA results. If there is a need of public interaction in the SEA then it can be implemented in this phase. SEA is a time-consuming process and it relatively new compared to other processes. Due to this it may face the problem of unavailability of baseline data, documentation, public interaction, etc. SEA technique also depends on quantitative data that may not be available for many cases (Bidstrup and Hansen, 2014). Moreover, there are many unavoidable uncertainties and boundary-setting issues associated with SEA.

3.3.4 Cost-benefit analysis

Cost-benefit analysis (CBA) (also known as benefit cost analysis) is a method to evaluate price (or expenses) and benefits of a project, which is further used to examine decisions (Fig. 3.5). The technique is particularly used at initial phases of a project to explore its viability and compare its

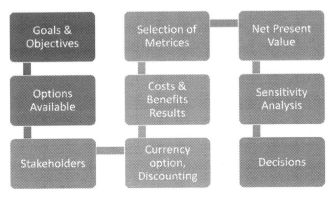

FIG. 3.5 Cost-benefit analysis.

expenses and advantages with other contesting projects. One of the most important characteristics of CBA is that it provides various options to decision makers to get maximum possible returns on the investment (Haveman and Weimer, 2001). In this way, CBA is an effective approach to explore all pros and cons of all alternatives to opt the best approach to get maximum possible benefits. In CBA, cost includes all expenses and investments in terms of monetary units; and benefits comprise all revenues obtained from a project. In fact, all costs and benefits associated with a project are expressed in terms of monetary units and adjusted for the time value. All these flows related to benefits and costs over specific period of time are defined in terms of net present value. The evaluation of cost in CBA includes direct, indirect, and intangible costs. Moreover, any other alternative investments and its associated costs can be included too. On the other hand, benefits include all types of returns that is, sales and revenues, intangible benefits, and any other advantages (Tol, 2001). Overall the process of CBA can be divided into different steps namely analyzing cost and benefits, assignment of monetary values to cost, assignment of monetary values to benefits, and, finally, comparison of cost and benefits. CBA can be divided into two approaches namely economic and social.

3.3.5 Travel cost analysis

Travel cost analysis (TCA) method is used to evaluate economic values of goods associated with ecosystems. It is often used to calculate the values of environmental goods and services that cannot be evaluated using market prices (Johnstone and Markandya, 2006). The estimation of economic values pertaining to recreational sites or ecosystems is carried out by this technique. The evaluation of travel cost comprises all economic costs and benefits of a recreational site. One assumption taken in this method is inclusion of cost for visiting the recreational site in travel cost. People who are willing to pay price to visit a recreational site is based on number of trips made at different travel costs. Therefore, the method tells about the willingness of people to pay cost on the basis of involvement of visitors. Moreover, it includes any variation in costs to access a recreational site, addition and removal of any recreational site, and variation in environmental quality or features in a site (Jones et al., 2017).

3.3.6 Social impact assessment

Social impact assessment (SIA) is a method to analyze the social impacts associated with plans, policies, projects, and other developments. The technique is used to study the social effects of planned development interventions; however, it can also be used to study the social impacts of any unexpected interventions such as demographic variations, natural disasters, etc. The objective of social impact assessment is to achieve sustainable development outcomes (Esteves et al., 2017). It also helps in promoting the adaptive management of plans, policies, and projects. The interconnection of economic, social, and biophysical impacts is considered in social impact assessment (Fig. 3.6).

3.3.7 Contingent valuation method

Contingent valuation method (CVM) is a technique used to evaluate economic values of various types of ecosystem and environmental services. This is one of the most popular methods to evaluate non-use-

FIG. 3.6 Social impact assessment.

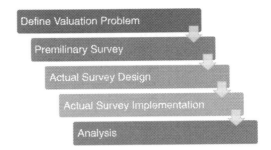

FIG. 3.7 Steps in contingent valuation method.

values; however, it can be used to estimate use values too. In CVM, the willingness of people to pay for a specific environmental service is considered (Whitehead and Haab, 2013). Moreover, the willingness of people to get compensation amount for leaving an environmental service is also included in many cases. Sometimes the utilities given by the nonmarket resources to the people may not have market values due their nonsaleable nature and they are typically enough to evaluate by price-based models (Ferreira and Marques, 2015.). These issues can be handled in CVM. The steps used in CVM are shown in Fig. 3.7. The process commences with the definition of a problem considered for evaluation and then proceeded to a preliminary survey that includes interaction with people. After this, actual survey is done, which is a relatively complex and time-consuming process. Finally it is implemented followed by the analysis of the results.

3.3.8 Hedonic pricing method

Hedonic pricing method (HPM) is used to evaluate quantitative values for various ecosystems and environmental utilities that affect the market prices directly. This method is well utilized to estimate the economic costs and benefits associated with environmental quality and amenities. It is well used to establish and analyze the relation between the characteristics and price of a good (Tyrvainen, 1997). To evaluate costs and benefits related to environmental aspect, HPM depends on the market deals and negotiations for various goods. So, the determination of market cost of a specific good is based on its attributes and domain of service (Arrondo et al., 2018), and also on an assumption that the goods are a combination of various characteristics and some of them cannot be sold individually due to lack of individual price.

3.3.9 Multicriteria analysis

Multicriteria analysis (MCA) represents a structured approach used to analyze overall possible alternatives and preferences and evaluate them under different criteria at the same time. In this methodology, preferable targets and goals are particularized and corresponding characteristics and indicators are recognized (Multi-criteria analysis: a manual. 2009). One of the important characteristics of MCA is that the assessment of indicators generally hinges on a quantitative analysis of various qualitative impact categories; however in many cases, assessment of indicators is not expressed in monetary terms. In MCA, various approaches are used to classify, compare, and select the most appropriate alternatives with respect to the given criteria. Based on the selected method, each criterion can be evaluated qualitatively or qualitatively (Caravaggio et al., 2019). MCA acts as a tool and can be effectively applied to the areas and sectors where single criterion–based methodologies are found ineffective, and important social and environmental impacts cannot be expressed in terms of monetary values.

3.3.10 Material intensity per service unit

Material intensity per service unit (MIPS) is a method to evaluate the environmental impacts associated with a product, process, or service. The method was introduced in 1992 for the idea of dematerialization on micro, meso, and macro level (Liedtke et al., 2014). For the sake of evaluation of environmental impact associated with manufacturing and services of a product, MIPS expresses the amount of material and resources used for the product or service (Ritthof et al., 2002). MIPS combines the overall materials to evaluate the total material intensity of a product or service, which is expressed as the ratio of total material input to the number of service units.

3.3.11 Analytic network process

Analytic network process (ANP) is basically a type of multicriteria decision-making methodology and effectively used to analyze different cases where complex interrelationships between decision elements and capability to implement qualitative and quantitative attributes are considered simultaneously (Kheybari et al., 2020). ANP is effectively used to handle the cases where decision problems cannot be structured as hierarchical due to the inner or outer dependencies and effects between and within the criteria and alternatives (Gencer and Gurpinar, 2007). It is also considered an indispensable tool to express the realization of a decision problem (Saaty TL., 2004); and has ability to explore indefinite interdependencies between various factors, subfactors, and options (Hosseini et al., 2013).

3.3.12 Environmental and sustainability rating systems

Environmental and sustainability rating systems (ESRS) is used to evaluate the performance of different projects in building and construction sector. It acts as an important tool that helps the decision makers during all phases of a project's life cycle. It also helps to promote interactions among the stakeholders during a project's life cycle to achieve the sustainability goals. There are so many sustainability and environmental rating systems used in building and construction sector. Some examples are BOMA 360, BREEAM, EDGE, Green Star, LEED, NABERS, etc.

3.4 Comparison of sustainability assessment methods

As discussed in previous section that there are so many methods used by the researcher/decision makers for sustainability assessment. It is also found that each method has its own effectiveness, suitability,

Table 3.1 Comparison of sustainability assessment methods.

S. No.	Criteria	LCA	SEIA	SEA	CBA	TCA	SIA	CVM	HPM	MCA	MIPS	ANP	ESRS
1	Cover all domains of Sustainability	High	High	High	Medium	High	High	Medium	Medium	High	Medium	High	High
2	Approach to alternative scenarios	Medium	Medium	Medium	High	Medium	High	Medium	High	High	Medium	Medium	Medium
3	Ability to handle scarcity of data	High	Medium	Medium	Medium	Medium	High	Low	Low	High	Low	Medium	Medium
4	Engagement of stakeholders in all phases	Low	Low	Medium	Medium	High	High	Medium	Medium	High	Low	High	High
5	Clear assessment	Low	Medium	Medium	Medium	High	Medium	High	Low	Medium	Low	High	Medium
6	Clear interpretation of results	Low	Medium	Medium	Medium	High	Medium	High	Low	Medium	Low	High	Medium
7	Manage interrelationships	Low	Low	Low	High	Medium	Low	Medium	High	High	Medium	High	Medium

and complexity. The suitability of a sustainability assessment methods depends on so many factors such as nature of problem, assumptions, scope, objectives, etc. Table 3.1 shows a comparison among different sustainability assessment methods on the basis of some important criteria namely coverage on all domains of sustainability, approach to alternative scenarios, capability of handling scarcity of data, involvement of stakeholders, clear assessment and interpretation of results, and management of interrelationships. The comparison has been shown with help of three options that is, high, medium, and low. "High" means the method strongly fulfill the criteria, while "low" represents the limitation of the method to deal with that criteria.

The comparison has been done on the basis of previous assessments carried out by the researchers found in the literature. Therefore, some uncertainties in the comparison can be expected; however, it gives a clear picture of benefits and limitations associated with different sustainability assessment methods. For an example, MCA is an effective tool to handle the complex cases that are difficult to handle with single criteria–based approach such as cost-benefit analysis, etc. One of the main benefits of MCA method is its ability to consider a variety of different criteria and contexts. Moreover, involvement of stakeholders in MCA is also beneficial as it further gives improved producers and conclusions. Consequently, it can be effectively used in association with participatory methods. It is also seen that MCA can deal effectively complicated policy matters as it improves structure and decision-making ability. In addition, the cases, where assigning of monetary values to prominent impacts is difficult, can be dealt using MCA. It also gives flexibility to the decision and policy makers to explore all economic, social, and environmental criteria. Various benefits and limitation of all these methods are discussed in Table 3.2. The benefit of using CVM instead of TCM is that CVM can also be used to estimate the use and nonuse values of an environmental good or service. CVM and HPM are quite useful when dealing with problems of ecosystem and environmental services. LCA is a quite popular approach to carryout sustainability assessment due to its comprehensive environmental assessment. MIPS is utilized to represent the quantity of material and resources consumed. ANP is a type of multi-criteria decision-making methodology that is effective in dealing with complex interrelationships between decision elements. ESRS is popular in sustainability assessment related to construction and building sector.

3.5 Conclusion

In the present time it becomes quite essential to think about alleviation of global issues and proceed toward sustainable development. The goal of sustainability is to think about fulfilling the present needs of the people with capability to produce and avail resources to fulfill the need in future. In this regard, sustainability assessment becomes a very important and influential exercise to make human activities more conscientious from an environmental point of view. A sustainability assessment method must have scope and objectives of assessment, appropriate sustainability indicators, effective assessment technique, and clear interpretation of results. Various sustainability assessment methods are available and it is quite complex to proclaim the best sustainability assessment method because every method has its own effectiveness, suitability and complexity, and limitations. It is also found from the literature that every method cannot evaluate all aspects of sustainability. However, it is quite expected that a sustainability assessment is able to produce clear and transparent results that can help decision and policy makers for attaining the goals of sustainable assessment.

Table 3.2 Benefits and limitations of sustainability assessment methods.

S. No.	Method	Merits	Demerits/limitations
1	Life cycle assessment (LCA)	• Comprehensive • Flexibility • Useful quantitative comparisons	• Requirement of extensive knowledge and data • Time consuming
2	Socioeconomic impact assessment (SEIA)	• Ability to evaluate social as well as environmental impacts • Participation of stakeholders and policy makers • Impacts are expressed in economic terms	• Nonmonetary impacts are not considered • Requirement of large statistical data
3	Strategic environmental assessment (SEA)	• Avoid conflicts • Promote socio-economic integration • Better analysis of large scale projects	• Involvement of uncertainties • Unclear definitions • Institutional difficulties
4	Cost-benefit analysis (CBA)	• Can categorize and prioritize resources for their effective and efficient utilization • Can assess projects that are beneficial for a nation • Comparison of various program	• Unable to convert many social benefits and costs in monitory units • Complex method
5	Travel cost analysis (TCA)	• Assessment of economic values of environmental goods associated with recreational sites • Cost effective	• Requirement of complicated statistical analysis • Time consuming
6	Social impact assessment (SIA)	• Promotes adaptive management of plans, policies, and projects • Consider interrelationship among economic, social and biophysical impacts • Allows transparencies in decision making process	• Unreliable results in many cases • Absence of system level integration • Inadequacy of public participation
7	Contingent valuation method (CVM)	• Handle use and nonuse values • Flexibility • Improves validity and reliability of results	• Involvement of hypothetical questions • Sources of bias may be present
8	Hedonic pricing method (HPM)	• Assessment of values based on market choice • Ability to deal with multiple interactions between goods and environmental quality • Versatile and relatively easy	• Requirement of large data • Expensive depending of data requirement
9	Multicriteria analysis (MCA)	• Explicit approach • Multiattribute decision analysis • Ability to provide preferences among different alternatives	• Extensive input data are required
10	Material intensity per service unit (MIPS)	• Deal with estimation of eco-efficiency of a product or service • Consider amount of materials used to produce good or service	• Difficult in dealing with several parameters such as emissions, waste, etc. associated with economy
11	Analytic network process (ANP)	• Established and reliable methodology • Provide consistency check • Able to deal with interrelationship among cost, comfort and safety	• Requirement of specific tools and Software • Complex
12	Environmental and sustainability rating systems (ESRS)	• Evaluation of construction and building sectors projects	• Generally focus on environmental issues. • Accuracy

References

Anand, S., Sen, A., 2000. Human development and economic sustainability. World Dev. 28 (12), 2029–2049.

Arrondo, R., Garcia, N., Gonzalez, E., 2018. Estimating product efficiency through a hedonic pricing best practice frontier. BRQ Bus. Res. Q. 21 (4), 215–224.

Baumann, H., Tillman, A.M., 2004. The Hitch-Hikers Guide to LCA: An Orientation in Life Cycle Assessment Methodology & Applications. StudentLitteratur AB, Sweden.

Berkhout, F., Hertin, J., 2000. Socio-economic scenarios for climate impact assessment. Glob. Environ. Change 10 (3), 165–168.

Bidstrup, M., Hansen, A.M., 2014. The paradox of strategic environmental assessment. Environ. Impact Assess. Rev. 47, 29–35.

Boustead, I., 1996. LCA-how it came about: the beginning in the UK. Int. J. Life Cycle Assess. 1 (1), 4–7.

Brandon, PS., Lombardi, P., 2011. Evaluating Sustainable Development in the Built Environment.

Caravaggio, N., Caravella, S., Ishizaka, A., Resce, G., 2019. Beyond CO_2: a multi-criteria analysis of air pollution in Europe. J. Clean. Prod. 219, 576–586.

Chauhan, M.K., Varun., Chaudhary S., Kumar, S., Samar, 2011. Life cycle assessment of sugar industry: a review. Renew. Sust. Energy Rev. 15 (7), 3445–3453.

Chiappori, P.A., Ekeland, I., 2006. The micro economics of group behavior: general characterization. J. Econ. Theory 130 (1), 1–26.

Esteves A.M., Factor G., Vanclay F., Gotzmann N., Moreira S., 2017. Adapting social impact assessment to address a project's human rights impacts and risks 67, 73-87.

Ferreira, S., Marques, R.C., 2015. Contingent valuation method applied to waste management. Resour. Conserv. Recycl. 99, 111–117.

Gencer, C., Gurpinar, D., 2007. Analytic network process in supplier selection: a case study in an electronic firm. Appl. Math. Model. 31 (11), 2475–2486.

Hasna, A.M., 2008. A review of sustainability assessment methods in engineering. Int. J. Environ. Cult. Econ. Soc. Sustain. 5, 1–12.

Haveman, R.H., Weimer, D.L., 2001. International Encyclopedia of the Social & Behavioral Sciences, 2845–2851.

Hondo H., Nishimura K., Uchiyama Y., 1996. Energy requirements and CO_2 emissions in the production of goods and services: application of an input-output table to life cycle analysis, CRIEPI report, Y95013, Central Research Institute of Electric Power Industry, Japan.

Hosseini, L., Tavakkoli-Moghaddam, R., Vahdani, B., Mousavi, S.M., Kia, R., 2013. Using the analytical network process to select the best strategy for reducing risks in a supply chain. J. Eng. Article ID 375628.

Iso.org (ISO/TC 323 - Circular economy).

ISO 14040, 1997. Environmental Management-Life Cycle Assessment-Principles and Framework.

ISO 14041, 1998. Environmental Management-Life Cycle Assessment-Goal and Scope Definition and Inventory Analysis.

ISO 14042, 2000. Environmental Management-Life Cycle Assessment-Life Cycle Impact Assessment.

ISO 14043, 2000. Environmental Management-Life Cycle Assessment-Life Cycle Interpretation.

Jimenez-Gonzalez, C., Kim, S., Overcash, M., 2000. Methodology for developing gate-to-gate life cycle inventory information. Int. J. Life Cycle Assess. 5 (3), 153–159.

Johnstone, C., Markandya, A., 2006. Valuing river characteristics using combined site choice and participation travel cost models. J. Environ. Manage. 80 (3), 237–247.

Jones, T.E., Yang, Y., Yamamoto, K., 2017. Assessing the recreational value of world heritage site inscription: a longitudinal travel cost analysis of Mount Fuji climbers. Tour. Manage. 60, 67–78.

Jorgensen, M.S., Remmen, A., 2018. A Methodological approach to development of circular economy options in businesses. Procedia CIRP 69, 816–821.

Kamali, F.P., Borges, J.A.R., Osseweijer, P., Posada, J.A., 2018. Towards social sustainability: screening potential social and governance issues for biojet fuel supply chains in Brazil. Renew. Sust. Energy Rev. 92, 50–61.

Kariyawasam, S., Wilson, C., Rathnayaka, M.I., Sooriyagoda, K.G., Managi, S., 2020. Conservation versus socio-economic sustainability: a case study of the Udawalawe National Park, Sri Lanka. Environ. Dev. 35, 100517.

Kerschner, C., 2010. Economic de-growth vs. steady state economy. J. Clean. Prod. 18 (6), 544–551.

Kheybari, S., Rezaie, F.M., Farazmand, H., 2020. Analytic network process: an overview of applications. Appl. Math. Comput. 367, 124780.

Kirchherr, J., Piscicelli, L., Bour, R., Kostense-Smit, E., Muller, J., Huibrechtse-Truijens, A., Hekkert, M., 2018. Barriers to the circular economy: evidence from the European Union (EU). Ecol. Econ. 150, 264–272.

Liedtke C., Bienge K., Wiesen K., Teubler J., Greiff K., Lettenmeier M., Rohn H., 2014. Resource use in the production and consumption system—the MIPS approach. Resources 3, 544–574.

Lopez, C.D., Carpio, C., Morales, M.M., Zamorano, M., 2019. A comparative analysis of sustainable building assessment methods. Sustain. Cities and Soc. 49, 101611.

Ludovico, D.D., Fabietti, V., 2018. Strategic environmental assessment, key issues of its effectiveness. the results of the speedy project. Environ. Impact Assess. Rev. 68, 19–28.

Malvestio, A.C., Fischer, T.B., Montano, M., 2018. The consideration of environmental and social issues in transport policy, plan and programme making in Brazil: a systems analysis. J. Clean. Prod. 179, 674–689.

Martin, A., Petersen, M., 2019. Poverty Alleviation as an economic problem. Camb. J. Econ. 43 (1), 205–221.

Moriguchi, Y., Kondo, Y., Shimizu, H., 1993. Analysing the life cycle impact of cars: the case of CO_2. Ind. Environ. 16, 42–45.

Multi-criteria analysis: a manual. 2009. (www.communities.gov.uk).

Nautiyal, H., Singal, S.K., Varun., Sharma A., 2011. Small hydropower for sustainable energy development in India. Renew. Sust. Energy Rev. 15, 2021–2027.

Nautiyal, H., Varun, 2012. Progress in renewable energy under clean development mechanism in India. Renew. Sust. Energy Rev. 16, 2913–2919.

Nautiyal, H., Varun, 2015. Clean development mechanism: a key to sustainable development. In: Thangavel, P., Sridevi, G. (Eds.), Environmental Sustainability. Springer, Cham, pp. 121–128.

Nautiyal, H., Shree, V., Khurana, S., Kumar, N., Varun, 2015. Recycling potential of building materials: a review. In: Muthu, S.S. (Ed.), Environmental Implications of Recycling and Recycled Products. Springer, Cham, pp. 31–50.

Nautiyal, H., Shree, V., Singh, P., Khurana, S., Goel, V., 2018. Life cycle assessment of an academic building: a case study. In: Muthu, S.S. (Ed.), Environmental Carbon Footprints. Elsevier, Amsterdam, pp. 295–315.

Nautiyal, H., Goel, V., Singh, P., 2019. Water footprints of hydropower projects. In: Muthu, S.S. (Ed.), Environmental Water Footprints, Environmental Footprints and Eco-design of Products and Processes. Springer, Cham, pp. 35–46.

Nautiyal, H., Goel, V., 2020. Sustainability assessment of hydropower projects. J. Clean. Prod. 265, 121661.

O'Neill, DW., 2012. Measuring progress in degrowth transition to a steady state economy. Ecol. Econ. 84, 221–231.

O'Neill, DW., 2015. The proximity of nations to a socially sustainable steady-state economy. J. Clean. Prod. 108, 1213–1231.

Poveda, CA., 2017. Assessment approaches, frameworks and other tools. In: Sustainability Assessment, 3–32.

Prieto-Sandoval, V., Jaca, C., Ormazabal, M., 2018. Towards a consensus on the circular economy. J. Clean. Prod. 179, 605–615.

Ramanathan, R., Geetha, S., 2012. Socio-economic impact assessment of industrial projects in India. Impact Assess. Proj. Apprais. 16 (1), 27–32.

Reinhardt, R., Christodoulou, I., Gasso-Domingo, S., Amante García, B., 2019. Towards sustainable business models for electric vehicle battery second use: a critical review. J. Environ. Manage. 245, 432–446.

References

Ritthof, M., Rohn, H., Liedtke, C., 2002. Calculating MIPS: Resource productivity of products and services. Wuppertal Institute for Climate. Environment and Energy at the Science Centre North Rhine-Westphalia.

Roseland, M., 2000. Sustainable community development: integrating environmental, economic, and social objectives. Progr. Plann. 54 (2), 73–132.

Saaty, TL., 2004. Fundametnals of the analytic network process—dependence and feedback in decision making with a single network. J. Syst. Sci. Syst. Eng. 13, 129–157.

Sala, S., Ciuffo, B., Nijkamp, P., 2015. A systemic framework for sustainability assessment. Ecol. Econ. 119, 314–325.

Sharma, A., Saxena, A., Sethi, M., Shree, V., Varun, 2011. Life cycle assessment of buildings: a review. Renew. Sustain. Energy Rev. 15 (1), 871–875.

Shree, V., Goel, V., Nautiyal, H., 2015. Carbon footprint estimation from a building sector in India. In: Muthu, S.S. (Ed.), The Carbon Footprint Handbook. Taylor and Francis, UK, pp. 239–258.

Socio-Economic Impact Assessment Guidelines, 2007. Mackenzie Valley Environmental Impact Review Board (reviewboard.ca).

Stinchcombe, K., Gibson, R.B., 2001. Strategic environmental assessment as a means of pursuing sustainability: ten advantages and ten challenges. J. Environ. Assess. Policy Manage. 3 (3), 343–372.

Sustainabilitydefinition.org.

Tol, RSJ., 2001. Equitable cost-benefit analysis of climate change policies. Ecol. Econ. 36 (1), 71–85.

Tyrvainen, L., 1997. The amenity value of the urban forest: an application of the hedonic pricing method. Landsc. Urban Plan. 37 (3-4), 211–222.

Unalan, D., Cowell, R., 2019. Strategy, context and strategic environmental assessment. Environ. Impact Assess. Rev. 79, 106305.

Use in the production and consumption system—the MIPS approach. Resources 3, 544–574.

Varun, Bhat IK., Prakash, R., 2008. Life cycle analysis of run-of river small hydro power development in India. Open Renew. Energ. J. 1, 11–16.

Varun, Bhat I.K., Prakash, R., 2009. LCA of renewable energy for electricity generation systems-a review. Renew. Sustain. Energy Rev. 13 (5), 1067–1073.

Varun, Prakash R., Bhat, IK., 2010. Life cycle energy and GHG analysis of hydro electric power development in India. Int. J. Green Energy 7 (4), 361–375.

Varun, Sharma A., Shree, V., Nautiyal, H., 2012. Life cycle environmental assessment of an educational building in Northern India: a case study. Sustain. Cities Soc. 4, 22–28.

Varun, Chauhan MK., 2014. Carbon footprint and energy estimation of the sugar industry: an Indian case study. In Muthu, S.S. (Ed.), Assessment of Carbon footprint in Different Industrial Sectors. Springer, Cham, pp. 53–79.

Varun, Sharma A., Nautiyal, H., 2016. Environmental impacts of packaging materials. In: Muthu, S.S. (Ed.), Environmental Footprints of Packaging. Springer, Cham, pp. 115–137.

Wackernagel, M., Rees, W.E., 1996. Our Ecological Footprint: Reducing Human Impact on the Earth. New Society Publishers, Gabriola Island.

Whitehead, J.C., Haab, T.C., 2013. Contingent valuation method. encyclopedia of energy. Nat. Resour. Environ. Econ. 3, 334–341.

www.ecosystemvaluation.org.

CHAPTER 4

Sustainability assessment of energy systems:
Indicators, methods, and applications

Imran Khan

Department of Electrical and Electronic Engineering, Jashore University of Science and Technology, Jashore, Bangladesh

4.1 Introduction

Sustainability issues tackled by the global energy sector are impractical for countries and their electricity authorities to cope with in isolation. New evolving energy policies, regulatory frameworks, and adopting related measures in relation to a sustainable energy sector is a vital goal for a nation's overall sustainable development. Regional as well as national sustainability issues must be taken into account to achieve global sustainable development. Hence, the assessment of an energy sector's sustainability is crucial.

Sustainability pays attention to maintaining a balance between economic needs, social prosperity, and measures to protect the environment. Using current technologies for energy extraction from fossil fuels or nonrenewable energy sources has significant economic, social, and environmental impacts. For instance, to limit the negative environmental impacts, that is, greenhouse gas (GHG) emissions from fossil fuel electricity generation (Khan, 2018), more efficient technologies need to be deployed, requiring huge capital costs, and this is not always an economically viable option for all countries, predominantly the developing economies.

On the other hand, renewable sources such as solar, wind, hydro are being considered as almost emission-free electricity-generation sources. However, the negative environmental impacts on renewable energy sources vary from one technology to another. For example, solar electricity generation is almost emission free. In contrast, the generation from geothermal energy emits considerable amounts of GHGs, depending on the location of the site. In New Zealand, for example, the geothermal power plants emit on average 99 g CO_2-e, which is 14 and 13 times higher than for wind and solar, respectively,[1] although emissions from renewable (Khan et al., 2018) sources are much lower than from nonrenewable (Khan, 2018) sources. Nevertheless, the cost of electricity generation from most renewable sources is higher than for nonrenewable sources. Based on available national energy resources, a sustainable energy system should be designed to attain the goal of global sustainable development. Therefore, a sustainability assessment of energy systems should be conducted through proper methods, so that appropriate steps can be initiated.

---

[1] https://ecotricity.co.nz/is-geothermal-really-renewable-geothermal-emissions-are-now-higher-than-coal-in-the-electricity-sector/ (accessed 24-Jul-2020)

Toward a sustainable future for all, the United Nations declared 17 sustainable goals,[2] of which affordable and clean energy is number seven. Being a fossil fuel–dominated energy sector, about 60% of GHGs is produced globally, and this is responsible for negative climate change.[3] To compare the GHG emission levels of renewable- and nonrenewable-dominated energy systems see, for example Khan, 2019a, 2019b). Hence, goal no. 7 is crucial for a sustainable future. To achieve this goal, a sustainability assessment of the energy sector is one of the preconditions.

The energy sector considers sustainability to be the integration of social, economic, and environmental aspects (Santoyo-castelazo and Azapagic, 2014). These three pillars of sustainability are linked together. The overall sustainability of any system depends on a healthy balance between these three pillars; which is achieved through the help of an institutional state, for example, the government (Khan, 2019).

4.1.1 Principle of sustainability

Generally, sustainability assessment is conducted for an issue that is multidisciplinary in nature. For instance, sustainability considers not only technical and economic factors but also environmental and social issues. In general, sustainability is a concept that involves many different indicators predominantly from the environment, economy, and society (Khan, 2020). Each of these criteria takes into account many different indicators (Khan, 2020).

A sustainable system can be defined as a well-balanced scheme or plan between many different indicators of the economy, society, and environment, which is able to serve present needs without hindering future necessities. Fig. 4.1 shows the concept of sustainability. Often, the economy is termed "profit", society as "people", and environment as "planet". Therefore, a system that can achieve profit (monetary or other), which underpins the development of the people and has no or minimal negative impact on the planet, could be characterized as a sustainable system. A balance between the three pillars must be maintained to obtain sustainable development.

4.1.2 Energy system and sustainability

Energy is an essential need for the global future. However, fossil fuel energy systems are responsible for global warming and negative climate change due to GHG emissions in tandem. To limit global warming below 2 °C is a target that was set at the Paris Climate Agreement in 2015, and this was adopted by 195 countries. To achieve this target, it is an urgent priority to reduce emissions from the energy sector. Thus, transition to a sustainable energy system is a timely demand for the world. Hence, a sustainability assessment for any energy system needs to be conducted to check its status on a sustainability index (SI). For example, Khan (2020) scaled this SI from 0 to 1. If a system obtains an SI value between 0.8 and 1.0 it is said to be a highly sustainable energy system (Khan, 2020). These SI values are illustrated in Fig. 4.2.

Many previous studies have assessed the sustainability of global, regional, and national energy systems (Abu-Rayash and Dincer, 2019, Shaaban et al., 2018, Ren and Dong, 2018, Atilgan and Azapagic,

---

[2] https://www.un.org/sustainabledevelopment/sustainable-development-goals/ (accessed 03-Jul-2020)
[3] https://www.undp.org/content/undp/en/home/sustainable-development-goals/goal-7-affordable-and-clean-energy.html (accessed 03-Jul-2020)

4.1 Introduction

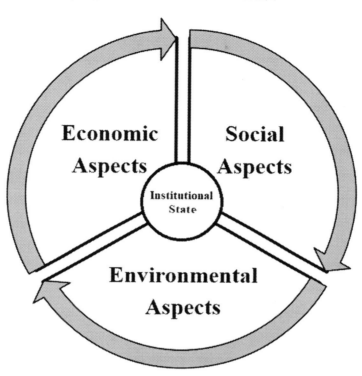

FIG. 4.1 Sustainability concept.

2017, Barros et al., 2015). For example, Maxim (2014) evaluated 13 electricity-generation technologies; and large hydro and coal systems were found to be the most and least sustainable technologies, respectively (Maxim, 2014). In terms of regional sustainability, Khan (2020) assessed the sustainability of the south Asia growth quadrangle region, consisting Bangladesh, Bhutan, India, and Nepal. It was found that this region's energy sector sustainability is dominated by India with a large share of generation fuel mix (Khan, 2020). On the other hand, many studies have conducted national electricity generation sustainability assessment. Thirteen electricity generation options in the United States were assessed and biopower and geothermal generation options were found to be more sustainable. In Bangladesh, solar and hydro were found to be the first and second most sustainable electricity generation technologies, respectively (Khan, 2019).

Although there are few other sustainability assessment methods such as "life cycle sustainability assessment" (Stamford and Azapagic, 2014), the majority of previous studies conducted sustainability assessment through the multicriteria decision analysis (MCDA) method. For this MCDA assessment, it is necessary to identify suitable indicators from the three pillars of sustainability. Therefore, in this chapter, different sustainability indicators and the MCDA method are discussed, along with an application of a method for sustainability assessment.

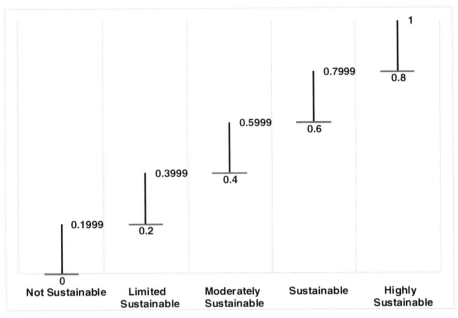

FIG. 4.2 Sustainability index value ranges (Data source: (Khan, 2020)).

The rest of the chapter is organized as follows. Section 4.2 describes the sustainability indicators related to the three pillars of sustainability for the energy sector. Section 4.3 discusses the available MCDA methods frequently used for energy sector's sustainability assessment. In Section 4.4, an MCDA method is applied to show the sustainability assessment procedure for the electricity sector. The final section concludes the chapter.

4.2 Sustainability indicators

Indicators can be defined as quantitative or qualitative measure of something. Sustainability indicators indicate the condition of specific criteria in evaluating the sustainability of a complete system. In energy system sustainability, there are two kinds of indicators: (1) quantitative, such as efficiency, capital costs of technology, and (2) qualitative, such as visual impact, social acceptability of any project. Indicators are essential as they underpin

- defining related policies
- decision making
- balancing the inputs
- identifying the inexplicit effects
- identifying the system's vulnerability
- sorting unpredicted correlation among different criteria.

To select an indicator for any impact assessment, it is always recommended to follow any one of the methods, either top-down or bottom-up. Policy makers decide which indicators to be selected for the impact assessment in the top-down approach. In this approach, sometimes the chosen indicators do not reflect the actual condition of the overall process. In contrast, the primary information from the grassroots level is used to select indicators in the bottom-up approach. The bottom-up approach usually works better than the top-down one for indicator selection in sustainability assessment. In addition, the literature studies also suggest that the bottom-up approach is more suitable for selecting sustainability assessment–related indicators than the top-down approach (Munier, 2005). Often, a hybrid approach, that is, both top-down and bottom-up, might work well than an individual approach (Chamaret et al., 2007).

Although there are many different indicators directly related to technological factors, such as efficiency, capacity factors, all these indicators can be broadly categorized under the three main pillars of sustainability, that is, economic, social, and environmental. For instance, the efficiency improvement of any technology requires additional costs, and is thus included as an economic indicator. Similarly, cultural factors could be integrated into the social indicator category. At the same time, in the literature, some indicators such as heat waves or noise were found to be either in the social or the environmental category (Atilgan and Azapagic, 2016). However, the most common indicators used for sustainability assessment of energy systems found in the literature are listed in Table 4.1.

There might be hundreds of indicators and it is difficult to select the most appropriate set for sustainability assessment of any energy system. Thus, it is necessary to select indicators based on some criteria. One such criterion might be the "what-why/how-whom" framework for indicator selection (Nathan and Reddy, 2011). This framework is depicted in Fig. 4.3.

At the first stage, it is necessary to identify the main objective of the assessment so that proper indicators can be selected that are more closely associated with the project goals. At the same time, the

Table 4.1 List of most common indicators for the energy system sustainability assessment found in the literature.

Criteria	Indicators	References
Economic	Mining and extraction cost, Pretreatment and enrichment cost, Transportation cost, Investment cost, Cost of fuel and CO_2, Operation and maintenance cost, Subsidies, Levelized cost of electricity, Efficiency, Resource availability and limitation, Capacity factor, Generation flexibility.	Khan and Kabir (2020), Barros et al., 2015, Khan (2020), Khan, 2020, Khan (2019), Klein and Whalley (2015), Maxim (2014)

(continued)

Table 4.1 (Cont'd)

Criteria	Indicators	References
Social	Employment generated (new job creation), Population displacement, Development of new areas, Risk of accident, Visual impact, Public health risk, Local economy development, Disturbance to existing social infrastructure and services, Social acceptability, Bird strike risk, Human rights and corruption, Energy security.	Barros et al. (2015), Khan (2019), Atilgan and Azapagic (2016), Klein and Whalley (2015), Maxim (2014), Khan and Kabir (2020)
Environmental	Global warming potential, Ozone depletion potential, Air toxicity, Water ecotoxicity, Acidification potential, Eutrophication potential, Water consumption, Land use, Abiotic depletion, Ecopoints of environmental impact, Noise, Heat wave, Smog, Bad odors, Local/regional/global impact, GHG emissions, Water use, Climate change, Photochemical smog, Human toxicity, Ecotoxicity, Solid wastes, Particulates, Effect on agriculture and seismic activity, River damage	Abu-Rayash and Dincer (2019), Barros et al. (2015), Khan (2020), Atilgan and Azapagic (2016), Klein and Whalley (2015), Maxim (2014), Khan and Kabir (2020)

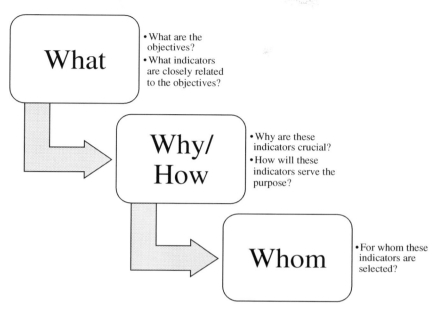

FIG. 4.3 What-why/how-whom framework for indicators selection.

number of indicators for the analysis needs to be selected in an optimal way so that they can represent the whole scenario of the analysis. Notably, very few indicators might not be able to convey the message, whereas a lot of indicators might dilute the purpose of the analysis.

In the second step, it is necessary to justify why the selected indicators are crucial for the analysis. The selected indicators must carry significant information related to the analysis. If the selected indicators are not key indicators, the analysis might result in different findings. Consequently, this will misguide decision makers in designing an appropriate policy. It is also essential to find an answer to the question, "How will the selected indicators serve the purpose of the analysis?" In addition, the indicators should be selected in such a way that information reliability can be ensured.

Finally, it is critical to answer, "To whom will the analysis and related results be communicated, and are they understandable by the target groups?" The indicators must be understandable by different target groups from different disciplines.

4.3 Sustainability assessment methods

Multicriteria analysis involves many different steps, and the main steps are illustrated in Fig. 4.4 (Opricovic and Tzeng, 2004). For energy-related multicriteria analysis, all these steps are also followed.

In the first step, the social, environmental, and economic criteria are selected along with proper indicators. All possible alternatives are then considered in the second step. Based on the selected criteria and indicators, these alternatives are evaluated with the help of a proper MCDA method in the third and fourth steps. Depending on the results of the MCDA method, the optimal solution is

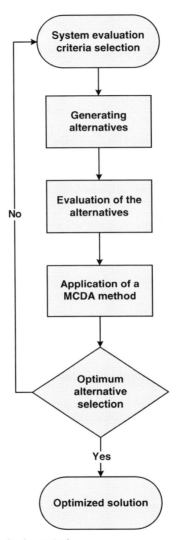

FIG. 4.4 Main steps involved in multicriteria analysis.

selected in the fifth step. In the final step, if the solution is not acceptable, the process starts again from the beginning.

After selection of the indicators, a proper MCDA method should be selected. A proper MCDA method underpins decision making for a multidisciplinary issue with conflicting interests or objectives in a properly defined way. There are many MCDA methods available in the literature, broadly categorized as elementary, unique synthesizing criterion, and outranking (Wang et al., 2017). However, these methods have advantages and limitations (Wu and Geng, 2014, Velasquez and Hester, 2013, Choudhary and Shankar, 2012).

4.3 Sustainability assessment methods

In this chapter, a few common MCDA methods comprising both unique synthesizing criteria and outranking criteria will be discussed.

4.3.1 Multiattribute Value Theory (MAVT)

In this method, partial value functions are determined first, and then weights are assigned to each criterion. The global value function is then calculated using Eq. (4.1) (Atilgan and Azapagic, 2016).

$$G(s) = \sum_{n=1}^{I} w_n v(s) \tag{4.1}$$

where:

$G(s)$—global value function that indicates the total sustainability value/score for the scenario s,
w_n—weight for criterion n for the sustainability indicator; this weight is assigned in accordance with the importance of the indicator,
$v(s)$—for scenario s on indicator n, it indicates the normalized value function reflecting the performance of the scenario,
I—total number of sustainability indicators.

The four basic steps involved in this method are (1) identification of the alternatives that are to be compared with each other, (2) identification of indicators, (3) for all alternatives, values need to be assigned to each indicator, and (4) calculation of a total score for each alternative through the application of a value function to all criteria. For an application of this MAVT method in energy system sustainability assessment, see Santoyo-castelazo and Azapagic (2014). The weighted sum method (WSM) and analytic hierarchy process (AHP) are the two most common methods of MAVT.

4.3.2 Weighted Sum Method (WSM)

The WSM method is one of the very common MCDA methods used for single-dimensional problems. If there are n criteria and N alternatives, the best alternative can be found through the performance score as depicted in Eq. (4.2) (Mateo, 2012, Wang et al., 2009).

$$S_{WSM} = \max_{i} \sum_{j=1}^{n} \alpha_{ij} w_j, \quad \text{for} \quad i = 1, 2, 3, \ldots N \tag{4.2}$$

where:

S_{WSM}—best alternative score
n—number of criteria
N—number of alternatives
α_{ij}—value of the ith alternative for jth criterion
w_j—importance weight for jth criterion

For a detailed energy-related sustainability assessment through the WSM method, see for example, Moreira et al. (2015) and Vo et al. (2017).

4.3.3 Analytic Hierarchy Process (AHP)

In the AHP method, a complex multidisciplinary problem is decomposed into a system of hierarchies. Compared to the WSM method, AHP is able to deal with multidimensional problems. The AHP method involves three main steps (Mateo JRSC 2012, Saaty, 1990): (1) computation of the weight for each criterion, (2) computation of the score of each alternative for each criterion, (3) obtaining the score for alternatives and ranking them.

For the first step, a pairwise comparison matrix (A) is created to compute the weights for different criteria. The importance of jth criterion related to kth criterion is represented by each entry (a_{jk}) to the matrix A. The entries to the matrix must satisfy $a_{jk} \times a_{kj} = 1$ and for all j $a_{jj} = 1$.

Conditions in identifying criteria importance:

- If $a_{jk} > 1$, criterion jth is more important than kth
- If $a_{jk} < 1$, criterion kth is more important than jth
- If $a_{jk} = 1$, both criteria have equal importance

Each entry to the normalized comparison matrix (A_{norm}) is computed as $\hat{a}_{jk} = \dfrac{a_{jk}}{\sum_{l=1}^{m} a_{lk}}$ and the criteria weight matrix (w) is calculated by averaging the entries on each row of the normalized comparison matrix through $w_j = \dfrac{\sum_{l=1}^{m} \hat{a}_{jl}}{m}$.

In the second step, the score matrix is obtained as $S = [s^{(j)}.....s^{(m)}]$, where $s^{(j)}$ is the jth column of matrix S, and $j = 1, ..., m$. Each entry of the S matrix s_{ij} represents the score of the ith option in relation to the jth criterion. Similar to the previous step, a pairwise comparison matrix ($B^{(j)}$) is also formed, in this case with entries as $b_{ih}^{(j)}$. The entries must satisfy $b_{ih}^{(j)} \times b_{hi}^{(j)} = 1$ and $b_{ii}^{(j)} = 1$ for all i. And the conditions are as follows:

- If $b_{ih}^{(j)} > 1$, ith option is more desired than the hth option
- If $b_{ih}^{(j)} < 1$, ith option is less desired than the hth option
- If $b_{ih}^{(j)} = 1$, both options are equivalent

In the final step, the global performance score of AHP is obtained as (Saaty, 1990)

$$G = S \times w \tag{4.3}$$

Notably, the AHP method is able to deal with (1) indicators having no values and/or weights and (2) indicators having both values and weights (Mulliner et al., 2016). For the former, a pairwise comparison matrix formation is used, and for the latter only the final step (Eq. 4.3) of numerical calculation is used. Here, only the former case is shown. An application of this AHP method in the energy sector can be found in Štreimikienė et al. (2016) and Chatzimouratidis and Pilavachi (2009).

4.3.4 Weighted Product Method (WPM)

The WPM method is similar to the WSM. The main difference is that in WPM multiplication needs to be conducted instead of addition. The WPM is also a dimensionless analysis, as it eliminates the units of measurement. A number of ratios are multiplied to each alternative to compare with others for each criterion. If A_M and A_N are two alternatives to be compared, the product needs to be calculated according to Eq. (4.4) (Mateo, 2012, Wang et al., 2009).

$$P\left(\frac{A_M}{A_N}\right) = \prod_{j=1}^{n}\left(\frac{a_{Mj}}{a_{Nj}}\right)^{w_j} \tag{4.4}$$

where:

n—number of criteria
a_{Mj}—Mth alternative's actual value with respect to jth criterion
a_{Nj}—Nth alternative's actual value with respect to jth criterion
w_j—weight of the jth criterion

Conditions:

- If $P\left(\frac{A_M}{A_N}\right) \geq 1$, the A_M is more desirable than A_N
- Best alternative $\geq$ all other alternatives

Being a dimensionless analysis method, one of the major advantages of this MCDA method is that it can handle both single and multidimensional problems. Although not in the energy field, a detailed step-by-step WPM method is employed in Supriyono and Sari (1977).

4.3.5 Technique for Order Preference by Similarity to Ideal Solution (TOPSIS)

This method finds the ideal (also known as "best") and anti-ideal (also known as "worst") solutions and compares the geometric distance between each alternative and the ideal alternative. The common steps in TOPSIS are as follows (Mateo JRSC 2012, Lozano-Minguez et al., 2011, Štreimikiene, 2013, Deveci et al., 2018):

Step 1: Creation of an evaluation matrix.

$$X = \left(x_{ij}\right)_{m \times n} \tag{4.5}$$

Where:

x_{ij}—intersection of each criterion and alternative
m—number of alternatives
n—number of criteria.

Step 2: Matrix normalization.
Using the normalization procedure, the normalized value of each intersection x_{ij} is calculated as

$$r_{ij} = \frac{x_{ij}}{\sqrt{\sum_{i=1}^{m} x_{ij}^2}}, \quad i = 1, 2, \ldots, m; \; j = 1, 2, \ldots, n \tag{4.6}$$

Thus, the normalization matrix is found to be

$$R = \left(r_{ij}\right)_{m \times n} \tag{4.7}$$

Step 3: Weighted normalized decision matrix formation.
The weighted normalized decision matrix can be written as

$$D_{ij} = r_{ij} \times w_j, \quad i = 1, 2, \ldots, m; \; j = 1, 2, \ldots, n \tag{4.8}$$

where w_j is the weight of the jth criterion and $\sum_{j=1}^{n} w_j = 1$, $j = 1, 2, \ldots, n$.

Step 4: Determination of the ideal and anti-ideal alternatives.
Ideal alternative can be found through

$$A_I = \left\{ \left\langle \min\left(D_{ij} | i = 1, 2, \ldots, m\right) | j \in J^- \right\rangle, \left\langle \max\left(D_{ij} | i = 1, 2, \ldots, m\right) | j \in J^+ \right\rangle \right\} \tag{4.9}$$

Anti-ideal alternative can be found through

$$A_{AI} = \left\{ \left\langle \max\left(D_{ij} | i = 1, 2, \ldots, m\right) | j \in J^- \right\rangle, \left\langle \min\left(D_{ij} | i = 1, 2, \ldots, m\right) | j \in J^+ \right\rangle \right\} \tag{4.10}$$

where

J^+—associated with the criterion with a positive impact and $J^+ = \{j = 1, 2, \ldots, n|j\}$
J^-—associated with the criterion with a negative impact and $J^- = \{j = 1, 2, \ldots, n|j\}$

Step 5: Determination of the distance.
Distance between the target alternative and the ideal condition (A_I)

$$d_{iI} = \sqrt{\sum_{j=1}^{n} \left(D_{ij} - D_{Ij}\right)^2}, \quad i = 1, 2, \ldots, m \tag{4.11}$$

Distance between the target alternative and the anti-ideal condition (A_{AI})

$$d_{iAI} = \sqrt{\sum_{j=1}^{n} \left(D_{ij} - D_{AIj}\right)^2}, \quad i = 1, 2, \ldots, m \tag{4.12}$$

Eqs. (4.11) and (4.12) are the popular classical Euclidean distance.
Step 6: Determining the similarity to the anti-ideal condition.

$$S_{iAI} = \frac{d_{iAI}}{\left(d_{iAI} + d_{iI}\right)}, \quad 0 \leq S_{iAI} \leq 1, \; i = 1, 2, \ldots, m \tag{4.13}$$

Conditions:
- If and only if the alternative solution has the ideal condition, $S_{iAI} = 1$
- If and only if the alternative solution has the anti-ideal condition, $S_{iAI} = 0$

Step 7: Preference order ranking.
The alternatives are ranked from ideal (best), that is, higher S_{iAI} value to anti-ideal (worst).
The TOPSIS method applied to energy-related problems can be found for example in Lozano-Minguez et al. (2011) and Brand and Missaoui (2014).

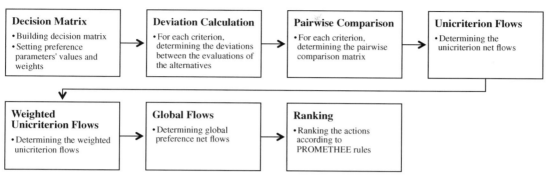

FIG. 4.5 Calculation steps for PROMETHEE method.

4.3.6 Preference Ranking Organization METHod for Enrichment of Evaluations (PROMETHEE)

This MCDA method uses outranking methodology in ranking different alternatives. In brief, this method involves three main steps: pairwise comparison, unicriterion flow computation, and aggregation of all flows into a global flow. Over time, this method has evolved from PROMETHEE I to VI. Thus, it is more convenient to mention the major steps rather than explaining each of them. In general, the steps are shown in Fig. 4.5. Further detail can be found in Greco et al. (2016) and Mateo JRSC (2012). For an application of the PROMETHEE method in the energy sector, see for example Strantzali et al. (2017) and Troldborg et al. (2014a, 2014b).

4.3.7 ELimination Et Coix Traduisant la REalite (ELECTRE)

The ELECTRE method involves two major processes in determining the rank of alternatives. These are (1) the creation of several outranking relations, that is, multiple criteria aggregation process, and (2) an exploitation process. The ELECTRE also has several types such as ELECTRE I to IV. Thus, similar to PROMETHEE, general steps are shown in Fig. 4.6. For the detailed method for ELECTRE, see for example Greco et al. (2016). An application of this method in the energy sector can be found in Catalina et al. (2011).

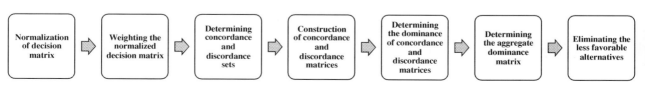

FIG. 4.6 Steps involved in ELECTRE method (Triantaphyllou, 2000).

4.3.8 VIseKriterijumska Optimizacija I Kompromisno Resenje (VIKOR)

The VIKOR method is able to provide a compromise solution for decision makers when it is difficult for them to be certain for any cases at the initial level of the system design. Previous studies have found that a compromise solution is closest to the ideal case and this solution is established through mutual concessions (Opricovic and Tzeng, 2004). The steps involved in this method are (Mateo JRSC 2012, Cristóbal, 2011, Alinezhad and Khalili, 2019) as follows:

Step 1: The best and worst values determination for all the criteria.

In terms of benefit, if the ith function represent a benefit, it can be written as $f_i^* = \max_j f_{ij}$ and $f_i^- = \min_j f_{ij}$ where f_i^* and f_i^- are the best and worst values for all the criteria function and $i = 1, 2, \ldots, n$.

Step 2: Computation the values of S_j and R_j.

$$S_j = \sum_{i=1}^{n} w_i \frac{\left(f_i^* - f_{ij}\right)}{\left(f_i^* - f_i^-\right)} \tag{4.14}$$

$$R_j = \max_i \left[w_i \frac{\left(f_i^* - f_{ij}\right)}{\left(f_i^* - f_i^-\right)} \right] \tag{4.15}$$

where w_i are the weights of the criteria, this expresses the decision makers preferences according to their relative importance and $i = 1, 2, \ldots, n$, and $j = 1, 2, \ldots, m$.

Step 3: Computation of Q_j.

$$Q_j = v \frac{\left(S_j - S^*\right)}{\left(S^- - S^*\right)} + (1 - v) \frac{\left(R_j - R^*\right)}{\left(R^- - R^*\right)} \tag{4.16}$$

where

v—weight of the strategy of the majority of criteria

$(1 - v)$—weight of the individual regret

$S^* = \min S_j$

$S^- = \max S_j$

$R^* = \min R_j$

$R^- = \max R_j$

$j = 1, 2, \ldots, m$.

Step 4: Ranking the alternatives.

The ranking would be three lists as per the values of S, R, and Q and they need to be in an ascending order.

Step 5: Proposing a compromise solution.

An alternative (A') would be ranked the best by the minimum value of Q if the following conditions are satisfied (Opricovic and Tzeng, 2004):

Condition 1: Acceptable advantage

$$Q(A'') - Q(A') \geq DQ \tag{4.17}$$

where

A''—the second alternative in the list obtained from the values of Q.

$DQ = \dfrac{1}{(J-1)}$, J is the number of alternatives.

Condition 2: Acceptable stability in decision making.

"Alternative A' must also be the best ranked by S and/or R. This compromise solution is stable within a decision making process, which could be: "voting by majority rule" (when $v > 0.5$ is needed), or "by consensus" $v \approx 0.5$, or "with veto" ($v < 0.5$)" (Opricovic and Tzeng, 2004).

If none of the condition is satisfied, the following set of composite solutions is proposed:

- If only Condition 2 is not satisfied, the composite solutions would be both A' and A'', or
- If Condition 1 is not satisfied, the composite solutions would be alternatives A', A'', ..., A^m; and A^m is determined by $Q(A^m) - Q(A') < DQ$ for maximum M (the positions of these alternatives are "in closeness") (Opricovic and Tzeng, 2004).

Examples of the application of the VIKOR method in an energy system can be found in Vučijak et al. (2013) and Cristóbal (2011).

4.3.9 COmplex PRoportional ASsessment (COPRAS)

The COPRAS method is becoming popular in the energy sector's multicriteria analysis due to its features. For instance, this method considers independent attributes, and the qualitative attributes are usually converted into quantitative ones, and this is a compensatory method. The method involves the following six steps (Chatterjee et al., 2011, Chatterjee and Bose, 2012, Büyüközkan et al., 2018):

Step 1: Development of the initial decision matrix.

$$X = \left[x_{ij} \right]_{n \times m}; \quad i = 1, 2, \ldots, n \quad \text{and} \quad j = 1, 2, \ldots, m \tag{4.18}$$

where

x_{ij}—the performance value of ith alternative on jth criterion
n—number of alternatives
m—number of criteria

Step 2: Normalized decision matrix formation.

To compare different criteria, normalization is conducted so that dimensionless values can be obtained. The normalization is conducted as

$$r_{ij} = \dfrac{x_{ij}}{\sum_{i=1}^{n} x_{ij}}; \quad j = 1, 2, \ldots, m \tag{4.19}$$

Step 3: Weighted normalized decision matrix formation.

$$a_{ij} = w_j \times r_{ij} \tag{4.20}$$

where $a_{ij} = \dfrac{w_j x_{ij}}{\sum_{i=1}^{n} x_{ij}}$ and w_j is the weight of the criteria $[w_1, w_2, \ldots, w_m]$.

Step 4: Calculation of maximizing and minimizing indexes.

$$\text{Maximizing index, } S_{+i} = \sum_{+j} a_{ij} \tag{4.21}$$

$$\text{Minimizing index, } S_{-i} = \sum_{-j} a_{ij} \tag{4.22}$$

where (+) indicates beneficial attributes or indicators and (−) indicates nonbeneficial or cost attributes or indicators. High scores in beneficial and low scores in nonbeneficial attributes are preferred.

Step 5: Calculating the overall performance value.

$$P_i = S_{+i} + \dfrac{\sum_{i=1}^{n} S_{-i}}{S_{-i} \sum_{i=1}^{n} \dfrac{1}{S_{-i}}} \tag{4.23}$$

Step 6: Final ranking of the alternatives.

The performance scores of the alternatives are ranked in a descending order. The highest rank will be assigned to the highest final score.

The application of the COPRAS method in the energy field can be found in Büyüközkan et al. (2018) and Chatterjee and Bose (2012).

4.3.10 Other methods

Many other MCDA methods are available in the literature for multicriteria decision-making in many different fields. For instance, additive ratio assessment (ARAS), weighted aggregates sum product assessment (WASPAS), stepwise weight assessment ratio analysis (SWARA), analytic network process (ANP). However, their application in the energy field is limited. and these methods are not discussed in this chapter. For further details of these methods, see for example Alinezhad and Khalili (2019).

4.4 Sustainability assessment: an application of COPRAS method

To better understand the sustainability assessment of energy systems it is more convenient to apply one of the MCDA methods. Hence, in this section, the COPRAS method is applied to assess the sustainability of common electricity generation technologies including renewable, nonrenewable, and nuclear. In terms of the indicators for the three pillars of sustainability, these are considered from two previous studies (Khan, 2019, Khan, 2020). These indicators are listed in Table 4.2. The generation technologies that have been taken into account for this sustainability assessment are gas, oil, coal, nuclear, hydro, solar, wind, and biomass. The indicators' value for respective technologies is listed in Table 4.3.

There are a number of reasons behind the selection of the indicators chosen for this analysis. First, both quantitative and qualitative indicators are selected. The economic and environmental indicators are the quantitative ones, whereas social indicators are qualitative. Notably, the qualitative indicators are translated into quantitative values by using a relative scale of very high to none (i.e., very high, high,

4.4 Sustainability assessment: an application of COPRAS method

Table 4.2 List of sustainable criteria and indicators used for this case study.

Criteria	Indicators [unit]	References
Economic	Efficiency (Eff.) [%]	Khan (2020)
	Capacity factor (C.F.) [%]	
	Levelized cost of electricity (L.C.O.E.) [USD/kWh]	
Environmental	Greenhouse gas emission (GHG.E.) [gCO$_2$-e]	Khan (2020)
	Land use (L.U.) [m^2/kWh]	
	Water consumption (W.C.) [L/kWh]	
Social	Odor [qualitative measure]	Khan (2019)
	Public health risk (P.H.R.) [qualitative measure]	
	New job creation (N.J.C.) [qualitative measure]	

moderately high, moderate, low, moderately low, very low, and none) for the eight technologies considered for this analysis (Khan, 2019). Second, the selected economic indicators are the most crucial parameters for electricity generation technologies. For instance, the LCOE covers all the costs involved in generating the electricity from capital to maintenance. Third, the environmental indicators are the very basic parameters in relation to electricity generation and its related impact on the environment. Finally, the social indicators include both positive and negative indicators. For example, odor and public health risks are the negative impacts of electricity generation technologies. On the other hand, new job creation is one of the positive impacts of electricity generation projects.

In terms of weight assignment for the COPRAS method, considering expert opinion is one solution. Generally, a numerical scale is defined by assigning numbers to the selected indicators by experts according to their importance to them. For instance, a scale of 1 to 10 could be used to assign points to each indicator. As an illustrative example, the weight assignment for the listed indicators is shown in Table 4.3. In actual analysis, these data must be collected through expert opinion surveys.

Taking into account all the steps listed in COPRAS method, the results are as follows: Table 4.3 presents the decision matrix (Step 1) for this analysis. Considering Steps 2 and 3, the weighted normalized matrix can be obtained, as shown in Table 4.4.

Table 4.3 Sustainability indicators' value used for this analysis (Data Source: (Khan, 2020, Khan, 2019)).

Generation technology (other)	Economic indicators			Environmental indicators			Social indicators		
	Eff.	C.F.	L.C.O.E.	GHG.E.	L.U.	W.C.	Odor	P.H.R.	N.J.C.
(Assumed experts' weight)	8	9	8	10	8	9	7	8	8
Gas	49	11.4	0.048	400	0.0002	0.36	6	4	6
Oil	40	7.8	0.054	705	0.0003	1	7	5	5
Coal	38	63.8	0.042	800	0.005	2	8	7	8
Nuclear	33	90.3	0.043	16	0.0001	2	3	8	3
Hydro	90	39.8	0.051	41	0.01	34	4	1	2
Solar	13	33.9	0.242	4	0.01	0.008	1	2	4
Wind	39	30	0.056	35	0.001	0.004	2	3	1
Biomass	30	33.9	0.054	30	0.001	1.16	5	6	7

Table 4.4 Weighted normalized decision matrix.

Technology	Economic indicators			Environmental indicators			Social indicators		
Gas	0.0472	0.0132	0.0260	0.0729	0.0021	0.0029	0.0507	0.0386	0.0579
Oil	0.0385	0.0090	0.0292	0.1285	0.0032	0.0082	0.0591	0.0483	0.0483
Coal	0.0366	0.0738	0.0227	0.1458	0.0536	0.0164	0.0676	0.0676	0.0772
Nuclear	0.0318	0.1045	0.0233	0.0029	0.0010	0.0164	0.0253	0.0772	0.0289
Hydro	0.0867	0.0460	0.0276	0.0074	0.1073	0.2796	0.0338	0.0096	0.0193
Solar	0.0125	0.0392	0.1312	0.0007	0.1073	6.58E-05	0.0084	0.0193	0.0386
Wind	0.0375	0.0347	0.0303	0.0063	0.0107	3.29E-05	0.0169	0.0289	0.0096
Biomass	0.0289	0.0392	0.0292	0.0054	0.0107	0.0095	0.0422	0.0579	0.0676

In Step 4, the maximizing and minimizing indices are calculated and these are listed in Table 4.5.

In Steps 5 and 6, the performance score is calculated as per Eq. (4.23) and the final rankings are listed in Table 4.6.

4.4.1 Results and discussion

To compare the ranks between different technologies, results are depicted in Fig. 4.7. For three different criteria the ranks are shown. With respect to the sustainability status, top-ranked economic performance was found for nuclear followed by hydro. In contrast, the least economically viable technologies were found to be solar followed by oil.

In terms of environmental ranking, wind was found to be the best technology followed by nuclear. The least technologies were found to be hydro and coal. On the other hand, for the social criteria, the best technology was found to be solar and the second best was hydro, whereas nuclear was the least technology for the social criteria. Overall, including all the three sustainability criteria, the COPRAS method revealed that wind is the most sustainable technology in this case, followed by nuclear and hydro. On the other hand, oil was found to be the least sustainable technology.

There are a few crucial factors that need to be taken into account for MCDA for any multidisciplinary issues. In this case, for instance, the selection of sustainability indicators plays a vital role toward the final result of the analysis. It is an essential requirement for any multicriteria analysis to

Table 4.5 Maximizing and minimizing indices.

Technology	Economic indicators		Environmental indicators		Social indicators	
	S+	S−	S+	S−	S+	S−
Gas	0.060429	0.026034	0	0.078051	0.057971	0.089372
Oil	0.047586	0.029288	0	0.140007	0.048309	0.107488
Coal	0.110502	0.02278	0	0.216012	0.077295	0.135266
Nuclear	0.136368	0.023322	0	0.020439	0.028986	0.102657
Hydro	0.132833	0.027661	0	0.394445	0.019324	0.043478
Solar	0.051784	0.131254	0	0.108149	0.038647	0.027778
Wind	0.072328	0.030373	0	0.017151	0.009662	0.045894
Biomass	0.068169	0.029288	0	0.025746	0.067633	0.100242

4.4 Sustainability assessment: an application of COPRAS method

Table 4.6 Final score and ranking obtained from the COPRAS method.

Technology	Economic		Environmental		Social	
	P_i	Rank	P_i	Rank	P_i	Rank
Gas	0.1059	6	0.0702	4	0.1148	6*
Oil	0.0880	7	0.0391	6	0.0955	7
Coal	0.1624	3	0.0253	7	0.1148	5*
Nuclear	0.1871	1	0.2681	2	0.0784	8
Hydro	0.1756	2	0.0138	8	0.1361	2
Solar	0.0608	8	0.0506	5	0.2215	1
Wind	0.1113	4	0.3195	1	0.1203	3
Biomass	0.1086	5	0.2129	3	0.1183	4

*Further decimal places are considered for this ranking (not shown here).

select the most important indicators. Otherwise, the obtained results would vary significantly. Thus, it is recommended to consult with relevant experts when selecting the indicators.

Another parameter for which the results would be significantly different is the number of sustainability indicators. Increasing the number of indicators does not ensure the optimal result; rather, it might skew the main findings. To check the impact of the number of indicators on the final result, the number of environmental and social indicators was increased to 4 and 6, respectively. The new indicators for environmental criteria were "water withdrawal" and for social these were "noise," "smog," and

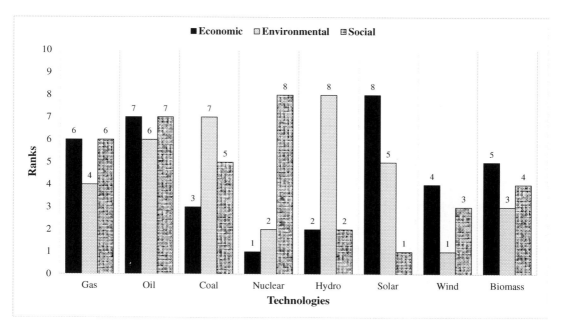

FIG. 4.7 Ranks obtained for different technologies using COPRAS method.

Table 4.7 Impact of experts' weight variation.

Assumed experts' Weight	Economic indicators			Environmental indicators			Social indicators		
	Eff.	C.F.	L.C.O.E.	GHG.E.	L.U.	W.C.	Odor	P.H.R.	N.J.C.
Used	8	9	8	10	8	9	7	8	8
New	9	8	10	10	6	9	7	10	9

"local economy development." The new social indicators consider both the positive (i.e., local economy development) and negative (i.e., noise and smog) indicators, as before. Although there were no changes observed for the top-ranked technology in each criterion, ranking was changed at other positions. For example, in environmental criteria, nuclear occupied second position, with a total of 9 indicators and it came third with 13 indicators. Biomass occupied the second position in environmental criteria for the latter case. Therefore, it is suggested that the number of indicators should be selected in the optimal way with the help of expert opinion.

Assignment of weights to the indicators is another vital factor for any MCDA method. If the weight assignment is not in accordance with expert opinion, the final result might be biased, and it would be difficult for decision makers to make proper decisions in relation to sustainable technology selection. For example, to check the impact of experts' weight assignment on final results, the assigned weights are changed, as listed in Table 4.7.

Although the first and second ranked technologies were not changed due to this weight change, impacts were observed for individual criterion. For instance, the economic indicators for gas and biomass were improved and degraded, respectively. Similarly, the social indicators for gas and coal were improved and degraded, respectively. On the other hand, no changes were observed for the environmental indicators as only the weight of "land use" was changed and the other two were kept unchanged (c.f., Table 4.7).

Sensitivity analysis for the result obtained through any method is another necessary part to be checked. In sensitivity analysis, the uncertainty of the result is checked through changing input parameters. If a small change in the input parameter changes the output or result significantly, the output obtained is proved to be not robust. Therefore, it is important to conduct sensitivity analysis for any output obtained through any MCDA method to check the method's robustness. For further detail on sensitivity analysis of MCDA method see Atilgan and Azapagic (2017), Catalina et al. (2011), Vo et al. (2017), and Ren, 92018).

The way in which the impacts of the number of indicator variations and weight variations were checked could be treated as the sensitivity analysis of the COPRAS method. Hence, it could be claimed that the COPRAS method is one of the robust MCDA methods, as the final best result did not vary due to the input parameter changes. For an MCDA method to be robust, it should meet some conditions: (1) the method should have little or no influence of scale and shape, (2) it should be able to adapt the input parameter change within certain limits, (3) the method should work independently irrespective of its application, and (4) it should incorporate the decision makers choice in evaluating the assessed criteria.

Although a single MCDA method is able to analyze and produce results to make decisions on energy issues, studies have found that the use of more than one method would be suitable for more optimal decision making. For example, Khan (2020) employed two MCDA methods to assess the sustainability status of Bangladesh's power generation expansion plan and found that results vary significantly from one method to another (Khan, 2020). Therefore, to make a more optimal decision, it is suggested that more than one MCDA method for energy sector sustainability assessment is used.

4.5 Conclusion

Sustainable development is one of the prerequisites for reducing negative climate change. One of the dominant sectors that could potentially contribute to achieving this sustainable goal is the energy sector. However, to make the energy system a sustainable one, it is necessary to assess its sustainability status (Khan, 2021). In this chapter, the sustainability concept for energy systems was discussed briefly. The many different indicators related to energy system sustainability were then presented with the help of the recent literature. Different MCDA methods frequently applied for the sustainability assessment of energy systems in the literature were presented. Finally, an MCDA method (COPRAS) was applied as an example to assess the sustainability status of different electricity generation technologies, and to show its application in the electricity generation sector.

Acknowledgement

The author would like to thank Md. Sahabuddin, Faculty Member, Jashore University of Science and Technology, Bangladesh for receiving comments on MCDA methods.

References

Abu-Rayash, A., Dincer, I., 2019. Sustainability assessment of energy systems: a novel integrated model. J. Clean. Prod. 212, 1098–1116. doi:10.1016/j.jclepro.2018.12.090.
Alinezhad, A., Khalili, J., 2019. New Methods and Applications in Multiple Attribute Decision Making (MADM). vol. 277. Springer Nature, Switzerland. doi:10.1007/978-3-030-15009-9.
Atilgan, B., Azapagic, A., 2016. An integrated life cycle sustainability assessment of electricity generation in Turkey. Energy Policy 93, 168–186. doi:10.1016/j.enpol.2016.02.055.
Atilgan, B., Azapagic, A., 2017. Energy challenges for Turkey: identifying sustainable options for future electricity generation up to 2050. Sustain. Prod. Consum. 12, 234–254. doi:10.1016/j.spc.2017.02.001.
Barros, C., Coira, M.L., Pilar, M., Cruz, D., Jos, J., 2015. Assessing the global sustainability of different electricity generation systems. Energy 89, 473–489. doi:10.1016/j.energy.2015.05.110.
Brand, B., Missaoui, R., 2014. Multi-criteria analysis of electricity generation mix scenarios in Tunisia. Renew. Sustain. Energy Rev 39, 251–261. doi:10.1016/j.rser.2014.07.069.
Büyüközkan, G., Karabulut, Y., Mukul, E., 2018. A novel renewable energy selection model for United Nations' sustainable development goals. Energy 165, 290–302. doi:10.1016/j.energy.2018.08.215.
Catalina, T., Virgone, J., Blanco, E., 2011. Multi-source energy systems analysis using a multi-criteria decision aid methodology. Renew. Energy 36, 2245–2252. doi:10.1016/j.renene.2011.01.011.
Chamaret, A., O'Connor, M., Récoché, G., 2007. Top-down/bottom-up approach for developing sustainable development indicators for mining: application to the Arlit uranium mines (Niger). Int. J. Sustain. Dev. 10, 161–174. doi:10.1504/IJSD.2007.014420.

Chatterjee, N.C., Bose, G.K., 2012. A COPRAS-F base multi-criteria group decision making approach for site selection of wind farm. Decis. Sci. Lett. 2, 1–10. doi:10.5267/j.dsl.2012.11.001.

Chatterjee, P., Athawale, V.M., Chakraborty, S., 2011. Materials selection using complex proportional assessment and evaluation of mixed data methods. Mater. Des. 32, 851–860. doi:10.1016/j.matdes.2010.07.010.

Chatzimouratidis, A.I., Pilavachi, P.A., 2009. Technological, economic and sustainability evaluation of power plants using the Analytic Hierarchy Process. Energy Policy 37, 778–787. doi:10.1016/j.enpol.2008.11.021.

Choudhary, D., Shankar, R., 2012. An STEEP-fuzzy AHP-TOPSIS framework for evaluation and selection of thermal power plant location: a case study from India. Energy 42, 510–521. doi:10.1016/j.energy.2012.03.010.

Cristóbal, J.R.S., 2011. Multi-criteria decision-making in the selection of a renewable energy project in Spain: the VIKOR method. Renew. Energy 36, 498–502. doi:10.1016/j.renene.2010.07.031.

Deveci, M., Canıtez, F., Gökaşar, I., 2018. WASPAS and TOPSIS based interval type-2 fuzzy MCDM method for a selection of a car sharing station. Sustain. Cities. Soc 41, 777–791. doi:10.1016/j.scs.2018.05.034.

Greco, S., Ehrgott, M., Figueira, J.R., 2016, second ed.Multiple Criteria Decision Analysis233 Springer, New York. doi:10.1007/978-1-4939-3094-4.

Khan, I., 2018. Importance of GHG emissions assessment in the electricity grid expansion towards a low-carbon future: a time-varying carbon intensity approach. J. Clean. Prod. 196, 1587–1599. doi:10.1016/j.jclepro.2018.06.162.

Khan, I., 2019. Power generation expansion plan and sustainability in a developing country: a multi-criteria decision analysis. J. Clean. Prod. 220, 707–720. doi:10.1016/J.JCLEPRO.2019.02.161.

Khan, I., 2019a. Temporal carbon intensity analysis: renewable versus fossil fuel dominated electricity systems. Energ. Source. Part A 41, 309–323. doi:10.1080/15567036.2018.1516013.

Khan, I., 2019. A Temporal Approach to Characterizing Electrical Peak Demand: Assessment of GHG Emissions at the supply SIDE and Identification of Dominant Household Factors at the Demand Side. University of Otago.

Khan, I., 2020. Data and method for assessing the sustainability of electricity generation sectors in the south Asia growth quadrangle. Data Br. 28, 1–8. doi:10.1016/j.dib.2019.104808.

Khan, I., 2020. Impacts of energy decentralization viewed through the lens of the energy cultures framework: solar home systems in the developing economies. Renew. Sustain. Energy Rev. 119, 1–11. doi:10.1016/j.rser.2019.109576.

Khan, I., 2020. Sustainability challenges for the south Asia growth quadrangle: a regional electricity generation sustainability assessment. J. Clean. Prod. 243, 1–13. doi:10.1016/j.jclepro.2019.118639.

Khan, I., Jack, M.W., Stephenson, J., 2018. Analysis of greenhouse gas emissions in electricity systems using time-varying carbon intensity. J. Clean. Prod. 184, 1091–1101. doi:10.1016/j.jclepro.2018.02.309.

Khan, I., Kabir, Z., 2020. Waste-to-energy generation technologies and the developing economies: a multi-criteria analysis for sustainability assessment. Renew. Energy 150, 320–333. doi:10.1016/j.renene.2019.12.132.

Khan, I., 2021. Sustainable Energy Infrastructure Planning Framework: Transition to a Sustainable Electricity Generation System in Bangladesh. In: Asif, M. (Ed.), Energy and Environmental Security in Developing Countries, 1st ed. Springer, Cham, pp. 173–198. https://doi.org/10.1007/978-3-030-63654-8_7.

Klein, S.J.W., Whalley, S., 2015. Comparing the sustainability of U.S. electricity options through multi-criteria decision analysis. Energy Policy 79, 127–149. doi:10.1016/j.enpol.2015.01.007.

Lozano-Minguez, E., Kolios, A.J., Brennan, F.P., 2011. Multi-criteria assessment of offshore wind turbine support structures. Renew. Energy 36, 2831–2837. doi:10.1016/j.renene.2011.04.020.

Mateo JRSC, 2012. VIKOR Eds. In: Cristobal, S, Ramon, J (Eds.), Multi Criteria Analysis in the Renewable Energy Industryfirst ed. Springer-Verlag, London, pp. 49–53. doi:10.1007/978-1-4471-2346-0_8.

Mateo JRSC, 2012. PROMETHEE Eds. In: Cristobal, S, Ramon, J (Eds.), Multi Criteria Analysis in the Renewable Energy Industryfirst ed. Springer-Verlag, London, pp. 23–32. doi:10.1007/978-1-4471-2346-0_5.

Mateo JRSC, 2012. TOPSIS. In: Cristobal, S, Ramon, J (Eds.), Multi Criteria Analysis in the Renewable Energy Industry1st ed. Springer-Verlag, London, pp. 43–48. doi:10.1007/978-1-4471-2346-0_7.

Mateo JRSC, 2012. AHP. In: Cristobal, S, Ramon, J (Eds.), Multi Criteria Analysis in the Renewable Energy Industryfirst ed. Springer-Verlag, London, pp. 11–17. doi:10.1007/978-1-4471-2346-0_3.

Mateo, J., 2012. Weighted sum method and weighted product method. In: Cristobal, S, Ramon, J (Eds.), Multi Criteria Analysis in the Renewable Energy Industryfirst ed. Springer-Verlag, London, pp. 19–22. doi:10.1007/978-1-4471-2346-0_4.

Maxim, A., 2014. Sustainability assessment of electricity generation technologies using weighted multi-criteria decision analysis. Energy Policy 65, 284–297. doi:10.1016/j.enpol.2013.09.059.

Moreira, J.M.L., Cesaretti, M.A., Carajilescov, P., Maiorino, J.R., 2015. Sustainability deterioration of electricity generation in Brazil. Energy Policy 87, 334–346. doi:10.1016/j.enpol.2015.09.021.

Mulliner, E., Malys, N., Maliene, V., 2016. Comparative analysis of MCDM methods for the assessment of sustainable housing affordability. Omega 59, 146–156. doi:10.1016/j.omega.2015.05.013.

Munier, N., 2005. Introduction to Sustainability: Road to a Better Future, first ed. Springer, Dordrecht, The Netherlands. doi:10.1007/1-4020-3558-6.

Nathan, H.S.K., Reddy, B.S., 2011. Criteria selection framework for sustainable development indicators. Int. J. Multicriteria. Decis. Mak. 1, 257–279. doi:10.1504/IJMCDM.2011.041189.

Opricovic, S., Tzeng, G.H., 2004. Compromise solution by MCDM methods: a comparative analysis of VIKOR and TOPSIS. Eur. J. Oper. Res 156, 445–455. doi:10.1016/S0377-2217(03)00020-1.

Ren, J., 2018. Sustainability prioritization of energy storage technologies for promoting the development of renewable energy: a novel intuitionistic fuzzy combinative distance-based assessment approach. Renew. Energy 121, 666–676. doi:10.1016/j.renene.2018.01.087.

Ren, J., Dong, L., 2018. Evaluation of electricity supply sustainability and security: multi-criteria decision analysis approach. J. Clean. Prod. 172, 438–453. doi:10.1016/j.jclepro.2017.10.167.

Saaty, T.L., 1990. How to make a decision: the analytic hierarchy process. Eur. J. Oper. Res 48, 9–26. doi:10.1016/0377-2217(90)90057-I.

Santoyo-castelazo, E., Azapagic, A., 2014. Sustainability assessment of energy systems : integrating environmental, economic and social aspects. J. Clean. Prod. 80, 119–138. doi:10.1016/j.jclepro.2014.05.061.

Shaaban, M., Scheffran, J., Böhner, J., Elsobki, M.S., 2018. Sustainability assessment of electricity generation technologies in Egypt using multi-criteria decision analysis. Energies 11, 1–25. doi:10.3390/en11051117.

Stamford, L., Azapagic, A., 2014. Life cycle sustainability assessment of UK electricity scenarios to 2070. Energy Sustain. Dev. 23, 194–211. doi:10.1016/j.esd.2014.09.008.

Strantzali, E., Aravossis, K., Livanos, G.A., 2017. Evaluation of future sustainable electricity generation alternatives : the case of a Greek island. Renew. Sustain. Energy Rev 76, 775–787 http://dx.doi.org/10.1016/j.rser.2017.03.085.

Štreimikiene, D., 2013. Assessment of energy technologies in electricity and transport sectors based on carbon intensity and costs. Technol. Econ. Dev. Econ 19, 606–620. doi:10.3846/20294913.2013.837113.

Štreimikienė, D., Šliogerienė, J., Turskis, Z., 2016. Multi-criteria analysis of electricity generation technologies in Lithuania. Renew. Energy 85, 148–156. doi:10.1016/j.renene.2015.06.032.

Supriyono, H., Sari, C.P., 1977. Developing decision support systems using the weighted product method for house selection. AIP. Conf. Proc. 2018. doi:10.1063/1.5042905.

Triantaphyllou, E., 2000. Multi-Criteria Decision Making Methods: A Comparative Study, first ed. Springer, Dordrecht. doi:10.1007/978-1-4757-3157-6.

Troldborg, M., Heslop, S., Hough, R.L., 2014. Assessing the sustainability of renewable energy technologies using multi-criteria analysis : suitability of approach for national-scale assessments and associated uncertainties. Renew. Sustain. Energy Rev 39, 1173–1184. doi:10.1016/j.rser.2014.07.160.

Velasquez, M., Hester, P., 2013. An analysis of multi-criteria decision making methods. Int. J. Oper. Res 10, 56–66.

Vo, T.T.Q., Xia, A., Rogan, F., Wall, D.M., Murphy, J.D., 2017. Sustainability assessment of large-scale storage technologies for surplus electricity using group multi-criteria decision analysis. Clean. Technol. Environ. Policy 19, 689–703. doi:10.1007/s10098-016-1250-8.

Vo, T.T.Q., Xia, A., Rogan, F., Wall, D.M., Murphy, J.D., 2017. Sustainability assessment of large-scale storage technologies for surplus electricity using group multi-criteria decision analysis. Clean. Technol. Environ. Policy 19, 689–703. doi:10.1007/s10098-016-1250-8.

Vučijak, B., Kupusović, T., MidŽić-Kurtagić, S., Ćerić, A., 2013. Applicability of multicriteria decision aid to sustainable hydropower. Appl. Energy 101, 261–267. doi:10.1016/j.apenergy.2012.05.024.

Wang, H., Duanmu, L., Lahdelma, R., Li, X., 2017. Developing a multicriteria decision support framework for CHP based combined district heating systems. Appl. Energy 205, 345–368. doi:10.1016/j.apenergy.2017.07.016.

Wang, J.J., Jing, Y.Y., Zhang, C.F., Zhao, J.H., 2009. Review on multi-criteria decision analysis aid in sustainable energy decision-making. Renew. Sustain. Energy Rev 13, 2263–2278. doi:10.1016/j.rser.2009.06.021.

Wu, Y., Geng, S., 2014. Multi-criteria decision making on selection of solar – wind hybrid power station location : a case of China. Energy Convers. Manag. J 81, 527–533.

CHAPTER 5

Sustainability measurement
Evolution and methods

Mariolina Longo, Matteo Mura, Chiara Vagnini, Sara Zanni
Department of Management, University of Bologna, Via Terracini, Bologna, Italy

5.1 Why measuring sustainability matters in the current business landscape

Sustainability is now part of everyday life. It is sufficient to analyze how many times the term *sustainability* appears in academic journals and nonspecialized newspapers to understand that its relevance has become a distinctive trait of our times. As highlighted by the international community, CO_2 emissions have grown significantly over the years, leading to greater concerns about their impacts, and demanding considerable determination of mitigation processes. Current society, future generations, and the planet do not have time left anymore for avoiding environmental catastrophic effects such as climate change, loss of biodiversity, and resource depletion. It is therefore topical to deal with sustainability and its implications for the effects on economy and everyday life.

In this context, the industrial sector plays a key role, as it is responsible for a third of total global greenhouse gas (GHG) emissions (Fischedick et al., 2014). Consequently, sustainability measurement in organizations has gained particular relevance both in the academic research and in the practice. Several theoretical frameworks for measuring sustainability have been proposed, and sustainability reporting and disclosure started to play a major role on executives' agenda. Organizations have reconsidered the whole process of value creation and to measure their performances accordingly. Instead of focusing on shareholders' value-maximization logics, companies have adopted a systemic approach to value creation, which involved measuring and disclosing their performances to stakeholders according to both an economic, social, and environmental bottom line (Neely et al., 2002; Elkington, 1997). At the same time, new reporting policies have forced companies to disclose their sustainability-related practices with proper tools and, on the other hand, consumers have become increasingly attentive and aware on sustainability practices, pushing companies to act accordingly.

However, measuring sustainability can be a tricky practice for current businesses. A comprehensive, standardized, and globally accepted framework for sustainability assessment at company level is still under debate (Searcy and Elkhawas, 2012), additionally existing frameworks (i.e., GRI, CDP, ASSET4) mostly focus on large and listed organizations, while small- and medium-sized enterprises are excluded. Furthermore, research on sustainability measurement tends to use a precise scale of analysis, which usually is either a lower scale that focuses on corporate logics for assessing sustainability, or an upper scale that defines sustainability measurement at the level of industrial ecosystems (Coenen et al., 2012; Martin and Sunley, 2006).

Despite a significant fragmentation, the field of sustainable measurement is evolving, and novel approaches adopting a systemic reasoning are being explored, as the case of *multiscalarity*.

5.2 The evolution of sustainability measurement research
5.2.1 Literature intellectual structure

Research in sustainability measurement has grown considerably over the past years, and it has explored a variety of issues, from sustainability disclosure to measurement in green supply chains, from the diffusion of environmental standards to the political use of sustainability metrics. Also, articles have been published in a variety of journals, and authors have used different terms, methods, and theoretical frameworks.

The comprehensive literature review on sustainability measurement has been conducted with the aims of bringing together different strands of the literature, identifying main contributions and unanswered questions, and mapping the evolution of sustainability measurement research by highlighting current and emerging trends (Mura et al., 2018). By analyzing 721 articles published from 1992 to 2016, authors revealed the breadth and rapid growth of research on sustainability measurement. The number of articles published on the topic has grown exponentially since 1992 (Fig. 5.1). In particular, the field has experienced an impressive growth over the last three years, as half the papers in the sample were published in 2013–2016. The top 20 journals cover 67.7% of the scientific production on the topic and consist of journals in general management (e.g., *Journal of Business Ethics*, *Business & Society*), accounting (e.g., *Accounting Auditing & Accountability Journal*, *Accounting Organizations, and Society*), operations management (*International Journal of Production Economics*), and engineering (e.g., *Journal of Cleaner Production*).

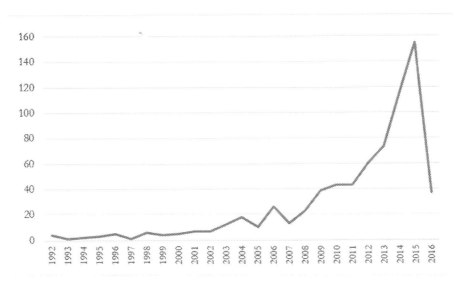

FIG. 5.1 The number of articles published on sustainability measurement from 1992 to 2016 (Adapted from Mura et al., 2018).

The literature on sustainability measurement is characterized by various research strands that can be grouped into nine main areas of research:

1. *Sustainability disclosure and performance.* Scholars dealing with this topic are interested in the types of information disclosed by companies, which is from purely environmental data on, for instance, GHG emissions to social information about employees and the society. Internal and external elements that characterize companies disclosing are investigated, as well as reasons that push them to do so, such as stakeholders' and institutional pressures, acquisition of legitimacy, and differentiation from competitors. Findings reveal that the disclosure of sustainability-related information is still less (Clarkson et al., 2011), and that large enterprises (LEs) disclose more than small and medium-sized enterprises (SMEs) (Luo et al., 2012). Recent studies, however, give more emphasis on differences between what companies disclose and their actual sustainability practices, and reveal that managers tend to selectively disclose only the information their stakeholders need (Depoers et al., 2016). These issues evolve together with demands to governance for standardized disclosing reports and a clear regulation on the matter.

2. *Determinants of sustainability disclosure.* From early articles to most recent ones, two main groups of papers can be distinguished. The first group analyses the drivers of sustainability disclosure, which are attributed to both organization's characteristics (e.g., size, strategic approach, board composition, and ownership) and external factors, such as stakeholders, country, institutions, communities, industry membership, and media exposure (Reverte, 2009). A second group of articles explores environmental regulation, which role is considered to have a positive effect on companies' disclosure decisions (Cowan and Deegan, 2011).

3. *Critical environmental accounting.* Papers in this cluster critically investigate the possibility to develop a dashboard of indicators able to measure sustainability at corporate level. Main concerns involve the unit and the scope of analysis of sustainability measurement, as well as the difficulty to adopt a systemic view of the field (Schaltegger and Burritt 2010). The critical perspective develops further by deepening the profound disconnection between sustainability reporting and current ecological issues (Milne and Gray, 2013; Tregidga et al., 2014). However, recent studies deepen positive situations of sustainable accounting and propose theoretical and practical methods for environmental accounting development (Bebbington and Larrinaga, 2014; Spence and Rinaldi, 2014).

4. *Sustainability metrics.* This cluster presents two lines of research. The majority of papers concentrate on the study of social and environmental indicators, and how these metrics are applied and reported externally, especially by multinational companies, whose reasons vary from differentiating themselves from competitors, to improving their performance and strategy (Roca and Searcy, 2012; Searcy and Elkhawas, 2012). A second stream of research focuses on environmental management accounting (EMA) and sustainability performance measurement systems (SPMS). In particular, authors provide insights on the integration of ecocontrol into management control and performance measurement systems (PMS) (Henri and Journeault, 2010), and its effect on organizational strategy (Gond et al., 2012).

5. *Sustainable operations and supply chain management.* In the first years, the number of authors approaching the topic is small and measurement is not central on the research, rather the theme was sustainability in operations and supply chain in broad terms (Caniato et al., 2012). However, from 2012 onwards, research expanded considerably. New studies analyze organizations' use

of sustainability measurement practices to develop sustainable operations and green supply, by assessing their environmental impact, suppliers' activity, and life cycle of products (Brandenburg and Rebs, 2015; Marshall et al., 2015).

6. *Carbon accounting.* The debate about carbon accounting and footprint is faced by scholars particularly in the first stage of sustainability measurement development. Topics range from the differences in the use of carbon accounting according to the level of analysis investigated (Stechemesser and Guenther 2012), to the metrics to account for carbon-related information (Schaltegger and Csutora, 2012). Authors also focus on the roles of environmental legislation, and political pressure, which are considered determinant to make corporations accountable for their carbon impacts. Nevertheless, the number of papers in this cluster reduces with time passing, probably due to the spread of optimism toward apparent positive results in relation to climate change issues.

7. *Diffusion of sustainability standards.* Early studies focus on the diffusion of sustainability standards worldwide and motivations that enhance it, such as stakeholders' pressure and political framework (Marimon et al., 2012; Reid and Toffel, 2009). By 2012, this cluster expands rapidly, and scholars take two research directions. One group of papers further explores determinants and outcomes of sustainability reporting and disclosure (Ceulemans et al., 2015), while a second group studies sustainability reporting consequences on companies (Vigneau et al., 2015).

8. *Assurance of sustainability reporting.* With a significant growth in the last years, this literature follows two strands of research. One group of papers explores the diffusion of assurance practices and how their use varies according to different organizations and industrial sectors (Segui-Mas et al., 2015). Another group links assurance practices with other organizational variables such as environmental performance and company reputation (Alon and Vidovic, 2015).

9. *Emerging clusters.* Three small clusters are emerging in the sustainability measurement literature. The first cluster analyses companies' reasons to disclose sustainability information, and gains insights on greenwashing, that is the corporates' behavior to provide a sustainable picture of themselves, despite not having any particular social or environmental engagement (Nurhayati et al., 2016). The second cluster investigates the level of diffusion of biodiversity accounting and reporting in different countries (Rimmel and Jonall, 2013), while the third cluster conducts the research on the effect of institutions and norms on sustainability standards (De Villiers and Alexander, 2014).

Highlights of the main areas of inquiry on sustainability measurement are summarized in Table 5.1.

5.2.2 Sustainability measurement: a broken compass

Despite sustainability measurement literature has grown exponentially over recent years, the cross-disciplinary nature of the subject and the lack of a sufficiently comprehensive conceptualization of the field have led to the creation of many separate, unconnected areas of inquiry. In particular, most studies have tended to consider only selected aspects of the measurement process and to concentrate on specific issues (e.g., sustainability reporting, carbon accounting, eco-efficiency, introduction of specific measures within organizations and supply chains). This has conducted to a quickly expanding but very fragmented field, with different theoretical perspectives, conceptualizations of the measurement process, and contributions to practice.

Table 5.1 Summary of the sustainability measurement literature clusters.

Cluster	Highlights
A. Sustainability disclosure and performance	Scholars investigate the type of information disclosed and determinants of companies' disclosure. Findings reveal that the discosure of sustainability-related information is still few (Clarkson et al., 2011), and that larger organizations disclose more than small- and medium-sized enterprises (Luo et al., 2012). Recent studies shed light on gaps between what companies disclose and their actual sustainability practices, and the lack of standardized disclosing reports (Depoers et al., 2016).
B. Determinants of sustainability disclosure	Two groups of research emerge. One group of articles analyses the drivers of sustainability disclosure at company level, which are attributed to both internal and external factors (Reverte, 2009). A second group explores the positive effect of environmental regulation on companies' decisions about disclosing (Cowan and Deegan, 2011).
C. Critical environmental accounting	Papers investigate the possibility to develop a set of indicators able to measure sustainability at corporate level, with concern about the unit of analysis, the scope, and the adoption of a systemic view of the field (Schaltegger and Burritt, 2010). The critical perspective on disconnection between sustainability reporting and current ecological issues develops further (Milne and Gray, 2013; Tregidga et al., 2014). Theoretical and practical methods for environmental accounting are explored (Bebbington and Larrinaga, 2014; Spence and Rinaldi, 2014).
D. Sustainability metrics	The majority of papers investigate the application of sustainability metrics, and how these are reported externally (Roca and Searcy, 2012; Searcy and Elkhawas, 2012). A second stream of research focuses on the integration of eco-control into performance measurement systems (Henri and Journeault, 2010), and its effect on organizational strategy (Gond et al., 2012).
E. Sustainable operations and supply chain management	Initially, measurement is not central on the research. Afterwards, new studies concentrate on the development of sustainable operations and green supply, and companies' assessment of their environmental impact, suppliers' activity, and life cycle of products (Brandenburg and Rebs, 2015; Marshall et al., 2015).
F. Carbon accounting	Topics range from the differences in the use of carbon accounting according to the level of analysis investigated (Stechemesser and Guenther. 2012), to the metrics to account for carbon-related information and the roles of environmental legislation and political pressure (Schaltegger and Csutora, 2012). The stream of the literature reduces with the spread of optimism toward apparent positive results in relation to climate change issues.
G. Diffusion of sustainability standards	Early studies focus on the diffusion of sustainability standards worldwide (Marimon et al., 2012). Recent studies are divided into two groups of research. One group further explores determinants and outcomes of sustainability reporting and disclosure (Ceulemans et al., 2015), while a second group sheds light on sustainability reporting consequences on companies (Vigneau et al., 2015).
H. Assurance of sustainability reporting	Two strands of research: the diffusion of assurance practices on different organizations and industrial sectors (Segui-Mas et al., 2015), and the link between assurance practices and other organizational variables (Alon and Vidovic, 2015).
I. Emerging clusters	Three small clusters are emerging in the sustainability measurement literature: greenwashing (Nurhayati et al., 2016), the diffusion of biodiversity reporting (Rimmel and Jonall, 2013), and institutionalization of sustainability reporting within organizations (De Villiers and Alexander, 2014).

Mura et al.'s (2018) analyses show the emergence of different subfields over the 1992–2016, and that while some literature areas have expanded significantly by proposing new insights and compelling findings (e.g., the integration of sustainability-related information in management control and PMS, and the assessment and management of green supply chains), other strands appear to be waning, as in the case of carbon accounting. While adopting different viewpoints could be beneficial, authors' findings show that studies conducted by scholars belonging to different academic communities tend to not talk to each other and, consequently, to overlap considerably.

Another reason for the lack of homogeneity in the sustainability measurement literature is strategic dispersion on how to manage sustainability processes within organizations (Roca and Searcy, 2012). This is mainly due to the lack of consideration of sustainability measurement in the general performance measurement and management (PMM) literature and vice versa, which has led to further duplication of efforts and missed opportunities in bringing together findings from the two areas. Moreover, on the practice, this deficiency has reflected on organizations struggle to measure environmental sustainability along with financial aspects (Chen et al., 2014), causing the so-called *decoupling effect*, which occurs when the metrics used to measure sustainability are actually decoupled from the PMS that companies use to actually measure their own—usually financial—performance.

The debate in the literature characterizing the need for creating homogeneity in measuring sustainability processes ended up with the creation of systems of sustainability reporting such as the Global Reporting Initiative (GRI) Standards, Thomson Reuters ASSET4, or Global Engagement Services (GES). However, the presence of more than one system of sustainability reporting measure does not contribute in the process of homogeneity, as they show great variations in content and scope (Searcy, 2012), and mostly focus on public companies or large organizations. These elements trigger negative behavioral consequences induced by sustainability measurement, such as *selective disclosure*. Given that there is no globally accepted standard available for measuring and reporting sustainability, and most companies report on their set of metrics assured by the third parties, managers usually selectively disclose only those key performance indicators that are more relevant to them. This has both positive and negative consequences. The pro is that, by selectively disclosing specific information, companies could understand which stakeholders are key for them, and what information really matters for those stakeholders. The con is that companies selectively disclose only those measures where they know they are performing better, thus hiding relevant data for a comprehensive assessment of their own performances.

5.2.3 Contribution to performance measurement and the management literature

The decoupling of sustainability measurement from the performance measurement literature has prompted authors of both disciplines to reflect differently on the benefits that would derive from the integration of the two fields on corporate strategy and performance. This has shed light on interesting insights. First of all, while the role of stakeholders in the measurement process has been a point of contention in performance measurement studies, for sustainability measurement research stakeholders play such an important role in the design, implementation, and use of sustainability measures that they should be considered an integral part of the measurement process, or even the focal point of studies. This is evident when considering research on the roles and effects of regulators and institutions that introduce standards and award certifications, as well as external auditors, rating agencies, and firms that assess suppliers' environmental practices and reporting.

Second, the presence of only technical aspects of measurement denounced by recent contributions in the PMM literature can be overcome by research in sustainability measurement, which offers both technical and social perspectives of measurement (Bititci et al., 2012; Chenhall et al., 2017). For example, the literature area on development and use of sustainability metrics (D) tends to embrace a systemic thinking of measurement by encompassing both social and technical aspects.

Third, instead of adopting a typical organizational point of view, the sustainability measurement literature has implications for research and practice in the wider PMM field in relation to how common measures could be established and how data could be shared effectively across the all value chains. For example, the area on sustainable operations and supply chain management (E) is mainly formed by articles on the assessment and evaluation of suppliers, and on the introduction of common sustainability measures by buyers and suppliers. In the area of carbon accounting (F), various authors discuss how carbon accounting could be used not only within, but also among organizations.

5.3 Methods and tools: the path toward sustainability measurement

The number of tools and approaches to develop a comprehensive system for measuring sustainability in organization is growing rapidly. The major areas of activity are summarized below and will be further explained in the following paragraphs:

1. Sustainability core issues and stakeholder mapping;
2. SPMS;
3. Sustainability reporting.

5.3.1 Sustainability core issues and stakeholder mapping

By learning from the sustainability measurement literature that stakeholders play a fundamental role in the measurement process, the search for stakeholders who matter the most to organization has become an essential activity for managers. There are internal tools helping them to assess the importance of their own stakeholders. The most used one is *stakeholder map*, which has been developed by Mitchell et al. (1997) and assesses stakeholders based on the following three key attributes:

- *Power:* It is the probability that one stakeholder has to influence the company's behavior.
- *Legitimacy:* It is assumed as a desirable social good. Activities of a stakeholder with legitimacy are perceived to be appropriate within a socially developed system of norms, values, beliefs, and definitions (Suchman, 1995).
- *Urgency:* It defines how fast a company has to react to its stakeholders' requests.

By linking the three different attributes, companies identify various classes of stakeholders based on the possession of one, two, or all three of the attributes, which can be grouped into the following three clusters (Fig. 5.2):

- *Latent stakeholders:* Only one of the stakeholder attributes (power, legitimacy, and urgency) is perceived by managers to be present. Organizations spend limited time, energy, and other resources to recognize those stakeholders' existence, and vice versa.

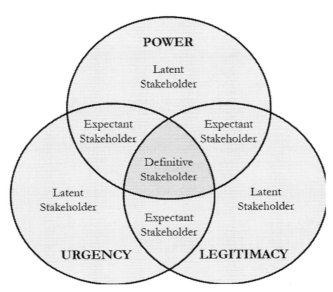

FIG. 5.2 Stakeholder typology based on the attributes of power, legitimacy, and urgency (Source: adapted from Mitchell et al., 1997).

- *Expectant stakeholders:* Any two of the three attributes are perceived by managers to be present. The level of engagement between managers and these expectant stakeholders is higher, as well as firm responsiveness to the stakeholder's interests.
- *Definitive stakeholders*: Stakeholders with all three of the attributes are core stakeholders for the company, which means managers give priority to competing those stakeholder claims.

An organization also needs to identify the most relevant sustainability priorities that are consistent with its business strategy. *Materiality matrix*, which is the result of the materiality analysis, is an important internal tool to identify which sustainability issues to focus on when taking into account also stakeholders' interests (see Box 5.1 for details on materiality analysis). For example, management of Enel Group identified 13 potential material topics of significance, such as energy distribution, customer focus, local communities, sustainability supply chain, decarbonization, etc. Then, these major topics were classified according to the relevance for the company itself and the importance for the company's stakeholders (Fig. 5.3). The result is a deeper investigation on sustainability issues for the company, along with the prioritization of real stakeholders' requests and engagement.

5.3.2 Sustainability performance measurement system

Increased strategic attention to environmental and social issues has triggered the demand for the integration of sustainability measures into PMS. Changes in the PMS are often attributed by both academics and practitioners to modifications on the balanced scorecard (BSC) and its performance perspectives. By accounting for sustainability issues, conventional BSC develops into a sustainability

5.3 Methods and tools: the path toward sustainability measurement

> **Box 5.1**
> **Materiality analysis**
>
> "Materiality" is the threshold beyond which an information or an indicator becomes important enough to be reported on.
>
> According to GRI Standards, material is a topic that reflects the reporting organization's significant economic, environmental, and social impacts, or substantively influences the assessments and decisions of stakeholders; materiality therefore has two dimensions, internal and external. Both dimensions should be assessed and evaluated during a process usually called "materiality analysis." To define its impacts, organization should take into account its values, policies, strategies, goals, core competencies. To understand its role on influencing stakeholders' decisions, the company should consider interests and expectations of its relevant stakeholders; main sectoral challenges and issues, as identified by peers and competitors; public policies and regulation.
>
> Direct engagement is the strength of a solid materiality analysis, as it guarantees more concrete, effective, and "customized" results. Regarding the external dimension, engagement requires a clear identification and prioritization of company's stakeholders, which can be involved through surveys, interviews, focus groups, with both traditional or digital methods and tools. Looking at the internal dimension, a committed top management should be engaged in identifying company's priorities between relevant economic, social, and environmental issues. Evidences emerging from a materiality analysis are often represented through a matrix, an impactful visual sign of what is more relevant for both the company and the stakeholders. The matrix is usually divided into four quadrants, and issues appearing in the upper right quadrant will be the focus of the reporting.
>
> The materiality matrix provides other significant information: not only the alignments (issues equally relevant for company and stakeholders), but also the misalignments should be carefully considered. They can point out areas of reputational risks (when an issue is more relevant for stakeholders than for the company), or suggest item in which the company should invest more in communication (in the opposite case).
>
> As a result of an accurate analysis, materiality matrix is a guide not only for sustainability measuring and reporting (material issues are those on which to focus the annual sustainability report), but also for strategic planning (material issues are an input for defining new strategies) and stakeholder engagement (material issues are the basis for planning future engagement).

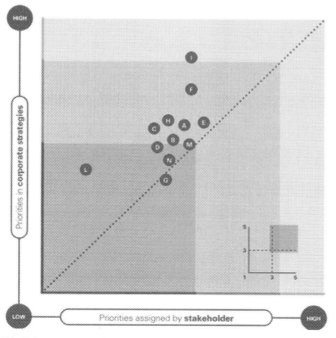

BUSINESS AND GOVERNANCE ISSUES

A Energy distribution
B Decarbonization of the energy mix
C Customer focus
D Ecosystems and platforms[1]
E Sound governance and fair corporate conduct
F Economic and financial value creation
N Innovation and digital trasformation

SOCIAL ISSUES

G Engaging local communities
H People management, development and motivation
I Occupational health and safety
L Sustainable supply chain

ENVIRONMENTAL ISSUES

B Decarbonization of the energy mix
M Environmental management

FIG. 5.3 Enel's materiality matrix (Enel Group's Sustainability Report, 2019).

BSC (SBSC) which explicitly integrates sustainability-related objectives and performance measures into its architecture. Sustainability objectives can be either integrated into the conventional four perspectives of the BSC (i.e., *learning and growth*, *costumer*, *internal process*, and *finance*), or added as a dedicated sustainability perspective complementing the four conventional BSC perspectives. Hansen and Schaltegger's (2016) analyze three architectural design combinations to integrate sustainability-related objectives into a company's PMM system. These are as follows:

1. *Add-on.* By means of stakeholder map and materiality analysis, management identifies core environmental and social objectives, which are, then, managed by adding one or more individual perspectives to the traditional BSC. Adding a performance perspective with environmental and social objectives is an important change to the existing PMS, as it allows to dedicate separately to sustainability-related aspects. However, some risks could emerge. First, the mere identification and addition of a new perspective in a consolidated BSC can fail if the link with the management is not strong. In the worst-case scenario, the sustainability perspective will only communicate with environmental or marketing department, and its activity will be decoupled from the rest of the business. Second, a new perspective can be easily removed if, for instance, it is not a priority for management anymore.
2. *Integration.* This second approach does not manage sustainability-related objectives emerging from stakeholder mapping and materiality analysis in a separate new perspective of BSC, rather it integrates them into the four traditional perspectives. Thus, the number of perspectives remains the same, while the amount of strategic objectives increases. The integration can be partial or full depending on whether sustainability goals are embedded into one or a few existing perspectives, or whether changes regard all four conventional perspectives, accordingly. Examples of sustainability integration go from extending the *learning and growth perspective* with green capabilities, to embedding innovation processes for sustainability into *internal process perspective*, and to accounting for product differentiation through sustainability-related features into the *costumer perspective*.
3. *Extension.* This last option applies simultaneously the previously mentioned design alternatives. Sustainability objectives are totally embedded into the standard four perspectives of BSC, and an additional sustainability perspective is introduced with the aim of managing environmental and social goals in the short term, but also in long time horizons.

Firms' consistent interest to improve their environmental and social performance has also led to the development of a large number of sustainability key performance indicators, which have been classified in different ways. A first two-dimensional perspective distinguishes between process-based measures and outcome-based measures (Delmas et al., 2013). Process-based indicators capture internal actions that firms establish to improve their environmental performance, such as dedication to environmental causes, refinement of environmental management systems, and managerial quality in general (see Box 5.2). On the other hand, outcome-based indicators consider impacts of these actions on actual environmental performance currently in place in the organization.

A more thorough assessment of the overall picture of a company's emissions is given by the GHG Protocol Corporate Standard, which categorizes GHG emissions at firm level into three "scopes." Scope 1 emissions are direct GHG emissions from sources that are owned or controlled by the company. Scope 2 emissions are indirect GHG emissions from the generation of purchased energy. Scope 3 emissions are all indirect GHG emissions that are not included in scope 2, and that occur in all downstream

> **Box 5.2**
> **Hera's approach to create and measure shared value**
> The practice of creating shared value (CSV), which is based on the interconnection between economic, social, and environmental values, has become fundamental in business, as highlighted by Porter and Kramer (2011). By agreeing with this idea, in 2016 Hera developed its own Corporate Social Responsibility approach with the aim of creating and measuring shared value. The creation of shared value in Hera occurs when business activities generate operating margins while meeting the drivers of the global agenda, that is the "calls to action" for change in specific fields defined in global, European, national, and local policies.
>
> Furthermore, the Group enhanced its sustainability reporting with new perspectives and tools, such as the CSV EBITDA, which measures the amount of Hera Group's consolidated EBITDA that has been generated by its shared-value business activities. The indicator is therefore aimed at highlighting the contribution from all the business activities that respond to the "calls to action" for change to the generation of EBITDA. Such "calls to action" have been summarized by the company in three drivers (i.e., Smart use of energy, efficient use of resources, and innovation and contribution to development) and nine impact areas.
>
> Hera Group's activities have been analyzed to identify those that are specifically consistent with the drivers and the impact areas for CSV. A full consistency can be considered for those activities that fully comply with the priorities defined by the "calls to action"; otherwise, a partial consistency occurs when only some of the activities evaluated for a specific business area meet the CSV requirements. In case of partial consistency, a specific KPI is identified to allow the calculation of the amount of the activity that is consistent with CSV and, consequently, to quantify its part of CSV EBITDA.
>
> In 2016, Hera's CSV EBITDA was about 33% of the overall consolidated EBITDA, while in 2019 the Group's CSV EBITDA amounted to € 422.5 million, that is 38.9% of the Group's total EBITDA. Currently, Hera's goal is to reach a CSV EBITDA of 42% by 2023.

and upstream activities of the value chain of the reporting company. Most of the largest companies in the world account and report the emissions from their direct operations (scopes 1 and 2), while an increasing amount of firms are adopting scope 3 as one of their major challenges. Indeed, although the last scope is optional, emissions along the value chain often represent a company's biggest impacts, and thus, measuring such interorganizations dynamics enables companies to focus their efforts on the greatest GHG reduction opportunities.

Besides sustainability key performance indicators, firms can also follow initiatives that provide a set of targets to reduce emissions. An example is the Science-Based Targets initiative, which supports companies with a defined pathway of goals in line with the objectives of the Paris Agreement by specifying how much and how quickly they need to reduce their emissions.

5.3.3 Sustainability reporting

Reporting to stakeholders in a transparent and public manner is fundamental for companies committed to sustainability, and there exists various channels of communication to do so. The most straightforward way for a company to release information is to use traditional corporate reports such as annual reports. Other available channels for disclosures of environmental and social practices are international sustainability measurement frameworks. The most well-known international set of corporate sustainability indicators are the 79 measures included in the GRI's G4 reporting guidelines. It has four sections: (1) company description and stakeholder engagement, (2) economic issues, (3), environmental issues, and (4) social issues, which is divided into labor practices and decent work, human rights, society, and product responsibility.

Other international sustainability measurement frameworks are Global Compact, ASSET4, and Carbon Disclosure Project (CDP). They propose indicators that move beyond the ones suggested by the GRI, for example by linking together different sustainability measures in cause and effect relationships (Searcy, 2012).

Although existing frameworks have provided interesting insights on reporting company's sustainability, they have guided organizations to report sustainability-related information using different metrics, as each of them has proposed their own set. Many organizations select indicators from different reporting bodies, which turns out in a fragmentary evaluation of sustainability-related performance and in a variety of reports that are not necessarily comparable (Domingues et al., 2017). Additionally, sustainability disclose is mainly a voluntary practice, and is generally run by public companies or large organizations. However, integrated sustainability reports for companies of different sizes are developing (for an example, see Box 5.3).

5.4 The future of sustainability measurement

Research on sustainability measurement has grown exponentially over the last two decades. By analyzing the extant and recent literature in the field, it is possible to highlight the dynamic evolution of sustainability measurement, and to identify emerging and future research trends.

Some literature areas have expanded significantly. The stream of sustainability metrics develops by providing greater focus on the integration of environmental and social indicators into organizational

---

Box 5.3
Sustainability Measurement and Management Laboratory (SuMM Lab)

The SuMM Lab acts as a potential permanent observatory aiming to explore the disclosure level of sustainability-related practices of small, medium, and large organizations through a structured database that contains online environmental and social sustainability-related data of companies coming from different industries (Longo and Mura, 2017). The project was launched in 2016 by Mariolina Longo and Matteo Mura with the funding provided by five large Italian companies.

By analyzing 712 articles published from 1991 to 2017 on sustainability measurement (Mura et al., 2018), the founders identified a final list of 69 KPIs, representing the following 11 thematic areas: environmental certifications, social certifications, energy, water, waste management, environmental impact, corporate social responsibility, supply chain, consumption, product innovation, and business model. Organizations were selected by stratifying the whole population of Italian companies by revenues. This process led to creation of a structured database containing 3,928 companies located in all Italian regions, with different sizes and operating in 32 different sectors. Data on sustainability-related practices were collected manually through secondary sources (i.e., organization's websites and sustainability reports published online) in two rounds (2017–2018 and 2019–2020) by the support of a pool of 80 students overall of the Management Engineering course from the University of Bologna.

SuMM Lab aims to overcome some limitations of the academic literature and existing international sustainability frameworks by developing a new measurement framework that is adapted to apply to both LEs and SMEs, and considering disclosure channels such as websites and journals in addition to annual reports. It also provides information on the contents of sustainability information disclosed by companies and possible drivers for their disclosure, which can be the baseline for future roadmaps that help organizations to cope with sustainable development and organizational processes. At the same time, the creation of an innovative database on companies' sustainability-related practices could be an opportunity for businesses and policy makers to understand contents and drivers of online sustainability disclosure. This could be a first step toward the plan of policies for industrial transition and the decarbonization of the economy, in line with the key characteristics of sustainable transition (Hansen and Coenen, 2015), and would constitute a fundamental knowledge base for the formulation of informed decisions and specific policies at local, national, and international level, as well as a tool for assessing their effectiveness.

processes such as resource planning, capital allocation, and performance evaluation. Studies in this cluster also explore the interplay of management control systems and sustainability control systems, and its effect on the integration of a sustainability perspective within organizational strategy. Other examples focus on sustainability disclosure, which evolve by exploring both nonfinancial and financial indices used for enhancing disclosure's credibility, and supply chain management area, which develops by adopting operation research methods to evaluate and manage green suppliers, sustainable practices, and life cycle of products. Finally, topics such as measuring a company's shared value, the diffusion of biodiversity reporting, and institutionalization of sustainability reporting within organizations appear as emerging trends.

At the same time, we need to take into account that other strands of the literature appear to be waning, as in the case of carbon accounting area, or are not significantly contributing to research, as determinates of sustainability disclosure cluster.

Nevertheless, the path toward a sustainable measurement is ongoing within companies. The number of tools and approaches to identify sustainability core issues is growing rapidly, as well as the diffusion of disclosure reports according to both an economic, social, and environmental bottom line. This process testifies how managers are becoming more and more inclined and sensitive to the issue, despite some gaps in the literature and the lack of integrated frameworks for sustainability assessment.

Future studies on sustainability measurement could explore the conceptualization of the measurement process, the links between strategy and performance measurement, and the behavioral effects of measurement practices.

Yet, what sustainability measurement needs more is to lose the "fragmented field" label, which might occur, perhaps, by abandoning the idea of using a precise scale of analysis in its studies, which usually is either a lower scale (corporate measurement), or an upper scale (ecosystems measurement). Treating the two dimensions separately is understandable if considering the complexity of the field, but it has undoubtedly contributed to fragment a phenomenon that needs, instead, to be addressed systematically. Further research and practical developments on sustainability measurement could greatly benefit by adopting a multiscalarity approach (Coenen et al., 2012; Martin and Sunley, 2006). This approach is based on the idea that changes originate partly as a result of interventions at various scales, from local to global levels of analysis. By assuming that sustainability is an ecological and societal concept that cannot be limited to specific boundaries, multiscalarity aims to create links between the different scales and to adopt a systemic reasoning that is equally applicable to both corporations and industrial ecosystems. Thus, despite it is still little explored, multiscalarity could have a significant impact on both the theory and the practice of sustainability measurement, which could be achieved through the integration of different—but related—literatures such as the economic geography and the ecological macroeconomics.

References

Alon, A., Vidovic, M., 2015. Sustainability performance and assurance: influence on reputation. Corp. Reputat. Rev. 18, 337–352.

Bebbington, J., Larrinaga, C., 2014. Accounting and sustainable development: an exploration. Account. Organ. Soc. 39, 395–413.

Bititci, U., Garengo, P., Dörfler, V., Nudurupati, S., 2012. Performance measurement: challenges for tomorrow. Int. J. Manag. Rev. 14, 305–327.

Brandenburg, M., Rebs, T., 2015. Sustainable supply chain management: a modeling perspective. Ann. Oper. Res. 229, 213–252.

Caniato, F., Caridi, M., Crippa, L., Moretto, A., 2012. Environmental sustainability in fashion supply chains: an exploratory case based research. Int. J. Prod. Econ. 135, 659–670.

Ceulemans, K., Molderez, I., Van Liedekerke, L., 2015. Sustainability reporting in higher education: a comprehensive review of the recent literature and paths for further research. J. Clean. Prod. 106, 127–143.

Chen, J.C., Cho, C.H., Patten, D.M., 2014. Initiating disclosure of environmental liability information: an empirical analysis of firm choice. J. Bus. Ethics 125, 681–692.

Chenhall, R.H., Hall, M., Smith, D., 2017. The expressive role of performance measurement systems: a field study of a mental health development project. Accounting. Organ. Soc. 63, 60–75.

Clarkson, P.M., Overell, M.B., Chapple, L., 2011. Environmental reporting and its relation to corporate environmental performance. ABACUS 47, 27–60.

Coenen, L., Benneworth, P., Truffer, B., 2012. Toward a spatial perspective on sustainability transitions. Res. Policy 41, 968–979.

Cowan, S., Deegan, C., 2011. Corporate disclosure reactions to Australia's first national emission reporting scheme. Account. Finan. 51, 409–436.

De Villiers, C., Alexander, D., 2014. The institutionalisation of corporate social responsibility reporting. Brit. Account. Rev. 46, 198–212.

Delmas, M.A., Etzion, D., Nairn-Birch, N., 2013. Triangulating environmental performance: what do corporate social responsibility ratings really capture? Acad. Manage. Perspect. 27, 255–267.

Depoers, F., Jeanjean, T., Jérôme, T., 2016. Voluntary disclosure of greenhouse gas emissions: contrasting the Carbon Disclosure Project and Corporate Reports. J. Bus. Ethics 134, 445–461.

Domingues, A.R., Lozano, R., Ceulemans, K., Ramos, T.B., 2017. Sustainability reporting in public sector organisations: exploring the relation between the reporting process and organisational change management for sustainability. J. Environ. Manage. 192, 292–301.

Elkington, J., 1997. Cannibals with Forks: The Triple Bottom Line of the 21st Century Business. Capstone, Oxford.

Enel Group, Sustainability Report 2019. http://www.sustainabilityreport2019.enel.com.

Fischedick, M., Roy, J., Abdel-Aziz, A., Acquaye, A., Allwood, J.M., Ceron, J.-P., et al., 2014. Industry. Climate change 2014: Mitigation of Climate Change. Contribution of Working Group III to the Fifth Assessment Report of the Intergovernmental Panel on Climate Change. Cambridge University Press, Cambridge, United Kingdom and New York, NY, USA.

Gond, J., Grubnic, S., Herzig, C., Moon, J., 2012. Configuring management control systems: theorizing the integration of strategy and sustainability. Manage. Account. Res. 23, 205–223.

Hansen, E., Schaltegger, S., 2016. The sustainability balanced scorecard: a systemic review of architectures. J. Bus. Ethics 133, 193–221.

Hansen, T., Coenen, L., 2015. The geography of sustainability transitions. Review, synthesis and reflections on an emergent research field. Environ. Innov. Soc. Transit. 17, 92–109.

Henri, J.F., Journeault, M., 2010. Eco-control: the influence of management control systems on environmental and economic performance. Account. Organ. Soc. 35, 63–80.

Longo, M., Mura, M., 2017. Assessing sustainability within organizations: the sustainability measurement and management lab (SuMM) [http://link.springer.com/chapter/10.1007/978-3-319-57078-5_33]. In: Campana, G. et al (Ed.), Sustainable Design and Manufacturing 2017, Smart Innovation, Systems and Technologies 68. Springer International Publishing. doi:10.1007/978-3-319-57078-5_33.

Luo, L., Lan, Y., Tang, Q., 2012. Corporate incentives to disclose carbon information: evidence from the CDP Global 500 Report. J. Int. Fin. Manage. Account. 23, 93–120.

Marimon, F., del Mar Alonso-Almeida, M., del Pilar Rodríguez, M., Alejandro, K.A.C., 2012. The worldwide diffusion of the global reporting initiative: what is the point? J. Clean. Prod. 33, 132–144.

Marshall, D., McCarthy, L., Heavey, C., McGrath, P., 2015. Environmental and social supply chain management sustainability practices: construct development and measurement. Prod. Plan. Control 26, 673–690.

Martin, R., Sunley, P., 2006. Path dependence and regional economic evolution. J. Econ. Geogr. 6 (4), 395–437.

References

Milne, M.J., Gray, R., 2013. W(h)ither ecology? The triple bottom line, the global reporting initiative, and corporate sustainability reporting. J. Bus. Ethics 118, 13–29.

Mitchell, R.K., Agle, B.R., Wood, D.J., 1997. Toward a theory of stakeholder identification and salience: defining the principle of who and what really counts. Acad. Manage. Rev. 22 (4), 853–886.

Mura, M., Longo, M., Micheli, P., Bolzani, D., 2018. The evolution of sustainability measurement research. Int. J. Manage. Rev. 20, 661–695.

Neely, A.D., Adams, C., Kennerley, M., 2002. The Performance Prism: The Scorecard for Measuring and Managing Business Success. Prentice Hall Financial Times, London.

Nurhayati, R., Taylor, G., Rusmin, R., Tower, G., Chatterjee, B., 2016. Factors determining social and environmental reporting by Indian textile and apparel firms: a test of legitimacy theory. Soc. Responsib. J. 12, 167–189.

Porter, M.E., Kramer, M.R., 2011. The big idea: creating shared value. how to reinvent capitalism—and unleash a wave of innovation and growth. Harv. Bus. Rev. 89 (1-2), 62–77.

Reid, E.M., Toffel, M.W., 2009. Responding to public and private politics: corporate disclosure of climate change strategies. Strateg. Manag. J. 30, 1157–1178.

Reverte, C., 2009. Determinants of corporate social responsibility disclosure ratings by Spanish listed firms. J. Bus. Ethics 88, 351–366.

Rimmel, G., Jonall, K., 2013. Biodiversity reporting in Sweden: corporate disclosure and preparers' views. Account. Audit. Account. J. 26, 746–778.

Roca, L.C., Searcy, C., 2012. An analysis of indicators disclosed in corporate sustainability reports. J. Clean. Prod. 20, 103–118.

Schaltegger, S., Burritt, R.L., 2010. Sustainability accounting for companies: catchphrase or decision support for business leaders. J. World Bus. 45, 375–384.

Schaltegger, S., Csutora, M., 2012. Carbon accounting for sustainability and management. Status quo and challenges. J. Clean. Prod. 36, 1–16.

Searcy, C., 2012. Corporate sustainability performance measurement systems: a review and research agenda. J. Bus. Ethics 107, 239–253.

Searcy, C., Elkhawas, D., 2012. Corporate sustainability ratings: an investigation into how corporations use the Dow Jones sustainability index. J. Clean. Prod. 35, 79–92.

Segui-Mas, E., Bollas-Araya, H.-M., Polo-Garrido, F., 2015. Sustainability assurance on the biggest cooperatives of the world: an analysis of their adoption and quality. Ann. Public Coop. Econ. 86, 363–383.

Spence, L.J., Rinaldi, L., 2014. Governmentality in accounting and accountability: a case study of embedding sustainability in a supply chain. Account. Organ. Soc. 39, 433–452.

Stechemesser, K., Guenther, E., 2012. Carbon accounting: a systematic literature review. J. Clean. Prod. 36, 17–38.

Suchman, M.C., 1995. Managing legitimacy: strategic and institutional approaches. Acad. Manage. Rev. 20, 571–610.

Tregidga, H., Milne, M., Kearins, K., 2014. (Re)presenting 'sustainable organizations. Account. Organ. Soc. 39, 477–494.

Vigneau, L., Humphreys, M., Moon, J., 2015. How do firms comply with international sustainability standards? Processes and consequences of adopting the global reporting initiative. J. Bus. Ethics 131, 469–486.

CHAPTER 6

Industrial sustainability performance measurement system—challenges for the development

Alessandra Neri

Department of Management, Economics and Industrial Engineering, Politecnico di Milano, Milan, Italy

6.1 Industrial sustainability

Sustainability is a central topic in current managerial, academic, and policy discourses. As present modes of production result in unsustainable socioeconomic and environmental consequences (Ansell and Cayzer, 2018), transformations at technological, managerial, organizational, and behavioral levels are needed to move toward a more sustainable industrial system (Blok et al., 2015). A large share of responsibility for the aforementioned unsustainable consequences is in the hands of the industrial sector (Mangla et al., 2017), leading to the need for long-term oriented changes in production modes, addressing both products and processes (Freire, 2018). The attainment of changes in current production modes cannot be possible without the transformation of industrial processes (Srnicek and Williams, 2015). From this specific perspective, *industrial sustainability* can be defined as the share of sustainability accounting for "all the actions referred to the production plant (and not just the production line)": industrial sustainability indeed requires "actions at the levels of material, product, process, plant, and systems of production as well as an integration of these actions into normal operations" (Trianni et al., 2017).

The enhancement of industrial sustainability–related performance is of fundamental importance for the overall improvement of sustainability; as it is hard to improve what it is not gauged and measured (Engida et al., 2018; Singh et al., 2012), the improvement of performance should itself necessarily pass from the measurement of the performance.

6.2 Industrial sustainability performance measurement

The measurement of performance is one of the most important steps toward their enhancement, and firms can approach the measurement activity with different aims and in different manners. Understanding how and why industrial firms measure their sustainability-related performance is of fundamental importance to shed light on possible issues related to the measurement of performance.

6.2.1 Why do firms measure industrial sustainability–related performance?

As sustainability has been proven to be a strong competitive factor for the industrial sector (Morioka et al., 2018), firms are encouraged to measure their sustainability-related performance. Firms might receive pressures from external and internal stakeholders to focus on social and environmental aspects of sustainability, besides the traditional economic one (Tsalis et al., 2018). The aim of the measurement of industrial sustainability–related performance can be thus twofold (Lozano and Huisingh, 2011): on the one hand, it allows the firm to assess its current state; on the other hand, it is a necessary step to communicate in an effective way to those outside of the firm the results and progresses toward enhanced sustainability.

External stakeholders exert strong pressures toward firms for them to enhance their performance and share the results achieved in a transparent way (Stacchezzini et al., 2016). It is therefore critical for firms to communicate their progress to external stakeholders, so as to improve their relationship with them (Fuente et al., 2017; Székely and Vom Brockem, 2017), while also protecting their reputation and legitimizing their operations (De Villiers et al., 2016).

Understanding and assessing the performance of the firm is vital for the industrial decision-makers aiming at fostering continuous improvement and performing benchmarking activities (Ghadimi et al., 2012; Howard et al., 2018).

The evaluation of internal performance is central for assessing the current state of the firm, recognizing where specific improving actions—also referred to as industrial sustainability measures (Trianni et al., 2017)—should be implemented so as to enhance the overall firm's performance (Cagno et al., 2019; Collins et al., 2016). The measurement of performance raises awareness and guides the decision-making process (Paju et al., 2010), allowing industrial decision-makers to identify which actions to adopt (Bhanot et al., 2017; Trianni et al., 2017) and subsequently to assess and monitor the effects and impacts of the adoption (Winroth et al., 2016).

Benchmarking activities allow a continuous comparison with peers and competitors, thus stimulating internal improvements (Ferrari et al., 2019). Notably, benchmarking activities should not address only the comparison with firms operating in the same sector but should entail also analyses based on different contextual factors, as size or geographical area (Apaydin et al., 2018; Siebert et al., 2018).

6.2.2 How do firms measure industrial sustainability–related performance?

The measurement of performance is carried out using performance indicators (Engida et al., 2018) that enable the performance measurement process, and support and drive industrial decision-makers in the attainment of targets and objectives (Globerson, 1985). When performance indicators are arranged in a set, they can be referred to as a performance measurement system (PMS) (Neely et al., 1995). PMSs help in properly evaluating performance and improving the management of the firm (Johnson and Schaltegger, 2016). Nonetheless, empirical analyses showed that firms are still relying on experience and sensitivity to measure their sustainability-related performance: although they consider a PMS as a fundamental tool they would be interested in adopting, they still do not have it (Cagno et al., 2019).

6.2.3 Focus of the present chapter

To foster the adoption of PMSs by firms for the measurement of industrial sustainability–related performance, it is of central importance to understand what characteristics an effective PMS should have, as well as the rationales for its development. The literature considered thousands of indicators for industrial sustainability (Hristov and Chirico, 2019; Rojas-Lema et al., 2020). Taking for granted that all the proposed indicators meet the characteristics suggested by Feng and Joung (2011), the real challenge lies in the selection of the right indicators to be used and to be included in a PMS, rather than in their development (Singh et al., 2014). From this perspective, several challenges arise toward an effective PMS development, as attention must be paid to the overall features that the PMS should cover (Garengo et al., 2005). This chapter aims at contributing to the extant literature by providing a holistic overview on the main aspects that should be considered for an effective development of a PMS. Particularly, the chapter will discuss specific aspects as the usefulness, completeness and balance, and usability of a PMS; the practical methods employed for the selection of indicators; and the contexts in which the PMS would be applied. A framework of PMSs will be then presented as a possible example tackling the discussed aspects, and suggestions for future streams of research will be offered.

6.3 Industrial sustainability PMS—toward an effective development

An effective PMS should entail several features in its development (Garengo et al., 2005). Five pivotal aspects have been identified in the literature, reported in Fig. 6.1.

The first aspect relates to the usefulness on the PMS, which should be aligned with the needs of internal and external stakeholders. The second aspect addresses the completeness and balance of the PMS in terms of appropriate consideration of all the sustainability-related areas. The third aspect concerns the usability and manageability, as the PMS should be straightforward in its use, also considering the number of indicators that the PMS should include. The fourth aspect refers to the selection of the performance indicators, which should be carried out objectively. The fifth aspect considers the context of application, understanding the specific characteristics of different contexts in terms of the industrial sector, firm's size, geographical area, awareness toward sustainability. All the above-mentioned features will be in-depth discussed in the following sections.

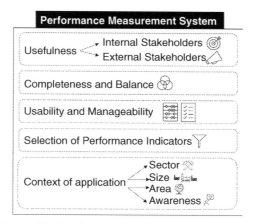

FIG. 6.1 Features to consider for an effective development of a PMS.

6.3.1 Usefulness to internal and external stakeholders

The indicators included in a PMS should be relevant for the specific firm and its goals (Lee and Lee, 2014). In the first instance, the industrial firm has two different options. On the one hand, the firm can adopt an already developed PMS: this would allow benchmarking but the system could not be properly applied in specific contexts and results excessively time-consuming (Hallstedt et al., 2015); on the other hand, the firm can develop its own PMS, based on the specific firm's needs and characteristics. Such a PMS would result tailored; nonetheless, it would threaten benchmarking activities and its development might be resource-intense (Staniškis and Arbačiauskas, 2009).

In general terms, the identification and use of the right indicators in the daily measurement of performance is limited (Bilge et al., 2014). Additional issues emerge trying to balance the requirements of the different stakeholders—particularly trying to be strategy aligned, fostering strategy development, and being focused on external stakeholders (Garengo et al., 2005). While it is important for a PMS to be tailored to a firm's needs to analyze internal performance and attain continuous improvement (Clarke-sather et al., 2011; Singh et al., 2016), standardized indicators also carry importance to communicate in-between firms and to conduct benchmarking activities (Ferrari et al., 2019). Additionally, the meeting of external stakeholders' requirements for communication and transparency strongly influences the indicators gauged by a firm—and thus potentially included in a PMS (Nordheim and Barrasso, 2007), focusing on an external perspective rather than on an internal one (De Villiers et al., 2016). For many firms, external pressures could be even perceived as the main driving force for sustainability performance evaluation (Staniškis and Arbačiauskas, 2009), with the risk to misrepresent the concept of industrial sustainability in favor of other related but conceptually different aspects, as social responsibility or environmental management (Gray, 2010; Montiel, 2008).

6.3.2 Completeness and balance according to a holistic perspective on industrial sustainability

The balance of different aspects and areas of interest in a traditional PMS is difficult (Garengo et al., 2005; Kitaw and Goshu, 2017) and specific further issues arise when trying to balance the different features deriving from industrial sustainability. The consideration of all the areas of industrial sustainability, often characterized by tradeoffs (Neri et al., 2018), could indeed lead to high complexity and heterogeneity (Tseng et al., 2020).

Particularly, two main issues should be considered tackling completeness and balance.

The first one is that available PMSs addressing industrial sustainability still appear unbalanced in terms of consideration of the different pillars of sustainability—environmental, social, and economic (Elkington, 1997). According to Trianni et al. (2019) and Van Schoubroeck et al. (2018): economic indicators still predominate both in theoretical developments and empirical applications; environmental-related indicators have been largely developed, also thanks to the growing importance of the concept of green/environmental sustainability (Tseng et al., 2020), nonetheless, there is still ample room for firms to address environmental-related indicators in a more structured manner; social-related aspects seem to lay behind both from a theoretical development and an empirical measurement perspective, although recent contributions are tackling this aspect (Feil et al., 2019; Mickovski and Thomson, 2017; Popovic et al., 2018).

The second issue is that PMSs should address sustainability in a holistic and integrated manner. Many PMSs still address industrial sustainability through compartmentalizations, separating the different areas constituting sustainability (Cagno et al., 2019; Lozano and Huisingh, 2011). Nonetheless, several mutual benefits and interdependencies can be tackled addressing all the pillars simultaneously (Cagno et al., 2018; Nehler and Rasmussen, 2016) and such an approach would help understand how the different performances are interrelated, and act in synergy to enhance overall sustainability (Amrina et al., 2016; Helleno et al., 2017).

6.3.3 Usability and manageability

The literature has largely discussed the right number of indicators that a PMS should include; nonetheless, the debate is still open. The usability and manageability of the PMS is an extremely important feature that might face a clash when trying to address the need for a complete and balance PMS. Indeed, as a high number of indicators would be necessary to cover as much as possible the topic in terms of breadth and depth (Garengo et al., 2005), the number of indicators included in the PMS should be nonetheless aligned with the actual human capacity to process information (Hubbard, 2009), thus ensuring clarity and simplicity (Garengo et al., 2005). The right threshold has not been identified yet, with suggestions ranging from 5 to 60 indicators (Collins et al., 2016; Globerson, 1985), although the literature agreed that a large number of indicators might divert from pursuing a focused strategy (Epstein and Widener, 2010; Medini et al., 2015). Empirical applications showed that the average number of indicators gauged within industrial firms varies from less than 20 up to 70, with a high variance and with interesting patterns emerging according to contextual factors, above all the size of the firm (Saeed and Kersten, 2020; Trianni et al., 2019). The number of considered indicators emerging from the empirical investigations is interestingly lower than the number of indicators included in recent developed PMSs (Agostinho et al., 2019; Jiang et al., 2018). In this way, some authors have already considered the need to shrink the number of indicators introduced in their theoretical PMSs to develop a more manageable set for empirical use (Ahmad et al., 2019; Taques et al., 2020). Notably, according to the case studies conducted by Cagno et al. (2019), manageability and handling of a PMS resulted the characteristics most requested by the investigated firms, above all for those firms that aim at moving from mere compliance with regulation to a more structured and systemic approach toward sustainability.

Some authors stated that a good solution could be the development of an index or composite indicator—a mathematical combination of other indicators able to simultaneously characterize all the pillars of sustainability (Garbie, 2015). Such an index could avoid bias and allow comparisons among firms with similar characteristics (Harik et al., 2015). Some shortcomings of the development of an index should be nonetheless underlined. First, a single multidimensional measure for sustainability can be difficult to obtain, given the complex interrelationship among dimensions (Kibira et al., 2009) and compensation effects among different areas should be avoided (Neri et al., 2018). Second, creating a single index means to assign a weight to the different parts; the literature showed that the procedure can be performed in different ways, as by assigning the same weight to all the indicators composing the index (Charmondusit et al., 2014); weighting the indicators according to the single firm's perspective—that is, according to the perspective of one or multiple industrial decision-makers (Garbie, 2014), customers (Ma and Kremer, 2016), or external stakeholders (Epstein and Widener, 2010); weighting according to different contextual factors of the firms, foremost the sector (Kocmanova and Docekalova, 2011). These approaches, nonetheless, have been deemed as subjective or leading to possible inconsistency (Calabrese et al., 2016; Feng and Joung, 2011). Last, some authors underlined the counterintuitiveness of using a single index for addressing a multifaced topic as industrial sustainability (Cayzer et al., 2017).

6.3.4 Selection of indicators

The literature showed that the selection of indicators is a huge challenge and a central part in the development of a PMS (Sloan, 2010). Different methods have been proposed: some authors based the selection on the literature occurrence rate (Veleva and Ellenbecker, 2001); some others suggested ways for weighting the indicators, as discussed in the previous section, mainly suffering from possible inconsistency (Calabrese et al., 2016; Madanchi et al., 2019). Additionally, if the weighting is performed only by industrial decision-makers, it may experience subjectivity (Callens and Tyteca, 1999); if it is performed only by external stakeholders or academics, its effectiveness could be limited (Salvado et al., 2015). A good compromise between the two options could be to adopt a mixed perspective, involving in the selection process industrial decision-makers, stakeholders, and academia (Li et al., 2012). Taking into account the complexity of different facets of industrial sustainability, multiple industrial decision-makers experienced with the areas of industrial sustainability within the same firm should be involved in the selection of indicators (Cagno et al., 2018). This approach, although necessary, might increase the complexity of the process as every industrial decision-maker has their priorities, interests, and perspectives on industrial sustainability, very often in conflict among each other (Frini and Benamor, 2018; Gong et al., 2018; Nicolăescu et al., 2015).

6.3.5 Context of application

The development of a PMS should also consider the context where it would be applied: the same indicators indeed might not be suitable for all the possible contexts of application (Rojas-lema et al., 2020). Firms featured by distinct contextual factors might pay attention to distinct aspects (Winroth et al., 2016) and require a different level of detail in terms of depth of indicators (Trianni et al., 2019). While trying to balance these specific aspects with conducting benchmarking activities, the inclusion and use of too many different PMSs could increase the difficulty to measure and compare performance in an effective manner (Christofi et al., 2012).

Contextual factors are generally referred to features as the sector, geographical area, and firm's size and they largely proved to influence the behavior of firms (Micheli and Cagno, 2010; Trianni et al., 2013). Along with them nonetheless, it is important to characterize firms according to other aspects. Concerning industrial sustainability, aspects related to how sustainability is addressed and managed within firms appeared important (Trianni et al., 2019), namely: the certifications held by each firm (May and Stahl, 2017); the presence within the firm of a specific manager in charge of sustainability (Jansson et al., 2017); the way sustainability is perceived and defined within the specific firm (De Oliveira et al., 2018; Held et al., 2018).

A particular focus must be reserved for small and medium enterprises (SMEs). SMEs constitute more than 99% of the European firms (Eurostat, 2018) and have consistent environmental and social impacts (Meng et al., 2018; Rojas-Lema et al., 2020). Being able to coordinate strategy and performance measurement is relevant for SMEs (Garengo et al., 2005) as they are still focused mainly on operational and economic aspects (Trianni et al., 2019), and usually independent, so that growth and innovation are played by owners (Ribeiro-Soriano, 2017). SMEs struggling to survive should be provided with a simple PMS focused on breadth rather than depth (Garengo et al., 2005). The general problems related to the measurement of performance in SMEs become even more evident focusing on industrial sustainability (Arena and Azzone, 2012). Although SMEs are pivotal for sustainability goals in the industrial sector (Singh et al., 2016), they are often not conscious of their impact (Feil et al., 2017); SMEs thus usually lack enough resources to measure performance effectively, as time, staff, money as well as competencies can constrain them (Jansson et al., 2017; Tremblay and Badri, 2018). When approaching the measurement of industrial sustainability performance, SMEs often require tailored PMSs and those designed for larger firms are difficult to apply in SMEs as the amount of data required might be unsustainable (Arena and Azzone, 2012; GRI, 2018).

The previous reasonings applied to SMEs could be of course extended to firms starting to introduce sustainability into their day-to-day activities, despite the size (Johnson, 2015; Witjes et al., 2015).

In both cases, not only a tailored PMS should be considered, but possibly, also a PMS able to adapt to the evolution of the firms according to increased awareness and competences related to industrial sustainability (Garengo et al., 2005).

6.4 A scalable framework for measuring industrial sustainability performance

The recently developed contribution by Cagno et al. (2019) is a valuable example of a PMS addressing the issues and challenges discussed.

The authors proposed an integrated scalable framework including three different levels of analysis, represented by three different industrial sustainability PMSs (ISPMSs). With the scalable framework, the authors tried to solve the dichotomy between the opposite needs for a standardized PMS (to compare across firms) and a tailored PMS (to allow for a firm's context to be taken into consideration, supporting the use of a standardized yet tailored system). Table 6.1 reports the framework proposed by Cagno et al. (2019).

The ISPMSs entail a decreasing numerosity of indicators—and thus different extent of depth, while aiming to adequately cover the sustainability pillars and their intersections—to an appropriate breadth. As the framework is scalable, firms can move among the different ISPMSs. Particularly the *Core* one, including 44 indicators, is thought for SMEs or firms with a low commitment or awareness toward industrial sustainability; moving to the *Intermediate* and then to the *Full*, including 76 and 104 indicators respectively, the IPMSs tackle the needs for a deeper level of detail required by larger firms or firms characterized by a higher level of awareness toward sustainability.

The three different levels of analysis, that is, the three different ISPMSs proposed, are thus appropriate to be applied in different contexts according to diverse contextual factors, and commitment and awareness toward sustainability. The ISPMSs are organized in areas related to the three pillars of

Table 6.1 The "Framework of Industrial Sustainability Performance Measurement Systems" proposed by Cagno et al. (2019). Adapted from Cagno et al. (2019).

Area	Category	Full ISPMS	Intermediate ISPMS	Core PMS
Economic	Investments	R&D investment	R&D investment	R&D investment
		Pollution prevention and control investment		
		Environment investment	Environment investment	Environment investment
		Energy efficiency investment		
		Safety investment	Safety investment	Safety investment
		Community investment		
		Ethics/philanthropy investment	Ethics/philanthropy investment	
	Costs and incomes	Operating cost	Operating cost	
		Overhead cost		
		Packaging cost		
		Production cost	Production cost	Production cost
		Set up cost		
		Inventory cost	Inventory cost	Inventory cost
		Labor cost	Labor cost	Labor cost
		Unit cost	Unit cost	Unit cost
		Maintenance cost	Maintenance cost	Maintenance cost
		Taxes		
		EHS fines	EHS fines	EHS fines
		Sales	Sales	Sales
		Market share	Market share	
		Revenues	Revenues	
		Profit	Profit	Profit
		Turnover		
	Production	Throughput	Throughput	Throughput
		New products	New products	New products
		Lead time	Lead time	Lead time
		Scrap	Scrap	
		Quality	Quality	Quality
		Mix flexibility		
		Volume Flexibility		
		DFx	DFx	
		Green product		
		IT level		

6.4 A scalable framework for measuring industrial sustainability performance

Social	Suppliers		Number of suppliers Local suppliers	Number of suppliers Local suppliers Certified suppliers	Number of suppliers Local suppliers
	Community		Community complaints Community projects Local employment Involvement of local community	Community complaints Community projects Local employment Involvement of local community	Community complaints Community projects
	Customers		Customer satisfaction Personalized products Services offered	Customer satisfaction Personalized products Services offered	Customer satisfaction Personalized products
	Employees		Number of employees Wage level Work satisfaction Involvement of employees Gender discrimination Ethnic group discrimination	Number of employees Wage level Work satisfaction Involvement of employees Discrimination	Number of employees Work satisfaction
			Safety training Environmental training	Training	Training
	OHS		Accidents Injuries Fatalities Near misses	Accidents Injuries Fatalities	Accidents Injuries Fatalities
			PPE Absenteeism Noise Dust Toxic substances OHS Administration Citations Safety expenditure	PPE Absenteeism Noise Dust Toxic substances OHS Administration Citations Safety expenditure	PPE Absenteeism
Environment	Water		Total water use Freshwater use Recycled water use Quality of water	Total water use Freshwater use Recycled water use Quality of water	Total water use Recycled water use
	Material		Total material use Recycled material use Hazardous material use	Total material use Recycled material use Hazardous material use	Total material use Recycled material use Hazardous material use

(continued)

Table 6.1 (Cont'd)

Area	Category	Full ISPMS	Intermediate ISPMS	Core PMS
	Energy	Toxic material use		
		Total energy use for production	Total energy use	Total energy use
		Renewable energy use for production	Renewable energy use	Renewable energy use
		Fuel use for production	Fuel use	Fossil fuel use
		Gas use for production	Gas use	
		Coal use for production	Coal use	
		Total energy use not for production		
		Renewable energy use not for production		
		Fuel use not for production		
		Gas use not for production		
		Coal use not for production		
	Air emissions	CO_2	CO_2	CO_2
		Other GHG	Other GHG	
		NO_X	NO_X	
		SO_2	SO_2	
		ODS	ODS	
		Metal emissions		
		Other emissions		
		Toxic emissions	Toxic emissions	Toxic emissions
	Waste	Hazardous solid waste	Hazardous solid waste	Hazardous solid waste
		Nonhazardous solid waste	Nonhazardous solid waste	Nonhazardous solid waste
		Hazardous liquid waste	Hazardous liquid waste	Hazardous liquid waste
		Nonhazardous liquid waste	Nonhazardous liquid waste	Nonhazardous liquid waste
		COD		
		BOD		
		Wastewater	Wastewater	
		Chemical waste		
		Waste disposed	Waste disposed	
		Waste recycled	Waste recycled	Waste recycled
		Energy recover		
		Material recover		
	Environmental management	Environmental accidents	Environmental accidents	
		Environmental fines	Environmental fines	
		Environmental certification	Environmental certification	
		Cost of compliance	Cost of compliance	

sustainability and then detailed into categories of performance and performance indicators. The selection of the indicators to include in each ISPMS has been done starting from the literature and applying different methods, in the order: literature frequency; evaluation of the content of information according to different perspectives, namely researchers, a panel of experts, and literature, thus avoiding the shortcomings deriving from the focus on only a single perspective.

Fig. 6.2 reports the main features of the framework by Cagno et al. (2019). For each ISPMS, Fig. 6.2 indicates the methodology for the selection of indicators to be included, the number of indicators included, and the coverage of the sustainability pillars and their intersections. Interestingly, as for this latter point, moving from the *Full ISPMS* to the *Core ISPMS*, the focus is on those indicators able to provide information on the intersection of, at least, two pillars.

6.5 Concluding remarks and future perspectives

This chapter provided an overview of the challenges associated with the effective development of a PMS for the measurement of industrial sustainability–related performance. The framework of Cagno et al. (2019) was presented as a possible framework to solve the challenges mentioned earlier in the chapter. The framework could nonetheless represent a base for additional studies focused on overcoming still existing limitation, addressing interesting aspects.

Particularly, the need for large empirical applications of theoretically developed PMSs could not be stressed enough. Such empirical studies would be of extreme relevance for understanding the extent to which the PMSs are appropriate according to different contextual factors as antecedents, while also understanding whether the different ways in which firms manage sustainability represent another antecedent or mediate and/or moderate the outcome (Aguinis et al., 2017).

As an industrial firm might evolve in terms of awareness toward sustainability, it would be interesting also to understand how a PMS should evolve as well so to adapt to the new needs of the firm. This point can be related to the presence of leading and lagging indicators within the PMS and to its dynamic adaptability (Garengo et al., 2005). Jointly, it would be of great relevance to understand the evolution of the PMS over time concerning the development by the firm of approaches to solve tensions and tradeoffs in performance measurement from both a short- and long-term perspective (Kim et al., 2019; Lévesque and Stephan, 2020), while also understanding the antecedents of the specific approaches (Slawinski and Bansal, 2012).

As sustainability is recognized as a strong competitive factor for the industry and as competitiveness is more and more at a system level instead of a single firm one (De Angelis et al., 2017; Shibin et al., 2017), opportunities for additional research arise. Especially for supply chains, the need to integrate the PMSs used for measuring sustainability by the different tiers of the supply chain, with a PMS for the overall system emerged as pivotal (Carter et al., 2019), complementing standard key performance indicators for the system with tailored ones for each firm (Neri et al., 2021).

Moreover, the relationship among PMSs aimed at measuring different areas of performance should be addressed both at a single firm and an industrial system level, as, for example, possible synergies between PMSs for industrial sustainability and quality management (Bastas and Liyanage, 2018; Machado et al., 2019).

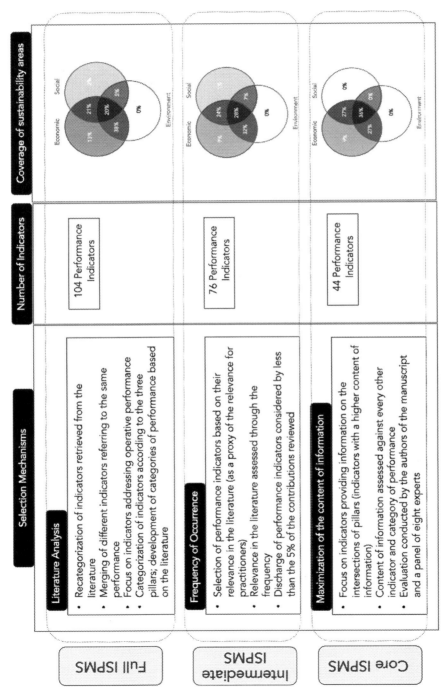

FIG. 6.2 Features of the "Framework of Industrial Sustainability Performance Measurement Systems" proposed by Cagno et al. (2019). For each ISPMS, the following are displayed: the selection mechanism; the number of indicators; the coverage of sustainability areas.

References

Agostinho, F., Richard Silva, T., Almeida, C.M.V.B., Liu, G., Giannetti, B.F., 2019. Sustainability assessment procedure for operations and production processes (SUAPRO). Sci. Total Environ. 685, 1006–1018. doi:10.1016/j.scitotenv.2019.06.261.

Aguinis, H., Edwards, J.R., Bradley, K.J., 2017. Improving our understanding of moderation and mediation in strategic management research. Organ. Res. Methods 20, 665–685. doi:10.1177/1094428115627498.

Ahmad, S., Wong, K.Y., Zaman, B., 2019. A comprehensive and integrated stochastic-fuzzy method for sustainability assessment in the Malaysian food manufacturing industry. Sustainability 11. doi:10.3390/su11040948.

Amrina, E., Ramadhani, C., Vilsi, A.L., 2016. A fuzzy multi criteria approach for sustainable manufacturing evaluation in cement Industry. Procedia CIRP 40, 620–625. doi:10.1016/j.procir.2016.01.144.

Ansell, T., Cayzer, S., 2018. Limits to growth redux: a system dynamics model for assessing energy and climate change constraints to global growth. Energy Policy 120, 514–525. doi:10.1016/j.enpol.2018.05.053.

Apaydin, M., Bayraktar, E., Hossary, M., 2018. Achieving economic and social sustainability through hyperconnectivity: a cross- country comparison. Benchmarking Int. J. 25, 3607–3627. doi:10.1108/BIJ-07-2017-0205.

Arena, M., Azzone, G., 2012. A process-based operational framework for sustainability reporting in SMEs. J. Small Bus. Enterp. Dev. 19, 669–686. doi:10.1108/14626001211277460.

Bastas, A., Liyanage, K., 2018. Sustainable supply chain quality management: a systematic review. J. Clean. Prod. 181, 726–744. doi:10.1016/j.jclepro.2018.01.110.

Bhanot, N., Rao, P.V., Deshmukh, S.G., 2017. An integrated approach for analysing the enablers and barriers of sustainable manufacturing. J. Clean. Prod. 142, 4412–4439. doi:10.1016/j.jclepro.2016.11.123.

Bilge, P., Badurdeen, F., Seliger, G., Jawahir, I.S., 2014. Model-based approach for assessing value creation to enhance sustainability in manufacturing. Procedia CIRP 17, 106–111. doi:10.1016/j.procir.2014.02.031.

Blok, V., Long, T.B., Gaziulusoy, A.I., Ciliz, N., Lozano, R., Huisingh, D., Csutora, M., Boks, C., 2015. From best practices to bridges for a more sustainable future: advances and challenges in the transition to global sustainable production and consumption. Introduction to the ERSCP stream of the Special volume. J. Clean. Prod. 108, 19–30. doi:10.1016/j.jclepro.2015.04.119.

Cagno, E., Neri, A., Howard, M., Brenna, G., Trianni, A., 2019. Industrial sustainability performance measurement systems: a novel framework. J. Clean. Prod. 230, 1354–1375. doi:10.1016/j.jclepro.2019.05.021.

Cagno, E., Neri, A., Trianni, A., 2018. Broadening to sustainability the perspective of industrial decision-makers on the energy efficiency measures adoption: some empirical evidence. Energy Effic 11, 1193–1210. doi:10.1007/s12053-018-9621-0.

Calabrese, A., Costa, R., Levialdi, N., Menichini, T., 2016. A fuzzy analytic hierarchy process method to support materiality assessment in sustainability reporting. J. Clean. Prod. 121, 248–264. doi:10.1016/j.jclepro.2015.12.005.

Callens, I., Tyteca, D., 1999. Towards indicators of sustainable development for firms. Ecol. Econ. 28, 41–53. doi:10.1016/S0921-8009(98)00035-4.

Carter, C.R., Hatton, M.R., Wu, C., Chen, X., 2019. Sustainable supply chain management: continuing evolution and future directions. Int. J. Phys. Distrib. Logist. Manag. 50, 122–146. doi:10.1108/IJPDLM-02-2019-0056.

Cayzer, S., Griffiths, P., Beghetto, V., 2017. Design of indicators for measuring product performance in the circular economy. Int. J. Sustain. Eng. 10, 289–298. doi:10.1080/19397038.2017.1333543.

Charmondusit, K., Phatarachaisakul, S., Prasertpong, P., 2014. The quantitative eco-efficiency measurement for small and medium enterprise: a case study of wooden toy industry. Clean Technol. Environ. Policy 16, 935–945. doi:10.1007/s10098-013-0693-4.

Christofi, A., Christofi, P., Sisaye, S., 2012. Corporate sustainability: historical development and reporting practices. Manag. Res. Rev. 35, 157–172. doi:10.1108/01409171211195170.

Clarke-sather, A.R., Hutchins, M.J., Zhang, Q., Gershenson, J.K., Sutherland, J.W., 2011. Development of social, environmental, and economic indicators for a small/medium enterprise. Int. J. Account. Inf. Manag. 19, 247–266. doi:10.1108/18347641111169250.

Collins, A.J., Hester, P., Ezell, B., Horst, J., 2016. An improvement selection methodology for key performance indicators. Environ. Syst. Decis. 36, 196–208. doi:10.1007/s10669-016-9591-8.

De Angelis, R., Howard, M., Miemczyk, J., 2017. Supply chain management and the circular economy : towards the circular supply chain. Prod. Plan. Control 29, 425–437. doi:10.1080/09537287.2018.1449244.

De Oliveira, A.C., Sokulski, C.C., Da Silva Batista, A.A., De Francisco, A.C., 2018. Competencies for sustainability: a proposed method for the analysis of their interrelationships. Sustain. Prod. Consum. 14, 82–94. doi:10.1016/j.spc.2018.01.005.

De Villiers, C., Rouse, P., Kerr, J., 2016. A new conceptual model of influences driving sustainability based on case evidence of the integration of corporate sustainability management control and reporting. J. Clean. Prod. 136, 78–85. doi:10.1016/j.jclepro.2016.01.107.

Elkington, J., 1997. Cannibals with Forks : The Triple Bottom Line of 21st Century Business. Oxford, Capstone.

Engida, T.G., Rao, X., Berentsen, P.B.M., Oude Lansink, A.G.J.M., 2018. Measuring corporate sustainability performance– the case of European food and beverage companies. J. Clean. Prod. 195, 734–743. doi:10.1016/j.jclepro.2018.05.095.

Epstein, M.J., Widener, S.K., 2010. Identification and use of sustainability performance measures in decision-making. J. Corp. Citizsh. 40, 43–73. doi:10.9774/GLEAF.4700.2010.wi.00006.

Eurostat, 2018. Statistics on Small and Medium-Sized Enterprises. **[WWW Document]**. URL https://ec.europa.eu/eurostat/statistics-explained/index.php/Statistics_on_small_and_medium-sized_enterprises (accessed 9.13.18).

Feil, A.A., De Quevedo, D.M., Schreiber, D., 2017. An analysis of the sustainability index of micro- and small-sized furniture industries. Clean Technol. Environ. Policy 19, 1883–1896. doi:10.1007/s10098-017-1372-7.

Feil, A.A., Schreiber, D., Haetinger, C., Strasburg, V.J., Barkert, C.L., 2019. Sustainability indicators for industrial organizations: systematic review of literature. Sustainability 11, 1–15. doi:10.3390/su11030854.

Feng, S.C., Joung, C.B., 2011. A measurement infrastructure for sustainable manufacturing. Int. J. Sustain. Manuf. 2, 204–221. doi:10.1504/IJSM.2011.042152.

Ferrari, A.M., Volpi, L., Pini, M., Cristina, S., García-Muiña, F.E., Settembre-Blundo, D., 2019. Building a sustainability benchmarking framework of ceramic tiles based on Life Cycle Sustainability Assessment (LCSA). Resources 8, 1–30. doi:10.3390/resources8010011.

Freire, P.A., 2018. Enhancing innovation through behavioral stimulation: the use of behavioral determinants of innovation in the implementation of eco-innovation processes in industrial sectors and companies. J. Clean. Prod. 170, 1677–1687. doi:10.1016/j.jclepro.2016.09.027.

Frini, A., Benamor, S., 2018. Making decisions in a sustainable development context: a state-of-the-art survey and proposal of a multi-period single synthesizing criterion approach. Comput. Econ. 52, 341–385. doi:10.1007/s10614-017-9677-5.

Fuente, J.A., García-Sánchez, I.M., Lozano, M.B., 2017. The role of the board of directors in the adoption of GRI guidelines for the disclosure of CSR information. J. Clean. Prod. 141, 737–750. doi:10.1016/j.jclepro.2016.09.155.

Garbie, I.H., 2015. Integrating sustainability assessments in manufacturing enterprises: a framework approach. Int. J. Ind. Syst. Eng. 20, 343. doi:10.1504/IJISE.2015.069922.

Garbie, I.H., 2014. An analytical technique to model and assess sustainable development index in manufacturing enterprises. Int. J. Prod. Res. 52, 4876–4915. doi:10.1080/00207543.2014.893066.

Garengo, P., Biazzo, S., Bititci, U.S., 2005. Performance measurement systems in SMEs: a review for a research agenda. Int. J. Manag. Rev. 7, 25–47. doi:10.1111/j.1468-2370.2005.00105.x.

Ghadimi, P., Azadnia, A.H., Mohd Yusof, N., Mat Saman, M.Z., 2012. A weighted fuzzy approach for product sustainability assessment: a case study in automotive industry. J. Clean. Prod. 33, 10–21. doi:10.1016/j.jclepro.2012.05.010.

Globerson, S., 1985. Issues in developing a performance criteria system for an organization. Int. J. Prod. Res. 23, 639–646. doi:10.1080/00207548508904734.

Gong, M., Simpson, A., Koh, L., Tan, K.H., 2018. Inside out: the interrelationships of sustainable performance metrics and its effect on business decision making: theory and practice. Resour. Conserv. Recycl. 128, 155–166. doi:10.1016/j.resconrec.2016.11.001.

Gray, R., 2010. Is accounting for sustainability actually accounting for sustainability…and how would we know? An exploration of narratives of organisations and the planet. Accounting. Organ. Soc. 35, 47–62. doi:10.1016/j.aos.2009.04.006.

GRI, 2018. Empowering small and medium enterprises: Recommendations for Policy Makers to Enable Sustainability Corporate Reporting for SMEs https://www.globalreporting.org/resourcelibrary/Empowering_small_business_Policy_recommendations.pdf.

Hallstedt, S.I., Bertoni, M., Isaksson, O., 2015. Assessing sustainability and value of manufacturing processes: a case in the aerospace industry. J. Clean. Prod. 108, 169–182. doi:10.1016/j.jclepro.2015.06.017.

Harik, R., El Hachem, W., Medini, K., Bernard, A., 2015. Towards a holistic sustainability index for measuring sustainability of manufacturing companies. Int. J. Prod. Res. 53, 4117–4139. doi:10.1080/00207543.2014.993773.

Held, M., Weidmann, D., Kammerl, D., Hollauer, C., Mörtl, M., Omer, M., Lindemann, U., 2018. Current challenges for sustainable product development in the German automotive sector: a survey based status assessment. J. Clean. Prod. 195, 869–889. doi:10.1016/j.jclepro.2018.05.118.

Helleno, A.L., De Moraes, A.J.I., Simon, A.T., Helleno, A.L., 2017. Integrating sustainability indicators and lean manufacturing to assess manufacturing processes: application case studies in Brazilian industry. J. Clean. Prod. 153, 405–416. doi:10.1016/j.jclepro.2016.12.072.

Howard, M., Hopkinson, P., Miemczyk, J., 2018. The regenerative supply chain: a framework for developing circular economy indicators. Int. J. Prod. Res. doi:10.1080/00207543.2018.1524166.

Hristov, I., Chirico, A., 2019. The role of sustainability key performance indicators (KPIs) in implementing sustainable strategies. Sustainability 11. doi:10.3390/su11205742.

Hubbard, G., 2009. Measuring organizational performance: beyond the triple bottom line. Bus. Strateg. Environ. 18, 177–191. doi:10.1002/Bse.564.

Jansson, J., Nilsson, J., Modig, F., Hed Vall, G., 2017. Commitment to sustainability in small and medium-sized enterprises: the influence of strategic orientations and management values. Bus. Strateg. Environ. 26, 69–83. doi:10.1002/bse.1901.

Jiang, Q., Liu, Z., Liu, W., Li, T., Cong, W., Zhang, H., Shi, J., 2018. A principal component analysis based three-dimensional sustainability assessment model to evaluate corporate sustainable performance. J. Clean. Prod. 187, 625–637. doi:10.1016/j.jclepro.2018.03.255.

Johnson, M.P., 2015. Sustainability management and small and medium-sized enterprises: managers' awareness and implementation of innovative tools. Corp. Soc. Responsib. Environ. Manag. 22, 271–285. doi:10.1002/csr.1343.

Johnson, M.P., Schaltegger, S., 2016. Two decades of sustainability management tools for SMEs: how far have we come?. J. Small Bus. Manag. 54, 481–505. doi:10.1111/jsbm.12154.

Kibira, D., Jain, S., McLean, C.R., 2009. A System Dynamics Framework for Sustainable Manufacturing, Proc. 27th Int. Conf. Syst. Dyn. Soc.

Kim, A., Bansal, P., Haugh, H., 2019. No time like the present: how a present time perspective can foster sustainable development. Acad. Manag. J. 62, 607–634. doi:10.5465/amj.2015.1295.

Kitaw, D., Goshu, Y.Y., 2017. Performance measurement and its recent challenge: a literature review. Int. J. Bus. Perform. Manag. 18, 381. doi:10.1504/ijbpm.2017.10007477.

Kocmanova, A., Docekalova, M., 2011. Corporate sustainability : environmental, social, economic, and corporate performance. ACTA Univ. Agric. Silvic. Mendelianae Brun. LIX, 203–208.

Lee, J.Y., Lee, Y.T., 2014. A framework for a research inventory of sustainability assessment in manufacturing. J. Clean. Prod. 79, 207–218. doi:10.1016/j.jclepro.2014.05.004.

Lévesque, M., Stephan, U., 2020. It's time we talk about time in entrepreneurship. Entrep. Theory Pract. 44, 163–184. doi:10.1177/1042258719839711.

Li, T., Zhang, H., Yuan, C., Liu, Z., Fan, C., 2012. A PCA-based method for construction of composite sustainability indicators. Int. J. Life Cycle Assess. 17, 593–603. doi:10.1007/s11367-012-0394-y.

Lozano, R., Huisingh, D., 2011. Inter-linking issues and dimensions in sustainability reporting. J. Clean. Prod. 19, 99–107. doi:10.1016/j.jclepro.2010.01.004.

Ma, J., Kremer, G.E.O., 2016. A sustainable modular product design approach with key components and uncertain end-of-life strategy consideration. Int. J. Adv. Manuf. Technol. 85, 741–763. doi:10.1007/s00170-015-7979-0.

Machado, M.C., Telles, R., Sampaio, P., Queiroz, M.M., Fernandes, A.C., 2019. Performance measurement for supply chain management and quality management integration. Benchmarking Int. J. doi:10.1108/bij-11-2018-0365.

Madanchi, N., Thiede, S., Sohdi, M., Herrmann, C., 2019. Development of a sustainability assessment tool for manufacturing companies. In: Thiede, S., Herrmann, C. (Eds.), Eco-Factories of the Future. Springer International Publishing. doi:10.1007/978-3-319-93730-4.

Mangla, S.K., Govindan, K., Luthra, S., 2017. Prioritizing the barriers to achieve sustainable consumption and production trends in supply chains using fuzzy Analytical Hierarchy Process. J. Clean. Prod. 151, 509–525. doi:10.1016/j.jclepro.2017.02.099.

May, G., Stahl, B., 2017. The significance of organizational change management for sustainable competitiveness in manufacturing: exploring the firm archetypes. Int. J. Prod. Res. 55, 4450–4465. doi:10.1080/00207543.2016.1261197.

Medini, K., Da Cunha, C., Bernard, A., 2015. Tailoring performance evaluation to specific industrial contexts - application to sustainable mass customisation enterprises. Int. J. Prod. Res. 53, 2439–2456. doi:10.1080/00207543.2014.974844.

Meng, B., Liu, Y., Andrew, R., Zhou, M., Hubacek, K., Xue, J., Peters, G., Gao, Y., 2018. More than half of China's CO_2 emissions are from micro, small and medium-sized enterprises. Appl. Energy 230, 712–725. doi:10.1016/J.APENERGY.2018.08.107.

Micheli, G.J.L., Cagno, E., 2010. Dealing with SMEs as a whole in OHS issues: warnings from empirical evidence. Saf. Sci. 48, 729–733. doi:10.1016/j.ssci.2010.02.010.

Mickovski, S.B., Thomson, C.S., 2017. Developing a framework for the sustainability assessment of eco-engineering measures. Ecol. Eng. 109, 145–160. doi:10.1016/j.ecoleng.2017.10.004.

Montiel, I., 2008. Corporate social responsibility and corporate sustainability: separate pasts, common futures. Organ. Environ. 21, 245–269. doi:10.1177/1086026608321329.

Morioka, S.N., Bolis, I., Evans, S., Carvalho, M.M., 2018. Transforming sustainability challenges into competitive advantage: multiple case studies kaleidoscope converging into sustainable business models. J. Clean. Prod. 167, 723–738. doi:10.1016/j.jclepro.2017.08.118.

Neely, A., Gregory, M., Platts, K., 1995. Performance measurement system design: a literature review and research agenda. Int. J. Oper. Manag. Prod. Manag. 15, 80–116. doi:10.1108/01443570010343708.

Nehler, T., Rasmussen, J., 2016. How do firms consider non-energy benefits? Empirical findings on energy-efficiency investments in Swedish industry. J. Clean. Prod. 113, 472–482. doi:10.1016/j.jclepro.2015.11.070.

Neri, A., Cagno, E., Di Sebastiano, G., Trianni, A., 2018. Industrial sustainability: modelling drivers and mechanisms with barriers. J. Clean. Prod. 194, 452–472. doi:10.1016/j.jclepro.2018.05.140.

Neri, A., Cagno, E., Lepri, M., Trianni, A., 2021. A triple bottom line balanced set of key performance indicators to measure the sustainability performance of industrial supply chains. Sustainable Production and Consumption 26, 648–691. doi:https://doi.org/10.1016/j.spc.2020.12.018.

Nicolăescu, E., Alpopi, C., Zaharia, C., 2015. Measuring corporate sustainability performance. Sustainability 7, 851–865. doi:10.3390/su7010851.

Nordheim, E., Barrasso, G., 2007. Sustainable development indicators of the European aluminium industry. J. Clean. Prod. 15, 275–279. doi:10.1016/j.jclepro.2006.02.004.

Paju, M., Heilala, J., Hentula, M., Heikkilä, A., Johansson, B., Leong, S., Lyons, K., 2010. Framework and indicators for a sustainable manufacturing mapping methodology, Proc. Winter Simulation Conference, 3411–3422. doi:10.1109/WSC.2010.5679031.

Popovic, T., Barbosa-Póvoa, A., Kraslawski, A., Carvalho, A., 2018. Quantitative indicators for social sustainability assessment of supply chains. J. Clean. Prod. 180, 748–768. doi:10.1016/j.jclepro.2018.01.142.

Ribeiro-Soriano, D., 2017. Small business and entrepreneurship: their role in economic and social development. Entrep. Reg. Dev. 29, 1–3. doi:10.1080/08985626.2016.1255438.

Rojas-Lema, X., Alfaro-Saiz, J.J., Rodríguez-Rodríguez, R., Verdecho, M.J., 2020. Performance measurement in SMEs: systematic literature review and research directions. Total Qual. Manag. Bus. Excell. 0, 1–26. doi:10.1080/14783363.2020.1774357.

Rojas-lema, X., Rodríguez-rodríguez, R., Verdecho, M., 2020. Total quality management & business excellence performance measurement in SMEs : systematic literature review and research directions. Total Qual. Manag. 0, 1–26. doi:10.1080/14783363.2020.1774357.

Saeed, M.A., Kersten, W., 2020. Sustainability performance assessment framework: a cross-industry multiple case study. Int. J. Sustain. Dev. World Ecol. 27, 496–514. doi:10.1080/13504509.2020.1764407.

Salvado, M.F., Azevedo, S.G., Matias, J.C.O., Ferreira, L.M., 2015. Proposal of a sustainability index for the automotive industry. Sustainability 7, 2113–2144. doi:10.3390/su7022113.

Shibin, K.T., Gunasekaran, A., Dubey, R., 2017. Explaining sustainable supply chain performance using a total interpretive structural modeling approach. Sustain. Prod. Consum. 12, 104–118. doi:10.1016/j.spc.2017.06.003.

Siebert, A., O'Keeffe, S., Bezama, A., Zeug, W., Thrän, D., 2018. How not to compare apples and oranges: generate context-specific performance reference points for a social life cycle assessment model. J. Clean. Prod. 198, 587–600. doi:10.1016/j.jclepro.2018.06.298.

Singh, R.K., Murty, H.R., Gupta, S.K., Dikshit, A.K., 2012. An overview of sustainability assessment methodologies. Ecol. Indic. 15, 281–299. doi:10.1016/j.ecolind.2011.01.007.

Singh, S., Olugu, E.U., Fallahpour, A., 2014. Fuzzy-based sustainable manufacturing assessment model for SMEs. Clean Technol. Environ. Policy 16, 847–860. doi:10.1007/s10098-013-0676-5.

Singh, S., Olugu, E.U., Musa, S.N., 2016. Development of sustainable manufacturing performance evaluation expert system for small and medium enterprises. Procedia CIRP 40, 609–614. doi:10.1016/j.procir.2016.01.142.

Slawinski, N., Bansal, P., 2012. A matter of time: the temporal perspectives of organizational responses to climate change. Organ. Stud. 33, 1537–1563. doi:10.1177/0170840612463319.

Sloan, T.W., 2010. Measuring the sustainability of global supply chains: current practices and future directions. J. Glob. Bus. Manag. 6, 92–107.

Srnicek, N., Williams, A., 2015. Inventing the Future: Postcapitalism and a World Without Work. Verso, London (UK).

Stacchezzini, R., Melloni, G., Lai, A., 2016. Sustainability management and reporting: the role of integrated reporting for communicating corporate sustainability management. J. Clean. Prod. 136, 102–110. doi:10.1016/j.jclepro.2016.01.109.

Staniškis, J.K., Arbačiauskas, V., 2009. Sustainability performance indicators for industrial enterprise management. Environ. Res. Eng. Manag. 2, 42–50. doi:10.5755/j01.erem.48.2.13.

Székely, N., Vom Brockem, J., 2017. What can we learn from corporate sustainability reporting? Deriving propositions for research and practice from over 9,500 corporate sustainability reports. PLoS One 12, 1–27. doi:10.1371/journal.pone.0174807.

Taques, F.H., López, M.G., Basso, L.F., Areal, N., 2020. Indicators used to measure service innovation and manufacturing innovation. J. Innov. Knowl. doi:10.1016/j.jik.2019.12.001.

Tremblay, A., Badri, A., 2018. A novel tool for evaluating occupational health and safety performance in small and medium-sized enterprises: the case of the Quebec forestry/pulp and paper industry. Saf. Sci. 101, 282–294. doi:10.1016/j.ssci.2017.09.017.

Trianni, A., Cagno, E., Neri, A., 2017. Modelling barriers to the adoption of industrial sustainability measures. J. Clean. Prod. 168, 1482–1504. doi:10.1016/j.jclepro.2017.07.244.

Trianni, A., Cagno, E., Neri, A., Howard, M., 2019. Measuring industrial sustainability performance: empirical evidence from Italian and German manufacturing small and medium enterprises. J. Clean. Prod. 229, 1355–1376. doi:10.1016/j.jclepro.2019.05.076.

Trianni, A., Cagno, E., Thollander, P., Backlund, S., 2013. Barriers to industrial energy efficiency in foundries: a European comparison. J. Clean. Prod. 40, 161–176. doi:10.1016/j.jclepro.2012.08.040.

Tsalis, T.A., Stylianou, M.S., Nikolaou, I.E., 2018. Evaluating the quality of corporate social responsibility reports: the case of occupational health and safety disclosures. Saf. Sci. 109, 313–323. doi:'0.1016/j.ssci.2018.06.015.

Tseng, M.-L., Chang, C.-H., Lin, C.-W.R., Wu, K.-J., Chen, Q., Xia, L., Xue, B., 2020. Future trends and guidance for the triple bottom line and sustainability: a data driven bibliometric analysis. Environ. Sci. Pollut. Res. Int. 1614-7499, 27 (27), 33543–33567. doi:10.1007/s11356-020-09284-0. 32572746.

Van Schoubroeck, S., Van Dael, M., Van Passel, S., Malina, R., 2018. A review of sustainability indicators for biobased chemicals. Renew. Sustain. Energy Rev. 94, 115–126. doi:10.1016/j.rser.2018.06.007.

Veleva, V., Ellenbecker, M., 2001. Indicators of sustainable production: framework and methodology. J. Clean. Prod. 9, 519–549. doi:10.1016/S0959-6526(01)00010-5.

Winroth, M., Almström, P., Andersson, C., 2016. Sustainable production indicators at factory level. J. Manuf. Technol. Manag. 27, 842–873. doi:10.1108/JMTM-04-2016-0054.

Witjes, S., Vermeulen, W.J.V, Cramer, J.M., 2015. Exploring corporate sustainability integration into business activities. Experiences from 18 small and medium sized enterprises in the Netherlands. J. Clean. Prod. 153, 528–538. doi:10.1016/j.jclepro.2016.02.027.

CHAPTER 7

Life cycle assessment: methods, limitations, and illustrations

Sara Toniolo, Lorenzo Borsoi, Daniela Camana
Department of Industrial Engineering, University of Padova, Padova, Italy

7.1 Introduction to the life cycle assessment (LCA) methodology

The second half of the twentieth century was characterized by a greater awareness on the importance of the protection of the environment and of the possible impacts associated with products and processes. Several organizations increased their interest in the application of methodologies aimed at a better understanding of the interactions that their products or activities had with the environment (Bjørn et al., 2018b). In the last years, LCA became one of the most important tools for environmental management. Probably it is not the best technique for every situation, but it has a fundamental characteristic: it is based on the life cycle thinking approach. According to this, every product, process, or activity must be seen from the perspective of its life cycle. This means that if there is the necessity of evaluating the environmental impacts of a product, its manufacturing is not enough, but its entire life cycle must be considered, from the extraction of the raw materials to its disposal or recycling. In other words, LCA is a method to assess the environmental impacts of goods and processes "from cradle to grave" quantitatively. According to Hellweg and Canals (2014) its goal is to identify which are the most effective interventions to improve the product system from an environmental point of view, avoiding the "burden shifting" phenomenon. In fact, it is possible that an improvement aimed to reduce a particular type of impact could lead to an increase in the severity of another type of impact. This is the case of biofuels: the use of these fuels instead of those of fossil origin leads to a decrease in the consumption of nonrenewable resources, but has a huge impact with regard to the exploitation of the soil (Hellweg and Canals, 2014). LCA does not focus exclusively on one type of impacts, but it is able to cover multiple environmental issues, such as water scarcity, climate change, eutrophication, depletion of resources, and land occupation and transformation (Bjørn et al., 2018a).

A well-known definition of LCA is given by the Society of Environmental Toxicology and Chemistry (SETAC). Namely, "LCA is a methodology to evaluate the environmental burdens associated with a product, process, or activity. It identifies and quantifies energy and materials used and waste released to the environment; it assesses the impact of energy, materials, and releases to the environment; it identifies and evaluates opportunities for environmental improvements. LCA embraces the entire life cycle of a product, process, or activity, encompassing extraction and processing of raw materials; manufacturing, transportation and distribution; use, reuse, maintenance; recycling and final disposal" (SETAC, 1993). Thus, it aims at assessing the environmental burdens through the identification and quantification of energy and materials consumed, waste produced, and possible environmental improvements at various points in the life cycle products, processes, and activities. Another recognized definition was

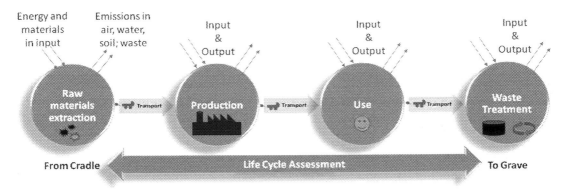

FIG. 7.1 Scheme of the life cycle thinking concept.

proposed by ISO, namely "LCA is a compilation and evaluation of inputs, outputs, and potential environmental impacts of a product system throughout its life cycle" (ISO, 2020a).

One of the main characteristics of LCA is that it has a quantitative nature. This means that it can be used to compare the environmental performance of different product systems. This quantitative nature is closely related to the fact that it is based on data. In fact, the flows that enter and exit the product system are obtained though measurements and the impacts are calculated by means of cause-effect models based on the proven causalities or empirical observed relationships (Bjørn et al., 2018a). Fig. 7.1 illustrates the life cycle thinking approach with reference to a life cycle application.

LCA has other fundamental characteristics. First of all, it must systematically and adequately examine the environmental aspects of the product systems under study. The degree of detail and the temporal extension are functions of the objective and system boundary. Moreover, each phase must be transparent.

Thanks to its characteristics, LCA is also used to inform decision makers in industry, government, or nongovernment organizations to select indicators of environmental performance, to implement eco-labeling, and to make environmental claims. One of the most important applications is the support on the marketing activities. In fact, LCA can also be used as a very effective tool of communication and its application can be an advantage with respect to competitors (Owsianiak et al., 2018).

The international regulatory reference for carrying out LCA studies is represented by the ISO standards belonging to the 14040 series, which is part of the group of ISO 14000 standards for the environmental management systems. These standards present general requirements that can be applicable to every type of product or activity. They were developed for the first time at the end of the nineties. Nowadays we rely on the following documents:

- ISO 14040:2006/AMD 1:2020 Environmental management—Life cycle assessment—Principles and framework—Amendment 1.
- ISO 14044:2006/AMD 2:2020 Environmental management—Life cycle assessment—Requirements and guidelines—Amendment 2.

ISO 14040 describes the principles and the framework, whereas ISO 14044 specifies the requirements and provides the guidelines. They provide information about the steps to develop an LCA study, namely (1) definition of the goal and scope of the LCA, (2) the life cycle inventory (LCI) analysis phase, (3) the life cycle impact assessment (LCIA) phase, and (4) the life cycle interpretation phase.

They also include a critical review of LCA, the limitations of LCA, the relationship between the LCA phases, and the conditions for the use of value choices and optional elements.

During the (1) definition of the goal and scope, the objectives of the study and the boundaries of the system are defined. In the (2) inventory analysis, inputs and outputs for each process are compiled; then they are summed across the whole system. In a typical LCA study, hundreds of emissions and resources are included in the calculation (Hellweg and Canals, 2014). During the (3) life cycle impact assessment, the data of the inventory are associated with specific environmental impacts and converted in common impact units to make them comparable. During the (4) interpretation phase, the inventory and impact assessment results are interpreted also with reference to the goal of the study. Fig. 7.2 outlines the four phases of the LCA methodology.

7.1.1 First phase

The first phase is the description of the goal and scope, which includes defining the objectives of the study and setting the methodological bases to develop the LCA. The goal definition must establish unambiguously what the intended application is, the reasons to carry out the study, and the type of public which it is intended for. In other words, it must identify the stakeholders to whom the results will be communicated and if the results are intended to be used in comparative assertion or if they will be disclosed to the general public (ISO, 2020a).

The scope instead must clearly define the extent, depth, and detail of the study. In the scope, the following items must be defined (ISO, 2020b):

- product system (or product systems in the case of comparative studies);
- function of the product system;
- functional unit;
- system boundary;
- allocation procedures;
- impact categories and methodologies of impact assessment and interpretation;
- data requirements and initial data quality requirements;
- assumptions;
- limitations;

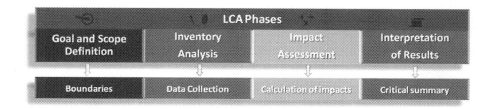

FIG. 7.2 Phases of the LCA methodology.

- type of critical review, if present;
- type and format of the report.

One of the most important items is the functional unit, as it allows quantifying the identified functions of the product system. It is the reference unit for the calculation of inputs and outputs and allows the comparison between different products in case of comparative studies (Toniolo et al., 2020). The product system and the system boundaries must be defined, determining what unit processes are included in the study. Considering only the manufacturing stage is not enough, each phase of the life cycle of the product under study must be considered, consistently with the life cycle thinking approach. Another important issue is the data quality. The description of the data quality is important to understand the reliability of the results and to interpret correctly. The data quality requirements specify the characteristics of the data needed for the study, they are (ISO, 2020a) as follows:

- time coverage;
- geographical coverage;
- technological coverage;
- data accuracy, completeness and representativeness;
- consistency and reproducibility of the methods;
- data source;
- information uncertainty.

7.1.2 Second phase

The second phase is called LCI and comprises three main steps, namely data collection, data elaboration, and application of allocation procedures.

During this phase, the increase of knowledge of the system can lead to an identification of new data requirements or limitations. This means that the data collection and elaboration procedures must change to satisfy the goal of the study (ISO, 2020a).

The main data required to conduct an LCA study are the following:

- consumption of materials, both new and secondary materials;
- consumption of resources such as energy and water;
- emission to air, water and soil;
- waste production.

The following items must be indicated as well:

- data sources;
- reference process;
- reference technology;
- geographical area;
- monitoring details;
- measuring methods;
- specific units of measurements.

Primary data, collected directly at the production sites, are preferable. If it is not possible to obtain them, secondary and tertiary data can be used as well. The former data are collected from the literature, technical manuals, or other studies, while the latter are obtained from estimations or technical coefficients. The collected data have to be validated by a calculation procedure. Moreover, data must be connected to the unit process and to the reference flow of the functional unit (ISO, 2020b).

One of the main issues LCA performers have to deal with are multifunctional processes. This particular problem occurs when a process of the product system delivers more product outputs and not only the one under study. This situation happens when the studied good is produced together with other coproducts. Multifunctionality constitutes a challenge in LCA, but it can be overcome by different types of solutions presented by ISO 14044. The first choice is simply trying to increase the modeling resolution by dividing the multifunctional process unit into new minor units. If this leads to a separation of the different products' flows, the problem is solved; otherwise the "system expansion" procedure is applied. This means that to ensure equal functionality, if we are dealing with two product systems, the second one is expanded to include the first system's secondary function, which is a function that has no relevance for the product's users. However, a complete functional equivalence is not obtained frequently, meaning that applying system expansion is not always feasible. As an alternative, ISO 14044 proposes allocation (Bjørn et al., 2018c).

Allocation consists in subdividing the inputs and outputs of the multifunctional process among the different product systems. When it is possible, allocation is performed in accordance with a causal physical relationship between the products. This is the case, for example, of an incineration plant that incinerates batteries and plastic waste: as plastic contains no cadmium, the emissions of this toxic metal can be entirely allocated to the batteries. In case a causal physical relationship does not exist, an allocation based on representative physical parameters is performed. However, it can be applied only if product and coproduct have similar functions, as the parameter must be able to represent a common function of the product systems. When no type of physical allocation can be applied, ISO standards propose an economic allocation, in which inputs and outputs of the multifunctional process are subdivided among the products based on their economic values (Bjørn et al., 2018c).

7.1.3 Third phase

The third phase is the impact assessment, during which the data of the inventory are associated with specific environmental impact categories, which are expressed by means of specific category indicators to quantify the impacts (ISO, 2020a).

The steps of this phase are as follows (ISO, 2020b):

- Classification: It relates the global or local impacts with the different input or output flows. The results of the inventory analysis are associated with the various impact categories.
- Characterization: Impacts are quantified within given categories by means of characterization factors derived from scientific models. This step allows the passage from a qualitative to a quantitative evaluation.

During the characterization step, Eq. (1) is considered (Goedkoop et al., 2013):

$$EP(j)i = Q \times EQ(j)i \tag{1}$$

where $EP(j)i$ is the environmental impact of substance i with reference to the impact category j, Q is the quantity of substance i, and $EQ(j)i$ is a factor representing the substance i contribution to the impact j. Different substances contributing to an environmental impact are aggregated considering their substance-specific effect and emissions or resources, which are grouped according to their impact categories expressed in a common impact unit, which permits the comparability of the results. For example, CO_2 and CH_4 emissions can both be expressed as CO_2-equivalent emissions thanks to the factor known as "global warming potential" developed by the Intergovernmental Panel on Climate Change (Hellweg and Canals, 2014).

Along with classification and characterization, which are mandatory steps, optional procedures also exist:

- Normalization: The relative importance of the indicators results is clarified, dividing the results by a reference value. This analysis allows the quantification of the contribution of each category.
- Weighting: Numerical factors are assigned to the impact categories results in accordance with their importance. Then, the results are multiplied by these factors and eventually aggregated in a single impact score. This procedure allows the comparison of single environmental effects.

7.1.4 Fourth phase

The fourth phase is the interpretation of the results obtained through the LCI and the life cycle impact assessment phases. The results are interpreted together with reference to the objective of the study (ISO, 2020b).

During this phase, it is recommended to identify the data that have a greater contribution to the result of a specific category indicator, through an analysis called, contribution analysis, which has the purpose of detecting and representing the most important data in an intuitive and effective way to focus attention on them. In this way the intervention priorities are identified, leading to effective and efficient improvement actions. During this phase, some controls must be done, such as completeness check and consistency check. The completeness check is performed to verify that all the needed data and information are available and complete. If this is not the case, there must be a justification, or the goal and scope of the study must be revised.

The consistency check, instead, is performed to determine whether assumptions, methods, and data are consistent with the goal and scope that have been defined.

In addition, a sensitivity analysis is needed to verify the consequent effects of variables' value on the output values (Laurent et al., 2020). The importance of this tool comes from the fact that the assumptions can affect the results of a study. In fact, according to ISO 14044, sensitivity analysis "tries to determine the influence of variations in assumptions, methods, and data on the results." An assessment of this type may be useful in reference to the choices relating to the following items: rules for allocation, system boundary definition, assumptions concerning data, selection of impact categories, assignment of inventory results to impact categories (classification), calculation of category indicators (characterization), normalized data, weighted data, weighting method, and data quality (ISO, 2020b). A sensitivity analysis can be conducted in different methods (Laurent et al., 2020). According to Heijungs and Kleijn (2001), perturbation analysis is one of the most prominent methods. It examines how much the outputs change if uniform, arbitrarily chosen, marginal perturbations in the input data are set. This analysis has two main purposes: the first purpose is to identify those input data and parameters that lead to relatively high deviations of the results, in opposition with the ones that do not considerably change the results noticeably; the second purpose is application-driven, as the information obtained from this analysis can be used as support for the development of products and processes (Laurent et al., 2020).

However, it is important to underline that the investigation of how impact results change, as a result of a change in input, should be related to the goal of the study (Laurent et al., 2020). According to Laurent et al. (2020), scenarios' analysis is an important auxiliary method. It is a method for predicting and analyzing possible future results, alternative to those of the main study, deriving from arbitrary decisions or the influence of external factors. It is also able to verify whether the assumptions made during the LCA study are relevant and valid under different scenario conditions, and to ensure these assumptions are well documented and transparent (Laurent et al., 2020).

Another important step in the conduction of an LCA study is the uncertainty analysis. Uncertainty is defined as "the discrepancy between a measured or calculated quantity and the true value of that quantity" (Finnveden et al., 2009) or as "the degree to which the quantity under study may be off from the truth" (Rosenbaum et al., 2018). We have to specify that different types of uncertainty exist: parameter uncertainty, model uncertainty, uncertainty due to choices, spatial and temporary variability, and variability between objects or sources and humans. As many input parameters to an LCA study are associated with uncertainties, this type of analysis can be considered of the fundamental importance. According to Laurent et al. (2020), the uncertainty analysis quantifies the influence of each individual uncertainty on the final result. In other words, it examines how they propagate through the modelling and impact assessment to result in overall uncertainty of the impact results (Laurent et al., 2020). The most widely used way for performing a quantitative uncertainty analysis is applying a Monte Carlo simulation (Laurent et al., 2020). In the Monte Carlo approach, the uncertainties of input parameters are represented by probability distributions that are characterized by a mean value and a standard deviation value. The LCA calculation is conducted a large number of times, considering for each iteration a different sample of values of the input parameters from their probability distribution. The resulting impact values are expressed as probability distribution (Laurent et al., 2020).

When quantitative methods cannot be applied, LCA practitioners should qualitatively discuss the uncertainties underlying how the data are modeled and how this modelling can influence the results and conclusions, critically evaluating the data sources, data quality, and data representativeness. The data relating to the most impactful processes, substances, or life cycle stages need special attention (Laurent et al., 2020).

7.2 International standards

It is important to highlight that LCA practitioners must not only comply with the standards described above, namely ISO 14040 and ISO 14044, but must also take into consideration other documents depending on the motivation that led them to use this methodology and the type of product or service under study.

For example, if the LCA methodology is applied with the purpose of publishing an Environmental Product Declaration (EPD), other specific standards must be considered. EPDs are LCA-based tools used to communicate the environmental performance of a product (Grahl and Schmincke 2007). A specific body, called the program operator or EPD operator, which conducts a type III environmental declaration programme, manages the elaboration process of EPDs (Ingwersen and

Stevenson, 2012). An EPD has to be created based on an appropriate set of specific rules, called Product Category Rule (PCR), which identifies and describes the process of preparing an EPD, making it comparable and verifiable (Butt et al., 2015). The programme operators produce the PCR document through a consultation process and involving the interested parties (ISO, 2000). Several PCRs exist for different sectors and are developed by program operators periodically. Besides the programme operators, also the European Committee for Standardization (CEN) has developed some European Standards to be used as PCRs recently. For instance, EN 15804:2012+A2:2019 provides core product category rules for all construction products and services and allows that EPDs of construction sector are derived, verified, and presented in harmonized way (CEN, 2019a). Other European Standards complementary to EN 15804 are EN 16810 for floor coverings (CEN, 2017a), EN 16783 that provides the rules for factory made and in-situ thermal insulation products (CEN, 2017b), EN 16757 for concrete and concrete elements for building and civil engineering (CEN, 2017c), EN 16485 for wood and wood-based products for use in construction (CEN, 2014), and EN 16908 for cement and building lime (CEN, 2017d). Other standards are EN 16887 for leather (CEN, 2017e), EN 17160 for ceramic tiles (CEN, 2019c), and EN 17213 for windows and doors (CEN, 2020). Table 7.1 presents a list of international standards and documents useful to apply the LCA methodology properly and to develop an EPD.

Table 7.1 List of international standards and documents related to LCA methodology.

Document	Year of publication	Title
ISO 14025	2006	Environmental labels and declarations—Type III environmental declarations—Principles and procedures
ISO/TS 14027	2017	Environmental labels and declarations—Development of product category rules
ISO 14040/AMD 1	2020	Environmental management—Life cycle assessment—Principles and framework–Amendment 1
ISO 14044:2006/AMD 2	2020	Environmental management—Life cycle assessment—Requirements and guidelines—Amendment 2
ISO/TR 14047	2012	Environmental management—Life cycle assessment—Illustrative examples on how to apply ISO 14044 to impact assessment situations
ISO/TS 14048	2002	Environmental management—Life cycle assessment—Data documentation format
ISO/TR 14049	2012	Environmental management—Life cycle assessment—Illustrative examples on how to apply ISO 14044 to goal and scope definition and inventory analysis
ISO/TS 14071	2014	Environmental management—Life cycle assessment—Critical review processes and reviewer competencies: Additional requirements and guidelines to ISO 14044
ISO/TS 14072	2014	Environmental management—Life cycle assessment—Requirements and guidelines for organizational life cycle assessment
ISO/WD TS 14074	Under development	Environmental management—Life cycle assessment—Principles, requirements and guidelines for normalization, weighting, and interpretation

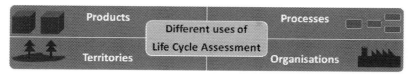

FIG. 7.3 LCA applications.

7.3 Applications

One of the key features of LCA is that it can be applied in a multitude of applications (Fig. 7.3). In particular, in the industrial sector LCA has five main purposes (Owsianiak et al., 2018): decision support in product and process development, marketing purpose, development of indicators for the products' environmental performance monitoring, choice of the supplier, and strategic planning. Moreover, according to Owsianiak et al. (2018), the same LCA can be performed for different purposes within the same organization. For example, it is not unusual that LCAs for product design have marketing purposes as well. LCA is typically used during the early stage of product and process design to assess and consequently improve specific product systems, supporting internal decision-making, such as for eco-design of products, process optimization, supply chain management, and strategic decisions (Hellweg and Canals, 2014). However, as we said, nowadays LCA is used in many other applications. In fact, many companies are using LCA results to highlight the phases of the product life cycle that generate the greatest impacts and how other companies across the value chain are addressing these impacts. This situation results in a better collaboration with the other actors. In other words, LCA can reveal whether collaboration with other actors in the value chain may be preferable to the intervention by the individual actor, if an improvement of the product system is needed (Hellweg and Canals, 2014). For example, LCA studies have demonstrated that the environmental impacts generated by washing clothes can be reduced by lowering the washing temperature. Actual improvements, however, can only be achieved thanks to the cooperation of different actors, such as cleaners and washing machines producers, which must respectively market detergents effective at low temperatures and devices that allow the selection of specific programs (Hellweg and Canals, 2014).

The construction sector turned out to be one of the most fertile grounds for an LCA study, and EPDs have become effective tools for communicating information on the environmental profiles of products related to this sector. Some applications show that energy use within the building is responsible for the highest contribution in most impact categories. There are studies that show that, in areas characterized by a cold climate, extra insulation material is able to decrease life cycle impacts, although these studies also demonstrate the existence of a tipping point at which further material use is not environmentally useful any longer. Other solutions that can be applied, according to the results of various LCA analyses, are the usage of renewable energy and the reduction of living space per person. However, it is underlined that the environmental impacts of a building also depend on the technology level, the local conditions, and the behavior of occupants (Hellweg and Canals, 2014).

Environmental performance communication and eco-labeling are not exclusive of the construction sector. In fact, as consumers' interest in the environment is increasing, communication to the public is becoming an important tool for gaining a competitive advantage. Moreover, eco-labels are increasingly used to make a product more appealing for these consumers (Owsianiak et al., 2018).

Other sectors in which LCA is widely applied are those of food and agriculture. The life cycle of a food product is usually divided into six stages: inputs production and transportation, cultivation, processing, distribution, consumption, and waste treatment. However, most of the studies are cradle-to-gate, focusing only on the first two stages, as the cultivation stage is usually responsible for the greatest contributions to many impact categories. The LCA studies that have been implemented during the last years agree on the fact that conventional and organic farming practices are equally valid. Instead, studies about local food showed that it is not necessarily the most sustainable answer, as transport is not often responsible for the decisive environmental impacts of a food product (Dijkman et al., 2018).

LCA is particularly suited to support decisions in waste management. In fact, it is able to assess whether it is preferable to avoid respecting the classic waste hierarchy, according to which reuse, recycling, and recovery are preferable to landfilling (Hellweg and Canals, 2014).

LCA is a widely used tool in the energy sector. For instance, if a new form of energy needs to be adopted, LCA can assess whether this change can actually bring environmental benefits before the money is invested. For instance, LCA is applied to calculate the impact associated with the utilization of biofuels that are widely considered as an environment-friendly source of energy, but thanks to LCA methodology, also the environmental burdens are highlighted. In the energy sector, LCA is also used to assess different scenarios of energy supply mixes to help design new sustainable energy systems (Hellweg and Canals, 2014).

In addition to biofuels, the use of electric vehicles (EVs) as an alternative to conventional ones has also been a hot topic in recent years, as it is considered a key technology to reduce greenhouse gases in the atmosphere. However, some aspects, such as the production of the battery system, might lead to shifting the problem to other life cycle phases or areas of impacts. LCA methodology is able to identify whether this is the case or not, even if many complain about the current lack of functional equivalence to ensure the comparability between two analyzed vehicles. In fact, despite this lack of methodological harmony, several studies agree on the fact that EVs lead to actual improvements only if they are powered by low-impact energy sources. This means that in regions where electricity is mostly obtained from fossil sources, the emissions of greenhouse gases may even increase due to transportation. Moreover, these studies have shown that the production phase of an EV has twice as much impact than that of a conventional vehicle. In particular, the production of the battery turned out to be responsible of almost a half of the total emissions of greenhouse gases during the EV's manufacturing stage. In addition, other types of impact are involved, due to the usage of copper, aluminum, and other rare earth metals to produce batteries and electric motors, whose mining may be harmful on a local level (Cerdas et al., 2018).

LCA is also applied as a tool to assess new technologies and promote proactive actions, such as with nanotechnology. However, almost all the LCA studies neglect nanoparticle emissions and their effect, excluding a potential source of impact for categories related to human and ecosystem health. Another problem is that a comparison between new and mature technologies needs to be updated because of upscaling and learning reasons (Hellweg and Canals, 2014).

It must be said that although LCA is typically used at the product level, applications at the corporate level are increasing. This means, that companies, especially the large enterprises, are increasing the interest to study the environmental performance of the whole organization or single plants, applying the so-called organizational LCA, which is a compilation and evaluation of the inputs, outputs, and potential environmental impacts of the activities associated with the organization adopting a life cycle perspective (Martínez-Blanco et al., 2015). However, nowadays this type of study is almost always limited to single impact categories. In fact, carbon footprint and water footprint are typically considered (Owsianiak et al., 2018).

Another challenging LCA methodology's application regards the so-called territorial LCA, which refers to the calculation of the environmental impacts associated with human activities in a territory and the reference flow is the relationship between a territory and a studied land planning scenario (Loiseau et al., 2013). In particular, two main approaches exist: territorial LCA can focus on the assessment of a given activity or supply chain of a specific territory (type A) or it can assess all production and consumption activities located in that territory (type B). As an example of territorial LCA of type A, there is a study of collective biogas plants in French West territories; instead, a territorial LCA analysis of type B was implemented in a French Mediterranean framework with the aim of providing a territory environmental baseline for 2010 (Loiseau et al., 2013; (Loiseau et al., 2018)).

7.4 Limitations

Although LCA is a very useful and widely used tool, it still has some disadvantages. For this reason, it is characterized by a continuous improvement, thanks to the publication of new updates and the development of new features. However, this reality, now as well as in the future, will continue to be studded with challenges to face and problems to solve. For instance, comprehensiveness may be considered a main strength of this tool; however, it is also a complication as the large amount of data require a modelling that is a simplification and generalization of the reality, leading to a divergence from the calculation of the actual impacts (Bjørn et al., 2018a).

It has been demonstrated that LCA results can have high uncertainties due to the large amounts of measured and simulated data. The modelling, thanks to which complex environmental interactions are represented in a simplified way, together with the assumptions and the choice of values can contribute to the total uncertainty as well. A question that emerges from this issue is how much uncertainty is acceptable. We have to underline that LCA is not always a tool to provide a single answer, but it is able to provide possible solutions to a problem, allowing its complete understanding (Hellweg and Canals, 2014). However, uncertainty in LCA studies remains one of the main problems on which action must be taken in the future.

Some studies have aimed at reducing uncertainties in LCA by assessing the life cycle of products and their impacts in a regionalized manner, as considering data that are specific for the product under study may increase the accuracy of the study. However, obtaining spatial data constitutes a challenge, as it is not frequent that companies know the whole supply chain and consumer phase. Moreover, even if regionalization makes LCA more relevant, it is also true that nowadays the main issue is how to perfectly match the regionalized impact assessment methods to regionalized emissions and resource consumptions (Hellweg and Canals, 2014). Another challenge is how the information displayed in product labeling can be communicated in a simple and understandable way without hiding uncertainties. Data gaps present a challenge as well, as for some impacts, such as biodiversity loss, the implementation of offsets may be difficult. In this context, a key role is held by the retailers who are in direct contact with both consumers and producers and can be a reliable source of information (Hellweg and Canals, 2014). Regarding the application of LCA as a supporting tool for managerial decisions, a challenge for future research will be to widen the system boundaries beyond waste treatment and recycling, so as not to miss potential improvements.

Another limitation is that LCA follows the "best estimate" principle. This means that LCA models are based on average information about the processes and do not contemplate possible rare events, regardless of how serious they may be (Bjørn et al. 2018a).

The final limitation we describe is the fact that LCA is able to evaluate when a product system is better than another, but it cannot assess whether a product is sustainable in general terms. This explains why LCA, even if it is one of the best tools in this field, cannot answer all the possible questions and, thus, cannot be applied in all the situations (Bjørn et al., 2018a).

References

Bjørn, A., Owsianiak, M., Molin, C., Laurent, A., 2018. Main characteristics of LCA. In: Hauschild, M.Z., Rosenbaum, R.K., Irvin Olsen, S. (Eds.), Life Cycle Assessment - Theory and Practice. Springer International Publishing AG, NY, pp. 9–16. ISBN 978-3-319-56474-6.

Bjørn, A., Owsianiak, M., Molin, C., Hauschild, M.Z., 2018. LCA history. In: Hauschild, M.Z., Rosenbaum, R.K., Irvin Olsen, S. (Eds.), Life Cycle Assessment - Theory and Practice. Springer International Publishing AG, NY, pp. 17–30. ISBN 978-3-319-56474-6.

Bjørn, A., Moltesen, A., Laurent, A., Owsianiak, M., Corona, A., Birkved, M., Hauschild, M.Z., 2018. Life cycle inventory analysis. In: Hauschild, M.Z., Rosenbaum, R.K., Olsen, Irvin, S., S. (Eds.), Life Cycle Assessment - Theory and Practice. Springer International Publishing AG, pp. 117–166. ISBN 978-3-319-56474-6.

Butt, A.A., Toller, S., Birgisson, B., 2015. Life cycle assessment for the green procurement of roads: a way forward. J. Clean. Prod. 90, 163–170.

CEN (European Committee for Standardization), 2014. EN 16485 Round and sawn timber. Environmental Product Declarations. Product category rules for wood and wood-based products for use in construction.

CEN (European Committee for Standardization), 2017. EN 16810 Resilient, textile and laminate floor coverings - Environmental product declarations - Product category rules.

CEN (European Committee for Standardization), 2017. EN 16783 Thermal insulation products - Product category rules (PCR) for factory made and in-situ formed products for preparing environmental product declarations.

CEN (European Committee for Standardization), 2017. EN 16757 Sustainability of construction works. Environmental product declarations. Product Category Rules for concrete and concrete elements.

CEN (European Committee for Standardization), 2017. EN 16908 Cement and building lime - Environmental product declarations - Product category rules complementary to EN 15804.

CEN (European Committee for Standardization), 2017. EN 16887:2017 Leather - Environmental footprint - Product Category Rules (PCR) - Carbon footprints.

CEN (European Committee for Standardization), 2019. EN 15804:2012+A2:2019 Sustainability of construction works - Environmental product declarations - Core rules for the product category of construction products.

CEN (European Committee for Standardization), 2019. EN 17160:2019 Product category rules for ceramic tiles.

CEN (European Committee for Standardization), 2020. EN 17213:2020 Windows and doors - Environmental Product Declarations - Product category rules for windows and pedestrian doorsets.

Cerdas, F., Egede, P., Herrmann, C., 2018. LCA of electromobility. In: Hauschild, M.Z., Rosenbaum, R.K., Irvin Olsen, S. (Eds.), Life Cycle Assessment - Theory and Practice. Springer International Publishing AG, NY, pp. 669–694. ISBN 978-3-319-56474-6.

Dijkman, T.J., Basset-Mens, C., Antón, A., Núñez, M., 2018. LCA of food and agriculture. In: Hauschild, M.Z., Rosenbaum, R.K., Irvin Olsen, S. (Eds.), Life Cycle Assessment - Theory and Practice. Springer International Publishing AG, NY, pp. 723–754. ISBN 978-3-319-56474-6.

Finnveden, G., Hauschild, M.Z., Ekvall, T., Guinée, J., Heijungs, R., Hellweg, S., Koehler, A., Pennington, D., Suh, S., 2009. Recent developments in Life Cycle Assessment. J. Environ. Manage. 91, 1–21.

Goedkoop, M., Heijungs, R., Huijbregts, M., De Schryver, A., Struijs, J., Van Zelm, R., 2013. ReCiPe 2008 Report I: Characterisation. Version 1.08. Ministerie van VROM, Den Haag, Nederland.

Grahl, B., Schmincke, E., 2007. The part of LCA in ISO type III environmental declarations. Int. J. Life Cycle Assess. 12, 38–45.

Hellweg, S., Milà i Canals, L., 2014. Emerging approaches, challenges and opportunities in life cycle assessment. Science 344, 1109–1113.

Heijungs, R., Kleijn, R., 2001. Numerical approaches towards life cycle interpretation five examples. Int. J. Life Cycle Assess. 6, 141–148.

Ingwersen, W.W., Stevenson, M.J., 2012. Can we compare the environmental performance of this product to that one? An update on the development of product category rules and future challenges toward alignment. J. Clean. Prod. 24, 102–108.

ISO (International Organization for Standardization), 2000. ISO 14020: Environmental labels and declarations — General principles.

ISO, 2020. ISO 14040:2006/AMD 1:2020. Environmental management – Life cycle assessment – Principles and framework. International Organization for Standardization, Geneva, Switzerland.

ISO, 2020. ISO 14044:2006/AMD 2:2020. Environmental management – Life cycle assessment – Requirements and guidelines. International Organization for Standardization, Geneva, Switzerland.

Laurent, A., Weidema, B.P., Bare, J., Lio, X., de Souza, D.M., Pizzol, M., Sala, S., Schreiber, H., Thonemann, N., Verones, F., 2020. Methodological review and detailed guidance for the life cycle interpretation phase. J. Ind. Ecol. 24, 1–18.

Loiseau, E., Roux, P., Junqua, G., Maurel, P., Bellon-Maurel, V., 2013. Adapting the LCA framework to environmental assessment in land planning. Int. J. Life Cycle Assess. 18, 1533–1548.

Loiseau, E., Aissani, L., Le Féon, S., Laurent, F., Cerceau, J., 2018. Territorial Life Cycle Assessment (LCA): what exactly is it about? A proposal towards using a common terminology and a research agenda. J. Clean. Prod. 176, 474–485.

Martínez-Blanco, J., Inaba, A., Quiros, A., Valdivia, S., Milà-i-Canals, L., Finkbeiner, M., 2015. Organizational LCA: the new member of the LCA family—introducing the UNEP/SETAC Life Cycle Initiative guidance document. Int. J. Life Cycle Assess. 20, 1045–1047.

Owsianiak, M., Bjørn, A., Laurent, A., Molin, C., Ryberg, M.W., 2018. LCA applications. In: Hauschild, M.Z., Rosenbaum, R.K., Irvin Olsen, S. (Eds.), Life Cycle Assessment - Theory and Practice. Springer International Publishing AG, NY, pp. 31–42. ISBN 978-3-319-56474-6.

Rosenbaum, R.K., Georgiadis, S., Fantke, P., 2018. Uncertainty management and sensitivity analysis. In: Hauschild, M.Z., Rosenbaum, R.K., Irvin Olsen, S. (Eds.), Life Cycle Assessment - Theory and Practice. Springer International Publishing AG, NY, pp. 271–322. ISBN 978-3-319-56474-6.

SETAC, 1993. Guidelines for Life - Cycle Assessment: A 'Code of Practice. SETAC, Brussels.

Toniolo, S., Tosato, R.C., Gambaro, F., Ren, J., 2020. Life cycle thinking tools: life cycle assessment, life cycle costing and social life cycle assessment. In: Ren, J., Toniolo, S. (Eds.), Life Cycle Sustainability Assessment for Decision-Making. Methodologies and Case Studies. Elsevier, Amsterdam, pp. 39–56. ISBN 9780128183557.

CHAPTER 8

Life cycle assessment for better sustainability: methodological framework and application

Aman Kumar, Ekta Singh, Rahul Mishra, Sunil Kumar
CSIR-National Environmental Engineering Research Institute, Nagpur, Maharashtra, India

8.1 Introduction

Sustainability evaluation is now an important aspect to be considered by the decision makers and environmental managers with the growing awareness about sustainability. To measure and assess sustainability, several different methods have already been developed (Wulf et al., 2019). Three sustainability aspects, that is, climate, community, and economy, incorporated with concepts have been widely adopted (Zijp et al., 2015). There has never been a greater need for a holistic environmental assessment tool. This need laid the foundation of Life Cycle Assessment (LCA). LCA is a method of analysis that records the environmental impacts of raw material procurement, via processing and use, to waste disposal of a product, process, or any human activity. The LCA emerged out of this need. The environmental effects of a substance are assessed over its whole life by LCA, which is a state-of-the-art tool. LCA can be well-defined as product system impact on environment including their designs and processes and the estimation and assembly of resource inputs and outputs, over its life cycle (ISO, 2006; Nwodo and Anumba, 2019).

What began as an approach for comparison of products environmental goodness (greenness) has grown into a systematic framework to provide industry and government with a clear empirical foundation for environmental sustainability. LCA offers a detailed view of the environmental implications of the modification or selection of goods or procedures and gives an accurate image of possible tradeoffs with respect to the environment. From one venue to another or from one medium to another, LCA is beneficial for addressing various cross-media concerns and preventing the transition of an issue.

Life Cycle Thinking stemmed from Life Cycle Sustainability Assessment (LCSA) that is a potential method to determine sustainability based on this principle. This method suggests that a system is considered from cradle to grave, for example, an organization, a service, or a product. They highlighted the significance of recognizing that life cycle of a product comprised various processes, specifically: the extraction of raw materials, processing, use and disposal of the product (Klöpffer, 2003). Crucially, this strategy allows "tradeoffs" in impacts between different processes to be detected and prevented and impacts to be transferred during various periods of time. The study also recommended that the evaluation of the various dimensions of sustainability be combined.

LCA has its weaknesses as well as its benefits, as seen in all complex evaluation methods. While a general structure for performing an evaluation is given by the International Standardization Organization (ISO), it leaves a lot to be interpreted. The LCA standard was developed in the ISO 14000 series after growing pressure to standardize methodologies in the late 1990s. The ISO 14040 and 14044 standards (ISO 14040, 2006; ISO 14044, 2006) are not designed to define each field details, only to define a general methodology in which the approach is used. LCA as a tool for assessment of environmental sustainability has attained popularity in recent years as demonstrated by the fast-growing number of publications and databases that support their execution (Guinee et al., 2011).

The aim of this chapter is to provide an overall idea of building LCA. The study also discusses as separate subsections, other applications of LCA. Hence, the scope is on having the entire structure of LCA, the method involved, stages, application areas, and limitations. The sections include an overview/background of LCA, the systematic analysis methodology, applications, limitations, and finally the conclusion.

8.2 LCA methodology

For estimating industrial systems, processes, and products, LCA is used as a method of "cradle-to-grave." The raw substance gathering from the earth for production is beginning of "cradle-to-grave," which finishes with returning of all substances to the earth. All stages of products life are evaluated by LCA from the viewpoint that the next operation is interconnected with previous one. LCA facilitates estimates of total combined effects on the environment from all levels of the product life cycle and enables a more environmentally sound selection of the process or path (Potrč Obrecht et al., 2020).

Over years, LCA has been growing the model of product orientedness to investigate the environmental effect of a broader outline, which explains on a social, economic, and environmental scale. Life Cycle Sustainability Analysis (LCSA) helps associate the informations gathered and research required to address sustainability issues in future (Raymond et al., 2020). LCA assists policy makers in selecting a technology, process, or product, which has the low environmental impact. To find optimal solutions with other aspects such as cost and performance data, this information can be used. The transfer of environmental impacts to another from one media and among different stages of lifecycle is identified by LCA (e.g., releases of air may be lowered but leads to more generation of wastewater, etc. by a new process).

Input from certain raw materials and energy in all stages of any product or technology would mandate: from acquirement to production, processes, and final dumping. Solid wastes, waterborne, atmospheric emissions may be generated by all phases of the lifecycle, just because material efficiency and the transformation of energy are never hundred percent. Losses and byproducts occur that can sometimes be very unwanted. LCA allows us to monitor all harmful and useful results (Potrč Obrecht et al., 2020; Raymond et al., 2020).

As standardized by ISO, the current LCA practice mainly follows the following four interconnected stages (ISO, 2006; Curran, 2013):

- Clearly define the scope and purpose of the analysis (containing the selection of a functional unit): Identify a technology/process/product; system boundaries and context development.
- Compile inventory of related environmental releases, input of material, and energy (Life Cycle Inventory (LCI) analysis): Quantify and recognize materials, water, and energy as inputs as well as outputs of environmental releases.
- Associated with the established releases and inputs, Life Cycle Impact Assessment (LCIA): Evaluate the possible environmental impacts and potential effects of ecology and human, and quantify metrics.

- Analyze the results to assist policy makers in making a more informed decision: To recommend or select a preferred technology, process, or product, compare data from impact assessment stages and inventory analysis.

8.3 Important aspects of LCA methodology

While LCA is specified by the ISO standard and offers a common structure to perform an evaluation, it left a lot to the interpretation of the practitioner. For seemingly the same product, the results of various LCA studies have been excoriated as a consequence. There is a great deal of uncertainty about what LCA can and cannot do, and how it fits into a sustainability strategy at the strategic level. This chapter discusses 10 essential aspects of LCA methodology that are frequently ignored or misunderstood. What aspects users should know when determining whether an LCA is conducted or when someone else evaluates an LCA are described in the following sections.

8.3.1 Goal setting and functional unit

The scope of the study limits and directs the effort to collect data by simplifying it. With the goal setting, the functional unit choice is linked, a specific aspect of LCA that distinguishes it from other approaches to assessment of environment. The service provided by the system defines the functional unit, and the system is then studied. After this the study objective is shaped. It is essential to correctly set the scale of the functional unit. If it is set small, its LCIA will show only a small share of the total equipment emission (frequently approximately infinitesimal) (Finnveden et al., 2009).

8.3.2 Assigning environmental burdens

A widely discussed topic continues to be coproduct allocation. In the form of a hierarchy, the standard of ISO offers a few directions that allow practitioners, if possible, for preventing distribution by either modelling (i.e., collecting more comprehensive data) the subprocesses involved in the development or extension of system limitations to involve further coproduct(s) processes (ISO, 2006). A major role is left to interpretation in how this can be done. There is a common agreement that an attractive way to deal with this almost intractable issue is to prevent allocation by subprocess modelling and system expansion. Both methods however, make the model bigger and more complex and require additional data that will be collected to finish this project. The collection of more information involves more time and effort, which questions the method practicality. Larger systems seem to be less simple, because more information is available than can be easily shared about how the data were achieved. Although the reactions to this would be more essential to sustainability and help decision makers to make wise choices through the model of subprocesses, the distribution may not always be preventable, particularly if it is not possible to acquire data easily for the subprocesses or for the extended framework.

8.3.3 Credit for avoided burden

To include the alternative output of exported functions, the boundaries are extended in a framework expansion approach. In the case where the demand for soybean oil is assumed to be displaced by corn oil, the use of energy reduces to 37,985 Btu/gal from 77,228 Btu/gal. The idea of device expansion seems rational on the surface; however, it is too troublesome. Not merely does the definition of a way of producing an alternate solution byproduct necessitate market data, it results in negative emissions as the system is attributed with burdens and compensated elsewhere (Ekvall et al., 2005).

Open-loop recycling is seen as a particular distribution requirement. The chosen solution appears to be economic allocation; and for capturing the activities of downstream recycling, it is considered the best way. But as this is a subjective concept, it is not easy to determine which procedure is most "fair" depending on how far the analysis is performed.

8.3.4 Consequential LCA

Consequential LCA is regarded as widening research boundaries to include the possible implications of a decision (Curran, 2013). In comparison to the conventional gasoline, Searchinger et al. (2008) reported that a 20% reduction in greenhouse gaseous emissions was achieved in an attributive study of US maize-based ethanol. Nevertheless, in a consequential study, they anticipated emissions to grow by 47% compared to gasoline to account for policies driven production increases, owing to changes in land use resulting from higher maize prices. The differentiation between attributive and consequential LCA explains the effect of specified definition on data and methodological choices for the LCIA and LCI phases (Finnveden et al., 2009). The logical difficulty of a consequential LCA is that it requires extra economic principles such as costs of minimal production, demand and supply elasticity, and so on. The consequential LCA is driven by economic relations descriptions in the models. In general, by deducing patterns of history in costs, consumption, and production, it seeks to represent complex economic relationships. This adds to the possibility that the final LCA results would be substantially affected by inadequate assumptions or other errors. To minimize this risk, it necessitates that valid explanations can be employed to validate various findings on various consequences.

8.3.5 Inventory data availability and transparency

A key obstacle for LCA practice is the absence of readily accessible inventory data. The process is made smoother by commercial software systems; however, it is not always clear how the data have been modeled. Publicly accessible databases, such as Australia's National Pollutant Inventory (NPI) and the Toxic Release Inventory (TRI) of the US EPA, are also government-sponsored. At no discount, they are conveniently accessible and usable. However, these sources cannot easily be used in most life cycle studies as these data are recorded for specific facilities or sites rather than as market averages for an area or country. Sometimes the outcomes must be assumed to represent a sector of the industry as a means to group them. Europeans have successfully developed publicly accessible databases via initiatives such as the Eco Invent Database and the European Commission Life Cycle Assessment Forum, most recently. In establishing a national inventory database, the United States has seen limited progress (National Renewable Energy Laboratory NREL, 2006).

8.3.6 Identifying data uncertainty

Uncertainty can be due to either errors or variations in data in the form of uncertainty. The analysis of uncertainty is the method of assessing data variability and effect on the final results. Ambiguity applies in both impact evaluation metrics and inventory data and could have an important impact on how results have been used when making decisions. Nevertheless, the real effects of insecurity have not yet been thoroughly examined in decision making (Christensen et al., 2020).

8.3.7 Distinguishing risk assessment

Risk assessment is a dynamic process necessitating knowledge and data to be applied across a wide variety of disciplines and activities, involving dose-response evaluation, exposure assessment, modelling, fate and transport, and source classification. LCA expands the space and time product system on the opposite side (Margni and Curran, 2012; US EPA, 2004). Sometimes, inventory is aggregated in a manner that limits awareness of the geographical position of pollution. Information on the temporal course of pollution (subsequent concentrations in the creating environment or some environmental effects that occur in the future) is also usually unaccompanied by the LCI details. With the inherent ambiguity in the modelling of environmental impacts, the effect of a very complex reality on simplified model is an impact measure, providing only an estimate of the individual affected quality status. LCA outputs and LCIA model are ideal for testing relative comparisons, but not suitable for actual risk assumptions (Finnveden et al., 2009).

8.3.8 Reporting quantitative and qualitative information

Discussions endorsed LCAs to publish both quantitative and qualitative information in the early 1990s. But, quantification results in the enthusiasm of LCIA models, the importance of qualitative knowledge is sometimes overlooked. It should be noted that not all environmental information can be measured. Also, not all impact data are accessible for such inventory data models. Present models of LCIA for health of human would, therefore, be capable to model the possible effect of the products comprising human health. Using nanotechnology, this is the case for goods manufactured. In those cases where the LCIA model cannot model inventory data, when the model is recorded and kept in the final analysis, the modeler risk losing essential input and output data.

The LCA report is different in length but usually is approximately 100 pages, following several LCI spreadsheets. Different simulation assumptions and engineering estimates must be implemented when performing the LCA (both within the LCI and LCIA). These decisions are often formed on the basis of the modeler or the person responsible for the report. Thus, almost every option and its impact on the decision must be mentioned clearly in the final outcome to specify in detail the conclusions drawn from the data. Due to differences in practice, all modelling assumptions and data sources should be expressed with the results (Scrucca et al., 2020).

8.3.9 LCA does not always state a "winner"

The LCA interpretation step involves assumptions employed in producing a clear understanding of the ambiguities, the results of an effect analysis, a review of the results of the inventory and the results, and in facilitating decision making, whether it involves the product being chosen, the process or the service being developed, etc. A simple "winner" between alternatives will very rarely be established by the results of an LCA. In certain situations, due to the ambiguity of the final outcome, it might not be possible to say that one option is better than the other. This does not really imply that resources are wasted or for policy makers, that LCA is not a feasible tool. The LCA technique also improves understanding of the environmental and health effects (local, regional, or global) of each option and the relative importance of individual forms of impacts in comparison with each alternative proposed in the report. The pros and cons of each option are more completely exposed by this knowledge.

8.3.10 LCA is an iterative process

The findings interpretation is based on the comparison and the placement of the data and results with previous results. The iterative design of the ISO system appears in this sense. We can return to better details if the uncertainties are too great. If sensitive analysis reveals that certain decisions are important, we can go backwards and do more refined analyses. In particular, the decision—to repeat previous steps before results—is taken to support the original objectives of the studies.

In particular, with regard to chemical engineering, the open literature can easily find examples of the applications of LCA in chemical processes, recommending that the profession generally accepts the term. LCA provides an ideal link between chemistry and environmental sciences. However, scientists should be knowledgeable of the need to maintain holistic assessments and avoid the need to set study boundaries too narrowly, such as concentrating on the production process, which is just one step in the entire life cycle. The incorporation of life cycle thinking into engineering curricula in universities would be beneficial to its application in practice (Burgess and Brennan, 2001).

8.4 Sustainability approach

While sustainable development is not a completely new concept, a widely adopted idea does not yet exist. The basic academic discipline also depends on definitions. The sustainability of three elements notion, specifically social, economic, and environmental is operationalized by almost half of the studies examined as dimensions, pillars, or spheres. This understanding of sustainability is often followed by the idea of "3 Ps": earth, people, and benefit. Sustainability is no separate key idea behind this three-dimensional definition, but to demonstrate that it is "a state in which social, economic, and environmental processes are in perfect balance" (Gencturk et al., 2016).

Sustainable development is one of the core principles of sustainability, described as the capacity of humanity to evolve to meet current requirements "without sacrificing the ability of future generations to meet their own needs." A quarter of the studies are quoted for this description from the Brundtland Study "Our Shared Future." It is also the sustainable concept provided by the UNEP/SETAC LCSA Guidance Document (Ciroth et al., 2011). The authors look at Brundtland in Jørgensen, et al. (2013) to establish the purpose of LCSA studies. Two main concepts to be discussed in LCSA are: poverty—taken in relation to the (Sen, 1985) approach to capabilities, and resources—defined to include natural, created, human, and social resources. Jørgensen et al. (2013) concluded, despite this different conceptualization of LCSA, that LC methodologies could not be captured in the full spectrum of sustainability. They therefore suggested a different approach to LCSA from that given by Kloepffer (2008):

$$LCSA = S\text{-}LCA_{modified}, LCA \text{ and } LC_{social\ capital}$$

$S\text{-}LCA_{modified}$ refers to the inclusion in S-LCA of issues such as poverty alleviation and created capital, while the definition of $LC_{social\ capital}$ is described as "a still unknown methodology of the life cycle" for addressing social capital (Jørgensen et al., 2013).

In one-third of the studies, the researchers do not describe the concepts of sustainable development used for operationalization, whereas others refer to several definitions. As per case study on their respective LCSA or sector involved, approximately 20% of the literature in this analysis establishes a specific concept. For instance, Moslehi and Arababadi (2016) considered "sustainability for decision-makers to be a (policy) design objective." Sustainability is described by Fawaz et al. (2016) as sustainable development, creating a constructed environment that is resource efficient and ecological.

The sustainability view of LCSA was addressed by Schaubroeck and Rugani (2017) and asserted that LCSA has a sustainable perspective, anthropocentric (and not self-centered, biocentric, or eccentric). They concluded that LCSA must focus on goodness of human, which consists not only welfare of human, but also happiness, because of the anthropocentric view. With respect to this reason, the other impact categories should always be evaluated. They suggest combining LCSA with integrated earth system models as a functional consequence. Schaubroeck (2018) also suggests the same kind of model.

8.5 Application of LCA

8.5.1 Sustainable cities

In association with each urban issue, it is not easy to describe sustainable cities. Waste, water, urban planning, mobility, green areas and habitats, food, electricity, and buildings are the key issues involved in urban sustainability, and their improvement is encouraged by various strategies. However, for the assessment of these methods, life cycle thinking (LCT) as a quantitative method is important. With association of each urban issue, the key sustainability strategies are compiled by LCT, including LCSA articles, social LCA (S-LCA), life cycle costing (LCC), and environmental LCA dealing with these strategies, as well as integrated assessments with combined instruments. The urban problems that accounted for a greater number of studies were water, waste, and houses. There is a need for methodological developments in the field of sustainability and social assessment that facilitate their implementation in urban areas (Cremer et al., 2020).

In addition, a thorough study of its subsystems is of interest, because it is under huge pressure of environment and because more than 50% of the world's population is provided with basic services. LCT studies may help to identify and prioritize urban sustainability strategies, especially for policy motives (especially to the local from the regional level). A spatial technique to preventing various forms of trade is important, that is: (1) a spatial approach; (2) life cycle (e.g., material extraction of urban elements); (3) indicators (e.g., global warming and local freshwater effects); (4) energy use affecting energy processes outside of the city's boundaries. Typically, these methods help to simulate natural systems in urban areas through sustainable fostering systems. It can be accomplished by direct methods, involving infrastructure of green urban, which can improve efficiency of resource in various urban problems, such as green areas (e.g., urban heat island), food (e.g., local production), urban planning (e.g., urban lungs), energy (e.g., thermal insulation), and water (e.g., management of stormwater). Contrarily, ecological structures are mimicked by indirect processes, such as renewable or recycling energy systems, to improve urban metabolism. Renaturalization policies, as seen in global initiatives, political agendas for sustainable cities could also be the subject of. On the one side, a range of urban concerns (e.g., food, urban planning, mobility, electricity) is partly assessed, although some of the solutions are not yet addressed, for example, increasing ITS use or addressing electricity affordability. The life cycle instruments methodological production, on the other hand, needs special attention at various levels. The first is to assist with the evaluation of urban complexity through integrated schemes combining life cycle tools and other methodologies, especially with questions that involve a broader evaluation (e.g., community, region, and economy), such as water or mobility. Second, sustainability evaluation may include a thorough study of its three dimensions, that is, the climate, society, and the economy. Present practices for LCSA and S-LCA must be updated for this purpose to strengthen their application in case studies. Finally, although some articles discussed major gaps in research, the researchers concentrated on global warming or energy use metrics, thus restricting the comprehension of such strategies for sustainability. Further research may encompass thematically and methodologically current research deficiencies. Urban issues with limited available information and life cycle information and overlaps between various issues should be focused. Three aspects can be focused on the methodological developments: (1) to reduce possible tradeoffs, encouraging multiindicators; (2) advancing the application of sustainable research from a holistic viewpoint (particularly including three sustainability dimensions), and (3) defining and developing best integrated modelling and evaluation schemes of urban issues. Life Cycle Instruments are usually revised by participants for optimizing the interpretation and communication of results of life cycles for decision and policy making processes other than these (Kameni Nematchoua et al., 2020).

8.5.2 Municipal solid waste management

The enormous increase in solid waste production worldwide requires the implementation of sustainable environmental waste management strategies. The LCA instrument will assist in addressing the call by quantifying environmental impacts (Fig. 8.1). For municipal solid waste management (MSWM), it assesses the environmental efficiency of the system that allows policy makers to choose the greatest management approach with minimal environmental effects. It helps to assess the evolution of time, geographical allocation, and methods used in LCA studies. LCA studies are divided into four categories in that countries where it is performed is based on the level of income—lower income, lower middle income, upper middle income, and higher income countries—to determine the dependency of the

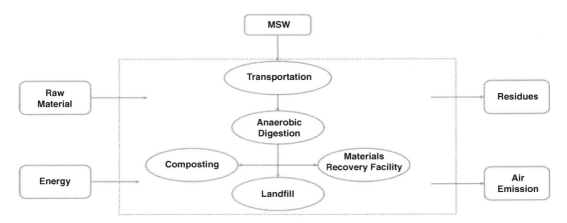

FIG. 8.1 System boundaries of LCA for municipal solid waste management.

publication of the studies and the economic situation of the region. The most common option for managing waste is integrated solid waste management. The MSWM system's life cycle costs and social effects have also been included in a very small number of reports. The results show that most of the studies on LCA are based in Asia and Europe. Shockingly, a single LCA report on MSWM has not been conducted by all countries in the world since 2013. The influence of growing gross domestic product (GDP) on the publication of LCA studies was also found to be insignificant, with the lack of economic constraints, time, and data as possible explanations. The government setting up of environmentally friendly policies and initiatives, together with the involvement of private, public, and non-governmental organizations through seminars and training courses, could assist enhancing the LCA applicability in the MSWM. (Khandelwal et al., 2019; Fernández-González et al., 2017; Harijani et al., 2017).

The distribution of LCA studies relating to the MSWM framework evaluated and the types of waste represented the environmental issues unique to Europe and Asia. The sustainability MSWM must be socially suitable, environmentally efficient, and economical; but a very limited number of studies have taken into account the LCC or the cost-benefit analysis of the MSWM system. Public, nongovernmental, and private organizations involvement through training courses and seminars could help to enhance the applicability of the LCA in the field of MSWM (Fullana et al., 2008). In addition, government understanding of providing adequate enticements with the introduction of more environmentally friendly initiatives and policies such as the UNEP-SETAC Life Cycle Initiative allows decision makers to improvise skills by providing materials and information platforms to disseminate best practices. It is not essential for each study to always have enough highest elements of the LCA methodology used in previous studies. The requirements for selection depend on different factors such as MSW structure, MSWM procedures, framework boundaries, environmental regulations, geographical scope, etc. Policy makers should approve LCA studies involving waste hierarchies and modified to local conditions on waste composition, treatment efficiency, energy mixes, etc. Thus, it may not be similar to 3R's (Reduce, Reuse, and Recycle) fundamental waste hierarchy. It is therefore recommended that policy makers in MSWM field recognize LCA as a method for evaluating "context-specific waste hierarchies" that are compatible with each MSWM framework's local conditions. The findings of the LCA studies would help the preparation and optimization of the ISWM scheme that could be significant in future waste management strategies development (Singh et al., 2021; Khandelwal et al., 2019).

8.5.3 Wastewater treatment

LCA is an instrument for measuring the impacts of a product, service, or operation from a cradle to a grave perspective. LCA was first introduced in the field of wastewater treatment (WWT) in the 1990s. It is obvious that LCA is an invaluable way of clarifying broad environmental implications in pursuit of ecological WWTs. The description of the functional unit, the structural limits, and the choice of the impact assessment methodology and the procedure followed to interpret the findings are heterogeneous. The discussion includes clear and emerging challenges for LCA applications in WWT: the paradigm shifts from pollutant removal to resorption recovery, the adjustment of LCA methodology to new target compounds, the emergence of regional influences, data quality improvement and data quality reduction, and greater compliance with ISO methodological requirements to achieve quality and quality.

Since 1995, there have been 45 published peer-reviewed international papers on WWT and LCA. The review of these papers demonstrated variability in the description of the functional and device limitations, the selection of the impact assessment methodology, and the procedure to interpret the results within ISO standards. Consequently, the standardized directives for WWT need to be established to ensure consistent application of the LCA methodology (ISO 14040, 2006; ISO 14044, 2006; Corominas et al., 2013).

Life cycle analysis has been common in the wastewater industry, although it was difficult for researchers and practitioners to synthesize findings from various studies in divisions between assumptions and methodology. Municipal wastewater management was based on LCAs with an emphasis on providing comprehensive guidelines for researchers and practitioners to perform LCA studies to inform wastewater management and infrastructure planning, design, and optimization (WWT plants, WWTPs; collection and reuse systems; based treatment technologies and policies), and to promote the implementation of new technology to advance treatment objectives and the sustainability of wastewater management. LCA methodology (1) provides informed purpose and scope descriptions, (2) selects functional unit and system limits, (3) selects the variables and their inventory sources, (4) identifies the best choice of impact evaluation methodologies, and (5) selects the appropriate methodology of data analysis and communication to determine the purpose and scope of a data analysis.

A collection of guidance on conducting LCA studies in the WWT sector is supported by state-of-the-art research findings. We provide (1) recommendations to facilitate the interpretation of the results, (2) selection of indicator guidance methodologies for impact assessment, (3) recommendation for inventory management, (4) guidance on functional unit selection and system limit definition, (5) methodological approach background information (attributional vs. consequences, process-based vs. input-output), and (6) LCA that can deal with (and not address) examples of questions. This review paved the way for the regular LCA usage to evaluate and enhance the environmental efficiency of wastewater systems based on a decade of study (Corominas et al., 2020).

8.5.4 Solar power

An opportunity is provided by applying the concept of net-zero energy (NZE) at urban scale to considerably decrease urban regions carbon footprint (Cellura et al., 2015). However, major challenges can be presented because of diversity in energy demand profile between different building types and excessive energy demands on the one hand, and small solar power potential in particular developments in high density on the other hand (Ortiz et al., 2014). There are various application, such as electricity generation, domestic hot water (DHW), and heating, utilizing solar energy (Finocchiaro et al., 2016). The life cycle environmental impact of implemented systems is usually overlooked by the analysis of NZE communities. LCA can evaluate whether a low impact operations stage can offset the environmental impact of the construction/material extraction or not (Guarino et al., 2020).

Guarino et al. (2020) studied the performance of two different urban energy systems in Calgary (Canada) for comparison with one large mixed-use community. Both systems examined consisted of a conventional energy–efficient system using heat pumps for hot water, heating, and cooling; the second design broadly implements solar thermal panels in combination with districts and seasonal heat storage systems for boilers. This analysis depends on the life Cycle Evaluation methodology and involves substitution of component, maintenance, on-site generation and use of energy, the system used, the manufacturing of systems, raw materials, and energy supplies, while exploring the system's performance with regard to the life cycle by using different ILCD 2011 indicators. From the perspective of all the indicators used in the study, the solar system is better than the conventional system. Land use and ozone depletion can be decreased by approximately 27% and 79.7%, respectively, while the remainder impact categories display a 39%–56% reduction. These approaches can be generalized to other existing programs with similar energy systems, weather constraints included in the thermal loads requirements, and operation. In addition, this study is based on assumptions and premises of real case studies in Canada, which further strengthens the soundness of the findings.

8.5.5 Agricultural strategic development planning

For the expansion of economy globally, the agriculture sector should be reflected as one of the most essential sectors, although it has significant role for the food safety. It is also responsible for substantial problems of environment such as emission of greenhouse gases at the same time. In most of the countries, a large percentage of total manufacturing benefit is represented by the agricultural sector and it is of great importance to large employers and the GDP. However, problems mostly linked with environment such as climate change, legislation, and protection of environment as well as social problems such as commercial margins, the increasing demand of food, and food security are being faced by this sector (Bonneau et al., 2017; Luque et al., 2017). With this in mind, the agricultural sector is developing, adopting, and implementing several new strategies to tackle many of these problems so that they become competitive, sustainable, and productive in market (Santucci et al., 2020). Hence, environmental footprint can be mainly minimized and controlled by these strategies that focus on providing tools, practices, and methods. The agri-sector also plays an important role in food safety, while at the same time it is considered the world's largest employer and covers about 40% of the world's population. It is also the primary source of income and job in many areas, especially rural. As per the United Nations (UN), there are approximately 500 million small farmers worldwide, who produce about 80% of their food, mainly in developing countries. At the same time, 75% of all crop diversity has been lost in the agricultural fields since the beginning of the twentieth century (UN, 2019). However, in the life cycle of food products, agricultural production is the most sustainable option that can be recognized with the help of LCA (Roy et al., 2009). This tool is being utilized increasingly to improve the environmental performance of goods, quarries, and services that include products of the agri-food sector and is regarded as a leading environmental impact assessment tool (Chen and Huang, 2019; Roy et al., 2009; Arzoumanidis et al., 2017). LCA is widely used for environmental analyses and improvements in agriculture and agri-food products (Fig. 8.2).

LCA was a helpful means of identifying and quantifying possible environmental consequences of apple juice, wine, or pepper pesto generation at three selected sites in Greece, North Macedonia, and

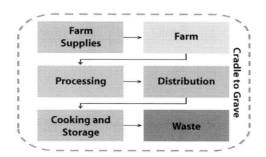

FIG. 8.2 System boundaries of LCA for agricultural strategic development planning.

Bulgaria in the context of the agricultural strategic development planning for the Balkan region. The three products have been described as the primary economic activities in the regions. Every product's entire production line is covered by LCA approach. Depending on LCA findings, which constitute six characterization factors for impact categories, proposals were made to minimize the footprint of apple orchard, wine, and pepper plot, along with manufacturing of apple juice, wine, and pepper pesto as the end product. The findings suggest that differences in culture and manufacturing have to be taken into account to optimize the environmental footprint. To enhance ecological quality of the products, decrease food losses and waste, and enhance crop yields, the whole strategy might also be beneficial for stakeholders, policy makers, and producers while simultaneously improving the three pillars of sustainability through strategic growth (Tsangas et al., 2020). Finally, LCA measured, monitored, and evaluated performance of environment from cultivation via distribution, and simultaneously promoted a strategy for the farming industry in any of the pilot areas of the Balkans.

8.5.6 Biofuels

Biofuels have often been seen as a benefit to global warming, but they have led to adverse environmental impacts and competition for a limited availability of natural resources. The environmental advantages and impacts of biofuels are evaluated by using a Thai Eco Scarcity methodology evaluation based on existing policies in Thailand. The findings after assessment illustrated that palm biodiesel and E85 (85% ethanol from cassava blended with 15% gasoline) have 43% and 95% more impacts than their counterparts for fossil fuels. A biodiesel and ethanol assessment showed a large number of tradeoffs between impacts on the environment, for example between fossil energy resources and other resources, freshwater and soil, and between global (greenhouse) and local (pesticides) consequences (Lecksiwilai and Gheewala, 2020). Also, it demonstrated the importance of using the concept of LCT to avoid problem shifts between impacts and life cycle stages. This is a key idea in policy making that has primarily matched greenhouse gas emission effects with unintentional effects on others. In the light of greenhouse gases and fossil energy resources, both biofuels will be much better than their fossil counterparts, as expected. However, the comparison with the fossil counterparts end up with better environmental profiles such as other categories, especially mineral resources, pesticide use, land and freshwater resources. Policy makers on a single scale can easily understand tradeoffs and comparative results using an LCA method.

In addition to identifying its advantages in GHG emissions and reducing fossil energy depletion, the comparable evaluation of fossil fuels and biofuels affirms the consequences of biofuel generation. These findings demonstrate the LCA benefits consider the entire product life cycle and all-important categories of effect, not just relevant one such as fossil fuel depletion and GHG emissions. The environmental benefits of using biofuels are typically GHG emissions because they do not take into account biogenic combustion and CO_2 emissions. The emission of GHG from both ethanol and biodiesel was lower than fossil fuel. It also confirmed the benefit of reduced fossil-fuel use. Contrarily, the impacts of the emissions of pollutants, intensive chemical utilization consumption, increased freshwater, and more agricultural land in biofuel processes can increase biofuels environmental impacts on fossil fuels. This may also be observed on comparing gasoline and conventional diesel along with E85 and biodiesel. When considering only energy resources and GHGs, the two biofuels are considerably better than their fossil equivalents; but, the evaluation is changed when other effects were seen, especially when it relates to pesticides and land use. This is an effective lesson for the inclusion of effects that are often not taken into account in policy making and therefore in evaluations. The Swiss Eco Factors 2013 evaluated same fuels. The Thai Eco Scarcity results also showed better outcomes. But, on using Swiss Eco factors 2013 with its fossil counterparts, both the biofuels showed worsened results. This evaluation strengthened the knowledge that the weighting factors obtained from different political priorities are important. More focus should be on increasing the quality of feedstock biomass and improving the content of oil extraction rate (OERs) and starch, instead of increasing production volumes to minimize environmental impact of biofuel production. It is also necessary to reduce chemical inputs, improve the water efficiency, and boost crop yield. The full environmental impact assessment using a single-scale impact evaluation method such as Thai Eco Scarcity makes it easy to understand the importance of the life cycle concept principle. The contrast of biofuels with fossil fuels exemplified the issues of changing from certain effects to other life cycle stages. To prevent single (or selected) problems leading to tradeoff between environmental impacts and unwanted negative effects, this vital result should be expressed. Throughout selection of suitable measures or policy formulation for prevention, this major concern should be regarded (Lecksiwilai and Gheewala, 2019; Gnansounou et al., 2009; Lecksiwilai and Gheewala, 2020).

8.6 LCA limitations and their probable solutions

There are several limitations as listed below:

- LCA accuracy and thoroughness rely heavily on data availability; data collection may be troublesome; therefore, it is important to be clear about assumption and uncertainty.
- Classic LCA will not decide what is the most performing or highly cost-effective technique, method, or product; hence, for a broad analysis of sustainability, LCA requires to be joined with social metrics, evaluation of technology, and analysis of cost.
- LCA does not usually quantify any specific actual effects contrary to conventional risk evaluation. LCA systems are appropriate for comparative evaluations, but not adequate for complete risk forecasts, when trying to create the link between potential impacts and a system.

LCA is an extensive task that needs multidisciplinary approach in economic and technical fields even for relatively smaller systems. LCA tasks are thus usually allocated to expert teams and can seldom be carried out with sufficient precision by one individual.

However, there are some strategies to overcome the limitations of LCA, which is given as follows:

- The most reliable solution is offered by extension of LCA. The social issues involvement and moving toward greater temporal resolution and spatial detail are developments in LCA (Ligthart et al., 2004).
- With respect to temporal and spatial information and other types of impacts inclusion, the most flexibility is offered by the development of a toolbox. There are no longer sets limits in the LCA rigid structure; in keeping with the relevant tool logic, every characteristic can be dealt (Suski et al., 2020).
- Hybrid analysis is the third strategy that is positioned in between the other two. In comparison to a toolbox and the extension of LCA, this strategy shows more consistency and more flexibility. Therefore, the strong points of the two strategies have been potentially combined by a hybrid analysis. The path to further discovery is more interesting than the already famous combination of process-LCA and input-output-LCA (Di Lullo et al., 2020).

8.7 Conclusion

Applying the LCA to every decision-making process offers a thorough understanding of the environmental impacts and human health of a process or product, which are traditionally not regarded. These useful data offer a way to take into account the sustainability and decisions impact, in particular those outside the site which are influenced directly by process or product selection. LCA is a method for better informing policy makers to take balanced decisions with the other criteria for decision making, such as costing and performance. This chapter gives an overview on all the aspects of LCA including the scopes, methodologies, applications, and finally the limitations.

References

Arzoumanidis, I., Salomone, R., Petti, L., Mondello, G., Raggi, A., 2017. Is there a simplified LCA tool suitable for the agri-food industry? An assessment of selected tools. J. Clean. Prod. 149, 406–425.

Bonneau, V., Copigneaux, B., Probst, L., Pedersen, B., 2017. Industry 4.0 in agriculture: focus on IoT aspects. Directorate-General Internal Market, Industry, Entrepreneurship and SMEs.

Burgess, A.A., Brennan, D.J., 2001. Application of life cycle assessment to chemical processes. Chem. Eng. Sci. 56 (8), 2589–2604.

Cellura, M., Guarino, F., Longo, S., Mistretta, M., 2015. Different energy balances for the redesign of nearly net zero energy buildings: an Italian case study. Renew. Sustain. Energy Rev. 45, 100–112.

Chen, Z., Huang, L., 2019. Application review of LCA (Life Cycle Assessment) in circular economy: From the perspective of PSS (Product Service System). Procedia Cirp. 83, 210–217.

Christensen, T.H., Damgaard, A., Levis, J., Zhao, Y., Björklund, A., Arena, U., Barlaz, M.A., Starostina, V., Boldrin, A., Astrup, T.F., Bisinella, V., 2020. Application of LCA modelling in integrated waste management. Waste Manage. 118, 313–322.

Ciroth, A., Finkbeier, M., Hildenbrand, J., Klöpffer, W., Mazijn, B., Prakash, S., Sonnemann, G., Traverso, M., Ugaya, C.M.L., Valdivia, S., Vickery-Niederman, G., 2011. Towards a live cycle sustainability assessment: making informed choices on products. UNEP/SETAC Life Cycle Initiative.

Corominas, L., Byrne, D., Guest, J.S., Hospido, A., Roux, P., Shaw, A., Short, M.D., 2020. The application of life cycle assessment (LCA) to wastewater treatment: a best practice guide and critical review. Water Res., 116058.

Corominas, L., Foley, J., Guest, J.S., Hospido, A., Larsen, H.F., Morera, S., Shaw, A., 2013. Life cycle assessment applied to wastewater treatment: state of the art. Water Res. 47 (15), 5480–5492.

Cremer, A., Müller, K., Berger, M., Finkbeiner, M., 2020. A framework for environmental decision support in cities incorporating organizational LCA. Int. J. Life Cycle Assess. 25 (11), 2204–2216.

Curran, M.A., 2013. Assessing environmental impacts of biofuels using lifecycle-based approaches. Manageme. Environ. Qual. Int. J 24, 34–52.

Curran, M.A., 2013. Life cycle assessment: a review of the methodology and its application to sustainability. Curr. Opin. Chem. Eng. 2 (3), 273–277.

Di Lullo, G., Gemechu, E., Oni, A.O., Kumar, A., 2020. Extending sensitivity analysis using regression to effectively disseminate life cycle assessment results. Int. J. Life Cycle Assess. 25 (2), 222–239.

Ekvall, T., Tillman, A.M., Molander, S., 2005. Normative ethics and methodology for life cycle assessment. J. Clean. Prod. 13 (13-14), 1225–1234.

Fawaz, A.N., Ruparathna, R., Chhipi-Shrestha, G., Haider, H., Hewage, K., Sadiq, R., 2016. Sustainability assessment framework for low rise commercial buildings: life cycle impact index-based approach. Clean Technol. Environ. Policy 18 (8), 2579–2590.

Fernández-Gonzalez, J.M., Grindlay, A.L., Serrano-Bernardo, F., Rodríguez-Rojas, M.I., Zamorano, M., 2017. Economic and environmental review of Waste-to-Energy systems for municipal solid waste management in medium and small municipalities. Waste Manage. 67, 360–374.

Finnveden, G., Hauschild, M.Z., Ekvall, T., Guinée, J., Heijungs, R., Hellweg, S., Koehler, A., Pennington, D., Suh, S., 2009. Recent developments in life cycle assessment. J. Environ. Manage. 91 (1), 1–21.

Finocchiaro, P., Beccali, M., Cellura, M., Guarino, F., Longo, S., 2016. Life cycle assessment of a compact desiccant evaporative cooling system: the case study of the "Freescoo". Sol. Energy Mater. Sol. Cells 156, 83–91.

Fullana, P., Frankl, P., Kreissig, J., 2008. Communication of Life Cycle Information in the Building and Energy Sectors: October 2008. Grup d'Investigació en Gestió Ambiental, Escola Superior de Comerç Internacional, Universitat Pompeu Fabra.

Gencturk, B., Hossain, K., Lahourpour, S., 2016. Life cycle sustainability assessment of RC buildings in seismic regions. Eng. Struct. 110, 347–362.

Gnansounou, E., Dauriat, A., Villegas, J., Panichelli, L., 2009. Life cycle assessment of biofuels: energy and greenhouse gas balances. Bioresour. Technol. 100 (21), 4919–4930.

Guarino, F., Longo, S., Vermette, C.H., Cellura, M., La Rocca, V., 2020. Life cycle assessment of solar communities. Solar Energy 207, 209–217.

Guinee, J.B., Heijungs, R., Huppes, G., Zamagni, A., Masoni, P., Buonamici, R., Ekvall, T., Rydberg, T., 2011. Life cycle assessment: past, present, and future. 45, 1, 90–96.

Harijani, A.M., Mansour, S., Karimi, B., Lee, C.G., 2017. Multi-period sustainable and integrated recycling network for municipal solid waste–A case study in Tehran. J. Clean. Prod. 151, 96–108.

ISO 14040, 2006. Environmental Management e Life Cycle Assessment-Principles and Framework: International Standard 14040. International Standards Organisation, Geneva.

ISO 14044, 2006. Environmental Management e Life Cycle Assessment-Requirements and Guidelines. International Standards Organisation, Geneva.

ISO, 2006. Environmental Management–Life Cycle Assessment–Principles and Framework. British Standards Institution, London, pp. 14040.

Jørgensen, A., Herrmann, I.T., Bjørn, A., 2013. Analysis of the link between a definition of sustainability and the life cycle methodologies. Int. J. Life Cycle Assess. 18 (8), 1440–1449.

Kameni Nematchoua, M., Sevin, M., Reiter, S., 2020. Towards sustainable neighborhoods in Europe: mitigating 12 environmental impacts by successively applying 8 scenarios. Atmosphere 11 (6), 603.

Khandelwal, H., Dhar, H., Thalla, A.K., Kumar, S., 2019. Application of life cycle assessment in municipal solid waste management: a worldwide critical review. J. Clean. Prod. 209, 630–654.

Kloepffer, W., 2008. Life cycle sustainability assessment of products. Int. J. Life Cycle Assess. 13 (2), 89.

Klöpffer, W., 2003. Life-cycle Based Methods for Sustainable Product Development, Springer, Cham.

Lecksiwilai, N., Gheewala, S.H., 2019. A policy-based life cycle impact assessment method for Thailand. Environ. Sci. Policy 94, 82–89.

Lecksiwilai, N., Gheewala, S.H., 2020. Life cycle assessment of biofuels in Thailand: implications of environmental trade-offs for policy decisions. Sustain. Prod. Consump.

Ligthart, T., Aboussouan, L., Van de Meent, D., Schönnenbeck, M., Hauschild, M., Delbeke, K., Struijs, J., Russel, A., Udo de Haes, H., Atherton, J., van Tilborg, W., 2004. Declaration of Apeldoorn on LCIA of non-ferrous metals. SETAC Globe 5, 46–47.

Luque, A., Peralta, M.E., De Las Heras, A., Córdoba, A., 2017. State of the industry 4.0 in the Andalusian food sector. Procedia Manuf. 13, 1199–1205.

Margni M., Curran MA, 2012. Life cycle impact assessment. In: Curran, MA, Salem, MA (Eds.), Life Cycle Assessment Handbook. Scrivener-Wiley Publishing, pp. 611.

Moslehi, S., Arababadi, R., 2016. Sustainability assessment of complex energy systems using life cycle approach-case study: Arizona State University Tempe campus. Procedia Eng. 145, 1096–1103.

National Renewable Energy Laboratory (NREL), 2006. U.S. Life Cycle Inventory Database. http://www.nrel.gov/lci (Accessed 19 November 2012).

Nwodo, M.N., Anumba, C.J., 2019. A review of life cycle assessment of buildings using a systematic approach. Build. Environ. 162, 106290.

Ortiz, J., Guarino, F., Salom, J., Corchero, C., Cellura, M., 2014. Stochastic model for electrical loads in Mediterranean residential buildings: validation and applications. Energ. Buildings 80, 23–36.

Potrč Obrecht, T., Röck, M., Hoxha, E., Passer, A., 2020. BIM and LCA integration: a systematic literature review. Sustainability 12 (14), 5534.

Raymond, A.J., Kendall, A., DeJong, J.T., 2020. Life Cycle Sustainability Assessment (LCSA): a research evaluation tool for emerging geotechnologiesGeo-Congress 2020: Biogeotechnics. American Society of Civil Engineers, pp. 330–339.

Roy, P., Nei, D., Orikasa, T., Xu, Q., Okadome, H., Nakamura, N., Shiina, T., 2009. A review of life cycle assessment (LCA) on some food products. J. Food Eng. 90 (1), 1–10.

Santucci, G., Martinez, C., Vlad-Câlcic, D. The sensing enterprise. https://www.theinternetofthings.eu/sites/default/files/[user-name]/Sensing-enterprise.pdf (Accessed 24 February 2020).

Schaubroeck, T., Rugani, B., 2017. A revision of what life cycle sustainability assessment should entail: towards modeling the net impact on human well-being. J. Ind. Ecol. 21 (6), 1464–1477.

Schaubroeck, T., 2018. Towards a general sustainability assessment of human/industrial and nature-based solutions. Sustain. Sci. 13 (4), 1185–1191.

Scrucca, F., Baldassarri, C., Baldinelli, G., Bonamente, E., Rinaldi, S., Rotili, A., Barbanera, M., 2020. Uncertainty in LCA: an estimation of practitioner-related effects. J. Clean. Prod., 122304.

Searchinger, T., Heimlich, R., Houghton, R.A., Dong, F., Elobeid, A., Fabiosa, J., Tokgoz, S., Hayes, D., Yu, T.H., 2008. Use of US croplands for biofuels increases greenhouse gases through emissions from land-use change. Science 319 (5867), 1238–1240.

Sen, A., 1985. A sociological approach to the measurement of poverty: a reply to Professor Peter Townsend. Oxf. Econ. Pap. 37 (4), 669–676.

Singh, E., Kumar, A., Mishra, R., Kumar, S., 2021. Eco-efficiency tool for urban solid waste management system: a case study of Mumbai, India. In: Kumar, S., Kalamdhad, A., Ghangrekar, M. (Eds.), Sustainability in Environmental Engineering and Science. Lecture Notes in Civil Engineering, vol 93. Springer, Singapore.

Suski, P., Speck, M., Liedtke, C., 2020. Promoting sustainable consumption with LCA–A social practice based perspective. J. Clean. Prod. 283, 125234.

Tsangas, M., Gavriel, I., Doula, M., Xeni, F., Zorpas, A.A., 2020. Life cycle analysis in the framework of agricultural strategic development planning in the Balkan region. Sustainability 12 (5), 1813.

UN Department of Economic and Social Affairs Disability. #Envision2030 Goal 2: Zero Hunger. Available online: https://www.un.org/development/desa/disabilities/envision2030-goal2.html (accessed on 23 November 2019).

US EPA, 2004. An Examination of EPA Risk Assessment Principles and Practices. EPA/100/B-04/00. US Environmental Protection Agency, Office of the Science Advisor, Washington, DC.

Wulf, C., Werker, J., Ball, C., Zapp, P., Kuckshinrichs, W., 2019. Review of sustainability assessment approaches based on life cycles. Sustainability 11 (20), 5717.

Zijp, M.C., Heijungs, R., Van der Voet, E., Van de Meent, D., Huijbregts, M.A., Hollander, A., Posthuma, L., 2015. An identification key for selecting methods for sustainability assessments. Sustainability 7 (3), 2490–2512.

CHAPTER 9

Life cycle sustainability dashboard and communication strategies of scientific data for sustainable development

Daniela Camana, Alessandro Manzardo, Andrea Fedele, Sara Toniolo
Department of Industrial Engineering, University of Padova, Padova, Italy

9.1 Introduction

In the last decades, business and policies are moving progressively inside the framework of sustainable development principles, by handling economic, social, and environmental issues together (WCED, 1987). All sustainability patterns—to be properly applied in their multifold dimensions (UN, 2015)—require cooperation among different stakeholders such as researchers, politicians, entrepreneurs, and citizens. Therefore, fair communication of scientific data is a key point for translating best scenarios of technical studies into reality via effective policies at every level (Sonnemann et al., 2017).

Consequently, a good sustainability assessment method must be scientifically robust and should evaluate multiple dimensions; at the same time, it must communicate results in a comprehensible way. Life cycle assessment (LCA) methodology (ISO, 2006a) is one of the leading environmental methods that has achieved an international consensus and is extensively used in many sectors (Yang et al., 2020), as it permits to investigate simultaneously many environmental impacts in the whole life cycle of products, processes, organizations, or policies. Life cycle thinking, in respect to other sustainability methods, has the strong advantage of examining expansively the whole supply chain, reducing burden shifting of impacts out of system boundaries, and reviewing and interpreting results with a fair technique (ISO, 2006b). Moreover, environmental analysis is recently combined with social studies and costs investigation in a multifactorial life cycle sustainability perspective (Toniolo et al., 2020). Hence, life cycle sustainability assessment (LCSA) may be technically advantageous for supporting interdisciplinary decision-making processes by integrating different results with a systematic approach (Visentin et al. 2020). The subsequent process of communication of data achieved needs to be simple and clear to all people involved. The life cycle sustainability dashboard (LCSD) might be helpful for this purpose as it gives a clear picture of positive and negative impacts in its comprehensive assessment of multiple aspects (Jesinghaus, 2009).

This chapter aims to provide some general highlights of communication strategies and to describe and comment the multifaceted tool of LCSD. The overall purpose is to give a contribution to the comprehensibility of the decision-making process in sustainable development policies.

The Section 9.2 begins illustrating how the ethical definition of sustainable development influences communication strategies. Subsequently, a brief description of the widespread method of LCSA and its communication approaches in the scientific literature are provided in Section 9.3. The multiform framework of the dashboard of sustainability (DoS) is illustrated in detail in Section 9.4 while methodology and case studies for LCSD are proposed in Section 9.5. Section 9.6 briefly describes other approaches, such as international guidelines, reports, goals, or standards. Final comments on sustainable development, life cycle sustainability, and communication strategies conclude the chapter in Section 9.7.

9.2 Ethical definition of sustainable development and communication strategies

The subjectivity of ethical choices in the definition of the sustainable development framework (WCED, 1987) is debated from a conceptual and substantial point of view (McIntyre et al., 2017). As known, the historical three main pillars of sustainability—economy, environment, and society—may be considered equally important or may be subjected to a preference ordination (Elkington, 2013). This evaluation may be done at theoretical level by governments, researchers, or other stakeholders leading to different definitions of sustainable development, mainly focused on economic expansion, social cooperation, or environmental protection (Giddings et al., 2002).

Similarly, the evaluation of priorities is present also inside each sustainability dimension, as each indicator or impact may be considered by decision makers more important, less important, or equally important than another one, in the same field. For example, in the environmental sector, is global warming more important than resources consumption? What costs are less significant in the supply chain? From a social perspective, work employment and children's education are equivalent?

Some methods have been proposed and mingled to approach and manage multifunctionalities: the multi criteria decision method (Ren and Toniolo, 2020), the multi-attribute value theory (Z. Guo et al., 2020), the scenario analysis (Atilgan and Azapagic, 2016; Ekener et al., 2018), and many others.

At a small scale, each researcher in every study makes analogous value choices, either explicitly or implicitly (Chiu et al., 2020); these assumptions lead to different weighing for indicators among the three dimensions and must be explicated as they affect results and outputs of stakeholders' communication. In fact, these valuations lead to attribute different mathematical weightings both at impacts inside the three areas and at impacts aggregated in the three pillars; these assessments finally generate different overall scores if a basic aggregation process is followed, as Fig. 9.1 easily illustrates.

Theoretically, with a provocative purpose, many other capitals that usually are considered within the three pillars might be considered separately, being complementary or substitutable (Avesani, 2020). Therefore health and safety concerns (Martin, 2017), financial equilibrium (Peña and Rovira-Val, 2020), technical development (Rodriguez et al., 2020), culture, individual freedom, governance, or many others might be added to the three-pillar concept of sustainable development (UN, 2015). In Fig. 9.2 a visual schematization of some possible different starting points of strategies is provokingly provided; in the figure, sustainability is not defined as a system with three independent and equivalent columns, but it is based on a different number of aspects, potentially weighting differently. As a result, different representations occur. Moreover, in a state of emergency, human health might become the prevalent value, leaving behind the other pillars of development; this is the case of the lockdown strategy during the COVID-19 pandemic, where economic issues of citizens and social rights have been temporarily overshadowed.

Usually, in the international debate, the triple bottom line (Janjua et al., 2020) is chosen, by including the financial issues in the economic perspective (Peña and Rovira-Val, 2020), the technical issues and the health in the environmental field (Martin, 2017; Mukherjee et al., 2020), and the general issues

9.2 Ethical definition of sustainable development and communication strategies

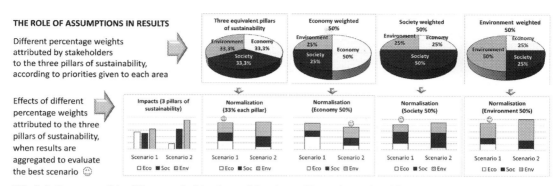

FIG. 9.1 Some possible different prioritizations of the three pillars of sustainability.

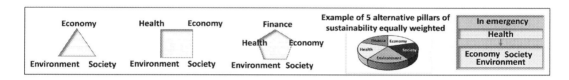

FIG. 9.2 The concept of sustainability, a provoking multidimensional scheme.

as freedom or inclusion of poor people in the social field (Huertas-Valdivia et al., 2020). Consequently, the three-dimensional representation of sustainability is the most widely used and internationally approved, but it is not the only one possible and graphically usable.

Moreover, in sustainability assessment, if some aspects are neglected, the risk is the burden shifting of impacts in time or space and the inconsistency of overall results for improving better conditions for all people and for resource decoupling (Kjaer et al., 2019). This is true not only among environmental issues, but also across economic and social concerns. For example, what is the overall gain of a scenario that allows the reduction in global carbon dioxide emissions if, in the same scenario, the local biodiversity decreases or the freedom of movement of the citizens is obstructed? On the other side, what are the economic advantages of improving industrial processes in terms of lower resource consumption if the financial market governs by itself the economic performances? Each alternative, in sustainability studies, is strictly connected to the others, and both rebound effects and consequences outside system boundaries should be always considered (Vivanco et al., 2018). However, time and space boundaries cannot be enlarged indefinitely, and the number of indicators cannot be infinite; therefore, assumptions and limitations are unavoidable.

Therefore, communication strategies in sustainability assessment, to be fair, should make explicit the definition of sustainable development and should highlight ethical choices behind each value preference and prioritised issues.

9.3 Life cycle sustainability

The LCSA is an increasingly used methodology (Visentin et al., 2020), with strong interdisciplinarity contents that range from engineering to social sciences and business accounting. LCSA is based on three parallel methods referring to the three pillars of sustainable development: LCA, life cycle costing (LCC), and social life cycle (S-LCA) (Sala et al., 2013b). LCA approach is well established and defined by International Standards ISO 14040 and ISO 14044 (ISO, 2006a; ISO 2006b), while LCC and S_LCA are not yet completely ruled and present areas of improvement (Toniolo et al., 2020; Yang et al., 2020) and debated open questions (Huertas-Valdivia et al., 2020).

Despite its different uses, the structure of LCSA is common and is visually schematized in Fig. 9.3 where complexity of data management is illustrated.

Each analysis—LCA, LCC, and S_LCA—can be performed independently, and results can be mingled and weighted at the end of the process (Zanni et al., 2020). Otherwise, each of the four steps of the study—goal and scope definition, inventory analysis, impact assessment, and interpretation of results—can be conducted with a comprehensive overview from the beginning of the study itself. At the state of the art, integration of results in a single score for the three dimensions of sustainability is given by about 50% of studies (Visentin et al., 2020).

LCSA case studies are mostly conducted in developed countries, and cover different categories such as products (Zhang et al., 2020), technologies (Z. Guo et al., 2020), service systems (López et al., 2020), and energy performances (Cerrato and Miguel, 2020).

As LCSA deals with a large amount of data and information, therefore a transparent communication strategy to interested parties is crucial for conscious evaluations at the end of the process.

9.3.1 Data report and illustration of results

Fair reporting on life cycle sustainability results is crucial to involve stakeholders and to support decision-making process among non-technical participants (Sonnemann et al., 2017). Tables and graphs with points, lines, or columns are typically used to share numerical information or quantitative data. Circles, rings, and spider diagrams help in a multifunctional assessment. Icons, arrows, labels, symbols, and colors supplement graphs, tables, or statements. Flow diagrams, together with defined space and time boundaries and images, complete the most widespread sustainability reporting options (Fig. 9.4).

Data sharing approaches in sustainability depend on researcher's thoughts, intended audience, contents to share, and tools used.

Table 9.1 summarizes some communication strategies applied in selected papers that used the LCSA methodology in the year 2020. As viewable, In LCSA studies many different methods have been utilized, many times mingled.

Many communication strategies have been developed in the years to help stakeholders' debate on priorities and scenarios. Among them, the life cycle sustainability triangle (LCST) and the LCSD are two well-established methods (Finkbeiner et al., 2010).

The LCST permits a visualization in a triangle of preferred options associated to weighting criteria (Hofstetter et al., 1999) and has been recently used also for life cycle engineering studies (Rodriguez et al., 2020). This scheme helps stakeholders to identify the best solution in terms of priorities given to the three issues considered, usually environment, society, and economy.

The LCSD is one flexible single tool for sustainability analysis (Visentin et al., 2020) that is described in detail in next paragraphs as it has advantages in many intended uses thanks to its multifield characteristics.

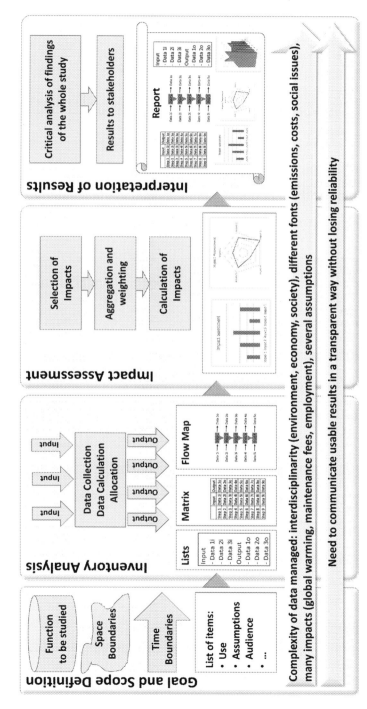

FIG. 9.3 Data management and communication in life cycle stages (adapted from ISO, 2006a).

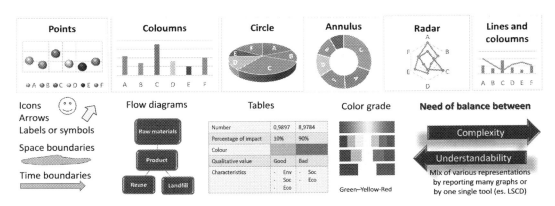

FIG. 9.4 Main communication strategies for scientific data reporting.

Table 9.1 Selected articles and communication strategies.

Topics and methodologies of selected articles using LCSA in 2020 (with fonts)	Communication strategies for data report
Seawater desalination technologies. Mathematical method to investigate cause-effect relationships. (Ren, 2020)	Tables, maps.
Chemical product design. Calculation of sustainability issues in design and case studies. (Zhang et al., 2020)	Tables, points, columns, flow diagrams.
Boxboard production. Case study using two multicriteria decision making methods. (Man et al., 2020)	Tables, dots and lines, flow diagrams.
Bio-composite supply chains. Comparative case studies by using input–output models. (Chen et al., 2020)	Tables, columns, flow diagrams.
Machine tools. The literature review of improvements for sustainability. (Feng and Huang, 2020)	Tables, circles, flow diagrams, figures.
Residential buildings. Development of key performance indicators and case study. (Janjua et al., 2020)	Tables, columns, spider diagrams, flow diagrams.
Industrial systems. Comparison of case studies using LCS index. (Xu et al., 2020)	Tables, lines, columns, vectors, flow diagrams, figures.
Pumped hydro energy storage. Comparative case study using MAVT and scenario analysis. (Z. Guo et al., 2020)	Tables, columns, flow diagrams.
Urban surfaces. Comparative case study using impacts and Sustainable Development Goals. (Henzler et al., 2020)	Tables, spider diagrams, flow diagrams, arrows, icons, symbols.
Resource-food-bioenergy systems. Case studies by optimization, modelling, and simulation. (M. Guo et al., 2020)	Tables, spider diagrams, flow diagrams, figures, maps.
Electricity, heat, and water supply in communities. Scenarios compared by MultiCriteria Decision Analysis. (Aberilla et al., 2020)	Tables, lines, flows, ternary diagrams, columns with dots, color scale.

9.4 The dashboard of sustainability, a tool for sharing results

The "Dashboard of Sustainability" (DoS) or "Digital Dashboard," in its original form, was developed with the aim of comparing specific and overall performances of various nations (Jesinghaus, 1999). The method proposed was quite robust and endorsed, as was settled after a peers' debate including data contribution of the United Nations Statistical Division, the World Bank, the Organization for Economic Cooperation and Development, and other international agencies (Hardi et al., 2002). The main methodological characteristics and advantages of the dashboard are shown in Fig. 9.5.

The name of the method derives from its more known visual output, that is like a car's dashboard (Hardi and Semple, 2000), with many circles, arrows, colors, and alarms, that provides information to the users, as an instrument panel.

Conceptually the dashboard is characterized by different circles, representing various areas of interest under study, that can be easily inserted and managed.

Usually, each circle is divided in concentric areas.

- In the middle the synthetic index analyzed can be reported.
- The inner ring is usually divided in three equal segments, according to the three pillars of sustainability. The methodology and the software are flexible; therefore, different adjustment of weights and other preferences of integration of multiple indicators are possible.
- The outer ring is divided in many sections, corresponding to all indicators selected for the study, grouped for each area of interest. When an indicator is considered more important than another, its section in the annulus of the circle is made bigger, proportionally to the weight given. If all indicators are considered equally important, all sections of the outer circle have the same dimensions.

The value of each indicator calculated or provided is normalized and scaled by a score (Hardi and DeSouza-Huletey, 2000) between 0 and 1000 points, where 0 is the worst results achieved (or achievable) in all scenarios for the indicator, while 1000 is the best one. A simple linear interpolation between two extreme cases permits to identify the position of intermediate cases for all indicators in all scenarios (Hardi and DeSouza-Huletey, 2000), as defined in the Eq. (9.1).

$$(DoS\,score)_i = 1000 \frac{(value_i - value_0)}{(value_{1000} - value_0)} \quad (9.1)$$

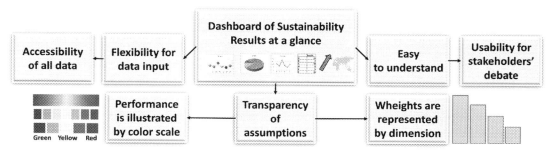

FIG. 9.5 Characteristics of the dashboard of sustainability (adapted from Jesinghaus, 2009).

where $(DoS\ Score)_i$ is the DoS score assigned to the indicator in a context i, $value_i$ is the indicator value for context i (intermediate); $value_0$ is the indicator with the worst value among all contexts; $value_{1000}$ is the indicator with the best value among all contexts.

For graphical representation, numbers obtained are transposed into colors. Therefore, the positive or negative performances of an indicator are consequently shown using a scale of at least seven colors, from dark red for bad performances (0 point) to yellow for averages to dark green for positive ones (1000 points) (Traverso et al., 2012b). The same colors interpretation and graduation applies to all the areas that aggregate results by arithmetic average in the dashboard. For each segment of the inner ring, the sum of the score is divided by the number of indicators. The overall score of the circle is the weighted sum of the pillars, divided for the number of the pillars. Therefore, bad sustainability hotspots are easily viewable in the overall picture, as they are dark red. At the same time, the parts of the circular annulus that are dark green present better performances. Consequently, performances of all indicators and aggregated factors are displayed immediately and visible at a glance.

Finally, an upper arrow located on a red-yellow-green rainbow provides an overall ranking score, while smaller needles may point to a value that represents the performances of each area studied.

The DoS cartogram may be customized, modified, and used for communicating different complex results not only for nations, but also for districts, cities, products, services, and organizations.

In fact, many features are available, added, and updated in the software developed by the Italian Joint Research Centre (Jesinghaus, 2009). The method is supported by a free-of-charge software available online and usable to compare performances based on different indicators, such as United Nations Commission of Sustainable Development Indicators (Jesinghaus, 2009), Millennium Development Goals (UN, 2005), Sustainable Development Goals (SDGs) (UN, 2012), or many others indicators directly insertable by the users.

In the software, results are available as pies, distribution of points, maps, or scatterplot. For a better readability, results of indicators may be shown as a tree of boxes of different colours, from red to green. If the dashboard is used, for example, for Millennium Development Goals applied to all nations, colors representing performances may be reported also in a geographic map for all countries investigated. Software permits also to deepen the study of interlinkages between indicators by showing them in a cartesian diagram to conduct sensitivity analyses, to manage outliers and data missing, to perform and develop new schemes, and to edit new types of dashboards.

The DoS allows stakeholders to be conscious of the meaning of assumptions, for example, the indicators selected, pillars considered, or importance of each choice.

Two examples of the use of the DoS with territorial indicators are briefly provided to show possible advantages and weak points.

The DoS has been utilised for the assessment of different scenarios for the Sicilian Town Master Plan for nine Italian provinces (Federico et al., 2006). As stakeholders' involvement in the discussion process has been considered important, outcomes of different policies have been evaluated and illustrated for a strategic environmental assessment not only to technicians and administrators, but also to citizens and entrepreneurs. For this purpose, the DoS has been revealed helpful as it has allowed among others, the prioritization of hierarchized weightings and the judgments and the understandability of quantitative and qualitative outcomes. However, the overall consistency of data has not been reached, stating different units of indicators and various methodologies used for each area.

The DoS has been also used for supporting the debate during the forum for the Local Agenda 21 in the Italian city of Padova (Scipioni et al., 2009). With the purpose of investigating the three pillars of sustainability with an overall study, 61 indicators have been selected in economic, social, and environmental areas. As the methodology is flexible, local changes have been promoted, by including these different indicators and by adding a time span evaluation perspective in the dashboard. The method has demonstrated to be reliable to sustain decision policies. However, two weak points have been delineated: the subjectivity of criteria used for the definition of the indicators and of their weights and the lack of possibility of making comparisons among different cities. Future improvements have been suggested, including sensitivity analyses on the effects of the change of indicators weight and on the outcomes using bottom up or top down data.

In general, the use of dashboards is considered important in sustainability assessment for its comprehensive approach, for communication tasks, and for collaborative decision support with a wide perspective in many sectors investigated, including logistics (Morana and Gonzalez-Feliu, 2015), agriculture (Barber et al., 2016), industries (Topor et al., 2017), and small and medium enterprises (Shields and Shelleman, 2020).

9.5 The life cycle sustainability dashboard

The use of the dashboards to present the results of the LCSA has been proposed to improve communicability of complex results in a straightforward and comprehensive way for no-expert stakeholders (Traverso et al., 2012b). The dashboard has been changed considering that each circle refers to a method (LCA, LCC, or S_LCA) or to a scenario compared by LCSA and each segment of the annulus is an impact calculated in LCA, LCC, or S_LCA. This choice permits to insert data in the software and to display results of different scenarios for the three life cycle approaches, in an aggregate way, by using weighting indices, if wished. This method is called LCSD.

Among different strategies for results communication, the LCSD has the advantage of summarizing different aspects in a unique complex and flexible tool (Finkbeiner et al., 2010). In fact, dashboards permit to manage together different indicators and various sustainability pillars throughout mathematical calculation and graphical picture (Jesinghaus, 2000). Moreover, LCSD summarizes many different representations of results in a unique picture.

The LCSD has been used in the years to compare economic, social, and environmental performances of products, processes, and technologies; four case studies from the literature are briefly described to illustrate research paths.

The first examined study compares impacts and identifies more sustainable strategies for extraction, cutting, finishing, and transport of the marble "Perlato di Sicilia" in Italy, with a life cycle thinking perspective (Capitano et al., 2011). In the research the comparison is made by considering different production processes, in two leading companies of the region. The assessment method CML-IA 2007 and its characterization factors are used for environmental assessment. Costs of extraction and production, fuel, waste disposal, and electricity are inputted for economic analysis. The workers' stakeholder category is chosen as most significant for social impacts, including indicators on total employees, women in administration, immigrants, limited contracts, unlimited contracts, health insurance, and annual health checks. Data fonts reported are databases, questionnaires, and surveys. Results achieved show that LCSA may help to address multiresults, but the optimum scenario for all categories is not clearly reachable (Capitano et al., 2011). Therefore, a visual strategy to communicate complex results to stakeholders, such as LCSD, is envisaged to assess weighting strategies and to make agreed decisions. However, a weak point of the LCSA outlined by authors is the difficulty of including reliable and comprehensive social factors into the methodology.

The applicability and practicability of both LCSA and LCSD are investigated with a case study on the assembly step of photovoltaic polycrystalline silicon modules in Italy and Germany, in different years (Traverso et al., 2012a). Environmental impact assessment is conducted using the Eco-indicator 99 model. LCC is performed including costs of photovoltaic cells, materials, equipment, electricity, and labor force. S-LCA is focused mainly on the workers group by analyzing discrimination, child labor, wages, working hours, social benefits, and health conditions. Data were directly collected by interviews, questionnaires, or Gabi and Sima-Pro databases. Comparative results are provided considering independently the three components of sustainability. LCSA and LCSD advantages outlined by the authors are the transparency of data output, the completeness of data available for expert audience, the possibility of communicating complex results by a color scale for each indicator making them easily readable, and the easiness of comparison among scenarios. However, one weak point outlined also by Traverso et al. (2012a) is the difficulty of individuating robust social indicators, problem that is still relevant in sustainability assessment (Huertas-Valdivia et al., 2020; Kloepffer, 2008). Moreover, the choice of indicators in all areas, the attribution to an area, and the weight of their importance remain subjective, also for theoretic reasons (Giddings et al., 2002). To face the subjectivity of prioritization, LCSD allows to give different weights to indicators and therefore permits sensitivity analysis, if properly used.

Another LCSD case study deals with the treatment and destination of byproducts (biological sludge and biogas) generated by treating domestic effluent in a wastewater treatment plant in South Brazil. Positive and negative characteristics of competing technologies in the phases of treatment and final destination are investigated by using LCA and LCC methodology, together with selected social indicators (Amaral et al., 2019). For the analysis and interpretation of results, four scenarios of management are compared by using some selected indicators: 8 for environmental impacts using the ReCiPe 2016 evaluation method (global warming, stratospheric ozone depletion, ozone formation, terrestrial ecosystems, terrestrial acidification, freshwater eutrophication, terrestrial ecotoxicity, freshwater ecotoxicity and human noncarcinogenic toxicity), 5 for economic issues (costs of energy, chemicals, workers, equipment acquisition, and maintenance), and 10 for health and social aspects (wages paid to workers, indoor noise, use of hazardous chemicals, indoor odor emission, biological risks, sludge N and P content, values of pathogens present in sludge, outdoor noise level, outdoor odor emission, and capacity to generate employment). As a result, the DoS permits to show many data in a visualized form, helping the decision-making process. Moreover, an overall score for the scenario, a sustainability index, can be calculated.

One recent research proposes the dashboard for evaluating impacts of soybean cultivation in Rio Grande do Sul state (Zortea et al., 2018). Boundaries for soybean collected include soil preparation, seed treatment and sowing, growing period, and harvest. LCA focuses on three categories by CML-IA midpoint methodology: global warming, acidification, and eutrophication. LCC includes feedstock costs, financial expenses, and infrastructure and maintenance costs. S_LCA considers workers, local community and society, and value chain actors as main stakeholders. Main data fonts used are questionnaires, material flow analyses, energy balances, cost analyses, and indicators on welfare, health, and safety. At the end of the analysis, the dashboard allows communication of complex data in a cumulative and understandable way, highlighting hotspots in the life cycle of the production process and helping the identification of more sustainable options. Nevertheless, even if the assessment

9.6 Other sustainability tools and communication strategies

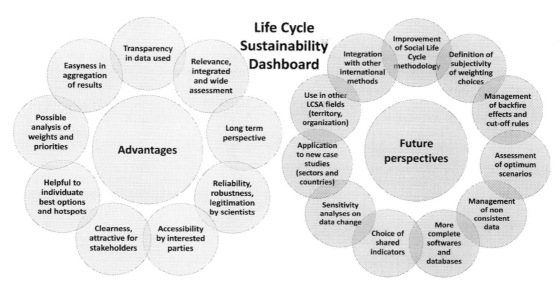

FIG. 9.6 LCSD advantages and future perspectives (comments from Jesinghaus, 2012; Sala et al., 2013a,b; UNEP, 2017; Valdivia et al., 2013; Visentin et al., 2020).

is carefully conducted, the identification of best alternatives is not risk-free as processes and activities are strongly interconnected and the optimization of one process might lead to a radical change of another one, with a backfire effect. In fact, all dimensions reveal to be strongly influenced among themselves.

General advantages of LCSD implementation, derived from methodology and applications, are provided in Fig. 9.6.

9.6 Other sustainability tools and communication strategies

Many instruments have been tested and numerous guidelines have been provided in the years to promote the sharing of data on sustainability among different nonexpert stakeholders, including governors, entrepreneurs, and citizens.

At institutional level, to meet the objectives of the 2030 Agenda, SDGs have been recently internationally adopted (UN, 2012). The official sharing of SDGs performance is widespread, declined at local, regional, national, and international level with recurrent reports open to the public debate. Agenda 21 is another common process for public decisions (Scipioni et al., 2009).

At industrial level, from the spread of the Corporate Social Responsibility Reporting (CSR) in the environmental and chemical sector in the late 1980s, many companies in different areas have started to account for their performance by different leaflets, annual reports, stakeholders' publications, etc. The main drivers have been the need of facing international competitiveness and targets of sustainability, by approaching environmental and social challenges worldwide, by involving stakeholder and social interested parties, by increasing the financial demand and the overall business performance (UNEP, 2019).

Various frameworks have been developed and updated: the methodology of the International Integrated Reporting Council, the Global Reporting Initiative, the AA1000 method of the Accountability Institute, the standards of the Sustainability Accounting Standards Board, the United Nations Global Compact, the guidelines, principles, and standards of the Organization for Economic Cooperation and Development, and many others (UNEP, 2019).

At product level, labels business-to-business or business-to-consumer, are also used for environmental and sustainability statements (ISO, 2006c, ISO, 2016, ISO, 2018). The presence of a great number of labels leads to a certain degree of confusion to citizens and to the difficulty of understanding which are truly consistent, and which are used only for greenwashing by companies. Some studies are also moving to a more comprehensive sustainable product declaration (Llatas et al., 2020).

At organizational level, some international standards are also applicable on management systems for all kinds of associations, in many fields, including social responsibility (ISO 2010). The common plan-do-check-act methodology permits to define targets and to assess trends in time. This approach of continuous improvement and recurrent analysis is useful also for studies on sustainability and in fact some standards cover different aims including for example sustainable procurement (ISO 2017a), sustainable events (ISO 2012), or sustainable communities (ISO 2017b).

9.7 Conclusions

The standing definition of sustainable development requires an ethical agreement (McIntyre et al., 2017) among governments, and finally among humans. In fact, the words "sustainable" and "development" may assume many facets (Sala et al., 2013a) and must be explicated. Moreover, if weak stakeholders are not included in the definition of the framework and left without voice in the policymaking process (Huertas-Valdivia et al., 2020), consequently the approach on sustainability will be decided by power people. On the other side, all levers of power must be properly included and measured, including for instance the international finance (Peña and Rovira-Val, 2020).

The abundance of existing methods on sustainability assessment (Sala et al., 2013a) and of communication strategies need to be unified and organized to avoid confusion and misinterpretation. Different reports might be mingled in terms of contents or form, grouping similar targets, by linking for example SDGs and GRI targets (Tsalis et al., 2020) or LCA impact categories and SDGs (Henzler et al., 2020). Every effort must be made to provide information characterized by reliability, relevance, clarity, transparency, and accessibility (UNEP, 2017).

In this context, sustainability measurement tools are multifield and cover with different degrees of completeness environmental, social, or economic aspects. However, it is difficult to find a holistic tool that covers the complexity of sustainability issues and that may be useful for managers, politicians, citizens (Rohan et al., 2018). LCSA can be a constructive methodology to investigate with a comprehensive approach the three dimensions of sustainability. LCSA must still be enforced to manage external impacts on different categories or dimensions, leakages and rebound effects (Mori and Christodoulou, 2012) in time and space, and must be more applied by case studies in different sectors and contexts. Moreover, the LCSD can help to visualize assumptions and to perform sensitivity analyses on priorities chosen. The same priorities that, in fact, are present in the definition of the "sustainable development" concept. This is a loop. What is sustainability? How should we measure it? How might we communicate sustainability? Fair answers are needed by humans.

Some authors suggest that it is not enough that human business moves in a sustainable development framework, but that the core human business should be the sustainable development itself (Gray and Bebbington, 2000). They propose a conceptual shift, a change of paradigm, in which companies are dedicated not to produce saleable goods and services, but to produce sustainable progress (Dyllick and Muff, 2016): this is indicated as "true sustainability." Therefore, sustainability is not more a way of producing, but becomes the object to be produced; this new horizon must be supported by all stakeholders, from companies to governments, from economic actors to citizens (Avesani, 2020). This conceptual shift of the center of the economy, from money to sustainable development, has massive economic, social, and environmental consequences and requires huge changes. Many researchers agree that a complete revolution in consumption patterns is the only one effective to solve the unsustainable depletion of hearth resources (Kjaer et al., 2019). The structural change is more and more felt also by scientists and politicians, and requires decoupling economic growth from environmental degradation, preserving social equilibrium (United Nations, 2019).

Scientific and reliable tools to assess and to communicate this concept of sustainability are needed, and LCSD may also be useful for this purpose.

References

Aberilla, J.M., Gallego-Schmid, A., Stamford, L., Azapagic, A., 2020. An integrated sustainability assessment of synergistic supply of energy and water in remote communities. Sustain. Prod. Consum. 22, 1–23. doi:10.1016/j.spc.2020.01.003.

Amaral, K.G.C.do, Aisse, M.M., Possetti, G.R.C., 2019. Sustainability assessment of sludge and biogas management in wastewater treatment plants using the LCA technique. Rev. Ambient. Água 14.

Atilgan, B., Azapagic, A., 2016. An integrated life cycle sustainability assessment of electricity generation in Turkey. Energy Policy 93, 168–186. doi:10.1016/j.enpol.2016.02.055.

Avesani, M., 2020. Sustainability, Sustainable Development, and Business Sustainability, Life Cycle Sustainability Assessment for Decision-Making. Elsevier Inc, Amsterdam. doi:10.1016/b978-0-12-818355-7.00002-6.

Barber, A.J., Manhire, J., Gasso-Tortajada, V., Oudshoorn, F., Sørensen, C., Moller, H., 2016. Benchmarking energy and water efficiency in New Zealand wine production: eco-verification and incentivising improvement using the New Zealand sustainability dashboard. Acta Hortic 1112, 411–417. doi:10.17660/ActaHortic.2016.1112.55.

Capitano, C., Traverso, M., Rizzo, G., 2011. Life Cycle Sustainability Assessment : an implementation to marble products. Life Cycle Manag. Conf. LCM.

Cerrato, M., Miguel, G.S., 2020. Life cycle sustainability assessment of the Spanish electricity: past, present and future projections. Energies 13. doi:10.3390/en13081896.

Chen, W., Oldfield, T.L., Cinelli, P., Righetti, M.C., Holden, N.M., 2020. Hybrid life cycle assessment of potato pulp valorisation in biocomposite production. J. Clean. Prod. 269. doi:10.1016/j.jclepro.2020.122366.

Chiu, A.S.F., Aviso, K.B., Tan, R.R., 2020. Differentiating ethical imperatives of the collective sustainability research community and the individual researcher. Resour. Conserv. Recycl. 160, 104928. doi:10.1016/j.resconrec.2020.104928.

Dyllick, T., Muff, K., 2016. Clarifying the meaning of sustainable business: introducing a typology from business-as-usual to true business sustainability. Organ. Environ. 29, 156–174. doi:10.1177/1086026615575176.

Ekener, E., Hansson, J., Larsson, A., Peck, P., 2018. Developing Life Cycle Sustainability Assessment methodology by applying values-based sustainability weighting - Tested on biomass based and fossil transportation fuels. J. Clean. Prod. 181, 337–351. doi:10.1016/j.jclepro.2018.01.211.

Elkington, J., 2013. The Triple Bottom Line: Does it All Add Up Enter the triple bottom line. doi:10.4324/9781849773348.

Federico, G., Lascari, G., La Gennusa, M., Rizzo, G., Traverso, M., 2006. An integrated and shared approach to sea of the regional town master plan of Sicily. Int. J. Sustain. Dev. Plan. 1, 287–302. doi:10.2495/SDP-V1-N3-287-302.

Feng, C., Huang, S., 2020. The analysis of key technologies for sustainable machine tools design. Appl. Sci. 10. doi:10.3390/app10030731.

Finkbeiner, M., Schau, E.M., Lehmann, A., Traverso, M., 2010. Towards life cycle sustainability assessment. Sustainability 2, 3309–3322. doi: 10.3390/su2103309.

Giddings, B., Hopwood, B., O'Brien, G., 2002. Environment, economy and society: fitting them together into sustainable development. Sustain. Dev. 10, 187–196. doi:10.1002/sd.199.

Gray, R., Bebbington, J., 2000. Environmental accounting, managerialism and sustainability: is the planet safe in the hands of business and accounting?. Adv. Environ. Account. Manag. 1, 1–44. doi:10.1016/S1479-3598(00)01004-9.

Guo, M., van Dam, K.H., Touhami, N.O., Nguyen, R., Delval, F., Jamieson, C., Shah, N., 2020. Multi-level system modelling of the resource-food-bioenergy nexus in the global south. Energy 197, 117196. doi:10.1016/j.energy.2020.117196.

Guo, Z., Ge, S., Yao, X., Li, H., Li, X., 2020. Life cycle sustainability assessment of pumped hydro energy storage. Int. J. Energy Res. 44, 192–204. doi:10.1002/er.4890.

Hardi, P., Desouza-Huletey, J.A., 2000. Issues in analyzing data and indicators for sustainable development. Ecol. Modell. 130, 59–65. doi:10.1016/S0304-3800(00)00202-7.

Hardi, P., Semple, P., 2000. The Dashboard of Sustainability—From a Metaphor to an Operational Set of Indices, Proc. Fifth International Conference on Social Science Methodology. Cologne, Germany 3–6 October 2000.

Hardi, P., Jesinghaus, J., O'Connor, J., 2002. The Dashboard of Sustainability: a measurement and communication tool. In: 378 ecological indicators 9 (2009), 364–380 Ninth Session of the Commission on Sustainable Development, 16–27 April. New York. UNCSD, New York.

Henzler, K., Maier, S.D., Jäger, M., Horn, R., 2020. SDG-based sustainability assessment methodology for innovations in the field of urban surfaces. Sustain 12, 1–32. doi:10.3390/su12114466.

Hofstetter, P., Braunschweig, A., Mettier, T., Müller-Wenk, R., Tietje, O., 1999. The mixing triangle: correlation and graphical decision support for LCA-based comparisons. J. Ind. Ecol. 3, 97–115. doi:10.1162/108819899569584.

Huertas-Valdivia, I., Ferrari, A.M., Settembre-Blundo, D., García-Muiña, F.E., 2020. Social life-cycle assessment: a review by bibliometric analysis. Sustain 12, 1–25. doi:10.3390/su12156211.

ISO, 2006a. ISO 14040:2006. Environmental Management—Life Cycle Assessment—Principles and Framework. International Organization for Standardization, Geneva, Switzerland.

ISO, 2006b. ISO 14044:2006. Environmental Management—Life Cycle Assessment—Requirements and Guidelines. International Organization for Standardization, Geneva, Switzerland.

ISO, 2006c. ISO 14025:2006 Environmental Labels and Declarations — Type III Environmental Declarations — Principles and Procedures. International Organization for Standardization, Geneva, Switzerland.

ISO, 2010. ISO 26000:2010. Guidance on Social Responsibility. International Organization for Standardization, Geneva, Switzerland.

ISO, 2012. ISO 20121:2012. Event Sustainability Management Systems — Requirements with Guidance for use. International Organization for Standardization, Geneva, Switzerland.

ISO, 2016. ISO 14021:2016 Environmental Labels and Declarations — Self-Declared Environmental Claims (Type II Environmental Labelling). International Organization for Standardization, Geneva, Switzerland.

ISO, 2017a. ISO 20400:2017. Sustainable procurement — Guidance. International Organization for Standardization, Geneva, Switzerland.

ISO, 2017b. ISO 37101:2016. Sustainable Development in Communities — Management System for Sustainable Development — Requirements with Guidance for Use. International Organization for Standardization, Geneva, Switzerland.

ISO, 2018. ISO 14024:2018 Environmental Labels and Declarations — Type I Environmental Labelling — Principles and Procedures. International Organization for Standardization, Geneva, Switzerland.

Janjua, S.Y., Sarker, P.K., Biswas, W.K., 2020. Development of triple bottom line indicators for life cycle sustainability assessment of residential buildings. J. Environ. Manage. 264, 110476. doi:10.1016/j.jenvman.2020.110476.

Jesinghaus, J., 1999. A European System of Environmental Pressure Indices. Vol. 1: Environmental Pressure Indices Handbook: The indicators. Part I: Introduction to the Political and Theoretical Background. Eurostat, EU Commission, Brussels, Belgium.

Jesinghaus, J., 2000. On the Art of Aggregating Apples & Oranges. Milan, Italy: Nota di Lavoro, No. 91. 2000. Fondazione Eni Enrico Mattei.

Jesinghaus, J., 2009. MDG Dashboard of Sustainability. Beyond GDP: Measuring Progress, True Wealth, and the Well-being of Nations. Virtual Indicator Exhibition p. 280-283. Office for Official Publications of the European Communities. doi: http://dx.doi.org/10.2779/54600.

Jesinghaus, J., 2012. Measuring European environmental policy performance. Ecol. Indic. 17, 29–37. doi:10.1016/j.ecolind.2011.05.026.

Kjaer, L.L., Pigosso, D.C.A., Niero, M., Bech, N.M., McAloone, T.C., 2019. Product/service-systems for a circular economy: the route to decoupling economic growth from resource consumption?. J. Ind. Ecol. 23, 22–35. doi:10.1111/jiec.12747.

Kloepffer, W., 2008. Life cycle sustainability assessment of products (with Comments by Helias A. Udo de Haes, p. 95). Int. J. Life Cycle Assess. 13, 89–95. doi:10.1065/lca2008.02.376.

Llatas, C., Soust-Verdaguer, B., Passer, A., 2020. Implementing Life Cycle Sustainability Assessment during design stages in building information modelling: from systematic literature review to a methodological approach. Build. Environ. 182, 107164. doi:10.1016/j.buildenv.2020.107164.

López, N.M., Sáenz, J.L.S., Biedermann, A., Tierz, A.S., 2020. Sustainability assessment of product-service systems using flows between systems approach. Sustain 12. doi:10.3390/SU12083415.

Man, Y., Han, Y., Liu, Y., Lin, R., Ren, J., 2020. Multi-criteria decision making for sustainability assessment of boxboard production: a life cycle perspective considering water consumption, energy consumption, GHG emissions, and internal costs. J. Environ. Manage. 255, 109860. doi:10.1016/j.jenvman.2019.109860.

Martin, T.M., 2017. A framework for an alternatives assessment dashboard for evaluating chemical alternatives applied to flame retardants for electronic applications. Clean Technol. Environ. Policy 19, 1067–1086. doi:10.1007/s10098-016-1300-2.

McIntyre, M.L., Caputo, T., Murphy, S.A., 2017. The inescapably ethical foundation of sustainability. Int. J. Bus. Gov. Ethics 12, 127–150. doi:10.1504/IJBGE.2017.086471.

Morana, J., Gonzalez-Feliu, J., 2015. A sustainable urban logistics dashboard from the perspective of a group of operational managers. Manag. Res. Rev. 38, 1068–1085. doi:10.1108/MRR-11-2014-0260.

Mori, K., Christodoulou, A., 2012. Review of sustainability indices and indicators: towards a new city sustainability index (CSI). Environ. Impact Assess. Rev. 32, 94–106. doi:10.1016/j.eiar.2011.06.001.

Mukherjee, C., Denney, J., Mbonimpa, E.G., Slagley, J., Bhowmik, R., 2020. A review on municipal solid waste-to-energy trends in the USA. Renew. Sustain. Energy Rev. 119, 109512. doi:10.1016/j.rser.2019.109512.

Peña, A., Rovira-Val, M.R., 2020. A longitudinal literature review of life cycle costing applied to urban agriculture. Int. J. Life Cycle Assess. 25, 1418–1435. doi:10.1007/s11367-020-01768-y.

Ren, J., 2020. Barriers Identification and Prioritization for Sustainability Enhancement: Promoting the Sustainable Development of the Desalination Industry, Life Cycle Sustainability Assessment for Decision-Making. Elsevier Inc, Amsterdam. doi:10.1016/b978-0-12-818355-7.00015-4.

Ren, J., Toniolo, S., 2020. Multi-criteria Decision-Making after Life Cycle Sustainability Assessment Under Hybrid Information, Life Cycle Sustainability Assessment for Decision-Making. Elsevier Inc, Amsterdam. doi:10.1016/b978-0-12-818355-7.00013-0.

Rodriguez, L.J., Peças, P., Carvalho, H., Orrego, C.E., 2020. A literature review on life cycle tools fostering holistic sustainability assessment: an application in biocomposite materials. J. Environ. Manage. 262, 110308. doi:10.1016/j.jenvman.2020.110308.

Rohan, U., Branco, R.R., Soares, C.A.P., 2018. Potentialities and limitations of sustainability measurement instruments. Eng. Sanit. Ambient. 23, 857–869. doi:10.1590/s1413-41522018170117.

Sala, S., Farioli, F., Zamagni, A., 2013a. Progress in sustainability science: lessons learnt from current methodologies for sustainability assessment: Part 1. Int. J. Life Cycle Assess. 18, 1653–1672. doi:10.1007/s11367-012-0508-6.

Sala, S., Farioli, F., Zamagni, A., 2013b. Life cycle sustainability assessment in the context of sustainability science progress (part 2). Int. J. Life Cycle Assess. 18, 1686–1697. doi:10.1007/s11367-012-0509-5.

Scipioni, A., Mazzi, A., Mason, M., Manzardo, A., 2009. The Dashboard of Sustainability to measure the local urban sustainable development: the case study of Padua Municipality. Ecol. Indic. 9, 364–380. doi:10.1016/j.ecolind.2008.05.002.

Shields, J.F., Shelleman, J.M., 2020. SME sustainability dashboards: an aid to manage and report performance. J. Small Bus. Strateg. 30, 106–114.

Sonnemann, G., Gemechu, E.D., Sala, S., Schau, E.M., Allacker, K., Pant, R., Adibi, N., Valdivia, S., 2017. Life cycle thinking and the use of LCA in policies around the world. In: Hauschild, M., Rosenbaum, R.K., Olsen, S. (Eds.), Life Cycle Assessment: Theory and Practice. Springer International Publishing, ISM-CyVi, UMR 5255, University of Bordeaux, Talence, France, pp. 429–463. doi:10.1007/978-3-319-56475-3_18.

Toniolo, S., Tosato, R.C., Gambaro, F., Ren, J., 2020. Life Cycle Thinking Tools: Life Cycle Assessment, Life Cycle Costing and Social Life Cycle Assessment, Life Cycle Sustainability Assessment for Decision-Making. Elsevier Inc, Amsterdam. doi:10.1016/b978-0-12-818355-7.00003-8.

Topor, D.I., Capusneanu, S., Tamas, A.S., 2017. Efficient green control (EGC) encouraging environmental investment and profitability. J. Environ. Prot. Ecol. 18, 191–201.

Traverso, M., Asdrubali, F., Francia, A., Finkbeiner, M., 2012a. Towards life cycle sustainability assessment: an implementation to photovoltaic modules. Int. J. Life Cycle Assess. 17, 1068–1079. doi:10.1007/s11367-012-0433-8.

Traverso, M., Finkbeiner, M., Jørgensen, A., Schneider, L., 2012b. Life cycle sustainability dashboard. J. Ind. Ecol. 16, 680–688. doi:10.1111/j.1530-9290.2012.00497.x.

Tsalis, T.A., Malamateniou, K.E., Koulouriotis, D., Nikolaou, I.E., 2020. New challenges for corporate sustainability reporting: United Nations' 2030 Agenda for sustainable development and the sustainable development goals. Corp. Soc. Responsib. Environ. Manag. 27, 1617–1629. doi:10.1002/csr.1910.

UNEP, 2017. Guidelines for Providing Product Sustainability Information Global Guidance on Making Effective Environmental, Social and Economic Claims, to Empower and Enable Consumer Choice. ISBN: 978-92-807-3672-4. http://hdl.handle.net/20.500.11822/22180 (Accessed 21 September 2020).

UNEP, 2019. Background to Sustainability REPORTING. - Enhancing the Uptake and Impact of Corporate Sustainability Reporting: A Handbook and Toolkit for Policymakers and Relevant Stakeholders. Section A. Handbook http://hdl.handle.net/20.500.11822/30663 (Accessed 21 September 2020).

United Nations, 2005. UN Millennium Project. Investing in Development: A Practical Plan to Achieve the Millennium Development Goals. United Nations, New York. (Accessed 21 September 2020) https://www.researchgate.net/publication/208574954_Investing_in_development_a_practical_plan_to_achieve_the_Millennium_Development_Goals.

United Nations, 2012. The Millennium Development Goals Report 2012 https://www.un.org/en/development/desa/publications/mdg-report-2012.html (Accessed 21 September 2020).

References

United Nations, 2015. Transforming Our World: the 2030 Agenda for Sustainable Development. United Nations, New York. USA https://sustainabledevelopment.un.org/post2015/transformingourworld (Accessed 21 September 2020).

United Nations, 2019. The Future is Now. Science for Achieving Sustainable Development. Global Sustainable Development Report 2019. / https://sustainabledevelopment.un.org/content/documents/24797GSDR_report_2019.pdf (Accessed 21 September 2020).

Valdivia, S., Ugaya, C.M.L., Hildenbrand, J., Traverso, M., Mazijn, B., Sonnemann, G., 2013. A UNEP/SETAC approach towards a life cycle sustainability assessment - Our contribution to Rio+20. Int. J. Life Cycle Assess. 18, 1673–1685. doi:10.1007/s11367-012-0529-1.

Visentin, C., Trentin, A.W., da, S., Braun, A.B., Thomé, A., 2020. Life cycle sustainability assessment: a systematic literature review through the application perspective, indicators, and methodologies. J. Clean. Prod. 270. doi:10.1016/j.jclepro.2020.122509.

Vivanco, D.F., Sala, S., McDowall, W., 2018. Roadmap to rebound: how to address rebound effects from resource efficiency policy. Sustain 10. doi:10.3390/su10062009.

WCED, 1987. Report of the World Commission on Environment and Development: Our Common Future. Our Common Future. Oxford University Press, Oxford https://sustainabledevelopment.un.org/milestones/wced (Accessed 21 September 2020).

Xu, D., Li, W., Dong, L., 2020. A composite life cycle sustainability index for sustainability prioritization of industrial systems. Life Cycle Sustain. Assess. Decis., 225–252. doi:10.1016/b978-0-12-818355-7.00011-7.

Yang, S., Ma, K., Liu, Z., Ren, J., Man, Y., 2020. Development and Applicability of Life Cycle Impact Assessment Methodologies, Life Cycle Sustainability Assessment for Decision-Making. Elsevier Inc, Amsterdam. doi:10.1016/b978-0-12-818355-7.00005-1.

Zanni, S., Awere, E., Bonoli, A., 2020. Life cycle Sustainability Assessment: An Ongoing Journey, Life Cycle Sustainability Assessment for Decision-Making. Elsevier Inc, Amsterdam. doi:10.1016/b978-0-12-818355-7.00004-x.

Zhang, X., Zhang, L., Fung, K.Y., Bakshi, B.R., Ng, K.M., 2020. Sustainable product design: a life-cycle approach. Chem. Eng. Sci. 217, 115508. doi:10.1016/j.ces.2020.115508 doi:.

Zortea, R.B., Maciel, V.G., Passuello, A., 2018. Sustainability assessment of soybean production in Southern Brazil: a life cycle approach. Sustain. Prod. Consum. 13, 102–112. doi:10.1016/j.spc.2017.11.002.

CHAPTER 10

Multicriteria decision-making methods for results interpretation of life cycle assessment

Ana Carolina Maia Angelo
Fluminense Federal University, Volta Redonda, RJ, Brazil

10.1 Introduction

Life Cycle Assessment (LCA) is a quantitative method to assess the environment performance of products, services, and processes for support decisions. According to ISO 14040, an LCA study is carried out through four iterative phases: (1) goal and scope definition, (2) inventory analysis, (3) impact assessment, and (4) interpretation of results (ISO, 2006). Depending on the goal and scope defined for the study, the LCA can comprise from one-impact category (e.g., in Carbon Footprint studies where only climate change is considered) to more than a dozen-impact categories.

Despite the international standard ISO and ILCD Handbook recommends the inclusion of all relevant impact categories in LCA (ISO, 2006; European Commission, 2010), in practice several LCA studies in different areas focus on only one or a few impact categories as seen, for instance, in Andrade and D'Agosto (2016) and Cui et al. (2010) in transportation area and in the waste management area as pointed out in the review carried out by Allesch and Brunner (2014). While dealing with few impact categories may facilitate the interpretation of LCA results, it also creates a risk of excluding relevant impacts from the assessment.

On the contrary, dealing with a large number of impact categories in comparative LCA studies results in tradeoffs, that is, conflicting results, because there is no the best overall environmental alternative that best performs in all impact categories. In other words, one alternative can be preferred in relation to some impacts, while another is preferred in relation to others. Therefore, the results from an LCA study can facilitate the decision-making process by providing an understanding of the pros and cons of each alternative analyzed, but determining the best alternative might be difficult. This is precisely where Multicriteria Decision Making (MCDM) can be useful.

MCDM is used to support decision making in problems involving several criteria and/or alternatives, various decision-makers can be also considered. As a tool for conflict management, MCDM can be very helpful for supporting environmental and sustainability problems, providing a consistent structure for decision-making process to deal with complex issues such as different stakeholders, a large amount of criteria from quantitative and qualitative nature, tradeoffs between criteria, and imprecise information (Matteson, 2014; Clímaco and Valle, 2016; Recchia, 2011).

In this chapter, an overview of the MCDM approach and its main methods are presented, as well as the powerful role of application of MCDM methods for LCA results interpretation. This chapter is structured as follows. This first section generally presents the topic to be addressed and the main purpose of the chapter. The next section presents an overview of the multicriteria approach followed by the MCDM basic process and its main methods. The third section discusses the integration of LCA and MCDM methods for improving support decision making and provides the reader some examples in different areas such as waste management, transportation mode, materials selection, and production systems comparison. Finally, the section four presents the main conclusions and references.

10.2 An overview of the multicriteria approach

Decision making can be a complex process seeking to establish satisfying solutions (compromise solutions) through a scientific base taking into account judgments of a specific decision maker or a group of decision makers. Even in cases where only one decision-maker is involved in the decision-making process, rarely it considers one criterion alone, which means the decision making is more often multicriteria than a monocriterion approach. In this sense, the multicriteria approach has been playing an important role for analyzing and structuring the decision-making process (Greco, Ehrgott, and Figueira 2016).

Multicriteria approaches originate from two Schools: the European Multicriteria Decision Analysis (MCDA), also called French School, and the American MCDM. While MCDA seeks to give recommendations by means of the introduction of outranking relations, MCDM tries to approach an ideal solution in terms of evaluating a discrete set of alternatives by multiattribute utility functions, linear or not (Greco, Ehrgott, and Figueira 2016). To simplify, the present chapter uses the acronym MCDM for both approaches.

10.2.1 The MCDM basic process

Three basic concepts are involved in the MCDM: (1) *alternative* or action that constitutes the object of the decision and involves a finite number of alternatives or infinite possibilities; (2) *criteria* that allow to evaluate and compare the alternatives; and (3) *problematic* that refers to the questions the decision aid wants to answer (Greco, Ehrgott, and Figueira 2016). The latter concept is an important issue in the decision-making process as it determines the most suitable MCDM method to fulfill the objective. All of these basic concepts are inputs of the MCDM basic process (Guitouni and Martel 1998), dealing with the comparison of alternatives in some criteria under a specific problematic to find a compromise solution (Fig. 10.1).

Structure the decision problem concerns the characterization of the decision-making situation, including the determination of the decision's emergency, the different alternatives and their consequences, the family of criteria, the quantity and the quality of information available, the problematic, the decision maker(s), and the stakeholders. This step is crucial as the formulation of the decision problem is often more important than its solution.

Model the preferences means the modeling of the decision-maker(s) preferences. It must be remarked that the assumptions about the preferences may affect the MCDM process and its solution, because the decision makers influence the decision-making process and they are also influenced by it (Guitouni and Martel 1998). The preferences can be addressed by four elementary binary relationships between two alternatives: indifference, weak or strong preference, and incomparability (Roy 1990).

Aggregate the alternative evaluations refers to the multicriteria aggregation procedure, which corresponds to the problematic concept defined in the first step (problem structuring). For instance, the choice problematic (P.α) covers the decision making oriented to select a single alternative, the sorting

10.2 An overview of the multicriteria approach

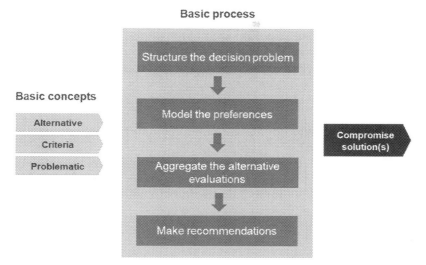

FIG. 10.1 The MCDM basic process.

problematic (P.β) relates to decisions lying on an assignment of each alternative to one category, the outranking problematic (P.γ) covers decisions with the objective to outrank the alternatives. Each problematic has a set of suitable methods and they are not the only possible ones (more information in Greco, Ehrgott, and Figueira 2016).

Make recommendations is the last step, which goes beyond the compromise solution as output. Although the core objective of MCDM is to support decision makers (e.g., managers, policy makers) to make better decisions, it must be notice that the complexity involved in the decision process and the limitations of problem structuring (e.g., insufficient information, uncertainty) may lead to not the better solution/alternative but to the compromise solution. In other words, the solution for the decision problem is not the optimal one but a satisfactory one (Guitouni and Martel 1998). In most cases, the process of knowledge construction is most relevant than the decision making per se (Greco, Ehrgott, and Figueira 2016).

10.2.2 MCDM methods classification

MCDM methods can be classified according to the aggregation procedure adopted to take into account all criteria analyzed. Besides the usual weighted sum, the MCDM methods can be classified into four groups (Polatidis et al. 2006; Greco, Ehrgott, and Figueira 2016) (Fig. 10.2).

The most traditional methods are those based on utility or value-function by single synthesizing criterion, in which formal rules mathematically structured reduce the criteria multiplicity into a unique criterion supposing that any alternative is comparable with every other. Several commonly used methods belong to this group, such as MAUT (Multiattribute Utility Theory), TOPSIS (Technique for Order Preference by Similarity to Ideal Solution), AHP (Analytic Hierarchy Process), and ANP (Analytic Network Process).

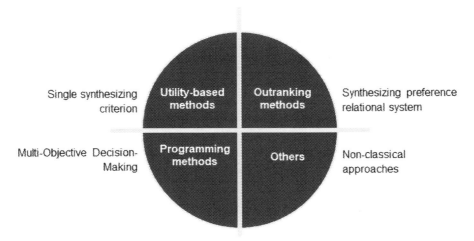

FIG. 10.2 MCDM methods classification.

The other approach is based on a synthesizing preference relational system, involving a pairwise comparison of the alternatives on each criterion supported by a well-structured mathematical rules based on discrimination and veto thresholds. PROMETHEE (Preference Ranking Organization Method for Enrichment Evaluation) and ELECTRE (Elimination and Choice Expressing Reality) belong to this group of outranking methods.

When the decision problem involves an infinite or a very large number of alternatives, Multiobjective Decision-Making methods are suitable. Comprising programming methods such as multiobjective optimization and goal programming, they are in general related to operational research approach to decision making (Azapagic and Perdan, 2005).

The other methods that are not in accordance with these approaches are the nonclassical approaches. In general, the nonclassical methods differ by their interactivity nature or use fuzzy theory to deal with uncertainties or imprecise information.

10.2.3 A brief description of the main MCDM methods

10.2.3.1 Analytic Hierarchy Process

AHP is the most widely applied multicriteria method in decision making (Vaidya and Kumar 2006). Developed by Saaty (1980), this method is based on the creation of a hierarchical structure of criteria, subcriteria (if they exist), and alternatives. Assuming comparability between alternatives, criteria and alternatives are compared in pairs to assess the relative preference among each other. The intensity of preference is defined through a relative measurement scale called Saaty's scale.

10.2.3.2 Technique for order preference by similarity to ideal solution

TOPSIS, also known as a reference point approach, is based on the concept that the alternative chosen should be the nearest to the ideal solution and farthest from the negative-ideal solution (Greco, Ehrgott, and Figueira 2016). In practice, TOPSIS compares a set of alternatives taking into account the distance of each alternative from the theoretic ideal solution and the negative one, being the most preferred alternative the one that is near the ideal alternative and far from the negative ideal one (Huang, Keisler, and Linkov 2011).

10.2.3.3 Preference Ranking Organization Method for Enrichment Evaluation

PROMETHEE family is composed of six methods based on positive and negative preference flows for each alternative, according to the selected criteria preferences (weights). PROMETHEE I deals with partial ranking, while PROMETHEE II deals with complete ranking. PROMETHEE III results in a ranking based on intervals and PROMETHEE IV, on continuous case. PROMETHEE V deals with constraints segmentation and PROMETHEE VI, with the representation of the human brain (Brans and Mareschal 2005).

10.2.3.4 Elimination and Choice Expressing Reality

ELECTRE family comprises six different methods. ELECTRE I is dedicated to choice problematic, aiming at reducing the size of a nondominated set of alternatives. ELECTRE IS is an improved form of ELECTRE I and uses an indifference threshold for modeling decision-makers preference. ELECTRE II outranks the alternatives from the best to worst option, using either strong or weak relationships between them. ELECTRE III allows the use of pseudocriteria and fuzzy outranking relations to deal with imprecise information. ELECTRE IV is similar to the ELECTRE III, but without the use of criteria weights. ELECTRE TRI deals with ordinal classification problems (sorting problematic) (Roy and Bouyssou 1993).

10.3 LCA and multicriteria methods integration

Before discussing about the integration of LCA and MCDM methods, it must be notice that both are tools to support the decision-making process. MCDM can be integrated to LCA in a variety way to support results interpretation. For instance, a review carried out by Zanghelini, Cherubini, and Soares (2018) pointed out the most common application of MCDM methods in LCA studies is at the Life Cycle Impact Assessment (LCIA) to assess the tradeoffs between different impact categories (midpoint or endpoint impact categories) or even between environmental impacts and others from the economic and/or social dimensions of sustainability, with the weighted sum and AHP as the most used MCDM methods.

These two methods and others from the single synthesizing criterion approach are considered compensatory methods as they allow some compensability among criteria represented by the tradeoffs, where a disadvantage on some criterion can be compensated by a sufficiently large advantage on another criterion (Rowley et al. 2012; Benoit and Rousseaux 2003; Guitouni and Martel 1998). On the contrary, the outranking methods might be more suitable to handle problems involving sustainability dimensions as they can be considered partially compensatory (Benoit and Rousseaux 2003; Rogers, Bruen, and Maystre 2000; Guitouni and Martel 1998) or even noncompensatory methods (Rowley et al. 2012). In fact, the review has pointed out an increase in the number of outranking methods applied in LCA studies due to their non/partial compensatory behavior. However, facing the multiplicity of MCDA methods none can be seem as the best method suitable to all decision-making problems (Guitouni and Martel 1998).

MCDM methods can be also applied in the in the Goal and Scope definition phase of LCA studies to define the impact categories representing the decision-makers' preferences (Zanghelini, Cherubini, and Soares 2018), and in the weighting procedure. Despite weighting in LCA is not recommended by ISO 14040:2006, MCDM is useful because it allows reflecting decision-maker or stakeholders' preferences in a consistent way through the well-structured MCDM methods. In this sense, AHP is commonly used in the weighting step due to its ability to convert subjective assessments of relative importance into a set of weights based on the judgment of knowledgeable and expert people (e.g., Halog and Manik 2011; Pires and Chang 2011; Myllyviita et al. 2012).

Considering the four LCA iterative phases (ISO, 2006): (1) goal and scope definition, (2) inventory analysis, (3) impact assessment, and (4) interpretation of results, Fig. 10.3 presents an integrated framework of LCA and MCDM.

Four examples are presented in the following, from which it is possible to verify the usefulness of LCA and MCDM integration in different areas, and multicriteria problematic and methods. For choice problematic, three examples are provided: one in the area of policy making with the case study of selection of municipal solid waste management (MSWM) option by applying a nonclassical MCDM method, other in material selection with the case study of sewer pipe materials comparison by applying a traditional single synthesizing criterion method, and the latter in the systems

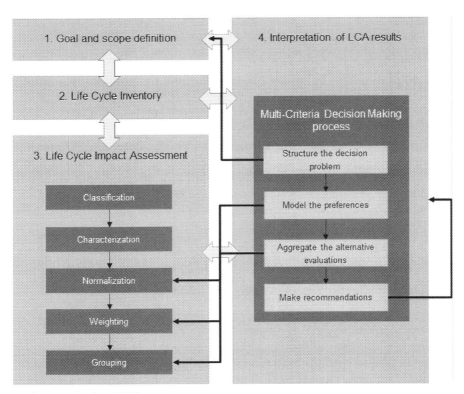

FIG. 10.3 LCA integrated with MCDM.

production by applying a traditional synthesizing preference relational system method to select the most sustainable poultry system production. For outranking problematic, an example of urban transport systems comparison by applying a partial/noncompensatory outranking method is provided. In all these examples, the MCDM basic process is highlighted for better understanding the integration approach.

10.3.1 Selection of MSWM option

An example of MCDM method integrated with LCA study can be found in Angelo et al. (2017), in which an interactive learning oriented multiattribute additive MCDM model using imprecise information, called VIP-Analysis (Dias and Climaco 2000), was applied to identify the most preferable MSWM alternative in terms of environmental performance. It is a nonclassical MCDM method for choice problematic and its main characteristic is that no precise values for weights are required, thus accepting imprecise information on decision-making preferences modeling. The objective of this study was to assess the potentialities of combining these two support decision tools. Regarding to the *decision problem structuring*, four solid waste management strategies (landfilling all MSW, 20% separate collection of organic wastes for anaerobic digestion (AD), 50% separate collection of organic wastes for AD, and AD of organic waste after material recovery facility (MRF)) were evaluated under 16 environmental impact categories. Thus, it was a multicriteria decision problem with four alternatives and 16 criteria (Table 10.1).

For *modeling of preferences*, as the MCDM method applied does not require values for weights, the criteria weighting was carried out by defining a ranking with the impact categories considering the same approach for weighting procedure used in the LCA method EDIP (Stranddorf et al.

Table 10.1 Alternatives and criteria used in the MSW strategy selection.

Alternatives (MSW strategy)	Criteria (LCA impact categories)	
	Preference direction	
	Minimize	*Maximize*
1. landfilling all MSW 2. 20% separate collection of organic wastes for AD 3. 50% separate collection of organic wastes for AD 4. AD of organic waste after MRF	1. Global warming potential 2. Ozone depletion potential 3. Particulate matter 4. Photochemical ozone formation 5. Freshwater eutrophication 6. Marine eutrophication 7. Acidification 8. Land eutrophication 9. Carcinogenic human toxicity 10. Noncarcinogenic human toxicity 11. Ionizing radiation 12. Ecotoxicity 13. Abiotic resources depletion	14. Nitrogen recovery 15. Phosphorous recovery 16. Potassium recovery (NPK-recovery)

2005), in which global warming potential ranked at the first position, followed by ozone depletion potential; noncarcinogenic human toxicity, photochemical ozone formation, NPK-recovery are in the third position, followed by ecotoxicity and abiotic resources at the fourth position; carcinogenic human toxicity, freshwater and marine eutrophication ranked at the fifth position, acidification at the sixth and, finally, particulate matter, ionizing radiation, and land eutrophication in the last (seventh) position.

In the *aggregation* step, the LCA results for each impact category were normalized obeying the preference direction (minimize or maximize) and aggregated through an additive value function under imprecise information and taking into account the minimax regret approach, resulting in values that enable pairwise confrontation of alternatives and dominated alternatives exclusion (Fig. 10.4). In the *making recommendations* step, as expected in terms of the Waste Hierarchy (European Commission 2005), the results pointed out the MSWM option of 50% separate collection of organic wastes for AD (scenario 3) as the most preferred waste management option. Contrary to the Waste Hierarchy, landfilling all MSW (scenario 1) seems a better option when compared to AD of organic waste after MRF (scenario 4).

Therefore the results pointed out the usefulness of applying MCDM method in LCA to improve the interpretation phase and make recommendations, especially in the solid waste management area, where various tradeoffs result concerning the environmental impact categories for which LCA is difficult to interpret and there is no alternative that completely optimizes all criteria together.

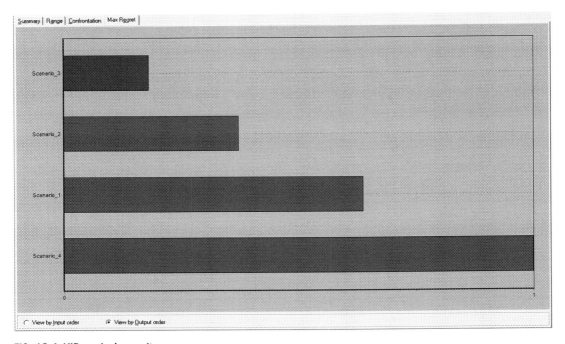

FIG. 10.4 VIP analysis results.

Source: Angelo et al. 2017.

10.3 LCA and multicriteria methods integration

Table 10.2 Alternatives and criteria used in the sewer pipe material selection.

Alternatives (sewer pipe material)	Criteria	
	Sustainable dimension	
	Environment (LCA indicators)	Economic (LCC indicators)
1. Concrete	1. Resource depletion	6. Initial cost
2. Polyvinyl chloride (PVC)	2. Energy consumption	7. Maintenance cost
3. Vitrified clay	3. Global warming	8. Disposal cost
4. Ductile iron	4. Acidification	
	5. Smog potential	

10.3.2 Selection of sewer pipe materials

A sustainable assessment was carried out by Akhtar et al. (2015) to evaluate and compare four typical materials for sewer pipes, taking into account LCA results and environment dimension and life cycle costing (LCC) for economic dimension. The authors applied AHP method to integrate the results and to provide decision-making recommendation for material selection among four alternatives and eight criteria with minimize preference direction (*decision problem structuring*) (Table 10.2).

The *modeling of preferences* was carried out through the application of Saaty scale that enables to determine relative weights for each criterion by pairwise comparison. It must be noticed that the study does not clearly define the decision makers in the process. The *aggregation* procedure obeys all steps of the AHP methodology, resulting in an overall priority values to select the best alternative.

In *making recommendations* step, although the authors have found PVC as the most sustainable material for sewer pipes, the AHP application suggested concrete pipes were slightly preferred than the PVC in the sensitivity analysis (by changing the weights of the environment and economic dimensions), indicating that AHP results can be strongly subjective (Fig. 10.5). In fact, the prioritization procedure in the AHP is very sensitive to the decision-maker preferences.

10.3.3 Selection of poultry production systems

Castellini et al. (2012) carried out a study for measuring the sustainability of different poultry production systems through a multicriteria approach. Regarding to *decision problem structuring*, production of three systems was evaluated under 24 criteria divided into four sustainability dimensions (Table 10.3 to select the most sustainable option by applying the ELECTRE I outranking method.

An interesting contribution of this study is that different stakeholders were considered in the decision-making process. In the *modeling preferences* step, a stakeholder consultation was carried out. Thirty people involved in the poultry production chain, divided into three different groups (10 per group)—scientists, consumers, and producers, were asked to determine importance co efficients for criteria weighting. As each stakeholder group has different points of view of the decision problem, the weighting procedure showed different results. While producers prioritized the economic indicators and quality, the opposite was found in the weights assigning by scientists who have given more importance to environment and social dimensions. The consumers' point of view is similar to the scientists.

Chapter 10 Multicriteria decision-making methods

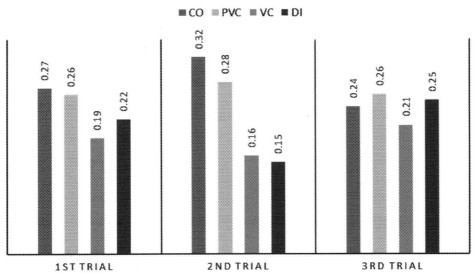

FIG. 10.5 Sensitivity analysis showing the priority vectors resulted from AHP.

Source: Akthar et al. 2015.

Table 10.3 Alternatives and criteria used in the poultry production systems comparison.

Alternatives (production system)	Criteria			
	Sustainable dimension			
	Environment	*Economic*	*Social*	*Quality*
1. Conventional 2. Organic 3. Organic-plus	1. Climate change 2. Land use 3. Ecotoxicity 4. Fossil fuels 5. EF/BC 6. ELR	7. Live weight at slaughtering 8. Feed conversion 9. Mortality rate 10. Net income 11. Revenue 12. Labor per production unit	13. Index of labor safety 14. Biodiversity birds 15. Moving (% budget time) 16. Foot pad lesions 17. Breast blister 18. H/L	19. Percentage of breast 20. Shear force 21. Fat content 22. Antioxidants 23. n-3 fatty acids 24. Oxidative stability

Abbreviations: EF/BC = sum of the ecological footprint of all activities/biocapacity; ELR = environmental loading ratio; H/L = heterophil/lynphocyte ratio of the birds; the environmental indicators comprise LCA results, ecological footprint indicators, and emergy analysis.

In the *aggregation of alternatives evaluations*, the performance of each alternative in each criterion was normalized and the alternatives were evaluated through pairwise comparison to establish the dominance relationship between them. Regarding the *making recommendation* step, the results indicated organic-plus as the most sustainable system production by taking into account the scientists and

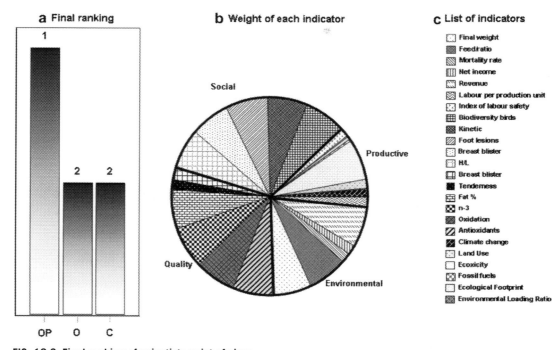

FIG. 10.6 Final ranking of scientists point of view.

Source: Castellini et al. 2012.

consumers' points of view (Figs 10.6 and 10.7). Also, the authors verified that the divergence rankings resulted from the different stakeholders' preferences modeling were in agreement with the current situation in poultry production systems, which corroborate with the usefulness of the MCDM method integration to environment and sustainable assessments.

10.3.4 Urban transport systems comparison

A study carried out by Angelo et al. (2019) compared metro and bus rapid transit (BRT) systems through a sustainable lifecycle analysis, taking the city of Rio de Janeiro as a case study focusing on BRT Transcarioca and the metro extension Line 4, both investments to receive Olympic and Paralympic games 2016. In the *decision problem structuring*, nine criteria of the three sustainable dimensions were evaluated with equal weights to identify the compromise solution between the BRT Transcarioca and the metro Line 4 by applying the outranking method ELECTRE III. It was chosen due to its noncompensatory characteristic, which turns ELECTRE III an appropriate MCDM method to handle environmental management problems (Govindan and Jepsen 2016) as well as those related to the multidimensional and complexity nature such as sustainable problems (Munda 2008).

164 Chapter 10 Multicriteria decision-making methods

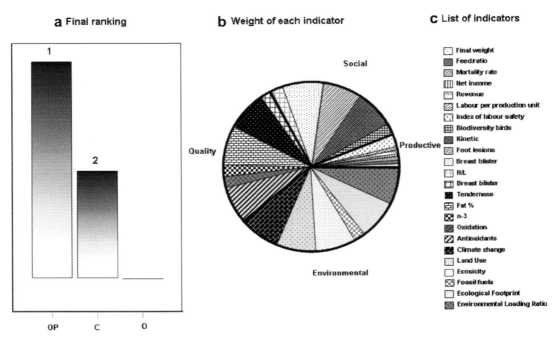

FIG. 10.7 Final ranking of consumers' point of view.

Source: Castellini et al. 2012.

The environmental criteria came from LCA results obtained by Martins and Angelo (2018) and the economic and social criteria from public reports (Table 10.4). All of them are minimizing criteria in terms of direction preference. In the *modeling preferences*, the criteria were considered equally important. The *aggregation* step obeyed the ELECTRE III aggregation procedure, in which no normalized values are required for alternatives pairwise comparison.

Regarding to *making recommendations*, the results of ELECTRE III indicated BRT was more sustainable than metro and an indifference between them in the sensitivity analysis, when the criteria-related investment and costs were removed from the analysis. In fact, BRT and metro were similar

Table 10.4 Alternatives and criteria used in the urban transport comparison.			
Alternatives (transportation mode)	**Criteria (environment LCA impact categories + economic and social indicators)**		
	Sustainable dimension		
	Environment	Economic	Social
BRT Transcarioca Metro line 4	Climate change Particulate matter Photochemical oxidant formation	Demand Investments on infrastructure Operational and maintenance costs	Perception of quality of service Travel time reduction Perception of transport expenditures reduction

from an environment perspective, and BRT has favorable performance in the majority of criteria from a social and economic perspective.

10.4 Discussion

Despite LCA is recognized as a broad-scope environment management tool to support decision, it is essentially a set of impact potential indicators from which decision makers have the challenge to deal with tradeoffs and to understand the implications of different choices/alternatives seeking a good decision making. In practice, LCA practitioners and decision makers appeal to optional elements under ISO Standards within an LCA study such as normalization and weighting.

The normalization procedure aims to obtain dimensionless LCIA results that can be multiplied by a set of weighting factors indicating the relative importance of the impact categories assessed in the LCA study. The normalized and weighted results thus can be summed up to a single-value score representing the overall impact (aggregation of results). Despite this may facilitate the interpretation of LCA results and therefore the decision making, weighting is a controversial step as it always involves value choices that influence the results and conclusions of the LCA study.

In this sense, coupling MCDM with LCA is successfully applied to support the interpretation of LCA results, providing a robust, rigorous, and transparent decision-making process. The integrated framework has been largely used especially in complex situations (e.g., sustainable assessments through Life Cycle Sustainability Assessment—LCSA). However, it must be remarked that MCDM can be complex due to the sophisticated analysis of qualitative and quantitative data, besides the information gathering and stakeholders' view consideration.

10.5 Concluding remarks

As discussed in this chapter, MCDM is a powerful approach for decision aiding, and integrated with LCA methodology, it facilitates the interpretation of results, thus helping managers make better decisions. Despite several LCA studies in different areas deal with only one or few impact categories, it is strongly recommended to include all relevant impact categories in the assessment in order to enhance all relevant aspects covered in the decision making. In this sense, there is a trend in integrating LCA and MCDM methods as it allows decision makers to deal with a large amount of diversified indicators in the same framework.

Some examples of studies that applied the integrated framework were presented, from which it is possible to observe the usefulness of MCDM methods not only for LCA results interpretation but also to model decision-makers' preferences under some decision problem. It must be remarked that there are a variety of MCDM methods, with specific application problematic and the choice of MCDM method should be consistent to the goal of the LCA study.

References

Akhtar, S., Reza, B., Hewage, K., Shahriar, A., Zargar, A., Sadiq, R., 2015. Life cycle sustainability assessment (LCSA) for selection of sewer pipe materials. Clean Technol. Environ. Policy. 17, 973–992. doi:10.1007/s10098-014-0849-x.

Allesch, A., Brunner, P.H., 2014. Assessment methods for solid waste management: a literature review. Waste Manag. Res. 32, 461e473.

Andrade, C.E., D'Agosto, M.A., 2016. Energy use and carbon dioxide emissions assessment in the lifecycle of passenger rail systems: the case of the Rio de Janeiro. J. Clean. Prod 126 (Abr), 526–536.

Angelo, A.C.M., Martins, I.D., Rodrigues, L.M., Cordeiro, M.C., Marujo, L.G., 2019. Comparative sustainable lifecycle analysis of bus rapid transit (BRT) and metro: a case study of Rio de Janeiro city, Proc 4th International Conference on Energy and Environment: bringing together Engineering and Economics. Guimarães, Portugal 16–17, May.

Angelo, A.C.M., Saraiva, A.B., Clímaco, J.C.N., Infante, C.E., and Valle, R. 2017. "Life cycle assessment and multi-criteria decision analysis: selection of a strategy for domestic food waste management in Rio De Janeiro." doi: 10.1016/j.jclepro.2016.12.049.

Azapagic, A., Perdan, S., 2005. An integrated sustainability decision-support framework Part II: Problem analysis. International Journal of Sustainable Development & World Ecology 12 (2), 112–131. doi:https://doi.org/10.1080/13504500509469623.

Benoit, V., Rousseaux, P., 2003. Aid for aggregating the impacts in life cycle assessment. Int. J. Life Cycle Assess. 8 (2), 74–82. doi:10.1007/BF02978430.

Brans, J.-.P., Mareschal, B., 2005. PROMETHEE methods. Multiple criteria decision analysis: state of the art surveys. Springer, New York, pp. 163–186.

Castellini, C., Boggia, A., Cortina, C., Dal, A., Paolotti, L., Novelli, E., Mugnai, C., 2012. A multicriteria approach for measuring the sustainability of different poultry production systems. J. Clean. Prod. 37, 192–201. doi:10.1016/j.jclepro.2012.07.006.

Clímaco, J.N., Valle, R., 2016. MCDA and LCSA - a note on the aggregation of prefer- ences. In: Kunifuji, S. (Ed.), Knowledge, Information and Creativity Support Systems pp 105e116, Advances in Intelligent Systems and Computing 416. Springer, Switzerland 2016.

Cui, S., Niu, H., Wang, W., Zhang, G., Gao, L., Lin, J., 2010. Carbon footprint analysis of the Bus Rapid Transit (BRT) system: a case study of Xiamen City. Int. J. Sustain. Dev. World Ecol 17 (4 (Jul), 8 pages.

Dias, L., Clímaco, J.N., 2000. Additive aggregation with interdependent parameters: the VIP-analysis software. J. Oper. Res. Soc. 51, 1070e1082.

European Commission, 2005. European Waste Framework Directive Avaiable at: http://eur-lex.europa.eu/legal-content/EN/TXT/?uri¼CELEX:32008L0098.

European Commission, 2010. International Reference Life Cycle Data System (ILCD) Handbook - General guide for Life Cycle Assessment - Detailed guidance, First edition. Publications Office of the European Union, Luxembourg March 2010. Translated by Luiz Marcos Vasconcelos. EUR 24708EN.

Govindan, K., Brandt Jepsen, M., 2016. ELECTRE : a comprehensive literature review on methodologies and applications. Eur. J. Oper. Res. 250 (1), 1–29. doi:10.1016/j.ejor.2015.07.019.

Greco, S., Ehrgott, M., Figueira, J.R., 2016. Multiple Criteria Decision Analysis - State of the Art Surveys Edited by Salvatore Greco, Matthias Ehrgott, and José Rui Figueira, Second ed. Springer-Verlag New York, New York. doi:10.1007/978-1-4939-3094-4 .

Guitouni, A., Martel, J.-.M., 1998. Tentative guidelines to help choosing an appropriate MCDA method. Eur. J. Oper. Res. 109 (2), 501–521. doi:10.1016/S0377-2217(98)00073-3.

Halog, A., Manik, Y., 2011. Advancing integrated systems modelling framework for Life Cycle Sustainability Assessment. Sustainability 3 (2), 469–499. doi:10.3390/su3020469.

Huang, I.B., Keisler, J., Linkov, I., 2011. Science of the total environment multi-criteria decision analysis in environmental sciences : ten years of applications and trends. Sci. Total Environ. The 409 (19), 3578–3594. doi:10.1016/j.scitotenv.2011.06.022.

ISO, 2006. ISO 14040 International standard, Environmental Management e Life Cycle Assessment e Requirements and Guidelines. International Organization for Standardization. Geneva, CH.

References

Matteson, S., 2014. Methods for multi-criteria sustainability and reliability assessments of power systems. Energy 71, 130e136.

Martins, I.D., Angelo, A.C.M., 2018. Avaliação Comparativa do Ciclo de Vida do Bus Rapid Transite (BRT) e Metrô: um estudo de caso do BRT Transcarioca e da Linha 4 do metrô do Rio de Janeiro. VI Congresso Brasileiro de Gestão do Ciclo de Vida, Brasilia, Brasil.

Munda, G., 2008. Social Multi-criteria Evaluation for Sustainable Economy. Springer-Verlag, Berlin Heidelberg.

Myllyviita, T., Holma, A., Antikainen, R., Lähtinen, K., Leskinen, P., 2012. Assessing environmental impacts of biomass production chains e application of Life Cycle Assessment (LCA) and Multi-Criteria Decision Analysis (MCDA). J. Clean. Prod. 29–30 (July 2012), 238–245. doi:10.1016/j.jclepro.2012.01.019.

Pires, A., Chang, N., 2011. Resources, conservation and recycling an AHP-Based Fuzzy Interval TOPSIS assessment for sustainable expansion of the solid waste management system in Setúbal Peninsula, Portugal. Resour. Conserv. Recycl. 56, 7–21. doi:10.1016/j.resconrec.2011.08.004.

Polatidis, H., Haralambopoulos, D.A., Munda, G., Vreeker, R., 2006. Selecting an appropriate Multi-Criteria Decision Analysis technique for renewable energy planning. Energ. Source Part B 1 (2), 181–193. doi:10.1080/009083190881607.

Recchia, L. (Ed.), 2011. Multicriteria Analysis and LCA Techniques: with Applications to Agro-engineering problems, Green Energy and Technology. Springer-Verlag, London; New York.

Rogers, M., Bruen, M., Maystre, L.-.Y., 2000. Electre and Decision Support: Methods and Applications in Engineering and Infrastructure Investment. Springer Science+Business Media New York Originally. doi:10.1007/978-1-4757-5057-7.

Rowley, H.V., Peters, G.M., Lundie, S., Moore, S.J., 2012. Aggregating sustainability indicators: beyond the weighted sum. J. Environ. Manage. 111, 24–33. doi:10.1016/j.jenvman.2012.05.004.

Roy, B., 1990. Decision-aid and decision-making. Eur. J. Oper. Res. 45, 324–331.

Roy, B., Bouyssou, D., 1993. Aide Multicritère à La Décision: Méthodes et Cas. Economica, Paris.

Saaty, T.L., 1980. The Analytic Hierarchy Process. McGraw-Hill, New York.

Stranddorf, H., Hoffmann, L., Schmidt, A., 2005. Impact categories, normalisation and weighting in LCA. Updated on selected EDIP97-data. Danish EPA Report 78.

Vaidya, O.S., Kumar, S., 2006. Analytic hierarchy process: an overview of applications. Eur. J. Oper. Res. 169 (1), 1–29. doi:10.1016/J.EJOR.2004.04.028.

Zanghelini, G.M., Cherubini, E., Soares, S.R., 2018. How Multi-Criteria Decision Analysis (MCDA) is aiding life cycle Assessment (LCA) in results interpretation. J. Clean. Prod. 172, 609–622. doi:10.1016/j.jclepro.2017.10.230.

CHAPTER 11

Composite sustainability indices (CSI); a robust tool for the sustainability measurement of chemical processes from "early design" to "production" stages

Mohammad Hossein Ordouei
Energy Research Center, University of Waterloo, Ontario, Canada

11.1 Introduction

Catastrophic industrial accidents, climate change due to anthropogenic activities, along with the discharge of polluting materials to the ecosystem, have far-reaching consequences for natural resources, human society, and the economy. The greenhouse gas (GHG) emission rate has been growing at an unprecedented rate for decades due to a variety of factors, including fossil fuel combustion (IPCC, 2013). Study shows that just in 2004 the global GHG emissions from the industries (excluding cement) and power plants were 5 and 10 Gigatons, respectively; in 2011 in the United States, the GHG emissions from both industries and power plants was about 53% of total emissions.

Chemical processes are among the leading contributors to both pollution and GHG emissions to the environment.

Market demand, availability of raw materials, site selection, easy access to transportation facilities, etc. are considered the crucial factors in a conventional process flow sheet design. This approach pursues financial accomplishment (e.g., optimum net present value, capital investment, return on investment, operations and maintenance costs, and marketability of final products). As a result, the designer may ignore the materials and energy impacts of the process on the environment and society. Therefore, the traditional process design has raised a conflict of interests between ecologists and regulators (whose interest is to protect the environment) versus stakeholders (who are fascinated by economic growth). Moreover, this trend ensues the generation of large amounts of waste and pollutants (EPA, 2012) and, consequently, an increase in the plant expenditure due to the installation and operation of control stations, such as waste treatment facilities, known as end-of-pipe treatment.

There is no simple solution to tackle these defies; however, there should be no doubt as to how the engineers must act. Sustainability is a modern approach that has surmounted the critical issue of the conflict of interests. Chemical engineers can produce a sorted process flow sheets comprising various production routes and equipment configurations to represent the production of certain chemical products. The quest is to pioneer screening methods that help separate unpromising processes at primitive design up to operating stages and reach sustainable chemistry objectives.

Holistic breakthroughs in reducing the GHG emissions, global warming, and process retrofitting mandate possessing strong chemical engineering competency. Process systems engineering can integrate various particular fields to scrutinize complex systems. Its viable techniques support revamping the existing processes and offering innovative solutions to battle sustainable development challenges, particularly cost reduction at enterprise and industry levels. Thus, sustainable development is a key concept to survive the environment, and *chemical engineers have considerable talent to achieve it* (Batterham, 2006). The objective of sustainable chemistry is to meet stakeholders', regulators', and society's demands by adhering to social responsibility, generating environmentally benign products, and low wastes, as well as saving the product's quality without compromising process profitability. Sustainable chemical processes not only curtail the vulnerability to climate change and the release of hazardous chemicals to the environment, but it also secures process profitability. The concept of sustainability and sustainability assessment techniques have been evolved during the last few years. The sustainability has become an inseparable part of the majority of human's activities from teaching (Carew and Mitchell, 2008) and engineering design, such as sustainable vehicle (van Lante and van Til, 2008), to the development of a framework for local sustainability indicators within a regional setting (Mascarenhas et al., 2010) and corporations' sustainability.

Unfortunately, there are many hurdles to implement a sustainable process design, mainly the shifting from the end-of-pipe treatment to the primitive design stage, due to the minimum available process data. Scientists and engineers have made several attempts in many different ways to ease this shift. Table 11.1 summarizes the recent essential methodologies. There are also a few outdated methodologies that are not listed in the table. Although the methods in this table have specific pros, they have one or a combination of the following cons: qualitative, inaccurate, time-intensive, score-based, dependent on the quality of training that a data collector may receive, demanding detailed process data, unappealing and challenging tasks.

As a general rule, the judgment of process performance based on a large number of indicators is challenging. The summarized frameworks in Table 11.1 necessitate the development of simple indices that account for a few sustainability factors and aggregating a set of the index into a composite index. The composite index enables a comprehensive assessment of the sustainability of a chemical process at early design and operating phases.

WAR algorithm (Hilaly and Sikdar, 1994 and 1995) and its concept of PEI balance (Young and Cabezas, 1999; Young et al., 2000), as well as Al-Sharrah's risk model (2007), consist of only an essential part of the sustainability assessment process, that is, environment and risk. Al-Sharrah's risk model (2007), however, needs a conceptual improvement to be useful at the initial stage of an intricate chemical process design together with the WAR algorithm and PEI balance.

The Composite Sustainability Indices (CSI) is among the best solutions to this significant exigence (Fig. 11.1); a complete study of the CSI methodology has been first pioneered by Ordouei (2014a) at the University of Waterloo, Ontario, Canada. The CSI is technically essential at all stages of chemical processes with the aim of pollution prevention and risk reduction. It requires a minimum amount of data to monitor and troubleshoot the sustainability performance of chemical processes. The succeeding sections discuss the elements and applications of the CSI methodology in more detail.

Table 11.1 The summary of the recent essential sustainability methodologies, descriptions, references, and disadvantages.

No.	Methodology	Description	Reference	Disadvantage(s)
		Sustainability methodologies		
1	Corporation sustainability	Strategic planning and external communications, supply chain, and decision making	Amimi and Bienstock, (2014)	Limited to operating businesses
2	Corporation SUSTAINABILITY	Local, regional, public, private, profit, nonprofit, industrial, agricultural, transportation	Ramos and Caeiro, (2010)	Limited to operating businesses
3	Dow Jones Sustainability Indices (DJSI)	Dow Jones introduced a corporate sustainability assessment to guarantee long-term benefits for stakeholders based on economic, environmental, and social developments.	Dow Chemical, 2021	Limited to operating businesses
4	FTSE4Good environmental leaders Europe 40 Index	A means for financers who seek European partnership in practical environmental management	FTSE4Good Environmental Leaders Europe 40 Index, (2020)	Limited to operating businesses
5	AIChE sustainability index (SI)	The SI collects data from the company's annual sustainability report, industrial performance rankings, government's pamphlets, and newsletters. The SI's output is aggregated to a scale from 0 to 7 on a spider chart	Cobb et al., (2007)	Limited to operating businesses
6	Life cycle assessment (LCA), known as cradle-to-grave analysis	The LCA is carried out by collecting the records of material and energy inputs and outputs, including emissions to the environment, transportation, handling, recycling, disposal, to calculate the potential environmental impacts of each step, as well as construing the outcomes for decision making.	EPA (2010)	Needs score of data, which is unavailable at the process design stage.
7	WAR algorithm and potential environmental impact (PEI) balance	The WAR algorithm employs the concept of pollution balance (similar to material balance) in a process flow sheet as a systematic method of waste reduction within a process. An amendment to the WAR algorithm is a PEI balance, which estimates the impacts of materials within a process on the environment. The PEI balance is an amendment to the WAR algorithm, which estimates the impacts of materials in a process on the environment.	Hilaly and Sikdar (1994, 1995); Young and Cabezas (1999); Young et al. (2000)	Not applicable

(continued)

Chapter 11 Composite sustainability indices (CSI)

Table 11.1 (Cont'd)

Sustainability methodologies

No.	Methodology	Description	Reference	Disadvantage(s)
8	Risk assessment of petrochemical plants	This method has four technical components, that is, frequency of accidents, chemical inventory, the toxicity of exposed chemicals, and plant size.	Al-Sharrah et al. (2007)	A sound methodology but needed modification; Limited to pure chemicals; multiplies risk by plant size (i.e., three).
9	Sugiyama et al.	Based on reaction routes, recycling configurations, operating conditions, as well as evaluation method may change by the economic and environmental assessment	Sugiyama et al. (2009)	They used Life Cycle Assessment (LCA) for environmental impact assessment.
10	Risk-based inherent safety index (RISI)	The RISI makes a relationship between the environmental consciousness of a process and declining the probability of explosions, toxic release, and similar accidents.	Rathnayaka et al. (2014)	The proposed equations for explosion damage radius (EDR) and fire damage radius (FDR) are independent of explosive mass, which is misleading. The proposed equations are proportional to the potential material hazard, which is not valid for some explosives such as ammonium nitrate. The explosion of 2700 tons of ammonium nitrate in Beirut port in August 2020 is strong evidence for the suggested disadvantages above.
11	Life cycle indexing system (LInX)	The LInX is claimed to be an accomplishment to the LCA method for the evaluation of chemical products and processes. It has four pillars; environment, health, and safety or EHS; cost; technical feasibility, and sociopolitical.	Khan et al. (2004)	The LInX is time-consuming and needs scores of information, which are not available at the primitive step of process design.
12	Environmental impacts assessment in separation processes	The EPA's methodology estimates the "energy" impacts of a process on the environment, and the materials impacts on the atmosphere, soil, water, and human are calculated based on the *WAR* algorithm.	Li et al. (2009)	The EPA's emission factors are the same for steam and electricity generation. In contrast, heat and power generations have different contributions to GHG emission due to the different efficiency factors of the boiler and steam turbine.
13	Emissions from power plants	This approach mandates data collections from the pilot plant and/or operating plants.	Hossain et al. (2011)	Needs score of data, which is unavailable at the process design stage; limited to power plants.

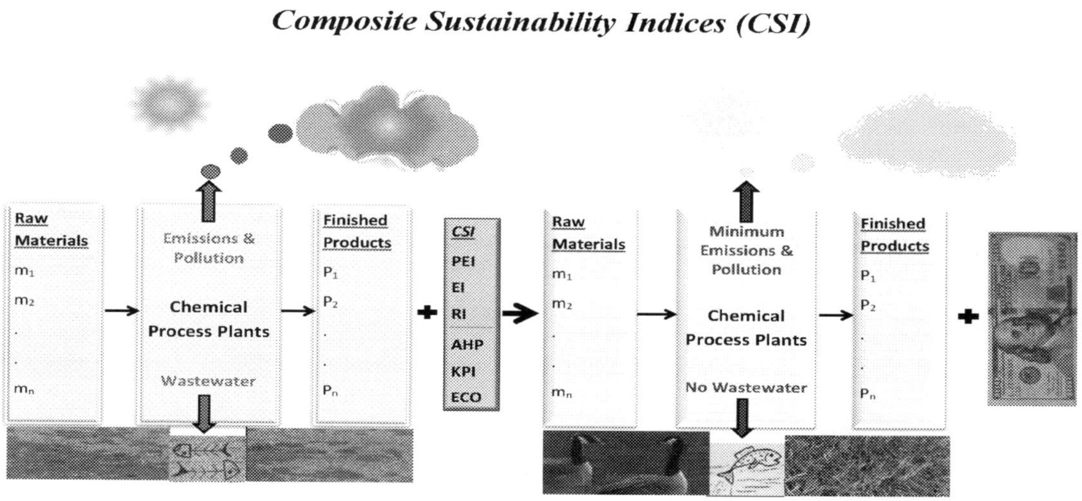

FIG. 11.1 Application of the CSI methodology on existing chemical process plants results in a reduction in emission and pollution discharge to the environment, as well as an increase in plant profitability.

11.2 The CSI methodology and applications

The invention of the CSI is a great movement toward the EPA's Sustainability objectives, which comprises three tenets: *nature, society,* and *economy*. It demonstrates that the lower impact on the environment and society a chemical process has, the more sustainable and profitable the process is. In other words, the higher the Key Process Indicator (KPI) value, the more sustainable chemical process. The CSI addresses the tenets as mentioned earlier by implementing three pillars in chemical processes:

- *Environment Index*: It quantifies the impacts of chemical compounds used within a process on *nature* in Potential Environmental Impacts (PEIs) per hour or PEI/h (Young and Cabezas, 1999). A definite advantage of the PEIs measurement over other methodologies is that it indicates that the impact of a small quantity of a substance on the environment may be higher than a significant amount of another material.
- *Energy Index:* It calculates the impacts of the steam and electricity used in a chemical process on *nature* and has a unit of kg CO_2/h (or kg NO_2/h or kg SO_2/h).
- *Risk Index*: It estimates the impacts of chemical substances used within a process on *society* in the unit of the number of affected people per year.

A conceptual decision model based on the Analytical Hierarchy Process (AHP) is then used to classify the above sustainability indicators and the corresponding weights (Saaty, 2008). The AHP gives a tangible result that can be aggregated to obtain the KPI of the chemical manufacturing plants under study. The proposed metrics by CSI can be applied to various process alternatives from conceptual design to operating phases to estimate the impacts of each process on the environment and society. It turns out that the CSI is essentially a powerful technical tool that avoids prejudice decision-making toward the process *economy* and warrants unbiased sustainability analysis. Thus, the economic and profitability studies of the processes need to be done separately and individually. In the case of concern with more than one sustainable chemical process, the CSI is competent to pinpoint the inherently safest and environmentally friendliest one. It has been demonstrated that the CSI methodology is coined to the profitability of a chemical process (Ordouei et al., 2016), and it is decisive at chemical manufacturing businesses pursuant to strategic planning toward the CSI method. As mentioned earlier, the output of the CSI approach is represented by a KPI as a measure of the *sustainability* performance of the corresponding process design. From the technical point of view, the CSI method is a robust screening tool that helps regulators, environmentalists, and investors to implement the *sustainability–profitability* analysis of the afore-mentioned process deigns.

The design of a chemical process instigates from a Design Basis, constituent of the quantity and composition of the raw materials entering the process battery limit, and the desired products as a result of market study. The process of interest can be as simple as a reaction taking place in a reactor and its piping around to a complex refinery plant designed by advanced process simulators, such as Aspen PLUS, Aspen HYSYS, CHEMCAD, PRO II, etc. There could be numerous process flow sheets design depending upon reactions involved and process routes. Hence, the evaluation of a process scheme in the conceptual design stage is based on the proposal offered by licensing corporations that are a connoisseur in particular processes. The proposal must specify minimum process data, such as block flow diagrams, the chemical compositions and the conditions of main material streams, and the energy streams data.

Each chemical compound of a material stream has several properties, such as flammability, melting and boiling points, viscosity, toxicity, and PEI. The CSI can be derived based on the materials' specific properties to be employed in the comparative analyses and screening of design proposals.

The CSI integrates the PEI concept, the improved risk assessment model, and the energy index model into a chemical process either at the conceptual design step or retrofitting of an operating chemical process. The PEI is an amendment to the WAR algorithm, advocated by the EPA, which employs eight impacts (Young and Cabezas, 1999; Young et al., 2000): human toxicity potential by ingestion (HTPI) and exposure (HTPE), terrestrial toxicity potential (TTP), aquatic toxicity potential (ATP), global warming potential (GWP), ozone depletion potential (ODP), photochemical oxidation potential (PCOP), and acidification potential (AP). The PEI uses the chemical compositions of the material streams to measure the corresponding impacts of each component. A risk evaluation model introduced by Al-Sharrah (2007) was developed by Ordouei et al. (2014b) and used in chemical processes for Inherently Safer Design (ISD). Given the characteristics of fossil fuels and energy generation/consumption in the forms of heat and power within a chemical manufacturing plant, Ordouei et al. (2018) presented a novel energy impacts assessment to estimate the GHG emissions to the environment.

11.2.1 WAste Reduction algorithm and potential environment impact balance

The WAR algorithm proposed the concept of pollution balance (similar to material balance) in a process flow sheet as a tool for a process designer to track a contaminant in an entire process plant (Hilaly and Sikdar; 1994, 1995). The WAR algorithm is a constructive tactic that can be used at the conceptual design stage of a chemical process and pledges an inherently less polluting process and notably reduces the cost of pollution control devices used in conventional process designs. An amendment to the WAR algorithm is the concept of Potential Environmental Impact (PEI) balance, which estimates the impacts of chemical pollutants within a process on the environment in PEI/h. The harmful effect of a chemical compound on the ecosystem as it were to discharge into the ecosystem is defined as the Potential Environmental Impact (Young and Cabezas, 1999; Young et al., 2000). The mathematical expression of the PEI balance is as follows:

$$\frac{dI_{System}}{dt} = I_{in}^{(t)} - I_{out}^{(t)} + I_{gen}^{(t)} \tag{11.1}$$

where

$\frac{dI_{System}}{dt}$ is the rate of change in the quantity of the PEI within the system (chemical process),

$I_{in}^{(t)}$ and $I_{out}^{(t)}$ are the rates of total PEIs into and out of the streams, respectively,

$I_{gen}^{(t)}$ is the rate of total PEIs generated or consumed by chemical reactions within the system.

At a steady state,

$$I_{in}^{(t)} - I_{out}^{(t)} + I_{gen}^{(t)} = 0 \tag{11.2}$$

Thus, the WAR algorithm methodology employs two main impact categories for the PEI assessment; each one includes four subcategories (totally eight environmental impact categories) that are as follows:

- *Global Atmospheric Impacts,* including GWP, ODP, PCOP, and AP.
- *Local Toxicological Impacts* contain human toxicity potential by ingestion and exposure both dermal and inhalation (HTPI and HTPE), TTP, and ATP.

The environmental impact categorization was first made by three groups (Heijungs et al., 1992): CML (Centre of Environmental Science), TNO (Netherlands Organization for Applied Scientific Research), and B&G (Fuel and Raw Materials Bureau). The WAR GUI software package (2008) was developed by the EPA to calculate all eight impact categories. Given a stream composition in a chemical plant, the WAR GUI estimates the individual and the total impact categories and presents the calculation results in the form of Table 11.2. Then, the entire raw material and product streams will be aggregated independently to use Eq. (11.2), which gives the total impacts generated.

11.2.2 Risk assessment index

Failure of a chemical manufacturing plant, mining processing, and refinery industries leads to fire, explosion, and life threats. It is also harmful to the plant, as well as an irreparable spoil for the company's reputation resulting in more economic shortfalls. Some of these tragedies follow: FlixBorough disaster in June 1974 due to the rapid combustion of cyclohexane claimed 28 lives, injured 38 people, and destroyed the entire plant; Mexico City tragedy in November 1984 due to leakage of the LPG supply line resulted in 650 deaths, injuring 7000 people, and affecting 39,000 more individuals; an explosion

Table 11.2 The calculated environmental impacts categories of a chemical stream in a process plant using WAR GUI software package (2008).

Stream compositions chemical, i	Environmental impacts categories								Total impacts of the component, i
	Local toxicological impacts				Global atmospheric impacts				
	HTPI	HTPE	TTP	ATP	GWP	ODP	PCOP	AP	
Chemical, 1									
Chemical, 2									
Chemical, 3									
...									
...									
Chemical, n									
Total impacts of each category									

in Kuwait's Mina Al-Ahmadi Refinery in June 2000 killed 4 people, wounded 49 others, and put the refinery out of operation for months. There are scores of disastrous incidents of this kind in chemical manufacturing industries everywhere in the world. One can find a score of the multimedia reports of such accidents in the U.S. Chemical Safety and Hazard Investigation Board's official website (USCSB, 2020), such as the BP America Refinery Explosion in 2005; Valero Refinery Propane Fire in Sunray, TX, in 2007; Husky Energy Refinery Explosion and Fire in 2018; and Philadelphia Energy Solutions (PES) Refinery Fire and Explosions in 2019. Other mishaps are (1) explosion at ExxonMobil Refinery in 2015 in Torrance, CA (USCSB, 2017); (2) explosion of a propylene gas tank at a Houston manufacturer, TX, which left two people killed and several houses in the plant vicinity damaged in January 2020 (Hanna et al., 2020); and (3) explosion caused by a failure at one of the blast furnaces of ArcelorMittal's steel mill located in northwestern Indiana in the United States in July 2020 (AP NEWS, 2020).

And these dreadful reports are going on and on because the majority of chemical compounds are inflammable, explosive, toxic, and corrosive, therefore inherently hazardous materials. It turns out that risk mitigation in the design step of a process is essential. Several attempts have been made in many different ways to impart risk evaluation methodologies at the primitive stage of process design, especially in the absence of detailed process data. Today, quite a few qualitative and quantitative methods for risk assessment of a chemical process are available. A score of risk assessment methodologies has been tabulated by Jafari et al. (2018), Athar et al. (2019), and Park et al. (2020). Table 11.3 represents risk evaluation methodologies used in industries (Ordouei et al., 2014b; Marhavilas et al., 2011). Unfortunately, most approaches are associated with one or more hindrances that make them unfavorable in the early design phase of designing chemical processes. These include hindrances that are qualitative, time-consuming, require detailed operating data, lack reliable data, and depend on the quality of data collected (e.g., training and experience of safety managers). Above all, the human risk analysis is the critical missing part in the majority of the existing methodologies.

Table 11.3 Major hazard and risk assessment methodologies used in industries (Ordouei et al., 2014b; Marhavilas et al., 2011).

No	Techniques	Formulation	Safety parameters	Application	Advantages	Disadvantages
1	Checklists		1. Historical record 2. Field inspection 3. Experienced individuals 4. Preestablished criteria	Equipment issues and human factors	1. A systematic approach based on questionnaire and checklist 2. Ensures that organizations are complying with standard practices 3. Easy application of the technique	1. Quality dependency on individual experts 2. Complex hazard sources identification 3. A supplement to another method 4. Qualitative information
2	What-if analysis		1. Boundaries of risk-related information 2. Problem identification (e.g., risk type, environmental impacts, economy) 3. Determination of subdivision of item 2 (e.g., location, tasks, subsystems) 4. Asking what-if questions	Equipment issues and human factors	1. Identifies hazards, hazardous situations, or specific accident events 2. Relatively easy to use/not expensive 3. Applicable to any activity or system 4. Most often is used to supplement techniques	1. Quality dependency on individual experts, documentation, and the experience of the review teams 2. Determines only hazard consequences 3. A loosely structured assessment 4. Qualitative
3	Safety audits (SA)		Operational safety programs	Administrative	1. Cheap 2. User friendly 3. Diagnoses equipment conditions or process procedures, which results in casualties, environmental impacts, or property loss	1. Not useful for the detection of hazard sources for technical installation 2. The outcome is a suggestion to management for various safety aspects of options 3. Qualitative
4	Task analysis (TA)		1. Worker's tasks 2. Interpersonal interaction 3. Human-machine interaction	Administrative	1. Detailed information can be provided 2. A well-structured picture of the work process can be built. 3. Safety-critical tasks can be identified	1. Time-consuming 2. Dependency on safety experts or production engineers 3. Qualitative

(continued)

Table 11.3 (Cont'd)

No	Techniques	Formulation	Safety parameters	Application	Advantages	Disadvantages
5	Sequentially timed event plotting (STEP)		Sequence of events	Operation	1. A valuable overview of the timing and sequence of events that contributed to the accident. 2. Plotting the sequence of events that contributed to the accident	1. Time-consuming 2. Qualitative
6	HAZOP		1. Pressure 2. Temperature 3. Flow rate 4. Equipment 5. Interlocks	Detailed engineering, operation, retrofitting	1. A systematic, documentary, imaginative methodology 2. Identifies deviations and causes of undesirable consequences 3. Recommends countermeasures to mitigate frequency and consequences of the deviations 4. Determines hazard causes and consequences 5. Very popular 6. Applicable to any system or procedure 7. Highly structured assessment relying on guide words to generate a comprehensive review	1. Expensive 2. Difficult to use 3. Requires a multidisciplinary team of experts 4. Time-consuming 5. Qualitative
7	Proportional-risk assessment (PRA)	$R = P.S.F$ R = Risk P = Probability S = Severity F = Frequency	1. Probability 2. Severity 3. Frequency	Operation, Construction	1. User friendly 2. Quantitative 3. Mathematical risk evaluation 4. Safe results, based on the recorded data of undesirable events or accidents 5. Incorporated in databases 6. Can be used in other risk-assessment techniques 7. Predicts hazards, unsafe conditions, and also prevents fatal accidents	1. Dependency on precisely recording the undesirable events 2. Time-consuming 3. Cannot be used at the early design stage 4. Dependency on safety experts or production engineers 5. Hard to find probability function

8	Decision matrix risk assessment (DMRA)	$R = S.P$ R = Risk P = Probability S = Severity	1. Probability 2. Severity	Operation	1. User friendly 2. Good data quality 3. Combination of risk analysis and risk evaluation 4. Predicts hazards, unsafe, and undesirable conditions 5. Prevents fatal accidents 6. Quantitative and graphical method 7. Facilitates prioritization and managing key risks	Dependency on safety experts or production engineers
9	Quantitative risk measure of societal risk (SRE)	$R = \{(S_k, F_k, N_k)\}$ S_k = kth accident scenario F_k = Frequency N_k = Consequence	1. Frequency 2. Severity	Operation	1. User friendly 2. Considers both public and worker risk 3. Contains a historical record of incidents 4. A quantitative and graphical technique 5. Encompasses the criteria for judging the tolerability of risk	1. Needs qualified safety managers to document the undesirable events 2. Time-consuming
10	Fault-tree analysis (FTA)		1. Equipment failures 2. Human errors 3. External events 4. Constructed from even and gates (AND/OR)	Operation	1. Models combinations of equipment failures, human errors, and external conditions causing an accident 2. Dependency on experts 3. Requires brainstorming meetings 4. Field inspections 5. Quantitative and qualitative 6. A highly structured method 7. Applicable for all type of risk-assessment 8. A practical root-cause analysis	1. Very complicated 2. Difficult to use 3. Time-consuming 4. Expensive 5. A system-level risk-assessment technique

As mentioned earlier, most chemicals are flammable, explosive, and toxic. Almost 70% of impacts are from explosion related to fire (Lees, 1996); however, both fire and explosion are less influential than toxicity based on the number of affected people (Belke, 2000). Toxic release to nature as a result of industrial accidents has adverse effects on the environment and human society. As a case at the point, the disaster in the Union Carbide Pesticide Plant on December 02, 1984, in Bhopal, India, was a massive warning to the industrial world. At least 30 tons of methyl isocyanate and some other poisonous gases were discharged into the shanty towns in the vicinity of the plant. And more than 600,000 people were exposed to the gas cloud that night. The estimates of the death toll as a result of this accident by government figures refer to an assessment of 15,000 killed over the years, and the plant is still abandoned (Taylor. 2014).

Koller et al. (2001) reviewed and classified the significant characteristics of 13 index methods and applied the theoretical concepts of 9 proper procedures at the conceptual design of 9 distinctive processes. They also suggested merging different techniques and using the history of previous incidents and accidents in the process plants to assess the risk in lack of information on equipment and plant. Tixier et al. (2002) also identified 62 safety risk analysis methods in industrial plants and identified the lack of human risk analysis in classical risk analysis. Then, they concluded there was no uniqueness of methods of accomplishment the risk analysis and, therefore, recommended combining of several methodologies.

Thus, Ordouei et al. (2014b) found it crucial to integrate risk to human society and chemical toxicity into their risk assessment model. They also justified the shortcomings pointed out by Keller and Tixier in the development of their risk assessment methodology. This model (Ordouei et al., 2014b) is a robust tool for ISD in chemical processes. The ISD is a systematic tactic to minimize the risks to the safety of a process plant, human, environment, and equipment during the design and operation of the process (Hendershot, 2011).

The proposed risk index by Ordouei et al. (2014b) is compliant with the ISD's four policies introduced by the Center for Chemical Process Safety (2009), which are as follows:

- *Intensification*: Minimization of the hazardous compound within a process plant.
- *Substitution*: Substitution of hazardous materials with benign compounds.
- *Moderation*: Handling and transporting of hazardous chemicals under reduced risk conditions (e.g., dilution, refrigeration, etc.).
- *Limitation*: Diminishing the probability of accidents and associated damages, for example, applying interlocking commands for process control.

Ordouei's (2014b) index mitigates the effect of obtaining unrealistic results for certain instances. It can be used at the primitive stage of chemical process designs, which are presented by the following equations:

$$(R.I)^P = \sum_i \sum_j M_j \times f_i \times H_i \times x_{i,j} \tag{11.3}$$

$$(R.I)^W = \sum_k \sum_l M_l \times f_k \times H_k \times x_{k,l} \tag{11.4}$$

where

- *R.I* is the Risk Index defined as the number of affected people per year. The *R.I* represents the maximum potential risks to society:

- The superscripts *P* and *W* stand for product and waste streams, respectively.
- The subscripts *i* and *k* denote the *chemical contents* in product and waste streams, respectively. As such, the superscripts *j* and *l* designate the product and waste *streams* within a process, respectively.
- M_j and M_l are chemical inventories in product and waste streams, respectively (the maximum 1 month of plant production in ton).
- f_i and f_k are frequency of accidents for chemical components *i* and *k* in the number of accidents per year.
- H_i and H_k are hazard effects in the number of people affected per ton of chemical components *i* and *k*.

Thus, the total risk $(R.I)^T$ is defined as the summation of the risks associated with product and waste streams as follows:

$$(R.I)^T = (R.I)^P + (R.I)^W \tag{11.5}$$

When both sides of Eq. (11.5) are divided by the annual plant capacity, a normalized risk index will be obtained, which is independent of the process size:

$$\frac{(R.I)^T}{\sum_s P_s} = \frac{(R.I)^P}{\sum_s P_s} + \frac{(R.I)^W}{\sum_s P_s} \tag{11.6}$$

where subscript *s* denotes the stream number of products, Eq. (11.6) enables process and safety engineers to compare two or more processes with different production capacities. The first term of the right-hand side of Eq. (11.6) represents the risks associated with product streams per ton, and the second term describes the risks associated with waste streams per ton. In the cases that all circumstances in design alternatives are identical, the term $\dfrac{(R.I)^P}{\sum_j P_j}$ would be eminent for the design ranking.

The Ordouei's (2014b) risk index has broad applications from small to large scales complex plants, such as chemical, refinery, and petrochemical processes. This risk model is one of the three pillars of the CSI methodology. This means that a chemical process endorsed by the CSI is inherently safe. A detailed method and databank for *f* and *H* factors have been presented by Ordouei et al. (2014b) and Al-Sharrah et al. (2007). To test the model, the chlorination of methane process has been studied, as shown in Fig. 11.2A,B. The risk index identified that the base model is inherently safer relative to the alternative process.

11.2.3 Energy impact index

A large portion of the expenses of each family and business goes to energy consumptions in their residential and commercial buildings and/or manufacturing plants. Energy generation from fossil fuels (e.g., coal, natural gas, and oil) in power plants in the form of heat and electricity causes GHGs emissions to the atmosphere and, consequently, global warming or climate change. Study reveals the following (Striebig et al., 2016):

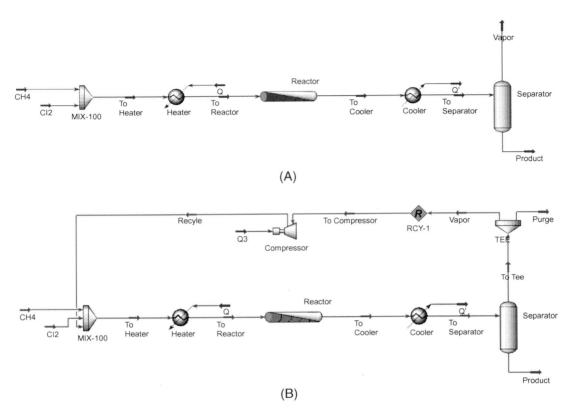

FIG. 11.2 Chlorination of methane.
(A) Original design without recycling. (B) Alternative design with recycling (Ordouei et al., 2014b).

- The CO_2 emission to the atmosphere has reached an amount of 165×10^9 metric tons (165 Gigatons) since the industrial revolution in the 1750s. Therefore, the main constituent of climate change is anthropogenic.
- About 2.2×10^9 tons (Gigatons) CO_2 per year is sunk to the planet's oceans, increasing the acidity of the oceans (i.e., a decrease in the ocean's PH), which is harmful to aquatic life.

In 2011, the highest source of GHG emissions in the United States was electricity production by 33% and the second one was the industrial sector by 20% (Striebig et al., 2016).

Thus, it is imperative to find technological solutions to battle climate change and also to provide a vital tool to estimate the impacts of GHGs emissions on the environment. Several methodologies have been proposed by scientists and engineers to fulfill this task, as summarized in Table 11.4. It can be seen from this table that the majority of the methodologies are merely applicable to premises (not industries), and the rest methods are either inaccurate or require a score of information and, therefore, not suitable at the process design phase.

Fortunately, a simple and quantitative energy efficiency methodology, which demands minimum data, such as fuel characteristics, does exist (Ordouei et al., 2018). The proposed method established a

Table 11.4 A summary of existing methodologies for energy efficiency calculations in different sectors.

No.	Methodology	Energy efficiency methodologies		
		Description	Reference	Disadvantage(s)
1	The Carbon Trust	It offers various conversion factors for GHG emissions from different sources within a business and public energy consumption, which is based on the annual interactive report published by the UK Government, Department for Business, Energy & Industrial Strategy. The calculation basis of the conversion factors is the energy type, bioenergy, passenger vehicle, etc.	The Carbon Trust, (2013); Department for Business, Energy & Industrial Strategy, (2013)	Only applicable to premises, not chemical industries.
2	Energy Usage Index (EUI)	The EUI is a methodology used for calculation of energy consumption within premises.	Oregon Department of Energy, (2013)	Not applicable to industries (including chemicals).
3	Energy Index (EI)	The EI uses software for energy index calculation in building per unit temperature per unit area. The lower the index, the more energy efficient the premise is.	Texas Instruments, (2013)	Not applicable to industries (including chemicals).
4	Energy efficiency index (ODEX)	ODEX is a useful method for energy savings related to operating industries, for example, chemical, steel, paper, nonferrous, cement, food, machinery, transport equipment, textile, and other nonmetallic plants. It is a weighted average of the contribution of the industrial category to the total energy consumption in a year.	Energy efficiency index (ODEX), (2012)	The ODEX does not give any information on gas emissions to the atmosphere and its impacts on the environment. Besides, it requires scores of information that are not readily available in the design stage.
5	Energy Development Index (EDI)	The International Energy Agency (IEA) presented four indicators for the evaluation of the energy function in human development as a tool to help estimate the Human Development Index proposed by the United Nations Development Programme (UNDP). The EDI is published annually by the World Energy Outlook for the international awareness and monitoring progress of particular countries from energy poverty toward modern energy access.	International Energy Agency (IEA), (2012)	It is neither applicable to industries (including chemicals), nor does it give any information on gas emissions to the atmosphere and its impacts on the environment.
6	Energy Efficiency Design Index (EEDI)	The EEDI is a technical index for the improvement of the energy efficiency of ships and vessels.	Germanischer Lloyd SE, (2013)	Not applicable to industries (including chemicals) and buildings.
7	WAR Algorithm	A software package developed by EPA (Section 11.2.1), which provides the users with an option to calculate the impact of the energy consumption in PEI/h.	WAR GUI, (2008)	It provides no information about NO_X and SO_X emissions. Neither does it distinguish between the efficiencies of steam and power generation from boilers and turbines, respectively.

relationship between heat and power generation/consumption and the emission rates (e.g., CO$_2$, NO$_2$, and SO$_2$) to the atmosphere. This methodology has broad applications in commercial and residential buildings, chemical processes, power plants, and other manufacturing industries, from design to operating stages. The energy efficiency index proposed by Ordouei is also one of the three pillars of the CSI methodology to estimate the impacts of energy used in an industry on the environment. A detailed study of this methodology has been provided by Ordouei et al. (2018). The index is shown below:

$$\dot{i}_e = \sum_k \sum_j \sum_i x_i \times \left\{ \frac{\dot{Q}_{i,k}}{\eta_k \times HV} \right\} \times \left(\frac{MW_j}{MW_i} \right) \quad (11.7)$$

where

η_k = Total efficiency of the process unit k (e.g., heat exchanger, incinerator, electromotor, etc.)
$\dot{Q}_i$ = The fraction of heat flow attributed to the component i in kJ/h (i = C, S, and N)
HV = The heating value of a fuel (kJ/kg fuel)
x_i = The mass fraction of component i
MW_j = The atomic weight of the component i (kg/kgmol). Also, the superscripts j denotes GHG emitted to the atmosphere (j = CO$_2$, SO$_2$, NO$_2$).
$\dot{i}_e$ = The rate of GHG emission, j, to the atmosphere in kg per hour due to either heat or power generation/consumption in the process unit k.

Eq. (11.7) estimates the CO$_2$, SO$_2$, and NO$_2$ generation as a result of fossil fuel combustion for both steam and electricity generation in thermal power plants. Dividing both sides of Eq. (11.7) by the annual plant capacity, a normalized energy efficiency index will be obtained, which is independent of the plant size:

$$\frac{\dot{i}_e}{\sum_s P_s} = \frac{\sum_k \sum_j \sum_i x_i \times \left\{ \frac{\dot{Q}_{i,k}}{\eta_k \times HV} \right\} \times \left(\frac{MW_j}{MW_i} \right)}{\sum_s P_s} \quad (11.8)$$

where subscript s denotes the stream number of products, Eqs. (11.7) and (11.8) are valid for an onsite (direct) power source in large-scale process plants, as the energy loss is negligible. However, if the *electricity* is received through off-site (indirect) power lines, a slight amount of electricity is dissipated into heat (about seven percent) due to the losses in the transmission and distribution grid (Striebig et al., 2016). Thus, the following waste factor has to be added to Eq. (11.8):

$$f_{trans.} = \frac{1}{(1-0.07)} = 1.075$$

to give Eq. (11.9):

$$\dot{i}_{e.c} = f_{trans.} \times \dot{i}_e = 1.075 \times \sum_k \sum_j \sum_i x_i \times \left\{ \frac{\dot{Q}_{i,k}}{\eta_k \times HV} \right\} \times \left(\frac{MW_j}{MW_i} \right) \quad (11.9)$$

where

i_e,c represents the corrected rate of GHG emissions.

$f_{trans.}$ is the correction factor for transmission loss from the power plant to the process plant.

Remember that Eq. (11.7) is valid for both steam and direct power generation/consumption, and Eq. (11.9) is valid merely for indirect power generation/consumption in chemical or power plants. Both Eqs. (11.7) and (11.9) have the following pros over the existing methodologies:

- Simple and user friendly for the estimation of GHG emission rates to the environment at any stage of process design and operation.
- Accurate, so that it calculates the emission rates of CO_2, SO_2, and NO_2 from fossil fuel combustion to the environment.
- Compatible with minimum available data, such as the characteristics of the fossil fuel, power consumption within the process.
- Applicable to commercial and residential buildings, mining and food processing, brewery, refinery, chemical, petrochemical, and power plants, as well as electric appliances in houses.

This index is even more accurate than that of EPA. To test this methodology, the process of chlorination of methane has been studied, as shown in Fig. 11.3A,B,C. The energy efficiency index determined the process of (Fig. 11.3B) as the most energy-efficient process, which has the least GHGs emissions to the environment.

11.3 Discussion

The CSI has a wide range of applications. The capability of the CSI model has been investigated by applying its components individually and combined with various case studies, as explained below:

1. The sole risk assessment indicator of the CSI was employed to identify the Inherently Safest Design (ISD) process for the *chlorination of methane*, as shown in Fig. 11.2A,B (Ordouei et al., 2014b). The risk assessment model indicated that the base design (Fig. 11.2A) is way less risky than the alternative process.
2. The energy intensity index of the CSI methodology was also individually used to determine the most energy-efficient process among other design arrays for the *chlorination of methane*, as shown in Fig. 11.3A,B,C (Ordouei et al., 2018). The proposed energy index method identified that Fig. 11.3B is the most energy-efficient process among all three designs.
3. The CSI approach has been used as a whole in research and product design. New products or property of new products are usually characterized either by lab experiments or software modeling. The CSI can then be applied to the product alternatives to ascertain the most sustainable goods. In a gasoline blending process; the octane number of three *gasoline blends* was determined utilizing Aspen HYSYS, followed by employing the CSI tool, which identified M5 blend (95% gasoline + 5% methanol) the most sustainable and profitable product (Ordouei et al., 2014c, 2015).
4. The CSI methodology can be extensively hired at the early design stage of a complex chemical process, such as a refinery plant. Ordouei et al. (2014d, 2016) have simulated two *hydrogenation*

186 Chapter 11 Composite sustainability indices (CSI)

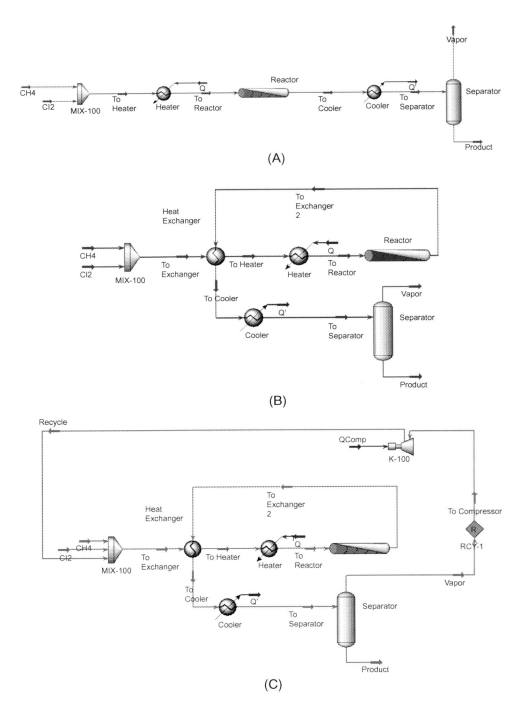

FIG. 11.3 Chlorination of methane.
(A) The chlorination design without a heat exchanger, (B) chlorination design with a heat exchanger, (C) chlorination design with a heat exchanger and recycling (Ordouei and Elkamel, 2017).

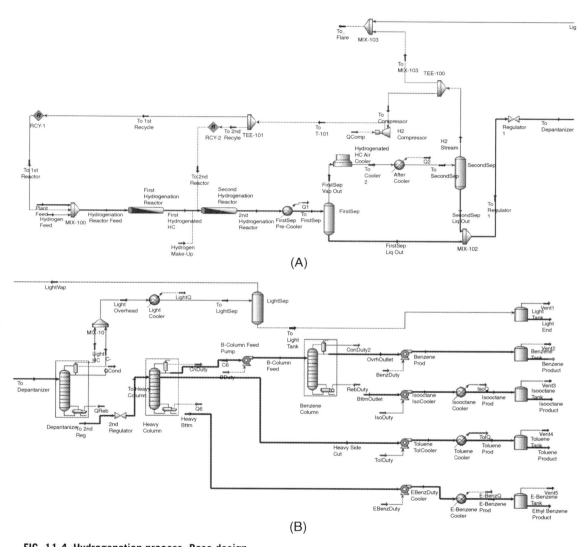

FIG. 11.4 Hydrogenation process: Base design.

(A), Reaction, and phase separation. (B) Purification. (Ordouei et al., 2016).

process plants (Figs. 11.4A,B and 11.5A,B), implemented the CSI methodology, and identified the base design (Fig. 11.4A,B) as the more sustainable and profitable process.

5. The *Cradle-to-cradle* concept (Fig. 11.6) has been implemented by Ordouei and Elkamel (2017) to design a *thinner recovery plant* in a typical auto industry through Aspen HYSYS as an alternative to the traditional painting unit. The CSI methodology was then utilized and revealed a significant environmental impact reduction in the alternative process and an escalation of the plant profitability due to sustainable retrofitting of the painting unit (Fig. 11.7A,B).

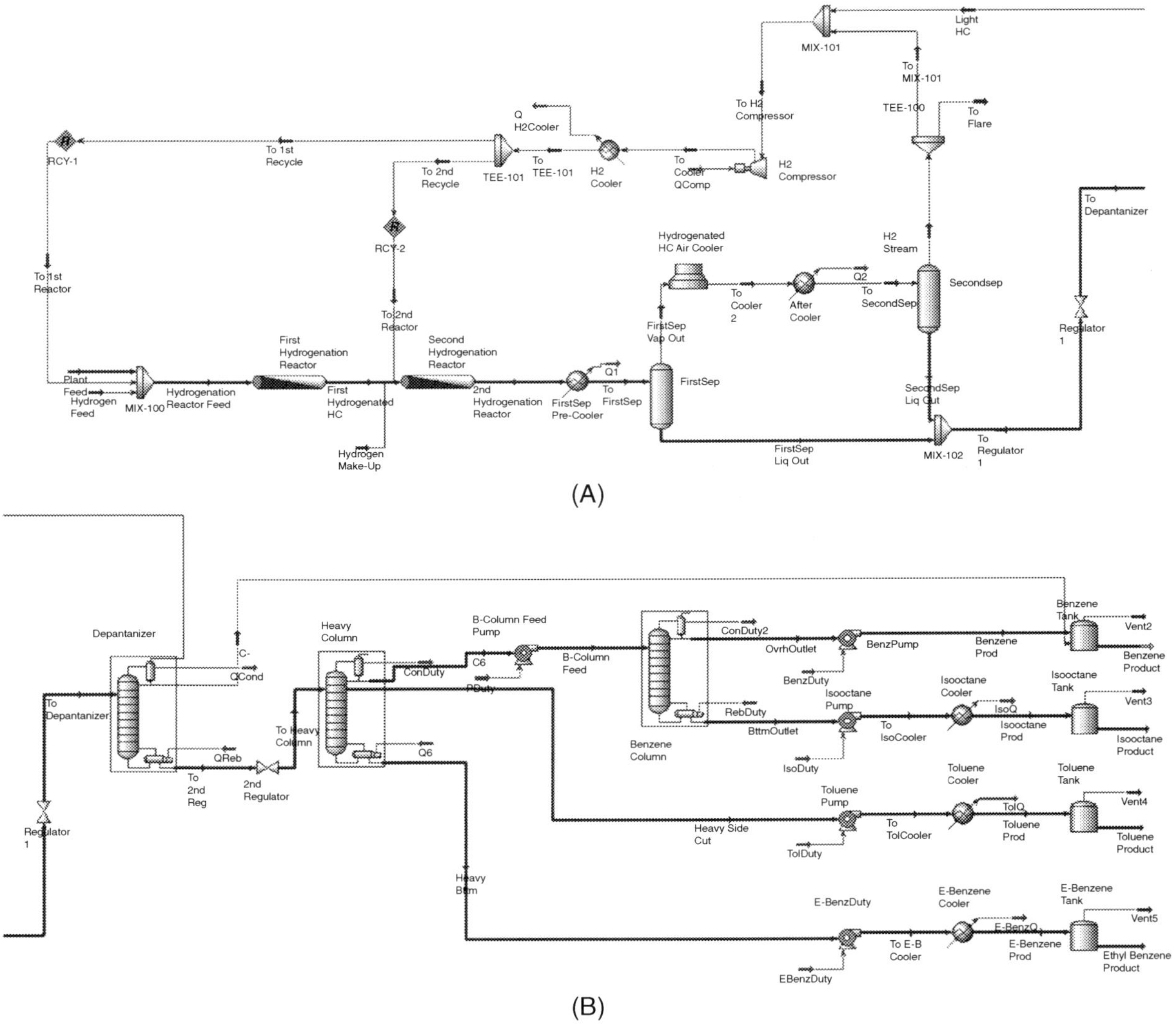

FIG. 11.5 Hydrogenation process: Alternative design.

(A) Reaction, and phase separation. (B) Purification. (Ordouei et al., 2016).

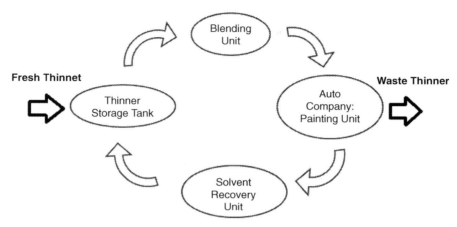

FIG. 11.6 Cradle-to-cradle integrated process (Ordouei and Elkamel, 2017).

6. The CSI has been successfully applied to the concept of Eco-Industrial Park (EIP) by Ordouei (2014a), as shown in Fig. 11.8A,B.
   a. In the EIP design, the hydrogen and CO_2 gases are delivered to the gasification process from Chlor-alkali and thermal power plants, respectively. The biomass from the farm and food wastes is transferred to the gasifier by a belt conveyer, which consumes negligible electricity power. A series of complex chemical reactions take place in the gasifier to generate large CO_2 content producer gas, which is the raw material for decarbonization processes. The application of the CSI (excluding energy index) indicates that the EIP is a sustainable process (Table 11.5), due to much lower risk and environmental impacts.
7. Life Cycle Assessment (LCA) is a world-wide accepted methodology. The LCA is a pathway from concept to research and development, from manufacturing to marketing and distribution. The CSI methodology is the right fit for the "product manufacturing" and "material processing" steps in the LCA.

From the above discussion, we learn that it is possible to add or remove more metrics to or from the CSI approach, depending on the nature of the product or process involved. For instance, in designing gasoline blends (Ordouei et al., 2014c, 2015), "octane number" and "mileage loss" metrics were added to the CSI, and the energy impact indicator was lifted from it due to consumption of the same amount of electric power in all of the blends.

It is also crucial to understand that the purity of a chemical product cannot be added to the CSI as a tenet, as the product purity affects the price, which is part of profitability analysis. In other words, in the CSI approach, purity has a relatively economic nature than a technical one.

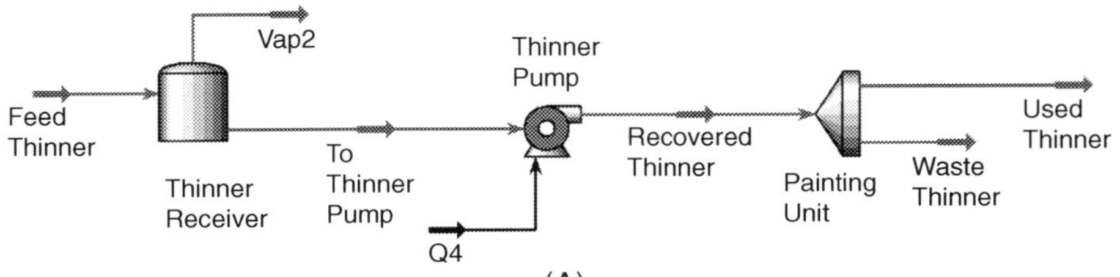

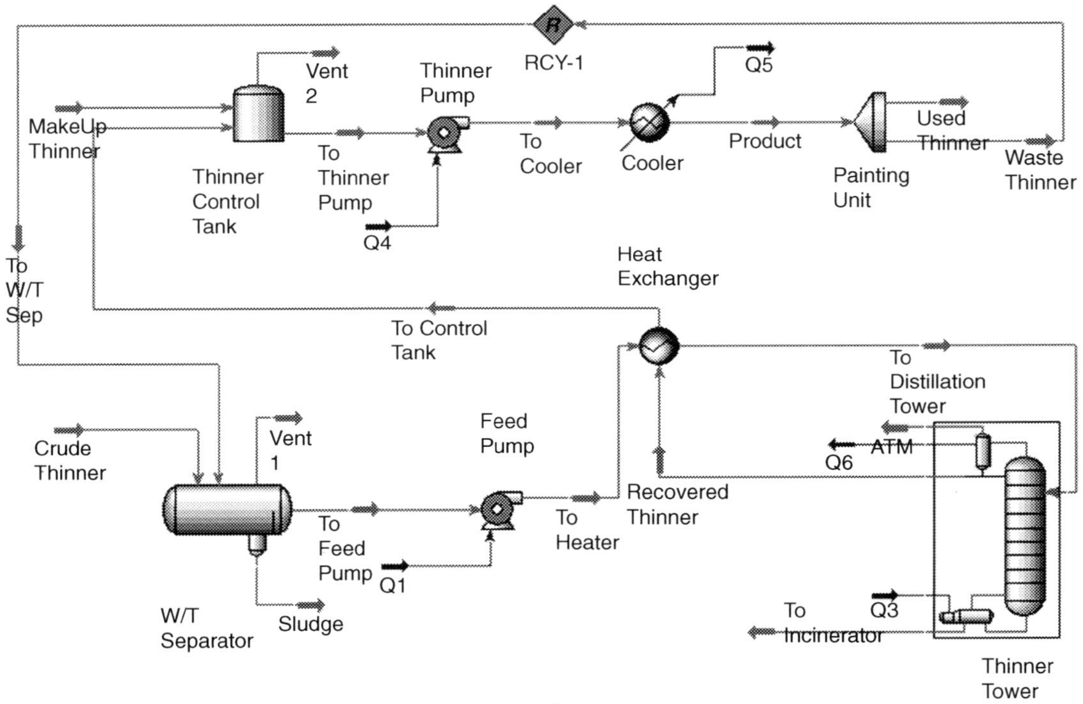

FIG. 11.7 A typical auto manufacturing factory.
(A) A traditional paint handling unit; (B) thinner recovery plant in a paint handling unit (Ordouei and Elkamel, 2017).

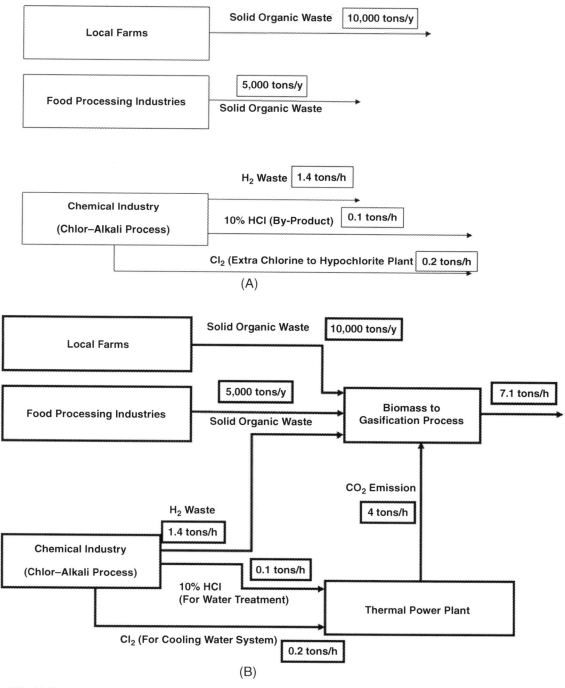

FIG. 11.8

(A) Waste stream scheme; (B) eco-industrial park (EIP) design for making producer gas (Ordouei et al., 2014a).

Table 11.5 The result of the CSI application on the EIP design.

Metrics	Waste streams scheme (Fig. 11.8A)	Eco-industrial park (EIP) design (Fig. 11.8B)
Potential environmental impacts, PEI/h	2280	10.4
Risk index, No. of affected people/year	1.4	0

11.4 Conclusions

We learned an essential lesson from the COVID-19 pandemic, which took the world plenty of time to be well prepared to fight it. Now, we have to intensify the battle against climate change by pushing the green technology to its cutting edge, supporting the scientists to find solutions, and share them with the world. The CSI presented in this chapter is a ground-breaking strategy that uses the existing technology and can thrust it to its limits. The CSI is a tactic for the measurement and justification of the sustainability performance of chemical processes and ranking them. It consolidates dynamically the complex data collected from design basis and process modeling and simulation into streamlined evidence that can be easily scrutinized. The CSI is also a robust means to minimize the pollutants and GHG emissions from chemical manufacturing industries to the ecosystem and diminish the risks associated with these industries to the society and the plants.

The CSI approach has other advantages, which make it stand out among different existing methodologies, which are as follows:

1. Applicable to following processes from research and conceptual design to production phases: fossil fuel, biofuel, paint, solvent, beverage, refinery, gas processing, petrochemical, biomass conversion, and mining processing industries.
2. Genuine and suitable at the initial stage of process design and minimum available data when the impacts of the decision making are exceptionally high.
3. Quantitative, simple, and user-friendly methodology.
4. Based on the engineering principles, standard tests, and reliable data presented by the EPA.
5. Delivers consistent and reliable results.
6. Capable of assessing and ranking the multiple process alternatives, in terms of energy efficiency, environmental consciousness, and inherent safety.
7. An appropriate strategy for LCA methodology as a manufacturing and processing module.

Reference

Al-Sharrah, G.K., Edwards, D., Hakinson, G., 2007. A new safety risk index for use in petrochemical planning. Trans IChemE 85 (B6), 533–540.

Amimi, M., Bienstock, C.C., 2014. Corporate sustainability: an integrative definition and framework to evaluate corporate practice and guide academic research. J. Clean. Prod 76, 12–19.

AP NEWS, 2020. No Injuries After Explosion Rocks NW Indiana Steel Mill. https://apnews.com/396131ae4df6d dd846c7bbe969c05d06 (Accessed July 18, 2020).

Athar, M., Shariff, A.M., Buang, A., 2019. A review of inherent assessment for sustainable process design. J. Clean. Prod 233, 242–263.

Batterham, R.J., 2006. Sustainability-the next chapter. Chem. Eng. Sci 61, 4188–4193.

Belke, J.C., 2000. Chemical Accident Risk in the U.S. Industry - A Preliminary Analysis of Accident Risk Data from U.S. Hazardous Chemical Facilities (the United States Environmental Protection Agency, Washington, DC, USA.

Carbon Trust Co., 2013. *Energy and Carbon Conversions 2013 Update*, Conversion Factors Fact Sheet.

Carew, A.L., Mitchell, C.A., 2008. Teaching sustainability as a contested concept: capitalizing on variation in engineering educators' conceptions of environmental, social, and economic sustainability. J. Clean. Prod 16, 105–115.

Center for Chemical Process Safety, 2009. Inherently Safer Chemical Processes: A Life-Cycle Approach, Second ed. John Wiley & Sons, Hoboken, NJ.

Cobb, C., Schuster, D., Beloff, B., Tanzil, D., 2007. Benchmarking sustainability. Chem. Eng. Prog. 103, 38–42.

Department for Business, Energy & Industrial Strategy. 2013. https://www.gov.uk/government/collections/government-conversion-factors-for-company-reporting (Accessed July 20, 2020).

Dow Chemical Co. 2021. https://corporate.dow.com/en-us/about/legal/public-policy/global/sustainability.html (Accessed April 19, 2021).

Energy efficiency index (ODEX). 2012. https://www.eea.europa.eu/data-and-maps/figures/energy-efficiency-index-odex-in-2 (Accessed July 20, 2020).

EPA, 2010. Science Inventory: Life-Cycle Assessment. US Environmental Protection Agency https://cfpub.epa.gov/si/si_public_record_report.cfm?Lab=NRMRL&TIMSType=&count=10000&dirEntryId=156704&searchAll=&showCriteria=2&simpleSearch=0&startIndex=20001 (Accessed July 8, 2020).

EPA, 2012. Waste Reduction Algorithm: Chemical Process Simulation for Waste Reduction. US Environmental Protection Agency https://www.epa.gov/chemical-research/waste-reduction-algorithm-chemical-process-simulation-waste-reduction (Accessed July 18, 2020).

FTSE4Good Environmental Leaders Europe 40 Index. 2020: https://www.ftserussell.com/products/indices/ftse4good (Accessed July 18, 2020).

Germanischer Lloyd, SE, 2013. *Rules for Classification and Construction, VI Additional Rules and Guidelines, 13 Energy efficiency*, Guidelines for Determination of the Energy Efficiency Design Index. Germanischer Lloyd SE, Hamburg.

Hanna, J., Alonso, M., Johnston, C., 2020. 2 People Killed in an Explosion at a Houston Manufacturer that Shook the City and Damaged Homes. CNN. https://www.cnn.com/2020/01/24/us/texas-houston-explosion/index.html (Accessed July 18, 2020).

Heijungs, R., Guinée, J.B., Huppes, G., Lankrejer, R.M., Udo de Haes, H.A., Sleeswijk, A.W., 1992. Environmental Life Cycle Assessment of Products Guide. Centre of Environmental Science, Leiden.

Hendershot, D.C., 2011. Inherently Safer Design; An Overview of key elements. Profess. Saf.

Hilaly, A.K., Sikdar, S.K., 1994. Pollution balance. A new methodology for minimizing waste production in manufacturing processes. J. Air Waste Manage. Assoc. 44, 1303–1308.

Hilaly, A.K., Sikdar, S.K., 1995. Pollution balance method and the demonstration of its application to minimizing waste in a biochemical process. Ind. Eng. Chem. Res. 34, 2051–2059.

Hossain, K., Khan, F., Hawboldt, K., 2011. IECP – an approach for integrated environmental and cost evaluation of process design alternatives and its application to evaluate different NOx prevention technologies in a 125 MW thermal power plant. Energy Sustain. Dev. 15 (1), 61–68.

International Energy Agency, 2012. World Energy Outlook. http://www.worldenergyoutlook.org/resources/energydevelopment/theenergydevelopmentindex/ (Accessed July 20, 2020).

IPCC (Intergovernmental Panel on Climate Change), 2013, IPCC 5th Assessment Report.

Jafari, M.J., Mohammadi, H., Reniers, G., Pouyakian, M., Nourai, F., Torabi, S.A., Rafiee Miandashti, M, 2018. Exploring inherent process safety indicators and approaches for their estimation: a systematic review. J. Loss Prev. Process Ind 52, 66–80.

Khan, F.I., Sadiq, R., Veitch, B., 2004. Life Cycle iNdeX (LInX): a new indexing procedure for process and product design and decision-making. J. Clean. Prod. 12 (1), 59–76.

Koller, G., Fischer, U., Hungerbühler, K., 2001. Comparison of methods suitable for assessing the hazard potential of chemical processes during early design phases. Trans IChemE 79 (Part B), 157–166.

Lees, F.P., 1996, Second ed Loss Prevention in the Process Industries: Hazard Identification, Assessment, and Control1. Butterworth-Heinemann, UK.

Li, C., Zhang, X., Zhang, S., 2009. Environmentally conscious design of chemical processes and products: multi-optimization method. Chem. Eng. Res. Des. 87, 233–243.

Marhavilas, P.K., Koulouriotis, D., Gemeni, V., 2011. Risk analysis and assessment methodologies in the worksites: on a review, classification, and comparative study of the scientific literature of the period 2000–2009. J. Loss Prev. Proc. Ind. 24, 477–523.

Mascarenhas, A., Coelho, P., Subtil, E., Ramos, T.B., 2010. The role of common local indicators in regional sustainability assessment. Ecol. Indic. 10, 646–656.

Ordouei, H.M., 2014. Integration of Safety, Pollution, and Risk Simple Indices into the Design and Retrofit of Process Plants Ph.D. Thesis). The University of Waterloo. Waterloo, Ontario, Canada.

Ordouei, M.H., Alhajri, I.H., Elkamel, A., Dusseault, M.B., 2014. Utility Assessment of Canadian Gasoline Blends, Proc: 6th International Conference on Chemical, Biological and Environmental Engineering (ICBEE 2014) September 8-9. Geneva, Switzerland.

Ordouei, M.H., Elkamel, A., 2017. New composite sustainability indices for cradle-to-cradle: case study on thinner recovery from waste paint in auto industries. J. Clean. Prod 166, 253–262.

Ordouei, M.H., Elkamel, A., Al-Sharrah, G, 2014b. New simple indices for risk assessment and hazard reduction at the conceptual design stage of a chemical process. Chem. Eng. Sci 119, 218–229.

Ordouei, M.H., Elkamel, A., Al-Sharrah, G, 2018. A new simple index for the estimation of energy impacts on environment. Environ. Eng.Manage. J. 17 (2)), 357–370.

Ordouei, M.H., Biglari, M., Mujiburohman, M., 2014d. A novel process design and simulation for hydrogenation plant in refineries. Energ. Source Part A 36 (22), 2474–2481.

Ordouei, M.H., Elkamel, A., Dusseault, M.B., Al-Hajri, I., 2015. New sustainability indices for product design employing environmental impact and risk reduction: case study on gasoline blends. J. Clean. Prod 108 (Part A), 312–320.

Ordouei, M.H., Elsholkami, M., Elkamel, A., Croiset, E, 2016. New composite sustainability indices for the assessment of a chemical process in the conceptual design stage: case study on hydrogenation plant. J. Clean. Prod 124, 132–141.

Oregon Department of Energy, 2013. Collecting Data for Energy Usage Index. http://www.oregon.gov/energy/cons/pages/sb1149/schools/eui.aspx (Accessed July 20, 2020).

Park, S., Xu, S., Rogers, W., Pasman, H., El-Halwagi, M.M., 2020. Incorporating inherent safety during the conceptual process design stage: a literature review. J. Loss Prev.Process Ind. 63, 104040.

Ramos, T.B., Caeiro, S., 2010. Meta-performance evaluation of sustainability indicators. Ecol. Indic. 10, 157–166.

Rathnayaka, S., Khan, F., Amyotte, P, 2014. Risk-based process plant design considering inherent safety. Saf. Sci. 70, 438–464.

Saaty, T.H., 2008. Decision making with the analytical hierarchy process. Int. J. Serv. Sci. 1 (1), 83–98.

Striebig, B.A., Ogundipe, A.A., Papadakis, M., 2016. Engineering Applications in Sustainable Design and Development. Cengage Learning, Canada.

Sugiyama, H., Fischer, U., Antonijuan, E., Hoffmann, V., Hirao, M., Hungerbuhler, K., 2009. How do different process options and evaluation settings affect economic and environmental assessments? A case study on methyl methacrylate (MMA) production process. Process Saf. Environ. Protect. 87 (6), 361–370.

Taylor, A., 2014. Bhopal: The World's Worst Industrial Disaster, 30 Years Later. *The Atlantic*. https://www.theatlantic.com/photo/2014/12/bhopal-the-worlds-worst-industrial-disaster-30-years-later/100864 / (Accessed July 18, 2020).

Texas Instruments. 2013. http://education.ti.com/en/us/products/computer_software/connectivity-software/ti-connect software/features/features-summary (Accessed July 20, 2020).

Tixier, J., Dusserre, G., Salvi, O., Gaston, D, 2002. Review of 62 risk analysis methodologies of industrial plants. J. Loss Prev. Process Ind 15, 291–303.

USCSB. 2017. Animation of 2015 Explosion at ExxonMobil Refinery in Torrance, CA. https://www.youtube.com/watch?v=JplAKJrgyew (Accessed July 18, 2020).

USCSB. 2020. https://www.csb.gov/videos/ (Accessed July 18, 2020).

Van Lante, H., van Til, J.I., 2008. Articulation of sustainability in the emerging field of nanocoatings. J. Clean. Prod. (16), 967–976.

WAR GUI (WAste Reduction Algorithm Graphical User Interface), 2008. Version 1.0.17. Program for the Reduction of Waste in Chemical Processes. Environmental Protection Agency (EPA). Download WAR GUI Software: https://www.epa.gov/chemical-research/waste-reduction-algorithm-chemical-process-simulation-waste-reduction (Accessed July 11, 2020).

Young, D.M., Cabezas, H., 1999. Designing sustainable processes with simulation: the waste reduction (WAR) algorithm. Comput. Chem. Eng. 23, 1477–1491.

Young, D., Scharp, R., Cabezas, H., 2000. The waste reduction (WAR) algorithm: environmental impacts, energy consumption, and engineering economics. Waste Manage. 20, 605–615.

CHAPTER 12

Sustainability assessment using the ELECTRE TRI multicriteria sorting method

Luis C. Dias

Univ Coimbra, CeBER, Faculty of Economics, Coimbra, Portugal

12.1 Introduction

Sustainable development is a top priority for countries and international organizations (European Union, United Nations, etc.). Firms are also aware of the importance of their impact on society and the environment, knowing that being more responsible helps their business (Carroll and Shabana, 2010). As one cannot manage what one cannot measure, sustainability assessment (or evaluation, no distinction is made in this chapter) is increasingly important. Assessment tools can be used for comparative assessments, for example, when choosing a new supplier, or when proposing a ranking of different companies. In such cases, a good assessment means the assessed entity is better than its rivals, but not necessarily good in absolute terms. Other sustainability assessment tools can be used in rating assessments in the form of classifications (e.g., A+, A, A-, B, etc.). In such cases, a good assessment means the assessed entity meets the requirements for the respective grade. This chapter deals with sustainability assessment from this classification/rating perspective, known as the *sorting problematic* (Roy, 1996).

Sustainability assessment is inherently multidimensional, as environmental, economical, and social aspects are considered. It is therefore particularly adequate to use Multicriteria Decision Analysis (MCDA), also known as Multicriteria Decision Aiding, Multicriteria Decision Making (MCDM), or Multiattribute Decision Making (MADM). MCDA (Belton and Stewart, 2002; Greco et al., 2016) explicitly acknowledges multiple and possibly conflicting criteria. It provides theoretically well-founded decision aiding models to aggregate multiple impact dimensions in a transparent and auditable way, balancing the multiple concerns of the decision-making entity (a person, a group, or an organization). At the time of writing this chapter, a search in the Scopus database for "multi-criteria" (including variants like "multicriteria" or "multiattribute") and "sustainability assessment" in title, abstract, and keywords yields over 600 publications, showing the potential use of MCDA in sustainability assessment. Among other, the interested reader might see the argumentation and framework by Munda (2005) and the reviews of Cinelli et al. (2014) and Diaz-Balteiro et al. (2017).

Conducting an MCDA sustainability assessment should begin with the clarification of its scope, namely answering the following structuring questions, not necessarily in this order:

- What will be assessed? (The assessed entities, called alternatives or actions in MCDA, which can be products, organizations, regions, etc., depending on the context).
- What will be the assessment criteria? (The set of assessment criteria, often called indicators or attributes).

- Whose preferences will be considered? (Identifying a decision maker, a panel of experts, stakeholder representatives, or other means to set parameters that reflect preferences).
- What type of result is sought? (Identifying the best alternative, a ranking, or a classification).

Clarifying these aspects is extremely important because all analyses that ensue are based on answers to these questions. Problem structuring methods (e.g., cognitive mapping, soft systems methodology) can help answering these questions, particularly when involving multiple stakeholders (Marttunen et al., 2017).

A second stage in MCDA is to characterize the performance of each entity on each criterion. Usually this characterization is based on a quantitative scale (monetary amounts, emissions, number of accidents, etc.). However, several MCDA methods, like the one presented in this chapter, also allow using qualitative scales. The latter can be used for criteria hard to quantify or having a subjective nature, as is the case of assessing perceived quality or aesthetics.

After answering the structuring questions and characterizing the performance of each entity on each individual criterion, an MCDA aggregation method can be applied to inform decision makers about conclusions at a global level. This chapter focuses on this aggregation stage of an MCDA analysis, assuming that the set of criteria has already been chosen and has been used to characterize the entities to be assessed. Many MCDA methods, mentioned in the following section, are available for this aggregation step. Section 12.3 presents the details of the ELECTRE TRI method (Yu, 1992) in particular. A pedagogic example is then analyzed in Section 12.4. Section 12.5 discusses different strategies to set the parameters of ELECTRE TRI, citing a few examples from the literature. Section 12.6 adds some conclusions about the use of ELECTRE TRI in sustainability assessment.

12.2 ELECTRE TRI in the MCDA panorama

MCDA/MCDM are umbrella terms that cover very distinct methods (Greco et al., 2016). Important distinctions can be made to properly situate the ELECTRE TRI method in this panorama. First, one can distinguish between optimization and evaluation. Multiobjective optimization methods deal with mathematical programming with multiple objective functions. The purpose of these methods is to set the values of decision variables subject to restrictions that define implicitly the feasible solutions. In contrast, MCDA evaluation or assessment methods consider a finite list of entities (the alternatives, in MCDA language) that are already known or will be known and can be listed. As this chapter deals with sustainability assessment, only the latter are addressed here.

The main feature distinguishing MCDA evaluation methods is the type of approach underlying the aggregation of the individual criteria. In particular, most MCDA aggregation methods belong to one of the following types:

1. Methods that derive a global performance value for each alternative;
2. Methods that measure how far each alternative is from given references;
3. Methods based on binary relations.

Methods of the first type can be further distinguished, according to whether the evaluation of one alternative is independent of the other alternatives. A global value for an alternative is obtained by aggregating its performances independently of the evaluation of the remaining alternatives in the well-known Multiattribute Value Theory (MAVT)/Multiattribute Utility Theory (MAUT) (Dias et al.,

2015; Keeney, 2006), but also in other methods such as the Ordered Weighted Average (OWA) (Yager, 1988) and fuzzy integrals (Grabisch and Labreuche, 2010). In contrast, the well-known Analytic Hierarchy Process (AHP) method (Saaty, 2008) is based on pairwise comparisons. This means that the global value obtained for each alternative aggregates how it compares relatively to the remaining alternatives, that is, the value would change (and even a rank reversal might occur) if a new alternative was added to the evaluated set (Forman and Gass, 2001). A popular method that also aggregates the performances of the alternatives into a single value is the Simple Additive Weighting or Weighted Sum Method (Yoon and Hwang, 1995), which requires that all criteria are measured using the same scale. Therefore, users of this method need to normalize the scales, which usually makes the evaluation of each alternative dependent of the other alternatives (see, e.g, Dias and Domingues, 2014).

As a second type, one finds methods that yield a global measure of how far each alternative is from one or more points in the multidimensional performance space. TOPSIS and VIKOR (Opricovic and Tzeng, 2004) are arguably the most popular methods in this class. These methods use two reference points based on the best and the worst performances among the set of alternatives. These references define an ideal point (a fictitious alternative with the best performances in all criteria) and the antiideal point (with the worst performances). The evaluation of each alternative expresses how far each alternative is from these two points (the closer to the ideal, the better). This evaluation thus depends on the ideal and antiideal points, which in turn depends on the set of alternatives being considered.

As a third type, one finds methods that build one or several binary relations, namely the so-called outranking methods, which include the well-known ELECTRE (Figueira et al., 2005) and PROMETHEE (Brans et al., 1986) approaches, among other. These methods do not need to convert and tradeoff the criteria scales, and thus they do not yield a global value for each alternative. When comparing an alternative to another alternative, the binary relation can be one of preference, indifference, or incomparability. ELECTRE and PROMETHEE are families of methods developed specifically for different needs: ELECTRE I/IS was developed to identify a single preferred alternative; ELECTRE II, III, and IV, as well as PROMETHEE I were developed to produce a (partial) ranking of the alternatives; and ELECTRE TRI was developed to sort the alternatives into predefined categories. PROMETHEE II computes a complete ranking from the global value of each alternative, and as such belongs to type 1. As in the case of AHP and other methods, the position of an alternative in outranking methods is relative to the other alternatives it is being compared to, which means it can be affected by changes in this set. ELECTRE TRI, detailed in the next section, is an exception, as the alternatives are compared to references that are external to the set of entities being assessed.

A recent MCDA taxonomy (Cinelli et al., 2020) focuses on additional issues, including (1) criteria structure (flat or hierarchical), (2) measurement scale (ordinal, interval, or ratio scale), (3) the presence of uncertainty or fuzziness, (4) degree of compensation, and (5) whether preferences are elicited directly or indirectly. The ELECTRE TRI method has different variants allowing any combination sought concerning aspects (1)–(5), making it a flexible tool for sustainability assessment.

12.3 ELECTRE TRI in detail
12.3.1 Origins and purpose

ELECTRE methods originated from the work of Bernard Roy and collaborators at SEMA consultancy and the Paris-Dauphine University, in France, from the 1960s onwards (Figueira et al., 2005). These methods are grounded on the notion of "outranking," a binary relation used in comparing alternatives: an alternative a is said to outrank another alternative b (denoted aSb) if the former is at least as good as the latter. In other words, aSb means that a is either preferred to b or a is as good as (indifferent to) b. The general philosophy of ELECTRE methods is as follows:

1. To obtain an outranking relation in a very cautious way, not demanding a rich type of information (by accepting ordinal scales and quantitative scales with imprecision), and not demanding normalization or scale conversions to a common unit. The conclusion that aSb should be supported by a large majority of the criteria and the absence of any serious shortcomings of a. A very poor performance on one criterion cannot be compensated by a very good performance on another criterion, thus not allowing full substitutability among criteria.
2. To use the obtained outranking relations to yield the required result, that is, identifying the best alternative (ELECTRE I/IS), ranking the alternatives (ELECTRE II/III/IV), or sorting the alternatives (ELECTRE TRI).

ELECTRE TRI was born in the context of a real-world application (credit granting decisions) and a PhD thesis (Yu, 1992) and has since been applied in many varied contexts (Govindan and Jepsen, 2016). In a sustainability assessment context, it can be used to sort the entities (products, supply chains, regions, etc.) according to predefined and ordered categories. The number of categories, as well as their characterization, is defined by the person or body responsible for the assessment, here named DM (as in decision maker). The labels associated with each category are also defined by the DM and can be seen as a sustainability rating (e.g., A, B, C, ..., or "Excellent," "Very Good," etc.).

12.3.2 Classification rules

Each entity under assessment is assigned to a category (rating) depending on how it compares to the references, called profiles, that define the categories. This comparison results from a multicriteria outranking relation (whose details are presented in Section 12.3.3), taking into account a set of n criteria $g_1,...,g_n$. Let $g_j(x)$ denote the assessment of an entity or profile x according to the jth criterion. The profiles are the boundaries that separate consecutive categories. Thus, the DM must define $k-1$ profiles $b^1, ..., b^{k-1}$ to define k categories $C^1, ..., C^k$}. A profile b^h is an n-dimensional vector of performance $(g_1(b^h),..., g_n(b^h))$ representing simultaneously the upper bound of category C^h and the lower bound of category C^{h+1} (Fig. 12.1). By convention, C^k denotes the best category. Each profile is at least as good as the lower profiles on every criterion, and consecutive profiles should be sufficiently different so that b^h is strictly preferred to b^{h-1} and no entity can be considered as indifferent to both b^h and b^{h-1} ($h = 2,...k-1$) (Yu, 1992).

The resulting category for an entity a, according to the most popular ELECTRE TRI variant (called "pessimistic variant") is the following (Yu, 1992):

- $a \rightarrow C^1 \Leftrightarrow \neg aSb^1$ (i.e., a is classified in category C^1 if a is not good enough to outrank b^1);
- $a \rightarrow C^h \Leftrightarrow aSb^{h-1} \wedge \neg aSb^h$, with $h \in \{2, ..., k-1\}$ (i.e., a is classified in category C^h if a is good enough to outrank b^{h-1}, but not to outrank b^h);
- $a \rightarrow C^k \Leftrightarrow aSb^{k-1}$ (i.e., a is classified in category C^k if a is good enough to outrank b^{k-1}).

12.3 ELECTRE TRI in detail

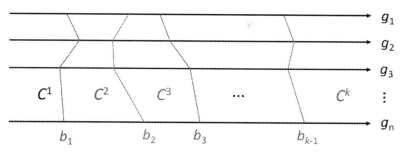

FIG. 12.1 Definition of categories using *n*-dimensional profiles.

The assessment of an entity a is independent of the assessment of other entities. In a sustainability assessment, many entities might be sorted in the same category, while, for instance, some category might be empty if no entity meets the requirements to be placed there. It is also noteworthy that not all entities need to be known at the outset: they can be assessed as they appear. The definition of whether an entity outranks a profile is presented next.

12.3.3 Valued outranking relations

To define whether an entity a outranks a profile b, ELECTRE looks at the advantage (disadvantage if negative) of a over b on the multiple criteria $g_1,\ldots,g_n$:

$$\Delta_j(a,b) = \begin{cases} g_j(a)-g_j(b), & \text{if } g_j(\cdot) \text{ is a maximization criterion (the more the better)} \\ g_j(b)-g_j(a), & \text{if } g_j(\cdot) \text{ is a minimization criterion (the less the better)} \end{cases} \quad (12.1)$$

The steps to be followed are:
1. *Determine the concordance of each criterion*

Concordance is given by an index that indicates how much $g_j(\bullet)$ agrees with aSb, on a scale from 0 (does not agree) to 1 (fully agrees) (Yu, 1992):

$$c_j(a,b) = \begin{cases} 0, & \text{if } \Delta_j(a,b) < -p_j \\ (p_j+\Delta_j(a,b))/(p_j-q_j), & \text{if } -p_j \le \Delta_j(a,b) < -q_j \\ 1, & \text{if } \Delta_j(a,b) \ge -q_j \end{cases} \quad (12.2)$$

This definition has two thresholds set by the DM separately for each criterion, allowing for imprecision in performance measurements:

- q_j (indifference threshold for the *j*th criterion): two levels of performance are considered to be indistinguishable if their difference is q_j or less.
- p_j (preference threshold for the *j*th criterion): one level of performance is considered to be undoubtedly better than another one if their difference is p_j or more (with $p_j \ge q_j$).

According to this definition, the jth criterion fully agrees with aSb if the performance of a is better or it is only slightly worse (up to the indifference threshold). The criterion does not agree with aSb if a is undoubtedly worse than b (a difference greater than the preference threshold). In the remaining intermediate situations, the criterion partially agrees with aSb. These parameters are optional and they can be set to zero (e.g., for a criterion assessed on an ordinal scale). Moreover, they can be expressed in relative terms (e.g., 1% as an indifference threshold) rather than constants. For more details about how to set these and the following parameters the reader may see Dias and Mousseau (2018).

2. *Determine the global concordance index*

Given the single-criteria concordance indices and a set of criteria weights, a global concordance index for aSb is computed as follows (Yu, 1992):

$$c(a,b) = \sum_{j=1}^{n} c_j(a,b) \times k_j \quad (12.3)$$

Here, k_j denotes the weight of criterion $g_j(\bullet)$. The weights can be interpreted as the number of votes allocated to each criterion considering the outranking conclusion must be supported by a sufficiently strong majority. The above expression assumes weights are nonnegative numbers and their sum is equal to 1 (100% of the votes).

3. *Determine the discordance (opposition) of each criterion*

ELECTRE allows rejecting aSb even if the majority of the criteria agree with this conclusion, when some criterion is strongly opposed to it. In other words, the DM can grant criteria the power of veto. Discordance is given by an index that indicates how much $g_j(\bullet)$ opposes to aSb, on a scale from 0 (does not oppose) to 1 (fully opposes) (Mousseau and Dias, 2004; Yu, 1992):

$$d_j(a,b) = \begin{cases} 0, & \text{if } \Delta_j(a,b) \geq -u_j \\ \left(-\Delta_j(a,b) - u_j\right) / \left(v_j - u_j\right), & \text{if } -v_j < \Delta_j(a,b) < -u_j \\ 1, & \text{if } \Delta_j(a,b) \leq -v_j \end{cases} \quad (12.4)$$

This definition has two other thresholds set by the DM separately for each criterion:

- u_j (nonopposition threshold for the jth criterion): the jth criterion does not oppose to aSb if a is worse than b but the difference does not exceed u_j ($u_j \geq p_j$). Originally, $u_j = p_j$ but setting these parameters to different values can add flexibility to the method (Mousseau and Dias, 2004).
- v_j (veto threshold for the jth criterion): the jth criterion completely opposes to aSb if a is worse than b with a difference of v_j or more ($v_j \geq u_j$).

According to this definition, the jth criterion can fully or partially veto the conclusion aSb. The parameter v_j defines a difference considered so large that it cannot be compensated on other criteria. The parameters v_j, u_j, p_j, and q_j can have different values for different profiles. For instance, the DM may define veto parameters only for the profiles associated with the best category, so that no entity can access that category if performance is too poor on one of the criteria.

4. *Determine the outranking credibility*

The overall credibility of aSb, summarizes the arguments in favor and against it (Yu, 1992):

$$\sigma(a,b) = c(a,b) \prod_{\substack{j \in \{1,\ldots,n\}: \\ dj(a,b) > c(a,b)}} \frac{1 - d_j(a,b)}{1 - c(a,b)} \quad (12.5)$$

Besides this expression other possibilities have been proposed, for example Mousseau and Dias (2004) suggest multiplying concordance by $\prod_j \left(1 - d_j(a,b)\right)$ or by $\left(1 - \max_j d_j(a,b)\right)$. Credibility increases with concordance and decreases with discordance. Moreover, if discordance is maximum then credibility becomes null.

The credibility can be seen as a fuzzy (valued) outranking relation, which can be transformed into a crisp relation by comparing it with a cutting level λ (Yu, 1992):

$$aSb \Leftrightarrow \sigma(a,b) > \lambda. \tag{12.6}$$

The cutting level λ represents the required level of majority (in the absence of opposition), and the DM can set it between 0.5 (weak majority) to 1.0 (unanimity). The higher λ is, the more difficult it will be for an entity to reach the top categories, thus allowing to control how demanding the assessment is.

12.3.4 Other variants

When ELECTRE TRI was proposed it had two variants, the pessimistic one presented above, and an optimistic one. The optimistic variant differs only in the classification rule (Yu, 1992):

- $a \to C^1 \Leftrightarrow b^1 Sa \wedge \neg aSb^1$;
- $a \to C^h \Leftrightarrow \left(b^h Sa \wedge \neg aSb^h\right) \wedge \left(aSb^{h-1} \vee \neg b^{h-1} Sa\right)$, with $h \in \{2, \ldots, k-1\}$;
- $a \to C^k \Leftrightarrow aSb^{k-1} \vee \neg b^{k-1} Sa$.

The optimistic classification is higher than the pessimistic one when for some profile b^h, neither aSb^h, nor $b^h Sa$ (the entity is said to be incomparable with the profile). Otherwise, the classifications will coincide.

More recently, Almeida-Dias et al. (2010) proposed the ELECTRE TRI-C method (calling the previous variants ELECTRE TRI-B), in which each reference profile is used to characterize each category (as a typical example) rather than limiting the categories. Each category can be characterized by more than one example in the ELECTRE TRI-nC variant (Almeida-Dias et al., 2012) and in the interactive method of (Rocha and Dias, 2008). The idea of using multiple profiles (as boundaries) to separate the categories has also been proposed as a new variant named ELECTRE TRI-nB (Fernández et al., 2017). Hierarchical versions of ELECTRE TRI have also been proposed, allowing to specify criteria as a hierarchy of main criteria and subcriteria (Corrente et al., 2016).

12.4 An illustrative example

The following pedagogic example illustrates a sustainability assessment using ELECTRE TRI, inspired by Domingues et al. (2015). In some country, policymakers (the DM) wish to tax passenger vehicles based on their sustainability, namely considering environmental (criteria g_1 and g_2), economic (g_3), and social concerns (g_4 and g_5):

- g_1: Greenhouse gas (GHG) emissions considering the vehicle's life cycle, including manufacturing, measured in $gCO2_{eq}/km$ (to be minimized: the less, the better);

- g_2: Recyclable mass of the vehicle, measured in percentage points relatively to the total mass (to be maximized: the more, the better);
- g_3: Costs for the user considering fuel (fossil, electricity, or other) and investment depreciation, measured in ¤/100 km, with symbol ¤ representing the country's currency (to be minimized);
- g_4: Emissions of particulate matter (PM) impairing health, stemming from tailpipe fumes and tire abrasion, measured in g/100 km (to be minimized);
- g_5: Qualitative assessment of the labor conditions in the places where the vehicle and its main parts are manufactured and assembled, on scale from 1 to 5 (to be maximized), noting that these numbers are just codes, for example, "1=Poor," "2=Passable," "3=Good," "4=Very Good," "5=Excellent."

Criteria g_1, g_3, and g_4 assume a service life of 15 years and 200,000 km (Domingues et al., 2015).

The DM decided to define three categories: "Malus," to bear a higher tax (C^1), "Neutral" (C^2), and "Bonus" to benefit from a tax bonus (C^3). It was thus necessary to define a profile b^1 separating C^1 from C^2, as well as a profile b^2 separating C^2 from C^3. The DM defined, for each criterion, the level it would require for the criterion to place the vehicle in each category. For criterion g_1 (GHG), the DM decided a vehicle would be "Malus" (C^1) if its life cycle emissions exceeded 200 gCO2$_{eq}$/km; it would be "Neutral" (C^2) if emissions did not exceed 200 gCO2$_{eq}$/km, but exceeded 100 gCO2$_{eq}$/km; and it would need to have emissions of 100 gCO2$_{eq}$/km or less to be "Bonus" (C^3) (this criterion is a minimization one). For criterion g_2, a maximization criterion, the DM stipulated that vehicles that can be recycled less than 75(%) should be C^1, from 75(%) onwards up to 90(%) the vehicle should be C^2, and for 90(%) or higher recyclability the vehicle would be able to reach C^3. For criterion g_3 the DM set 30¤/100 km as the boundary between C^1 and C^2 and set 20¤/100 km as the boundary between C^2 and C^3. Similarly, the DM defined the reference performances for criteria g_4 and g_5 depicted in Table 12.1, along with the parameters associated with each criterion.

The DM set all indifference thresholds $q_j = 0$, implying that if a vehicle a is worse than the required performance in a profile b^h on some criterion, then this criterion will not fully agree with the outranking statement aSb^h. In the criteria measured on cardinal scales (g_1 to g_4) the preference thresholds are positive, meaning that if a is worse than b^h on some criterion by a difference smaller than p_j, then this criterion will partially agree with the outranking statement aSb^h, and thus will contribute with part of its weight for the concordance coalition. For instance, $p_1 = 10$ and $g_1(b^1) = 200$, which means that g_1 will fully support placing a vehicle in the second category if emissions are 200 gCO2$_{eq}$/km or lower,

Table 12.1 ELECTRE TRI profiles, weights, and thresholds (example).

	g_1: GHG↓ (gCO2$_{eq}$/km)	g_2: Recyclable↑ (%)	g_3: Cost↓ (¤/100 km)	g_4: PM↓ (g/100 km)	g_5: Labor↑ (1–5)
b^2	100	90	20	4	4
b^1	200	75	30	10	2
k_j	0.3	0.2	0.2	0.15	0.15
q_j	0	0	0	0	0
p_j	10	5	2	5	0
u_j	10	5	2	5	0
v_j	90	55	22	25	3

Legend: ↓ Criterion to be minimized; ↑ Criterion to be maximized will partially support with a concordance of 5/10 (50%) if the vehicle is 5 $gCO2_{eq}$/km worse that the requirement, and will not give its support if the vehicle is 10 $gCO2_{eq}$/km worse than the requirement. In the qualitative criterion g_5, the DM set $p_j = 0$, meaning that this criterion does not support aSb^h if the labor conditions have a grade lower than required by the profile.

The DM set veto thresholds for all the criteria and followed the original ELECTRE TRI formulation placing $u_j = p_j$. In the first criterion, for instance, $u_1 = 10$ and $p_1 = 90$ means that if a vehicle a is worse than b^h on some criterion by a difference larger than 10 $gCO2_{eq}$/km, then the criterion will oppose to aSb^h with a discordance index that increases 1/80 per each $gCO2_{eq}$/km above 10. This means that, for instance, the discordance with aSb^1 of a vehicle with emissions of 270 $gCO2_{eq}$/km would be 0.75, implying a very low credibility $\sigma(a,b^1)$. In the qualitative criterion g_5, the chosen numerical coding together with thresholds $u_5 = 0$ and $p_5 = 3$ means that a disadvantage of one level yields a discordance of 1/3, a disadvantage of two levels yields a discordance of 2/3, and a disadvantage of three levels yields a discordance of 1.

Finally, the DM defined the criteria weights and the cutting level representing the majority to be required. The DM established that the environmental dimension should have 50% of the weight, with GHG emissions more important than recyclability, setting a weight of 0.3 for g_1 and a weight of 0.2 for g_2. Then, the DM established the two social criteria together should have more weight (30%) than the economic criterion, and set $k_3 = 0.20$ and $k_4 = 0.15$, and $k_5 = 0.15$. This means that g_2 is as important as g_3, and g_4 is as important as g_5. Moreover, g_1 alone is as important as a coalition of g_4 plus g_5. The cutting level λ was set to 0.65, implying for instance that the two environmental criteria alone do not have sufficient weight to support an outranking, needing the support of another criterion.

After specifying the ELECTRE TRI parameters the DM can now assess existing vehicles as well as new ones appearing in the future. For instance, a manufacturer was planning to introduce a new car (a_x) with relatively high life cycle emissions of 210 $gCO2_{eq}$/km in GHG (g_1) and 13 g/100km in PM (g_4), but exhibiting a total cost of 15¤/100 km (g_3) and 95% recyclability (g_2). Labor conditions in its manufacturing are Passable (level 2 in g_5).

Table 12.2 summarizes how a_x compares with the first profile, b^1. On criterion g_1 it fails to meet the required value with a disadvantage of $200 - 210 = -10$, equal to $-p_j$, which is sufficiently large for concordance to be null, but not so large for discordance to occur. On criteria g_2 and g_3 it is better than the required value, and on g_5 it meets the required value exactly: these three criteria agree with the outranking. a_x fails to meet the required value on criterion g_4 but the disadvantage is small: the criterion partially agrees with $a_x Sb^1$. The global concordance, considering the weights set by the DM, is 0.61, and the credibility of the outranking has the same value, as discordance is null. Yet, $\sigma(a_x,b^1)$ does not attain the required majority set at $\lambda = 0.65$. Therefore, a_x does not outrank b^1 and is classified as C^1.

Table 12.2 Classification of a_x.

	$g_1 \downarrow$	$g_2 \uparrow$	$g_3 \downarrow$	$g_4 \downarrow$	$g_5 \uparrow$		
a_x	210	90	15	13	2		
b^1	200	75	30	10	2		
Δ_j	-10	15	10	-3	0		
c_j	0	1	1	0.4	1	$c(a_x,b^1) = 0.61$	
d_j	0	0	0	0	0	$d^{max}(a_x,b^1) = 0$	$\sigma(a_x,b^1) = 0.61$

The conclusion that a_x is classified as C^1 depends of course on the parameter values. A sensitivity analysis can be conducted to find how much each parameter could change without leading to a different conclusion. Such analyses are usually performed changing one parameter at a time. For instance, the stability interval for the cutting level is $\lambda \in]0.61, 1.00]$. If $\lambda \leq 0.61$ then a_x would reach C^2, and as $\lambda = 0.65$ is not far from 0.61, it might be considered a sensitive parameter. The stability interval for each weight, changing one at a time and adjusting the remaining ones proportionally is $k_1 \in]0.25, 1.00]$, $k_2 \in [0.00, 0.28[$, $k_3 \in [0.00, 0.28[$, $k_4 \in]0.00, 1.00]$, and $k_5 \in [0.00, 0.24[$. Therefore, the classification of a_x is relatively sensitive to the weights of the first three criteria, with a value not far from the respective stability interval limits.

The manufacturer did not want to sell a vehicle classified as C^1, and for this reason it decided to introduce the vehicle in this market with a cleaner powertrain technology. This new model, a_y, has much better life cycle emissions of 101 gCO_{2eq}/km in GHG (g_1) and 5 g/100 km in PM (g_4), and it still is 90% recyclable (g_2). However, it has an increased cost of 15¤/100 km (g_3) and the labor conditions in its manufacturing are considered to be poor (level 1 in g_5).

Table 12.3 summarizes how a_y compares with the profiles. Only the last criterion does not fully support $a_y S b^1$. The weights of the remaining four criteria yield a concordance of 0.85 and discordance in the last criterion is much lower, not affecting the global credibility $\sigma(a_y, b^1) = 0.85 > \lambda$. Thus, a_y is good enough to outrank b^1. Next, one checks if it is good enough to outrank b^2. Criteria g_2 and g_3 fully agree with $a_y S b^2$. Criteria g_1 and g_4 agree to a large extent with $a_y S b^2$, as the disadvantage of a_y is rather small. Only criterion g_5 does not agree with the outranking and actually it is strongly opposed, as labor conditions are "poor" and a "very good" was sought. As a veto threshold was attained, discordance is maximum and credibility is null. Hence, $a_y S b^1$ and $\neg(a_y S b^2)$, that is, a_y is C^2. To reach category C^3 the manufacturer must improve on criterion g_5. This poor performance cannot be compensated on the other criteria, even if performance on those remaining criteria improved.

For this vehicle, a sensitivity analysis reveals that the stability interval for the cutting level is $\lambda \in]0.85, 1.00]$. As $\lambda = 0.65$ is far from 0.85, it might be considered an insensitive parameter. The stability interval for each weight, changing one at a time and adjusting the remaining ones proportionally

Table 12.3 Classification of a_y.

	g_1 ↓	g_2 ↑	g_3 ↓	g_4 ↓	g_5 ↑			
a_y	101	90	19	5	1			
b^1	200	75	30	10	2			
Δ_j	99	15	11	5	−1			
c_j	1	1	1	1	0	$c(a_y, b^1) = 0.85$		
d_j	0	0	0	0	1/3	$d^{max}(a_y, b^1) = 1/3$	$\sigma(a_y, b^1) = 0.85$	
	g_1 ↓	g_2 ↑	g_3 ↓	g_4 ↓	g_5 ↑			
a_y	101	90	19	5	1			
b^2	100	90	20	4	4			
Δ_j	−1	0	1	−1	−3			
c_j	0.9	1	1	0.8	0	$c(a_y, b^2) = 0.79$		
d_j	0	0	0	0	1	$d^{max}(a_y, b^2) = 1$	$\sigma(a_y, b^2) = 0$	

is $k_j \in]0.00, 1.00]$ (for $j = 1,...,4$), and $k_5 \in [0.00, 0.35]$, that is, a_y changes to C^1 only if $k_5 > 0.35$, and it never reaches C^3 due to discordance on criterion g_5. Therefore, the classification of a_y in C^2 is quite insensitive to changing one of the weights.

12.5 Setting the parameter values

Setting the parameter values of ELECTRE methods in general is discussed at length by Dias and Mousseau (2018). Depending on the situation, several strategies are possible. These are illustrated next based on a few case studies.

1. *Elicitation of precise parameter values*

Silva et al. (2015) carried out an ELECTRE TRI assessment of the sustainability of dairy farms in the Northwest of Portugal. The dairy farms were assessed separately in environmental criteria and socioeconomic criteria. The environmental assessment included criteria "Storage capacity of manure," "Area of application of manure in the soil," "Incorporation in excess of nitrogen in the farm," "Production of greenhouse gases," "Storage structures near water lines," "Individualized collection of rainwater," and "Animal well-being." In turn, socioeconomic criteria included "Producer´s age," "Time dedication," "Professional development," "Successors," "Dependence on external income," "Investment strategies," and "Future perspectives." In total, 1705 farms were assessed in three categories in each assessment: "Not sustainable," "Barely sustainable," and "Sustainable."

This study can be considered as a typical ELECTRE TRI application concerning how the parameter values were set. Three experts were involved in the study and the parameters (weights, profiles, thresholds) were set based on a dialogue with these experts. Then, using a specially developed software integration in a geographic information system (Silva et al., 2014), results could be shown as interactive maps on a computer (Fig. 12.2).

2. *Inference of parameter values*

Sánchez-Lozano et al. (2014) built an ELECTRE TRI model to assess the suitability of different sites for siting solar farms in the Spanish region of Murcia. After excluding the clearly unsuitable areas (protected sites, roads, rivers, etc.), the potential sites were assessed quantitatively on 10 criteria: "Agrological capacity," "Slope," "Orientation," "Plot area," "Distance to main roads," "Distance to power lines," "Distance to towns or villages," "Distance to electricity substations," "Solar radiation," and "Average temperature." The final result was a classification in four categories, from "Poor capacity (C^1)" to "Excellent capacity (C^4)."

Again, the ELECTRE TRI parameters were set with the involvement of domain experts, but now using an indirect elicitation approach for the weights and cutting level. Indirect approaches, also called disaggregation methods, infer the parameters from examples provided by the DM. In the case of ELECTRE TRI, the DM classifies a few entities as an example, allowing to infer parameters such as weights (Dias and Mousseau, 2018). This is illustrated in Fig. 12.3. If the DM states a_x is a good example for category C^3 and a_y is a good example for category C^1, then all parameter values leading to different classifications of these examples can be excluded, retaining only the set of parameter vectors compatible with the examples. Inference is usually performed through an optimization program that minimizes an error function to yield a vector of parameter values inside the compatible set.

In this case, 20 land plots were used as examples, which were classified by an expert, one at a time. As Fig. 12.3 shows, defining one example limits the possible categories for other examples, for

208 Chapter 12 Sustainability assessment using the ELECTRE TRI

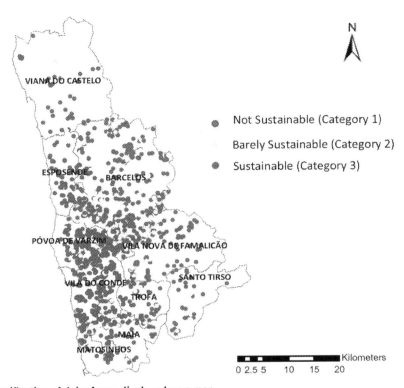

FIG. 12.2 Classification of dairy farms displayed on a map.

Credit: (Silva et al., 2014)

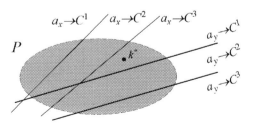

FIG. 12.3 Classification examples constrain the parameter space.

Classifying a_x in C^3 and simultaneously classifying a_x in C^3 leads to the darker shaded region. Vector k^* is a central vector representing this region.

Credit: N/A

example, $a_x \to C^1$ is no longer possible if $a_y \to C^3$. After classifying only 7 plots, the parameter region compatible with these examples became so small that the remaining 13 plots had their category defined. Then, a central vector for the weights and the cutting level was used (for details, see Sánchez-Lozano et al., 2014), allowing to show the classifications on a map (Fig. 12.4).

12.5 Setting the parameter values 209

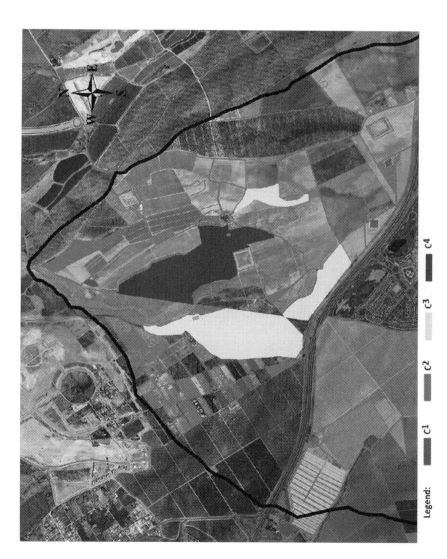

FIG. 12.4 Suitability classification of 20 plots of land (7 provided as examples and 13 inferred from those examples).

Credit: Adapted from (Sánchez-Lozano et al., 2014)

This case shows that as an alternative to a direct elicitation, one can ask for examples of results the DM seeks, and then infer a classification model. This can be done in an interactive way, allowing the DM to revise earlier examples (Mousseau et al., 2006).

3. *Direct constraints on the parameter values*

Another elicitation strategy is illustrated by Covas et al. (2013), who assessed over 4000 parishes in Portugal concerning their suitability to locate a sustainable data center on behalf of a telecommunications company. Separate assessments were made in four dimensions: Environmental (criteria related with renewable energy, free cooling, and protected areas), Economic (investment costs, operational costs, and attractiveness for the company's customers), Risks (natural hazards, human hazards and crime), and Social (attractiveness to employees, availability of skilled labor, technical support and other services).

In the same way that Fig. 12.3 illustrates how a classification example reduces the region of parameter values compatible with it, a comparison of criteria weights can have the same effect. For instance, a DM can state that criterion g_1 should be no less important than another criterion g_2, implying that only the parameter vectors having $k_1 \geq k_2$ are considered. In this application, the DM (the data center team of the company) made several of these comparisons, stating that some criteria should have the same weight and that some criteria should have more weight than some other ones, without indicating numerical values. It was then possible to compute, using optimization (Dias et al., 2002), the compatible classifications for each parish (e.g., Fig. 12.5). As the parameter space region was not reduced to a single vector, the assessment can result in an interval of categories, which can be considered robust relatively to the information provided.

This type of robustness analysis (Dias, 2007) can be seen as an inversion of perspective relatively to a sensitivity analysis. A sensitivity analysis, as illustrated in Section 12.4, focuses on a given result

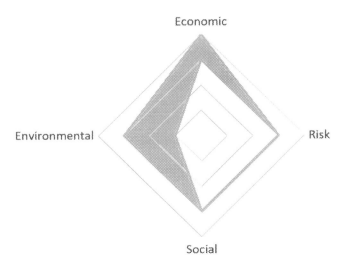

FIG. 12.5 Suitability of one parish to locate a sustainable data center: it is classified as C^3 in terms of the risk and social dimension, C^3 or C^4 in the economic dimension, and C^1, C^2, or C^3 in the environmental dimension.

Credit: N/A

and finds ranges for the parameter values compatible with that result (usually varying one parameter at a time). In contrast, a robustness analysis focuses on a given range of values for the parameter values (usually allowing these to vary simultaneously) and finds the results and conclusions that can be obtained from such ranges.

Another study that used direct constraints on the parameter values is the environmental life cycle assessment of vehicles by Domingues et al. (2015). Rather than eliciting weights, a set of bounds were used to ensure no criterion would have much more weight than another one. For each vehicle assessed, the allowed variation of the parameter values leads to an interval of categories. If an exact category is sought, besides the approaches used in cases (1) and (2), the DM might use the best classification under a "benefit of doubt" perspective or use the worst classification under a "cautionary" perspective.

4. *Stochastic analyses*

A stochastic analysis can be used to replace or complement the strategies outlined above. It consists in running the assessment multiple times (usually thousands) using randomly generated parameter values, and then producing statistics about the results obtained (Tervonen et al., 2007). For instance, Dias et al. (2018) combined a robustness analysis and a stochastic analysis for an assessment of policies to foster smart grids in Brazil. Eight policies were qualitatively assessed on seven criteria by a panel of stakeholders and experts: "Environment and human health," "Flexibility and capabilities of the electricity system," "Security of supply," "Fairness and efficiency of the electricity markets," "Financial benefit to the market agents," "Economic and social benefit to the country," and "Feasibility and adoption." Four categories were defined: "Uninteresting (C^1)," "Wait and see (C^2)," "Priority (C^3)," and "Maximum priority (C^4)."

The inputs of the same stakeholders and experts were used to define constraints on the criteria weights. Then, the best and worst possible categories for each policy were computed considering the constraints as in case (3). One policy was robustly sorted into C^4, two other policies were sorted into C^3, and the remaining policies could be either C^3 or C^4. For the latter cases, a stochastic analysis computed the probability of each classification using random parameter values. For instance, one of the policies had probability of 0.916 for C^3 and probability of 0.084 for C^4, meaning that C^4 would occur for only a few specific parameter values.

12.6 Conclusion

This chapter presented a method for sustainability assessment that does not lead to a number, but to a classification or rating based on predefined and ordered categories. The examples presented show it can be applied in a large variety of contexts. The following characteristics make ELECTRE TRI an interesting method to consider in sustainability assessment:

- It considers multiple criteria or indicators, which can have a rather different nature (environmental, economic, social, technical).
- It is not demanding in terms of requirements. The criteria can be assessed on quantitative or qualitative scales, and the scales can be used directly without need for normalization.
- It uses criteria weights that reflect the importance attached to each criterion, rather than reflecting monetarization or tradeoff rates. Indeed, the method rejects full substitutability, not allowing a

poor performance on one dimension to be compensated by a good performance on some other dimension.
- The assessment of one entity is independent of other entities being assessed, rather than the result of a comparison or "tournament" among those entities. This means that entities can be assessed as they come, even before knowing the entire set. It also means the assessments of the entities do not change if a new entity appears. Moreover, thousands of entities can be assessed without difficulty.
- The many variants of ELECTRE TRI and different parameter elicitation strategies allow choosing the most suitable approach for each situation.

Being an MCDA approach, ELECTRE TRI acknowledges the legitimacy of the DM to build the assessment model according to its views or the views of the participating actors (experts, stakeholders, or even the general public). It is therefore a tool that does not replace the DM by an automated and supposedly objective analysis. This can be a cause of concern as the results will to some extent depend on the choice of the method and the also on the parameter values defined by the DM.

Regarding the choice of the method, the DM could compare the results to those obtained by a different MCDA method. One should not forget, however, that parameters such as criteria weights have a different meaning from one method to another (Choo et al., 1999; Roy and Mousseau, 1996). This means that one should not simply use the same weights for methods interpreting these parameters in different ways. In such cases, the parameter elicitation should start from the beginning and will likely lead to a different weights vector. Moreover, although in general a very good entity will have a very good assessment independently of the method used, the assessment of those entities with a mix of good and poor performance on the different criteria can change from one method to another, due to the different characteristics of MCDA methods. The validity of the results therefore rests mainly on the coherence between the requirements of the DM, the characteristics of the chosen MCDA method, and the elicitation process.

Regarding setting the parameter values, the flexibility of ELECTRE TRI entails using several parameters for each criterion and its algebraic expressions, although elementary, are not so familiar as the popular weighted sum. Nevertheless, this concern can be alleviated by elicitation strategies based on incompletely defined preferences, as illustrated in cases (2) to (4). Such strategies can be based on examples that build on past experience or on reasonable constraints that still leave much freedom for the parameter values to vary.

As Section 12.3.4 shows, ELECTRE TRI has been evolving, with new variants being proposed to address specific requirements. In parallel, a recent research stream has proposed ways of deriving a ranking of the assessed entities based on ELECTRE TRI principles and parameters (Dias et al., 2018; Rolland, 2013). The use of ELECTRE TRI in sustainability assessments, among other applications, is thus expected to grow and to keep improving in the future.

Acknowledgments

The author gratefully acknowledges the collaboration with colleagues from the Energy for Sustainability Initiative of the University of Coimbra and the support of Fundação para a Ciência e Tecnologia—FCT through projects CENTRO-01-0145-FEDER-030570 and UIDB/05037/2020.

References

Almeida-Dias, J., Figueira, J.R., Roy, B., 2010. ELECTRE TRI-C : a multiple criteria sorting method based on characteristic reference actions. Eur. J. Oper. Res. 204 (3), 565–580. doi:10.1016/j.ejor.2009.10.018.

Almeida-Dias, J., Figueira, J.R., Roy, B., 2012. A multiple criteria sorting method where each category is characterized by several reference actions: the Electre Tri-nC method. Eur. J. Oper. Res. 217 (3), 567–579. doi:10.1016/j.ejor.2011.09.047.

Belton, V., Stewart, T.J., 2002. Multiple Criteria Decision Analysis: An Integrated Approach. Kluwer.

Brans, J.P., Vincke, P., Mareschal, B., 1986. How to select and how to rank projects: the Promethee method. Eur. J. Oper. Res. 24 (2), 228–238. doi:10.1016/0377-2217(86)90044-5.

Carroll, A.B., Shabana, K.M., 2010. The business case for corporate social responsibility: a review of concepts, research and practice. Int. J. Manage. Rev. 12 (1), 85–105. doi:10.1111/j.1468-2370.2009.00275.x.

Choo, E.U., Schoner, B., Wedley, W.C., 1999. Interpretation of criteria weights in multicriteria decision making. Comput. Ind. Eng. 37 (3), 527–541. doi:10.1016/S0360-8352(00)00019-X.

Cinelli, M., Coles, S.R., Kirwan, K., 2014. Analysis of the potentials of multi criteria decision analysis methods to conduct sustainability assessment. Ecol. Indic. 46, 138–148. doi:10.1016/j.ecolind.2014.06.011.

Cinelli, M., Kadziński, M., Gonzalez, M., Słowiński, R., 2020. How to support the application of multiple criteria decision analysis? Let us start with a comprehensive taxonomy. Omega 96, 102261. doi:10.1016/j.omega.2020.102261.

Corrente, S., Greco, S., Słowiński, R., 2016. Multiple criteria hierarchy process for ELECTRE tri methods. Eur. J. Oper. Res. 252 (1), 191–203. doi:10.1016/j.ejor.2015.12.053.

Covas, M.T., Silva, C.A., Dias, L.C., 2013. Multicriteria decision analysis for sustainable data centers location. Int. Trans. Oper. Res. 20 (3), 269–299. doi:10.1111/j.1475-3995.2012.00874.x.

Dias, L., Mousseau, V., Figueira, J., Clímaco, J., Clímaco, J., 2002. An aggregation/disaggregation approach to obtain robust conclusions with ELECTRE TRI. Eur. J. Oper. Res. 138 (2), 332–348. doi:10.1016/S0377-2217(01)00250-8.

Dias, L.C., 2007. A note on the role of robustness analysis in decision-aiding processes. In: Roy, B., Ali Aloulou, M., Kalaï, R. (Eds.), *Robustness in OR-DA, Annales du LAMSADE, No. 7*. Université-Paris Dauphine, pp. 53–70.

Dias, L.C., Domingues, A.R., 2014. On multi-criteria sustainability assessment: spider-gram surface and dependence biases. Appl. Energy 113, 159–163. doi:10.1016/j.apenergy.2013.07.024.

Dias, L.C., Silva, S., Alçada-Almeida, L., 2015. Multi-criteria environmental performance assessment with an additive model. In: Ruth, M. (Ed.), Handbook of Research Methods and Applications in Environmental Studies. Edward Elgar Publishing Limited, pp. 450–472.

Dias, L.C, Antunes, C.H., Dantas, G., de Castro, N., Zamboni, L., 2018. A multi-criteria approach to sort and rank policies based on Delphi qualitative assessments and ELECTRE TRI: the case of smart grids in Brazil. Omega 76, 100–111. doi:10.1016/j.omega.2017.04.004.

Dias, L.C., Mousseau, V., 2018. Eliciting multi-criteria preferences: ELECTRE models. In: Luıs, C., Dias, A. Morton, Quigley, J. (Eds.), Elicitation - The Science and Art of Structuring Judgement, 349–375. doi:10.1007/978-3-319-65052-4_14 Vol. 261.

Diaz-Balteiro, L., González-Pachón, J., Romero, C., 2017. Measuring systems sustainability with multi-criteria methods: a critical review. Eur. J. Oper. Res. 258 (2), 607–616. doi:10.1016/j.ejor.2016.08.075.

Domingues, A.R., Marques, P., Garcia, R., Freire, F., Dias, L.C., 2015. Applying multi-criteria decision analysis to the life-cycle assessment of vehicles. J. Clean. Prod. 107, 749–759. doi:10.1016/j.jclepro.2015.05.086.

Fernández, E., Figueira, J.R., Navarro, J., Roy, B., 2017. ELECTRE TRI-nB: a new multiple criteria ordinal classification method. Eur. J. Oper. Res. 263 (1), 214–224. doi:10.1016/j.ejor.2017.04.048.

Figueira, J., Mousseau, V., Roy, B, 2005. ELECTRE methods. In: Figueira, J., Greco, S., Ehrgott, M. (Eds.), Multiple Criteria Decision Analysis: State of the Art Surveys. Springer, Cham, pp. 133–153.

Forman, E.H., Gass, S.I., 2001. The analytic hierarchy process — an exposition. Oper. Res. 49 (4), 469–486.

Govindan, K., Jepsen, M.B., 2016. ELECTRE: a comprehensive literature review on methodologies and applications. Eur. J. Oper. Res. 250 (1), 1–29. doi:10.1016/j.ejor.2015.07.019.

Grabisch, M., Labreuche, C., 2010. A decade of application of the Choquet and Sugeno integrals in multi-criteria decision aid. Ann. Oper. Res. 175 (1), 247–286. doi:10.1007/s10479-009-0655-8.

Greco, S., Ehrgott, M., Figueira, J.R., 2016. Multiple Criteria Decision Analysis - State of the Art Surveys. Springer-Verlag, NY.

Keeney, R.L., 2006. Using preferences for multi-attributed alternatives. J. Multi-Criteria Decis. Anal. 14, 169–174. doi:10.1002/mcda.

Marttunen, M., Lienert, J., Belton, V., 2017. Structuring problems for multi-criteria decision analysis in practice: a literature review of method combinations. Eur. J. Oper. Res. 263 (1), 1–17. doi:10.1016/j.ejor.2017.04.041.

Mousseau, V., Dias, L., 2004. Valued outranking relations in ELECTRE providing manageable disaggregation procedures. Eur. J. Oper. Res. 156 (2), 467–482. doi:10.1016/S0377-2217(03)00120-6.

Mousseau, V., Dias, L.C., Figueira, J., 2006. Dealing with inconsistent judgments in multiple criteria sorting models. 4OR 4 (2), 145–158. doi:10.1007/s10288-005-0076-8.

Munda, G., 2005. Measuring sustainability: a multi-criterion framework. Environ. Dev. Sustain. 7 (1), 117–134. doi:10.1007/s10668-003-4713-0.

Opricovic, S., Tzeng, G.-H., 2004. Compromise solution by MCDM methods: a comparative analysis of VIKOR and TOPSIS. Eur. J. Oper. Res. 156 (2), 445–455. doi:10.1016/S0377-2217(03)00020-1.

Rocha, C., Dias, L.C., 2008. An algorithm for ordinal sorting based on ELECTRE with categories defined by examples. J. Glob. Optim. 42 (2), 255–277. doi:10.1007/s10898-007-9240-3.

Rolland, A., 2013. Reference-based preferences aggregation procedures in multi-criteria decision making. Eur. J. Oper. Res. 225 (3), 479–486. doi:10.1016/j.ejor.2012.10.013.

Roy, B., 1996. Multicriteria Methodology for Decision Aiding. Kluwer Academic.

Roy, B., Mousseau, V., 1996. A theoretical framework for analysing the notion of relative importance of criteria. J. Multi-Criteria Deci. Anal. 5 (2), 145–159. doi:10.1002/(SICI)1099-1360(199606)5:2<145::AID-MCDA99>3.0.CO;2-5.

Saaty, T.L., 2008. Decision making with the analytic hierarchy process. Int. J. Serv. Sci. 1 (1), 83. doi:10.1504/IJSSCI.2008.017590.

Sánchez-Lozano, J.M., Henggeler Antunes, C., García-Cascales, M.S., Dias, L.C., 2014. GIS-based photovoltaic solar farms site selection using ELECTRE-TRI: evaluating the case for Torre Pacheco, Murcia, Southeast of Spain. Renew. Energy 66, 478–494. doi:10.1016/j.renene.2013.12.038.

Silva, S., Alçada-Almeida, L., Dias, L.C., 2014. Development of a web-based multi-criteria spatial decision support system for the assessment of environmental sustainability of dairy farms. Computer. Electron. Agric. 108, 46–57. doi:10.1016/j.compag.2014.06.009.

Silva, S., Alçada-Almeida, L., Dias, L.C., 2015. Multi-criteria sustainability classification of dairy farms in a Portuguese regionAssessment Methodologies: Energy, Mobility And Other Real World Application. Imprensa da Universidade de Coimbra., pp. 343–364. doi:10.14195/978-989-26-1039-9_15.

Tervonen, T., Lahdelma, R., Almeida Dias, J., Figueira, J., Salminen, P., 2007. SMAA-TRI. In: Linkov, I., Kiker, G.A., Wenning, R.J. (Eds.), Environmental Security in Harbors and Coastal Areas: Management Using Comparative Risk Assessment and Multi-Criteria Decision Analysis. Springer, Netherlands, pp. 217–231.

Yager, R.R., 1988. On ordered weighted averaging aggregation operators in multicriteria decisionmaking. IEEE Trans. Syst. Man Cybern. 18 (1), 183–190. doi:10.1109/21.87068.

Yoon, K., Hwang, C.-L., 1995. Multiple Attribute Decision Making: An Introduction. Sage Publications.

Yu, W., 1992. Electre Tri: Aspects Méthodologiques et manuel D'utilisation. Document du Lamsade no. 74. Université de Paris–Dauphine.

CHAPTER 13

Sustainability improvement opportunities for an industrial complex

Rahul Singh Yadav[a], Dilawar Husain[b], Ravi Prakash[a]
[a]*Department of Mechanical Engineering, Motilal Nehru National Institute of Technology, Allahabad, Uttar Pradesh, India*
[b]*Department of Mechanical Engineering, Maulana Mukhtar Ahmad Nadvi Technical Campus, Malegaon Nashik, Maharashtra, India*

13.1 Introduction

The Micro, Small, and Medium Enterprises (MSME) sector in India consumes energy equivalent to about 35 million tons of oil equivalent annually, which is about 30% of the energy consumption by large industries. This sector accounts for 28% of GDP and consists of 63 million industries across the country (BEE 2018). The Government of India promotes entrepreneurship by mentoring, nurturing, and facilitating startups throughout their life cycle (Startup India 2016). For this purpose, industrial estates and complexes have been promoted and developed by the government to provide the enabling infrastructure. However, inadequate provision of required utilities remains a bottleneck for the prospective entrepreneurs.

Various techniques that can reduce the significant amount of energy consumption (such as Solar photovoltaic (PV) system Hong et al. 2017) as well as improve the indoor thermal comfort of an industry are centralized water chiller for air conditioning (Chow et al. 2004; Hamza and Safwatb, 2016), turbo ventilators for better ventilation, and solar light tube for daylight harvesting (Thakkar 2013; Wang et al. 2015), etc. The eco-industrial development practices involves physical exchanges between one another for raw material and wastes/byproducts and sharing/consolidating/coordinating the management of utilities and infrastructures such as water supply, energy utilization, pollutant emissions, and distributions (Gibbs and Deutz, 2007; Yang and Feng, 2008).

This study aims to present a sustainable design of energy systems and utilities for an industrial complex, which consists of the major energy and water infrastructural facilities based on the principle of energy reduction through efficient methods and utilization of renewable energy (mainly, solar energy) for reduced greenhouse gas emissions. It also helps to minimize the demand of bioproductive lands (in gha) during the operation of an industrial complex. The case study presented is that of an industrial complex located at a government engineering institute that is the Motilal Nehru National Institute of Technology (MNNIT), Allahabad, Uttar Pradesh, India. Through energy, cost, and emission analysis of the different energy systems and utilities, the feasibility of the proposed design for this industrial complex has been examined.

13.2 Methodology

13.2.1 Design of the various systems and utilities

Considering the different requirement and with the aim of reducing energy consumption through utilization of renewable energy and more efficient systems, the selected systems and utilities are depicted in Fig. 13.1 and their design methodology is as follows.

13.2.1.1 Rooftop solar photovoltaic system

As the total energy needs of the industrial complex are met by electricity from the grid, the replacement of source of electricity with the grid-connected Solar PV–based electricity would make the industrial complex self-sufficient for its energy requirements. In this study, RETScreen 4 software has been used to assess the feasibility of the rooftop solar PV system.

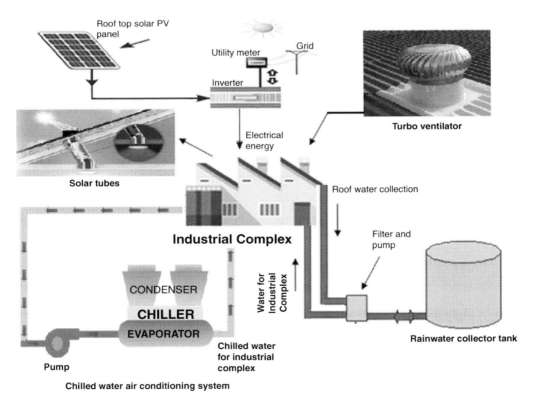

FIG. 13.1 Proposed sustainable systems for the Industrial Complex.

13.2.1.2 Rainwater harvesting system

In India, industries mainly utilize groundwater for processing activities, which can cause the groundwater level to deplete. Rooftop rainwater harvesting system (RHS) may provide approximately 1.4 m³ water per m² of catchment area in India. For RHS design, "A Practical Guide on Roof top Rain Water Harvesting" report (Kalimuthu, 2016) has been used.

13.2.1.3 Solar day-lighting system

Solar daylight harvesting is an efficient method for reducing the artificial lighting energy consumption during daytime either fully or partially in the industrial sheds. For the calculation of solar light tubes along with the conventional lighting, eQuest 3.65 software has been used.

13.2.1.4 Turbo ventilators

Turbo ventilators are powered by the wind to create effective ventilation for different industries. Number of air changes per hour for a space depends on the type of the facility. Most of the sheds in the industrial complex comes under the space type: factory (light).

The number of turbo ventilators (N) required can be calculated as

$$N = (\text{Volume of shed} \times \text{Number of air changes}) / (60 \times \text{Exhaust capacity of ventilator}) \tag{13.1}$$

The standard suggested air changes per hour for light factory are 10–20. Air changes per hour for the industrial sheds have been taken as 12. A standard turbo ventilator of size 24 inches for a roof height of 20 ft and exhaust capacity of 1972 cfm (cubic feet per minute) manufactured by the company Turbo Ventilators India (TVI 2019) was selected.

13.2.1.5 Chilled water air conditioning system

Chilled water air conditioning (using water cooled condensers) is a sustainable way to minimize operational energy for maintaining thermal comfort as compared to window air-conditioners (ACs) (with air-cooled condensers). To calculate the amount of electrical energy that can be saved using chilled water air conditioning, simulation has been performed on eQuest 3.65 software for the industrial complex.

13.2.2 Ecological footprint (EF)

The environmental impact was assessed through potential electrical energy reduction with commensurate reduction in CO_2 emissions. The expression for evaluating EF (in gha) due to carbon emissions is provided as Eq. (13.2):

$$\text{EF due to carbon emissions} = P_c \{(1 - S_{oc})/Y_c\}.ei \tag{13.2}$$

Where, P_C is annual CO_2 emissions, S_{oc} is the fraction of annual oceanic CO_2 sequestration and Y_C is the annual rate of carbon uptake/hectare of forestland at world average yield. The term ei here represents equivalence factor of CO_2 absorption land (1.28 gha/ha, Lin et al. 2016).

13.3 Case study

13.3.1 Survey of MNNIT Industrial Complex

The MNNIT Industrial Complex consists of 21 sheds, of which 9 sheds are in active use. Practically all the utilities (power, cooling, heating, ground water pumping) are dependent on grid electricity with high carbon emissions. For the design of sustainable utility systems, these active sheds were surveyed based on a questionnaire. The data collection included manpower and output products, energy, water consumption, and general information about the shed such as area of plots, machinery, and equipment in use. Fig. 13.2 shows the layout of the Industrial Complex.

13.3.2 Data collection

The relevant data for the nine active sheds were collected to estimate energy and water consumption of the industries along with requirement of appliances and manpower. Total energy and water consumption and the current cooling capacity requirement were calculated for all the industries. From the data collected from all the active sheds, following total loads were estimated: 17,312 kWh of electricity per month, cooling load of 22 TR, water consumption of 45,850 L per month, and diesel consumption of 425 L per month. Also, total roof area available in all 21 sheds is 3581 m^2.

The annual carbon emissions from the existing energy systems and utilities of the industrial complex are evaluated as approximately 260 tCO_2. The above-mentioned energy and utility data were collected from the 9 active industrial sheds, and it is extrapolated to all 21 industrial sheds in the simulation model. Thus for the case when all 21 sheds are actively used, the annual carbon emissions are expected to be nearly 550 tCO_2. The sustainable utility systems and their designs proposed in this study are aimed at reducing such carbon emissions by adopting renewable energy and energy-efficient technologies.

13.4 Results and discussion

The results obtained for the case study are presented as follows.

13.4.1 Rooftop solar PV system

Total area suitable for solar panel installation in all 21 sheds is roughly 3581 m^2 of which around 50% has been considered (other 50% accounting for other installations). The simulation results show that 512 MWh of electricity can be generated annually. The greenhouse gas emissions are reduced from 608 tCO_2/year to 139 tCO_2/year giving a total reduction of 469 tCO_2/year (76.9%).The payback period for the system is about 6.6 years and the system can fulfill 100% of the power requirement of the whole industrial complex by utilizing the roof area available with a surplus generation of 28 MWh annually, which can be sold to the grid. Also, the operational EF reduction from the solar electricity generation is 201.51 gha/year.

13.4 Results and discussion

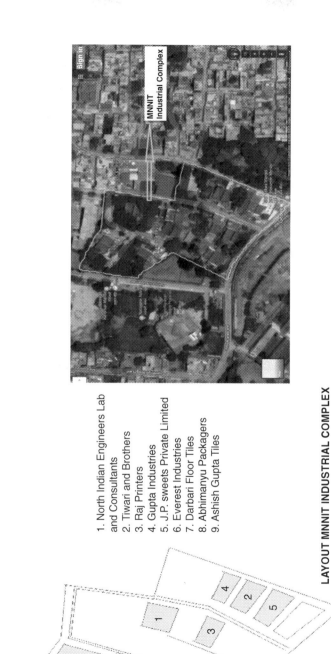

1. North Indian Engineers Lab and Consultants
2. Tiwari and Brothers
3. Raj Printers
4. Gupta Industries
5. J.P. sweets Private Limited
6. Everest Industries
7. Darbari Floor Tiles
8. Abhimanyu Packagers
9. Ashish Gupta Tiles

LAYOUT MNNIT INDUSTRIAL COMPLEX

FIG. 13.2 Layout of MNNIT Industrial Complex.

Table 13.1 Solar panel specification and results.				
Components	Specifications		Cost (INR/kW)	Total Cost (INR)
Poly-Si (Company: EMMVEE)	Rated power at Standard Temperature Conditions (STC)	320 W	32,000	9,408,000
	Module efficiency	16.41%		
	Open circuit voltage—V_{oc} (Volts)	45.7		
	Short circuit current—I_{sc} (Amps)	9.06		
	Dimensions	1970 mm × 990 mm × 35 mm		
	Operating temperature range	−40 °C to 8. °C		
	No. of panels	918		
	Total panel area	1792 m²		
	Total system capacity	294 kW		
Balance of System (BOS)	Inverter (capacity = 294 kW)		30,000	8,820,000
	Spare parts, transportation, training, and commissioning		6000	1,764,000
	Total INITIAL Cost			19,992,000 (approx. 20 million)
Annual maintenance cost	Parts and labor		28	8232
Periodic cost	New inverter installation (~ after 12 years)		30,000	88,200,00
Credit	End of project life			588,000

The technical specifications and cost for the system are presented in Table 13.1.

13.4.2 Rainwater harvesting system

The proposed RHS system capacity (1000 m³) is based on the annual water requirement of the MNNIT industrial complex and the annual rainfall received at Prayagraj city, Uttar Pradesh, India. The proposed RHS system may also recharge the groundwater level of the region.

The EF of the RHS system has been estimated to be 11.68 gha and annual average EF of the system is about 0.584 gha (i.e., 20 year life cycle of the system). The detailed specification showing size, materials, and cost of the RHS for the industrial complex has been presented in Table 13.2.

13.4.3 Solar day-lighting System

The simulation is performed considering a total of four solar light tubes per shed as per the specifications given above for all 21 sheds. Lighting load of 0.96 W/ft² for the workspace has been considered. Fig. 13.3A and B show the prepared model of the sheds with and without solar light tubes in eQuest 3.65 software.

Table 13.2 Rainwater harvesting potential and cost of materials for the system.

Rainwater potential	Annual rainfall × catchment area × run-off coefficient (0.6) = 2270 m³	Roof area = 3852 m²; annual rainfall at Allahabad = 982 mm Run-off coefficient (concrete) = 0.6–0.8 (Pacey and Cullis (1989))				
Component	*Specifications*	*Materials required (for four tanks)*				
Tank	$V = t \times n \times q$; V = Volume of the tank (L); t = length of dry season(days); $q \times n$ = consumption, (L/day) V = 250 days × 3570 L/day = 892,500 L = 892.5 m³ 4 Tanks of 250 m³ each	Material	Quantity	Unit cost (INR)[a]	Cost (INR)	The total cost of the four tanks is around INR 1.14 million.
		Brick	24,000	7/brick	168,000	
		Cement	21,178.4 kg	350/50 kg pkt	148,248	
		Steel rod	644.38 kg	45/kg	289,97	
		Aggregate	34 m³	4000/50 ft³	96,055	
		Gravel	160 m³	3500/50 ft³	395,523	
		Sand	30 m³	3500/50 ft³	74,160	
		4" PVC pipes, 20 feet in length	120 No.	600/20 ft	72,000	
		Concrete Mixer	40.4 h	500/h	20,200	
		Vibrator	40.4 h	400/h	16,160	
Manpower	272 labor days	Skilled = INR 500/day = INR 68,000 Unskilled = INR 300/day = INR 40,800				
Material transportation	Cost = INR 3/km ton (400 tons)	Total cost = INR 12,230 Total weight = 401,782 kg				

[a]Costs are obtained from market survey

The simulation results show that the annual energy consumption without skylights or by conventionally lighting the workspace (area lighting) is 20,970 kWh. On performing the simulation with the light tubes, the annual energy consumption was reduced to 12,870 kWh which is 38.6% reduction in electrical energy consumption for lighting annually. The reduction in electrical energy consumption is 8100 kWh annually, which saves carbon emissions of 8.9 tCO_2 per year and the commensurate operational EF reduction is 3.187 gha/year. The cost of purchase and installation for 1 light tube is INR 5800 and therefore cost of 84 light tubes for 21 sheds amounts to INR 487,200. Cost of the electricity saved by day light harvesting annually is INR 64,800, providing a payback period of 7.5 years.

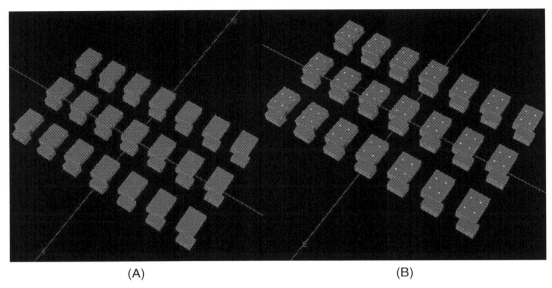

FIG. 13.3 Industrial sheds.

(A) without solar light tubes, (B) with solar light tubes.

13.4.4 Turbo ventilators

Based on the size of the shed and other parameters, two turbo-ventilators (based on Eq. 13.2) would be required in each shed with a total of 42 for all the sheds to achieve the required ventilation. The total cost (purchase and installation) of 1 turbo-ventilator amounts to INR 4657. Hence, the total cost of 42 turbo-ventilators for all the sheds is INR 195,615. Also, if exhaust fans of same capacity are used of 400 W power rating, the annual electricity consumption would be 61,320 kWh costing INR 490,560 for all 21 sheds; which is totally saved by replacing the exhaust fans. The payback period for turbo-ventilators is 4.7 months. By saving 61,320 kWh of electricity annually, 68.06 tCO_2 emissions are reduced with commensurate reduction in operational EF as 24.13 gha/year.

13.4.5 Chilled water air conditioning system

Annual electricity consumption for the baseline design (using split ACs for cooling) containing nine sheds is 209 MWh, which is almost equal to the yearly electricity consumption obtained from the survey of nine active sheds thereby validating the simulated results. For the analysis of the whole industrial complex, the model is extended for all the 21 sheds. The annual electricity consumption obtained in the extended model on performing the simulation for all sheds is obtained as 487 MWh. Also, the electricity consumption for space cooling in the baseline extended model (which has split ACs) is 107 MWh/year.

The chilled water system of 24 TR cooling capacity will be required. On replacing the split ACs with the chilled water system, the annual electricity consumption for cooling reduced from 107 MWh to 80 MWh. Hence, annually 27 MWh (i.e., 25%) of cooling electricity consumption can be saved by replacing the individual split ACs with the centralized chilled water system. Simultaneously, it leads

Table 13.3 Economic and environmental benefits of proposed systems.

Proposed systems	Electrical grid energy saving (kWh/year)	Emission reduction (tCO$_2$/year)	Payback period (year)	Operational EF reduction (gha/year)
Rooftop solar PV system	512,000	469	6.6	201.51
Rainwater harvesting system	Marginal	Marginal	-	Marginal
Daylight harvesting system	8100	8.9	7.5	3.187
Turbo-ventilators system	61,320	68.06	0.40	24.134
Chilled water air conditioning system	27,000	29.97	-	10.626

to an annual reduction of 29.97 tCO$_2$. The operational EF reduction due to reduced CO$_2$ emissions is evaluated as 10.6 gha/year.

13.4.5.1 Comparative assessment of proposed systems

The electrical energy saving, emission reduction, and operational EF reduction due to the proposed sustainable utility systems have been presented in Table 13.3. The combined emission reduction potential and operational EF reduction potential are evaluated as 575 tCO$_2$ and 240 gha/year, respectively.

From this comparative assessment, it is observed that energy efficiency measures alone may reduce the existing annual carbon emissions (~ 600 tCO$_2$) by nearly 18%. In addition, if rooftop solar PV is further employed as proposed, then the major annual carbon emissions may be offset leading to a "near net-zero carbon" industrial complex.

13.5 Scope of future work

Various studies have suggested different tools to measure industrial sustainability in recent times. However, there seems to be no definite criterion on which industrial clusters or complexes may be compared. Looking at the Sustainable Development Goals (SDGs) to be achieved by 2030, the SDG 9 relates to industry, innovation, and infrastructure (SDGs, 2015). Hence, there is a strong need to enhance sustainability in the industrial sector not only at the level of individual industries but also as a cluster of industries or industrial complexes. In such complexes, various industries exist not only in close geographical proximity but may also operate in a symbiotic mode through exchange of materials, energy, water, and wastes. As several such industrial complexes are already operating in different locations and regions, a question arises if they can be compared on the basis of sustainability to set benchmarks and learn from best practices for improved performance in the industrial sector.

Dong et al (2013) reported that the total industrial carbon footprint of Shenyang Economic and Technological Development Zone was 15.29 Mt in 2007, of which onsite emissions, upstream emissions, and downsides emissions accounted for 44.57%, 55.40%, and 0.02%, respectively. Budihardjo et al (2013) reported that the EF of an industrial zone of Central Java, Indonesia is about 3755 gha. However, such studies only concentrated on environmental sustainability of industrial clusters and did not cover the economic and social aspects of sustainability.

Pandey and Prakash (2018, 2020) used the concept of Industrial Sustainability Index (ISI) to compare various industries (similar or dissimilar) based on three major attributes of sustainability, that is economic, social, and environmental. The simplicity of this indicator enabled comparing the process/manufacturing industries rather than product-based analyses as suggested by initial indicators. This ISI is a simplified tool, which represents the socioeconomic benefit of any type of industry per unit of its carbon emissions. Carbon emissions could be direct and indirect. The direct carbon emissions may be due to the inherent nature of the production process, while the indirect emissions are caused due to energy consumption in the production process. The ISI as proposed assesses social, economic, and environmental goals of any type or types of industries (i.e., small, medium, or large scale). The ISI tool was also employed to examine the impact of industrial symbiosis on sustainability (Pandey and Prakash, 2019). They observed that industrial symbiosis is helpful in improving sustainability only in conditions where energy-efficient technologies are employed.

The concept of ISI may also be used to quantitatively assess the economic, social, and environmental sustainability of industrial clusters or complexes as illustrated in Fig. 13.4 and elaborated next.

The evaluation of the ISI for an industrial complex may be carried out using Eq. (13.3):

$$ISI = \frac{(RVA) \times (EMP)}{CO_2 \text{ emissions}} \qquad (13.3)$$

Here, the term "RVA" represents the resource value addition from an industrial complex (i.e., the difference of the total annual economic values of material, water and energy outputs [products], and that of inputs). It is represented here as million INR per year. The limitation of RVA using Indian currency (INR) can be overcome if the RVA is expressed in US dollars with purchasing power parity (i.e., PPP $). The use of purchasing power parity can make the RVA units universal in nature rather than being country-specific.

The term "EMP" represents the total number of persons employed by the industrial complex in a year. This refers to manpower with full-time employment (FTE, i.e., 8 h per day). The manpower employed for less than 8 h per day can be proportionately converted to the equivalent FTE. For example, employment for 4 h per day would be equivalent to 0.5 FTE.

According to Eq. (13.3), the employment count is proportional to the ISI index. It means those industries providing more employment achieve higher ISI values. Hence, the industrial clusters with high ISI values also help to reduce the GINI coefficient of the region or country by supporting employment and reducing income inequality for achieving social sustainability. More than economic security, employment leads to human dignity. Rather than seeing labor as a factor of production alone, the focus

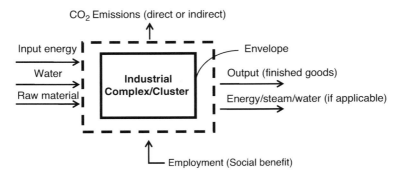

FIG. 13.4 Sustainability analysis for an industrial complex.

should be on the human dignity of every worker by allowing every single individual to use his/her physical and mental capacity to prosper.

In case an industrial complex is a net-zero (or negligible) emission complex, the quantitative evaluation of ISI may be carried out by assigning a nominal value of 1 tCO_2 emissions. For cases where an industrial complex is carbon negative, an appropriate value of less than 1 but greater than zero for carbon emissions may be assigned in the denominator of the ISI expression. This would facilitate in achieving a higher numerical value of ISI for carbon negative industrial complexes as compared to zero carbon or carbon-positive industrial complexes.

13.6 Conclusions

The installation of the designed utility systems can make the industrial complex more efficient in terms of energy consumption and also leads to reduced carbon emissions. All the proposed systems were found to be cost effective. From the results obtained it is found that a total of 512 MWh electricity can be generated annually by rooftop solar PV system which can fully power the industrial complex. The surplus solar electricity available can be sold to the grid or potentially utilized to power the centralized cooling system.

The results also show that that RHS is capable of providing water throughout the year and therefore reduces the huge burden of the industrial complex on groundwater aquifers. The rainwater potential, for the available roof area of the industrial complex, is more than its total annual water consumption. Daylight harvesting can save the annual electricity consumption for artificial lighting by 38.6% with commensurate reduction in carbon emissions. The installation of turbo-ventilators can effectively replace exhaust fans, thereby saving significant carbon emissions. The chilled water air-conditioning system was found to be more energy efficient than the individual window/split ACs, with significant potential of carbon and operational EF reduction.

All the proposed utility systems, if employed, can potentially offset the entire annual carbon emissions from the industrial complex thereby leading to a "net-zero carbon" industrial complex. Such sustainable utility systems may also be adopted for other industrial complexes in the country to achieve the goal of green industry.

Short Biography of the Authors

Rahul Singh Yadav completed his Master's studies at the Department of Mechanical Engineering at the Motilal Nehru National Institute of Technology, Allahabad, Uttar Pradesh, India. His academic interests lie in the domain of renewable energy systems.

Dilawar Husain has completed his PhD from the Department of Mechanical Engineering at the Motilal Nehru National Institute of Technology, Allahabad, Uttar Pradesh, India. He is currently working as an Assistant Professor with the Department of Mechanical Engineering, Sandip University, Nashik, Maharashtra, India. His academic interests include life cycle analysis, renewable system technologies, and sustainability.

Dr Ravi Prakash is currently working as a Professor with the Department of Mechanical Engineering at the Motilal Nehru National Institute of Technology, Allahabad, Uttar Pradesh, India. His academic interests include energy management, life cycle energy analysis, and sustainability. He is a Fellow of the Institution of Engineers (India) and a member of the International Solar Energy Society.

References

Budihardjo, S, Hadi, S.P., Sutikno, S., Purwanto, P., 2013. The ecological footprint analysis for assessing carrying capacity of industrial zone in Semarang. J. Hum. Resour. Sustain. Stud. (1), 14–20. doi:10.4236/jhrss.2013.12003.

Bureau of Energy Efficiency, BEE, 2018. Annual Report 2017-2018, Ministry of Power. Government of India, New Delhi, India https://beeindia.gov.in/content/annual-report, (Accessed November 9, 2019).

Chow, T.T., Fong, K.F., Chan, A.L.S., Yau, R., Au, W.H., Cheng, V, 2004. Energy modelling of district cooling system for new urban development. Energy Build. 36, 1153–1162. doi:10.1016/j.enbuild.2004.04.002.

Dong, H., Geng, Y., Xi, F., Fujita, T., 2013. Carbon footprint evaluation at industrial park level: a hybrid life cycle assessment approach. Energy Policy 57, 298–307. doi:10.1016/j.enpol.2013.01.057.

Gibbs, D., Deutz, P., 2007. Reflections on implementing industrial ecology through eco-industrial park development. J. Clean. Prod. 15 (17)), 1683–1695. doi:10.1016/j.jclepro.2007.02.003.

Hamza M.H., Safwatb, H. (2016), Proposed district cooling plant for the British University in Egypt Campus, In: Y. Bahei-El-Din and M. Hassan (Eds.) Advanced Technologies for Sustainable Systems, Springer, Cham doi:10.1007/978-3-319-48725-0_12.

Hong, T., Lee, M., Koo, C., Jeong, K., Kim, J., 2017. Development of a method for estimating the rooftop solar photovoltaic (PV) potential by analyzing the available rooftop area using hillshade analysis. Appl. Energy 194, 320–332. doi:10.1016/j.apenergy.2016.07.001 http://www.erpublications.com/uploaded_files/download/download_14_06_2013_19_37_02.pdf, (Accessed November 15, 2019).

Kalimuthu, A., 2016. A Practical Guide on Roof top Rain Water Harvesting. WASH Institute, New Delhi, India.

Lin, D., Hanscom, L., Martindill, J., Borucke, M., Cohen, L., Galli, A., Lazarus, E., Zokai, G., Iha, K., Eaton, D., Wackernagel, M., 2016. Working Guidebook to the National Footprint Accounts. Global Footprint Network (GFN), Oakland https://www.footprintnetwork.org/content/documents/National_Footprint_Accounts_2016_Guidebook.pdf, (Accessed December 3, 2019).

Pacey, Arnold Cullis, Adrian, 1989. Rainwater Harvesting: The Collection of Rainfall and Runoff in Rural Areas. Intermediate Technology Publications, London, England, pp. 55 pg.

Pandey, A.K., Prakash, R., 2018. Industrial sustainability index and its possible improvement for paper industry. Open J. Energy Effic. 7, 118–128. doi:10.4236/ojee.2018.74008.

Pandey, A.K., Prakash, R., 2019. Impact of industrial symbiosis on sustainability. Open J. Energy Effic. **8**, 81–93. doi:10.4236/ojee.2019.82006.

Pandey, A.K, Prakash, R., 2020. Opportunities for sustainable improvement in aluminium industry. Eng. Rep. doi:10.1002/eng2.12160.

Startup India, 2016. Ministry of Commerce and Industry. Government of India, Delhi, India https://www.startupindia.gov.in/content/sih/en/about_us/about-us.html, (Accessed December 14, 2019).

Sustainable Development Goals (SDGs), 2015. United Nation Development Programme https://www.undp.org/content/undp/en/home/sustainable-development-goals.html, (Accessed December 15, 2019).

Thakkar, V.N., 2013. Experimental study of tubular skylight and comparison with artificial lighting of standard ratings. Int. J. Enhanc. Res. Sci. Technol. Eng. 2 (6), 1–6.

Turbo Ventilators India (TVI): Industrial Ventilating Solutions (2019), Dehradun, India. http://www.turboventilatorsindia.com/features-specification-ventilators.html, (Accessed December 15, 2019).

Wang, S., Jianping, Z., Lixiong, W., 2015. Research on energy saving analysis of tubular daylight devices. Energy Procedia 78, 1781–1786. doi:10.1016/j.egypro.2015.11.305.

Yang, S., Feng, N., 2008. A case study of industrial symbiosis: Nanning Sugar Co., Ltd. in China. Resour. Conserv. Recycl. 52 (5)), 813–820. doi:10.1016/j.resconrec.2007.11.008.

CHAPTER 14

Coupled life cycle assessment and data envelopment analysis to optimize energy consumption and mitigate environmental impacts in agricultural production

Ashkan Nabavi-Pelesaraei[a], Zahra Saber[c], Fatemeh Mostashari-Rad[d], Hassan Ghasemi-Mobtaker[b], Kwok-wing Chau[e]

[a]*Department of Mechanical Engineering of Biosystems, Faculty of Agriculture, Razi University, Kermanshah, Iran*
[b]*Department of Agricultural Machinery Engineering, Faculty of Agricultural Engineering and Technology, University of Tehran, Karaj, Iran*
[c]*Department of Agronomy and Plant Breeding, Sari Agricultural Sciences and Natural Resources University, Sari, Iran*
[d]*Department of Agricultural Biotechnology, Faculty of Agricultural Sciences, University of Guilan, Rasht, Iran*
[e]*Department of Civil and Environmental Engineering, Hong Kong Polytechnic University, Hung Hom, Kowloon, Hong Kong*

14.1 Introduction

Different forms of direct (animal power, fuels, electricity, and human labor) and indirect energy (fertilizer, herbicide, and machinery), and also renewable and nonrenewable energy forms, are important in crop production. The consumption of various energy forms in crops has escalated throughout the years, due to increasing global demand for food (Kosemani and Bamgboye, 2020). Excessive use of energy resources, however, exacerbates the agricultural sector' environmental impacts. Agriculture sector accounts for 10%–12% or so of all greenhouse gases (GHG) with carbon dioxide (CO_2) emissions, according to previous reports on GHG emissions, as a contamination indicator (Hosseini-Fashami et al., 2019). As a powerful and popular tool in the environmental assessments, life cycle assessment (LCA) plays a key role in environmental auditing, risk evaluation, environmental impacts, and environmental performance assessment. Throughout the life cycle of a service or/and product, LCA is exposed to all its environmental impacts, and this is one of the unique features of LCA method. Therefore, all environmental effects are considered in all stages of production. As a feasible and identification management tool, LCA is used to compare different environmental options' performance and help option selection. Researchers have recently applied LCA management tools to make decisions in assessing environmental impacts of various agricultural products as well as examining different production systems from an environmental perspective (Abeliotis et al., 2013). Improving energy efficiency to diminish environmental effects on agroecosystems will be beneficial, given the connection between environmental impacts and energy consumption. Furthermore, the goal of accomplishing sustainable production is to protect fossil resources, save production costs, and reduce environmental pollution (Nabavi-Pelesaraei et al., 2017c). In this case, there are various methods to reduce environmental impact and energy optimization. Among these, one of the popular methods for this purpose is data envelopment analysis (DEA) (Mohammadi et al., 2013). As a powerful method in the literature, DEA is more appropriate for performance evaluation activities than traditional econometrics method, involving simple ratio and regression analysis (Zhu, 2014). In fact, DEA is considered as a statistical method using linear programming techniques to convert inputs into outputs, with the aim of assessing comparable products or organizations. In this method, each decision making unit (DMU) is free to select any input and output combination for enhancing its relative efficiency. The efficiency score or relative efficiency is actually the total weighted output to the total weighted input's ratio (Zhu, 2014). In this method, linear programming is used to compute the relative efficiency. As a result of DEA, the relative efficiency is the efficiency score assigned to a DMU. In fact, this energy efficiency is analyzed as a nonnegative value estimated on the linear relationships among inputs and outputs of the DMUs under assessment. Compared to similar DMUs, it actually specifies how efficient a DMU is in producing outputs at certain levels, according to the input amount (Mardani et al., 2017).

Relationships among input and output energies in agricultural systems considering environmental aspect are very complicated. In most items, optimization effects are focused on energy only but in this chapter, we investigate energy optimization by DEA and its effect to reduce environmental life cycle impacts as a multiobjective decision-making approach. In fact, joining DEA and LCA has been expanded comprehensively, and various elements are described here. Furthermore, some examples are expressed for better recognition of this optimization of energy-environmental aspects of agricultural products and finally, the interpretation of results are investigated.

14.2 Data collection

As farm conditions such as soil type, weather, operations, and equipment used, etc. vary in different case studies, studies should focus on specific areas in a country, such as a city. It is possible that in a research work, the pattern of energy consumption in a country is examined. In that case, the collection of data must be carried out separately in various country areas. In case studies, descriptions of information related to planting systems of the studied crops, involving statistics of planting areas, annual yields, etc. are effective to better understand the importance of those crops. Furthermore, it is possible to compare and generalize the results better when referring to geographical locations such as latitude and longitude, altitude, and type of soil under study (Nabavi-Pelesaraei et al., 2021). As shown in Fig. 14.1, describing the geographical location of case studies can be proper for identifying the surveyed areas for a clearer evaluation of the analysis.

In the data collection stage, the number of farmers (or in other words, the statistical population) should be determined. If the number of statistical population is limited and accessible, the census method should be used to collect data. But in most cases, the large number of statistical population makes the census work time-consuming, costly, and impossible. Therefore, in these circumstances, the sampling method is usually proposed, which examines a certain number of the statistical population. As the number of samples should be large enough so that their properties and behavior can be generalized to the whole community, various methods have been proposed in statistics to determine the optimal

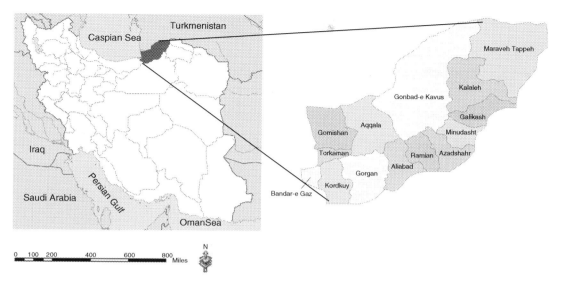

FIG. 14.1 Sample diagram of a case study in the north of Iran.

sample size. To determine the sample size, the following three fundamental formulas of sampling method are applied:

1. *Cochran method*: In this method, Cochran emphasized that in the case of limited population, the sample size can be slightly reduced. To compute the sample size under these conditions, he proposed the following formula (Cochran, 1977):

$$n = \frac{\frac{z^2 pq}{d^2}}{1 + \frac{1}{N}\left(\frac{z^2 pq}{d^2} - 1\right)} \quad (14.1)$$

Here, d represents the ratio deviation of allowable error from the average population ($= 0.05$), p denotes the computed attribute proportion inside the population ($= 0.5$), q is $1-p$ ($= 0.5$), n shows the involved sample size, z denotes the reliability coefficient ($= 1.96$ at 95% confidence level), and finally N represents the ratio of the number of milling factories to the target population.

2. Yamane (1967) presented another simple formula to replace Cochran formula. He suggested the sample size of $p = 0.5$ at 95% confidence level as follows:

$$n = \frac{N}{1 + N(e^2)} \quad (14.2)$$

where e denotes the precision level (95%) and N represents the size of population.

3. *Neyman allocation*: Categorized samples are employed with this sample allocation method. Maximizing the examination accuracy for specific sample sizes is the main objective of this method. By applying the allocation method, the best sample size of stratum h is presented as follows (Mohammadi and Omid, 2010):

$$n = \frac{\left(\sum N_h S_h\right)}{N^2 D^2 + \sum N_h S_h^2} \quad (14.3)$$

Table 14.1 Summary of previous works on the use of energy and environmental emissions in agro-ecosystems from sample size viewpoint.

Research work	Method of sample size	Computed rate of sample size
Cetin and Vardar (2008)	Neyman	95
Samavatean et al. (2011)	Cochran	136
Ozkan et al. (2011)	Neyman	85
Banaeian et al. (2011)	Cochran	25
Hemmati et al. (2013)	Cochran	50
Kazemi et al. (2015)	Yamane	30
Amid et al. (2016)	Neyman	70
Taki et al. (2018)	Neyman	120 irrigated and 90 rainfed farms
Aydın and Aktürk (2018)	Cochran	55
Soni et al. (2018)	Yamane	55 paddy rice–wheat and 48 paddy rice–potato systems
Fathollahi et al. (2018)	Cochran	25 alfalfa farms and 21 corn silage
Unakıtan and Aydın (2018)	Neyman	169
Valdivieso Pérez et al. (2019)	Cochran	50
Elahi et al. (2019)	Neyman	360
Naderi et al. (2020)	Cochran	91
Musafiri et al. (2020)	Cochran	299

where $D^2 = d^2/z^2$; z shows the coefficient of reliability (1.96 representing 95% reliability); d is the precision where $(x- \overline{X})$; S_h^2 shows the h stratification's variance; S_h represents h stratification's standard deviation, N_h presents the population number in h stratification; N depicts the holding number in the target population, and finally n shows the involved sample size.

Table 14.1 presents the sample size descriptions in some research works.

As a research tool, a questionnaire is applied to collect data after determining the sample size. In fact, a questionnaire refers to a set of questions to gather some sort of data from farmers in agricultural studies. Items such as fully computerized, adaptive computerized questionnaires with administration, paper, pencil, and face-to-face way, etc. are among different types of questionnaires based on Mellenbergh (2008). The first one is some type of questions given on the computer and next one sorts some of questions in the computers, in relation to the previous questions' answers. The second one is a kind of questionnaire in which interviewees must answer the questions with the help of the computers. The third one is a kind of questionnaire in which the items can be written on a piece of paper. The last is a questionnaire that consists of some sort of oral questions asked by an interviewer.

Questionnaires in energy studies and LCA in agriculture sector involve two main parts. The first part involves general information about the crop type, production method, producer specifications, meteorological information, and the farm used characteristics, comprising of farm size, soil structure, irrigation systems, harvesting method, and materials, etc. The second part addresses the consumption of all inputs separately for each operation. For example: What agricultural machineries have been used in tillage operations? What is the weight of the agricultural machineries? How much is the machineries effective life? How many labors are needed for this operation? What is the type and amount of fuel consumed? etc.

Table 14.2 summarizes a sample questionnaire for data gathering in agricultural studies.

Table 14.2 Summary of a sample questionnaire for agricultural and livestock products.

Questionnaire No: ….
Company Name: ….
Data: ….
The cultivation area or the number of fattening, broiler, calf, etc.: ….
Production duration: ….
Seed weight or mean weight of broiler, fattening, calf, etc. at the starting point of period: ….
Total output per FU (100 m^2, ha, 1000 broiler, 1000 calf, etc.) [a]: ….

A. Inputs:

1. *Labor*

Fixed labors' number in production duration: ….
Daily labors' number in production duration: ….
Working hours per day: ….
Gender of workers: ……

2. *Machinery*

Machinery operation used in production duration: ….
Machinery types applied in production duration: ….
Total machinery weight: ….
Economic life of machinery (h): ….

3. *Fuels*

Diesel fuel applied (Gasoline, petrol, etc.): ….
The total consumption of fuel in production duration: ….
Total engine lubricant consumption in production duration: ….
Total fuel consumption for other purposes (heating, etc.): ….

4. *Chemicals*

Types of chemical fertilizers and its analysis (N, P_2O_5, K_2O, etc.): ….
Total chemical fertilizers weight consumption from each type: ….
Chemical biocides used: ….
The amount of biocides (weight) from each type: ….

5. *The weight of farmyard manure:* ….

6. *Others energy Source*

Electricity consumption rate in each implementation: ….
Total electricity consumption in production duration: ….
Total natural gas consumption: ….

7. *Water*

Water supply source: ….
The amount of water consumed: ….

8. *Other inputs:* ….

B. Outputs weight: …

[a] *The measurement unit depends on the energy use and production' scale.*

A summary of the process of gathering requirement information to start energy-LCA studies in agricultural systems is demonstrated in Fig. 14.2.

14.3 Energy in agriculture
14.3.1 Energy analysis

Energy analysis is performed to quantify the energy input needed for producing specified goods or services and the energies produced through the harvesting of products (Yilmaz and Aydin, 2020). All energy inputs that are transformed throughout the whole stages of production systems, or in the life cycle, are referred to as input energies (Pelletier et al., 2011). These agricultural operations involve seedbed preparation, plowing, planting, manual weeding, fertilizing, spraying, and harvesting leading to some activities that bring energy use to the product. Some important inputs consist of agricultural machinery, human labor, fertilizers, fuels, electricity, biocides, and seed. Output energy involves the final product, which varies according to agricultural ecosystems. In wheat agricultural production, as an instance, the output energy involves wheat yield and straw; however, broiler yield is referred to output energy in poultry production system.

All inputs and outputs of the agroecosystems have the standard coefficients accompanied by standard energy, because of their limitation. Hence, each input and output's energy values can be computed by multiplying the energy coefficient with the pertinent physical value. Table 14.3 outlines the coefficients of each energy input attained by previous research works.

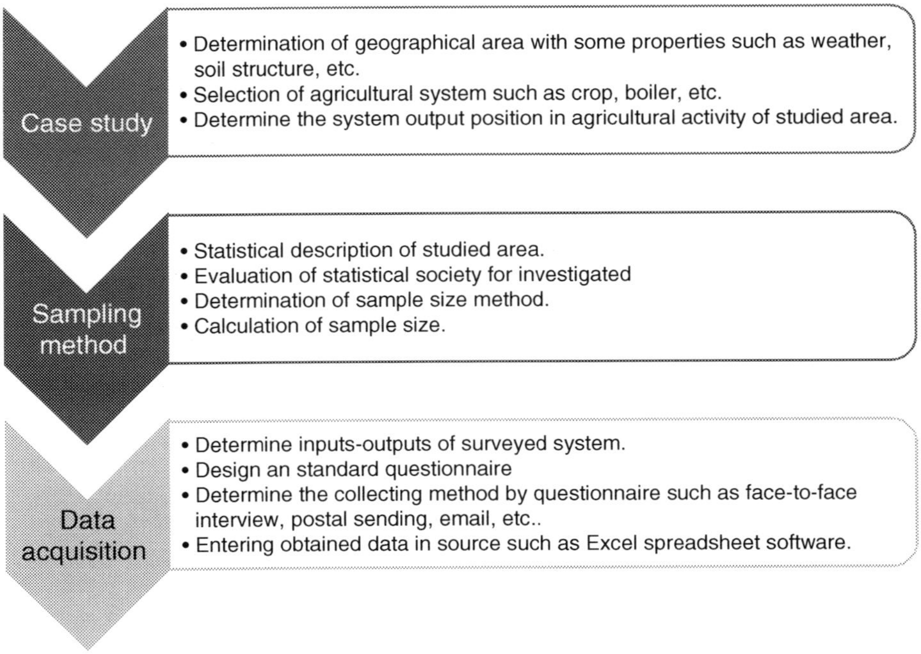

Case study
- Determination of geographical area with some properties such as weather, soil structure, etc.
- Selection of agricultural system such as crop, boiler, etc.
- Determine the system output position in agricultural activity of studied area.

Sampling method
- Statistical description of studied area.
- Evaluation of statistical society for investigated
- Determination of sample size method.
- Calculation of sample size.

Data acquisition
- Determine inputs-outputs of surveyed system.
- Design an standard questionnaire
- Determine the collecting method by questionnaire such as face-to-face interview, postal sending, email, etc..
- Entering obtained data in source such as Excel spreadsheet software.

FIG. 14.2 Flowchart of data gathering for energy-LCA researches in agricultural production.

Table 14.3 Coefficients of energy in different operation of agriculture.

Inputs used (unit)	Energy equivalent (MJ/unit)	References
1. Human labor (h)	1.96	Beheshti Tabar et al. (2010)
2. Machinery		
(a) Tractor and self-propelled (kg)	9–10	Hatirli et al. (2005)
(b) Implement and machinery (kg)	6–8	Hatirli et al. (2005)
(c) General machinery (kg)	142.70	Kaltsas et al. (2007)
(d) machinery applied (h)	62.70	Mobtaker et al. (2010)
3. Fuel		
(a) Diesel fuel (l)	56.31	Kizilaslan (2009); Mostashari-Rad et al. (2019)
(b) Gasoline (l)	46.3	Hosseinzadeh-Bandbafha et al. (2017)
(c) Oil (l)	36.7	Hosseinzadeh-Bandbafha et al. (2017)
(d) Natural gas (m^3)	49.5	Hosseinzadeh-Bandbafha et al. (2017)
(e) Lubricant (l)	43.80	Zentner et al. (2004)
4. Chemical fertilizers (kg)		
(a) Nitrogen	66.14	Bakhtiari et al. (2015)
(b) Phosphate (P_2O_5)	12.44	Nabavi-Pelesaraei et al. (2018)
(c) Potassium (K_2O)	11.15	Erdal et al. (2007); Nabavi-Pelesaraei and Amid (2014)
(d) Zinc (Zn)	8.40	Kazemi et al. (2015)
(e) Sulfur (K_2O)	1.1	Mousavi-Avval et al. (2011a)
(f) Ferrum (Fe^{2+})	6.3	Nabavi-Pelesaraei et al. (2014)
5. Farmyard manure (kg)	0.3	Kizilaslan (2009)
6. Biocides (kg)		
(a) Herbicides	85	Pishgar-Komleh et al. (2012)
(b) Insecticides	199	Nabavi-Pelesaraei et al. (2018); Ozkan et al. (2004)
(c) Fungicides	92	Nabavi-Pelesaraei et al. (2018); Ozkan et al. (2004)
7. Electricity (kWh)	11.93	Jekayinfa et al. (2013)
8. Feed		
(a) Concentrate (kg)	13.6	Hosseinzadeh-Bandbafha et al. (2017)
(b) Maize silage (kg)	10.41	Hosseinzadeh-Bandbafha et al. (2017)
(c) Dry alfalfa (kg)	10.92	Hosseinzadeh-Bandbafha et al. (2017)
(d) Barley (kg)	15.28	Hosseinzadeh-Bandbafha et al. (2017)
(e) Soybean meal (kg)	12.06	Atilgan and Hayati (2006)
(f) Dicalcium phosphate (kg)	10	Atilgan and Hayati (2006)
(g) Fatty acid (kg)	9	Heidari et al. (2011)
(h) Vitamins and minerals (m^3)	1.59	Heidari et al. (2011)
9. Irrigation (m^3)	0.63	Hatirli et al. (2006)
10. Nylon (kg)	17.91	Kitani, 1999
11. Transportation (t.km)	4.5	Nabavi-Pelesaraei et al. (2017b) (2017a)
12. Seed (kg)	As production system' output	

They are not in relation to the kind of crop; however, they have fixed values. As an instance, both phosphate and electricity have specific coefficients for conversion to the energy consumption's equivalent in various output type. So, each input value would be an important difference for agroecosystems in the use of energy. Furthermore, the findings are compared with other research works using the coefficient of standard energy to estimate the consumption amount of energy. Moreover, input and output energies can be computed by multiplying each item to its energy equivalent presented in Table 14.3.

The machinery's energy tantamount is depicted in the following equation (Mousavi-Avval et al., 2017b):

$$ME = \frac{ELG}{TC_a} \tag{14.4}$$

where C_a shows the effective capacity of field (ha/h) computed by Hatirli et al. (2005), T represents the machineries' economic life (h), G is the machines' mass (kg), L shows the useable life of machines (years), E depicts the production energy of machines (GJ/kg year) from Table 14.3 and finally ME shows the machines' energy (GJ/ha).

$$C_a = \frac{SWE_f}{10} \tag{14.5}$$

where E_f is the field efficiency, W represents the working width (m), and S shows the working speed (km/h) for a machine's economic life of y years.

14.3.2 Energy indices and forms

Evaluating input and output energies alone cannot indicate the efficiency or inefficiency of energy consumption in the agricultural sector. Hence, for this purpose, energy indices are used to analyze input-output energy, as introduced below for each of them.

1. *Energy use efficiency* is computed as crop yield energy divided by all inputs. Energy use efficiency also assesses the efficiency of systems in applying the support energy provided by crop husbandry (Alluvione et al., 2011).
2. *Energy productivity* describes how much yield is achieved in each agricultural system per MJ of energy consumption (Jat et al., 2020).
3. *Specific energy* indicates how much energy is consumed per kilogram of final product. Indeed, specific energy is the opposite of the energy productivity (Kaveh et al., 2020).
4. *Net energy*, which is energy gain, is a very important index related to the difference of all energy inputs and removed energy outputs. Indeed, net energy is an essential parameter when the arable land availability for plant cultivation is limited (Alluvione et al., 2011).

The use of energy in agroecosystems may be investigated by two approaches. The first approach is a kind of inserting energy to system, which divides into two groups including the usage of direct energy—referring to inputs transformed at the supply chain node of concern, such as the energy inputs of field-level to dry crops or to power agricultural machineries—and the usage of indirect energy, referring to the cumulative energy required in upstream processes in connection with the delivery and production such inputs (Pelletier et al., 2011). Another approach is based on conservation of energy resources, which is divided into two categories including renewable energy as energy collected from renewable resources, replenished on a human timescale naturally (Ali et al., 2020), and nonrenewable energy obtained from resources that will not be replenished or will run out during our life process. Most nonrenewable energy resources are fossil fuels involving coal, petroleum, and natural gas (Aydoğan and Vardar, 2020).

14.4 Life cycle assessment

LCA refers to a method that compares and evaluates various services or products in the field of their environmental impacts throughout the life cycle from the extraction of raw materials to the end of a service or product life (ISO, 2006). To supply decision supports is the important purpose of LCA in the value chain of the product. The total impacts of environment are manifested by the functional unit (FU) (Saber et al., 2020). LCA is considered as a very suitable technique to assess the environmental effects of the ecosystems, increasingly applied to evaluate the product's ecological sustainability (Roy et al., 2009). Indeed, LCA is the most useful method and comprehensive tool available to prevent problem-shifting from one stage of life cycle to the other stage and it is able to analyze all potential environmental burdens throughout the life cycle of product (ISO, 2006) involving the supply chain and downstream processes (Finnveden et al., 2009). The results of LCA can be applied to make decision for producers, policy makers, and consumers in choosing sustainable production processes and products (Meier et al., 2015). In total, LCA comprises the following four steps (Kaab et al., 2019a, 2019b):

1. Scope and goal description
2. Life cycle inventory (LCI) analysis
3. Life cycle impact assessment (LCIA)
4. Interpretation of final results.

14.4.1 Scope and goal definition

The ultimate definition is created by the organizations, such as trade group, industry, environmental offices, or companies, commissioning LCA. In fact, product systems basically involve the systems function as a basis to determine FU. In agricultural production, physical amount of yield (1 kg, 100 kg, 1 t, 10 t, etc.) is usually used as FU. Furthermore, to define the system boundaries is another main step in LCA. The quality and accessibility of data would be essential matters in life cycle inventory (LCI) (Zanghelini et al., 2018). A sample system boundary in agricultural production is demonstrated in Fig. 14.3.

14.4.2 Life cycle inventory

LCI step comprises making flow inventory of the service or product. In fact, it involves inputs, energy, crude substances along with all of emissions to soil, water, and air. The flow model of system is actually made by data related to the required input and output to build the model, which are collected from all activities within the system boundary (Bianco and Blengini, 2019). LCI is categorized into two groups, that is, indirect emissions (off-farm, off-orchard, and background) and direct emissions (on-Farm, on-orchard, and foreground), for agricultural production life cycle. In this regard, indirect emissions

236 Chapter 14 Coupled life cycle assessment and data envelopment analysis

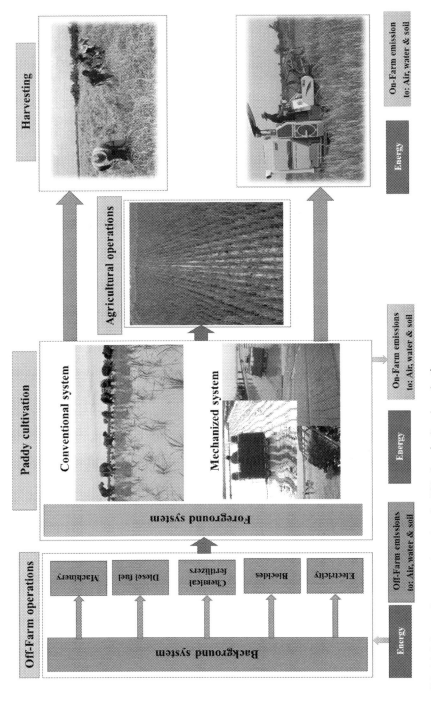

FIG. 14.3 A sample system boundary for LCA of agricultural production.

involve environmental impacts of production process for all inputs used in the agriculture systems (Mostashari-Rad et al., 2021), containing agricultural machinery, fertilizers, lubricating oil, diesel fuel, electricity, biocides, etc. (Ghasemi-Mobtaker et al., 2020). The number of each required inputs can increase or decrease with respect to agricultural crops. Furthermore, outputs of the agricultural systems involve yield. Outputs can increase or decrease depending on agricultural crop production. Direct emissions, on the other hand, are in relation to environmental impacts obtained by the consumption of inputs to air, water, and soil. Direct emissions can consist of several items, based on agricultural system and its products. However, important indirect emissions in agricultural systems are outlined as follows:

1. Direct emissions to soil and air owing to diesel fuel consumption in agricultural machinery (EcoInvent, 2019) with related coefficients as presented in Fig. 14.4.
2. Direct emissions of fertilizers to air and water (IPCC, 2006).
3. Direct emissions of biocides that can be achieved by complementary methods such as PestLCI 2.0 model (Fantin et al., 2019).
4. Direct emissions of human labor activity to air (Mousavi-Avval et al., 2017a).

Fig. 14.5 shows a sample of coefficients of the above-mentioned items for production of horticultural crops.

It should be noted that the above items are common emissions and many emissions related to different inputs can be added into direct emissions. For example, compost consumption has several emissions to air, soil, and water. An example of LCI for agricultural production is shown in Table 14.4.

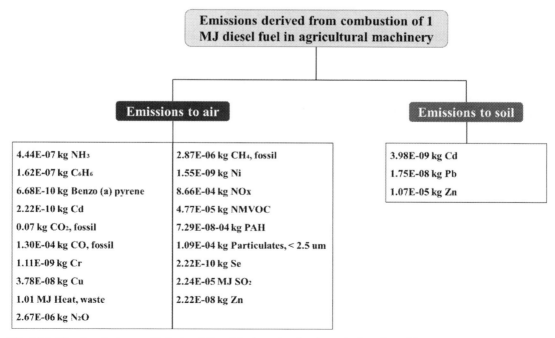

FIG. 14.4 Direct emission coefficients of diesel fuel combustion in agricultural machinery.

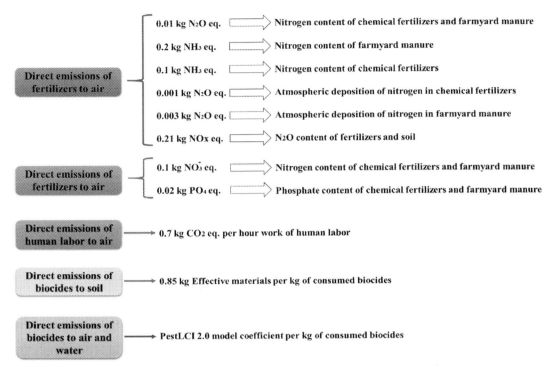

FIG. 14.5 Coefficients of direct emissions related to use fertilizers, biocides, and human activity for production of horticultural crops.

Table 14.4 A sample LCI report of a crop production system based upon 1 ha.		
Item	Unit	Amount
A. *Indirect emissions*		
1. Agricultural machinery	kg	3.60
2. Chemical fertilizers	kg	
(a) Nitrogen		64.96
(b) Phosphate		75.46
(c) Potassium		132.70
3. Farmyard manure	kg	1755.08
4. Biocides	kg	
(a) Pesticides		2.71
(b) Fungicides		1.92
5. Diesel fuel	kg	104.70
6. Gasoline	kg	185.05
7. Electricity	kWh	288.42

(continued)

Item	Unit	Amount
B. Direct emissions		
1. Emissions by diesel fuel burning to air		
(a) NH_3	kg	3.05×10^{-03}
(b) C_6H_6	kg	1.11×10^{-03}
(c) Benzo (a) pyrene	kg	4.58×10^{-06}
(d) Cd	kg	1.52×10^{-06}
(e) CO_2, fossil	kg	475.18
(f) CO, fossil	kg	0.89
(g) Cr	kg	7.62×10^{-06}
(h) Cu	kg	2.59×10^{-04}
(i) N_2O	kg	0.02
(j) Heat waste	MJ	6918.07
(k) CH_4, fossil	kg	0.02
(l) Ni	kg	1.07×10^{-05}
(m) NO_x	kg	5.94
(n) Non-methane volatile organic compound (NMVOC)	kg	0.33
(o) Polycyclic aromatic hydrocarbon (PAH)	kg	5.00×10^{-04}
(p) Particulates, < 2.5 μm	kg	0.75
(q) Se	kg	1.52×10^{-06}
(r) SO_2	kg	0.15
(s) Zn	kg	1.52×10^{-04}
2. Emissions by diesel fuel burning to soil		
(a) Cd	kg	2.73×10^{-05}
(b) Pb	kg	1.20×10^{-04}
(c) Zn	kg	0.07
3. Emissions by gasoline burning to air		
(a) C_4H_6	kg	1.59×10^{-07}
(b) C_2H_4O	kg	3.13×10^{-06}
(c) C_3H_4O	kg	3.77×10^{-07}
(d) C_6H_6	kg	3.80×10^{-06}
(e) CO_2	kg	0.57
(f) CO	kg	0.04
(g) CH_2O	kg	4.81×10^{-06}
(h) CH_4	kg	2.51×10^{-04}
(i) N_2O	kg	1.66×10^{-05}
(j) NO_x	MJ	0.01
(k) PAH	kg	6.85×10^{-07}
(l) Particulates, > 2.5 μm, and < 10 μm	kg	7.22×10^{-05}
(m) C_3H_6	kg	1.05×10^{-05}
(n) C_7H_8	kg	1.67×10^{-06}
(o) SO	kg	1.36×10^{-04}
(p) Volatile organic compounds (VOC)	kg	7.72×10^{-04}
(q) C_8H_{10}	kg	1.16×10^{-06}

(continued)

Table 14.4 (Cont'd)

Item	Unit	Amount
3. Emissions by fertilizers to air		
(a) N_2O	kg	1.64
(b) NH_3 derived from pure nitrogen in chemical fertilizers	kg	7.89
(c) NH_3 derived from pure nitrogen in farmyard manure (FYM)	kg	9.63
4. Emission by atmospheric deposition of fertilizers to air		
(a) N_2O derived from pure nitrogen in chemical fertilizers	kg	0.10
(b) N_2O derived from pure nitrogen in FYM	kg	0.12
5. Emissions by fertilizers to water		
(a) NO_3^-	kg	13.90
(b) PO_4^{3-}	kg	1.89
6. Emission by N_2O of fertilizers and soil to air		
(a) NO_x	kg	0.39
7. Emission by heavy metals of fertilizers to soil		
(a) Cd	mg	3609.36
(b) Cu	mg	168,618.40
(c) Zn	mg	436,267.36
(d) Pb	mg	361,308.35
(e) Ni	mg	14,496.59
(f) Cr	mg	51,467.96
(g) Hg	mg	56.46
8. Emission by human labor to air		
(a) CO_2	kg	366.49
9. Emissions by biocides to air		
(a) Diazinon	kg	0.28
(b) Captan	kg	0.19
10. Emissions by biocides to water		
(a) Diazinon	kg	0.13
(b) Captan	kg	0.10
11. Emissions by biocides to soil		
(a) Diazinon	kg	2.30
(b) Captan	kg	1.63
C. Output		
1. Yield	kg	4040.14

14.4.3 Life cycle impact assessment

LCIA can provide some sort of conclusions for creating better business decisions. It helps in categorizing and assessing environmental impacts with what matters to the concerned party and turning out to the sustainability matter such as global warming or human health (Rattanatum et al., 2018). In this step, the systems' environmental impacts are evaluated to decrease a series of results in terms of an inventory table within the objectives framework of the evaluation (Lasvaux et al., 2016). According to ISO 14042,

LCIA can be conducted by utilizing four stages: (1) impact categories' choice and classification, (2) characterization, (3) normalization, and (4) weighting (Nabavi-Pelesaraei et al., 2019). Herein, some methods are applied to implement these steps, including CML 2 baseline 2000 (Kouchaki-Penchah et al., 2017), IMPACT 2002+ (Jolliet et al., 2003), Eco-indicator 99 (Dreyer et al., 2003), EDIP2003 (Hauschild and Potting, 2004), ReCiPe2008 (Owsianiak et al., 2014), and ReCiPe2016 (Huijbregts et al., 2017), etc. It should be noted that SimaPro software is employed to implement LCA analysis.

In recent years, three main LCIA methods have been used for LCA of agricultural production systems including CML2 baseline 2000, IMPACT 2002+, and ReCiPe2016, which are illustrated as follows.

1. *CML2 baseline* 2000: This method is an update of the CML 92 method and includes more advanced models. There are 10 impacts categories in this method, which are described in details in Fig. 14.6.
2. *IMPACT* 2002+: Four different impact categories including resource, climate change, human health, and ecosystem standard are involved in this method. In fact, IMPACT 2002+ endpoint analysis tool is damaged (Jolliet et al., 2003). This method can perform environmental burdens' analysis for four influencing areas along with 15 midpoint impact categories, involving carcinogens and noncarcinogens, global warming, earthly acid/ nutrient, terrestrial ecotoxicity (TE), aquatic

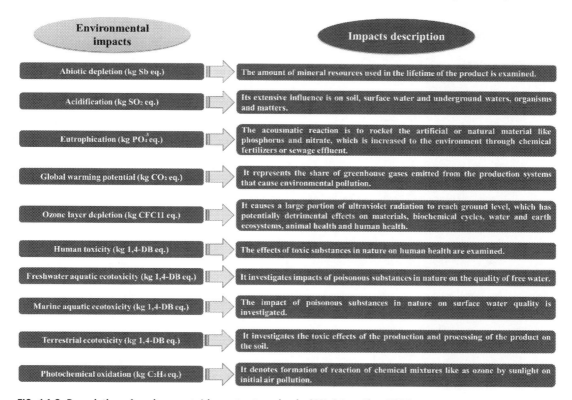

FIG. 14.6 Description of environmental impact categories in CML 2 baseline 2000.

eutrophication, aquatic acidification, aquatic ecotoxicity, ionizing radiation, ozone layer depletion (OLD), respiratory inorganics and respiratory organics, mineral extraction, nonrenewable energy, and land occupation. Impacts on various effect categories are evaluated in the inventory, followed by a damage assessment with reference to sources classes, climate change, human health, and ecosystem (Chen et al., 2010). The relationships between endpoint damage categories and midpoint impact categories under IMPACT 2002+ are presented in Fig. 14.7.

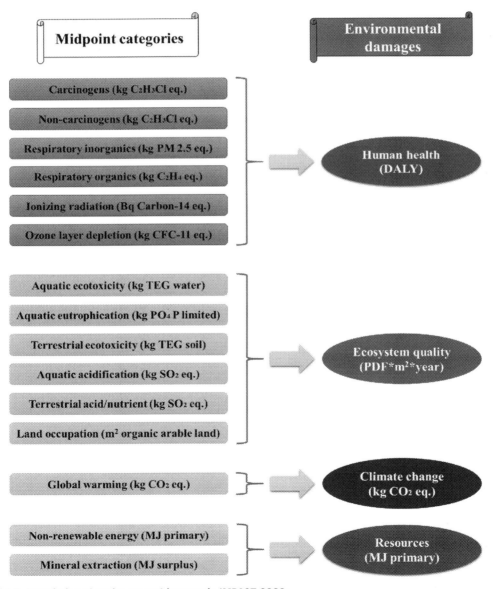

FIG. 14.7 Description of environmental impacts in IMPACT 2002+.

3. *ReCiPe*2016: There are some models for various impact categories, particularly in photochemical ozone formation, land and water use, and fine particulate matter creation. In fact, models are chosen practically and ReCiPe2016 is applied as a result of its ability to integrate some sort of impact categories (Huijbregts et al., 2017). There are three protection areas involving resource shortage, human health, and ecosystem quality retained for the implementation in ReCiPe2016, and in fact, all these protection areas were already chosen in ReCiPe2008 (Saber et al., 2021). Moreover, three protection areas exist in Endpoints levels. As an index in relation to human health, disability-adjusted life year (DALY) presents a period of time particularly for a personal disability in years associated with an incident or illness or even a missing period. Besides, the species year represents the ecosystem quality unit. United States dollars of 2013 (USD2013) denotes the resource scarcity's unit, demonstrated additional expenses in relation to the prospective mineral extraction and fossil resources (Huijbregts et al., 2017). Fig. 14.8 presents interrelationships among three endpoint protection areas and 17 impact categories of midpoints involved in ReCiPe2016.

14.4.3.1 Normalization

Normalization refers to the estimation of the magnitude of the category indicator results, in relation to some reference information (ISO, 2006). In this regard, external normalization references is selected; as an instance, the reference system data are independent from the environmental characteristics of the research work and accordingly do not vary based on the study system (Hélias et al., 2020). Ideally, external normalization data need a comprehensive inventory of resources and emissions consumption of the reference area throughout the reference time (such as the world or regional emissions for 1 year). Indeed, normalization provides the plausibility evaluation of the order of result magnitude and simplifies the result interpretation by presenting them with an appropriate unit, such as person equivalents (Pizzol et al., 2017). Furthermore, it is suggested as an elementary level for a potential weighting of environmental effects and assembling all these effects into a single reference system.

14.4.3.2 Weighting

Weighting method is of a major analysis mechanism in recognizing the important issues and dimension to evaluate. It serves a researcher to demonstrate research results after the characterization step by using a common reference (Mostashari-Rad et al., 2020).

14.5 Data envelopment analysis

One of the suitable methods in evaluating and determining a certain production unit's relative efficiency is DEA, in which each of the production unit is referred to DMU (Angulo-Meza et al., 2019). This method considers the computation of relative efficiency of a DMU group which applies various inputs to create output (Li et al., 2018). Hence, inefficiency levels and reasons of the units are determined by applying this technique (Hosseinzadeh-Bandbafha et al., 2018). Further, DEA method is used in agricultural studies to determine the efficient and proper fields and to compute the consumption of optimal energy on the basis of the agricultural crop producers' outputs.

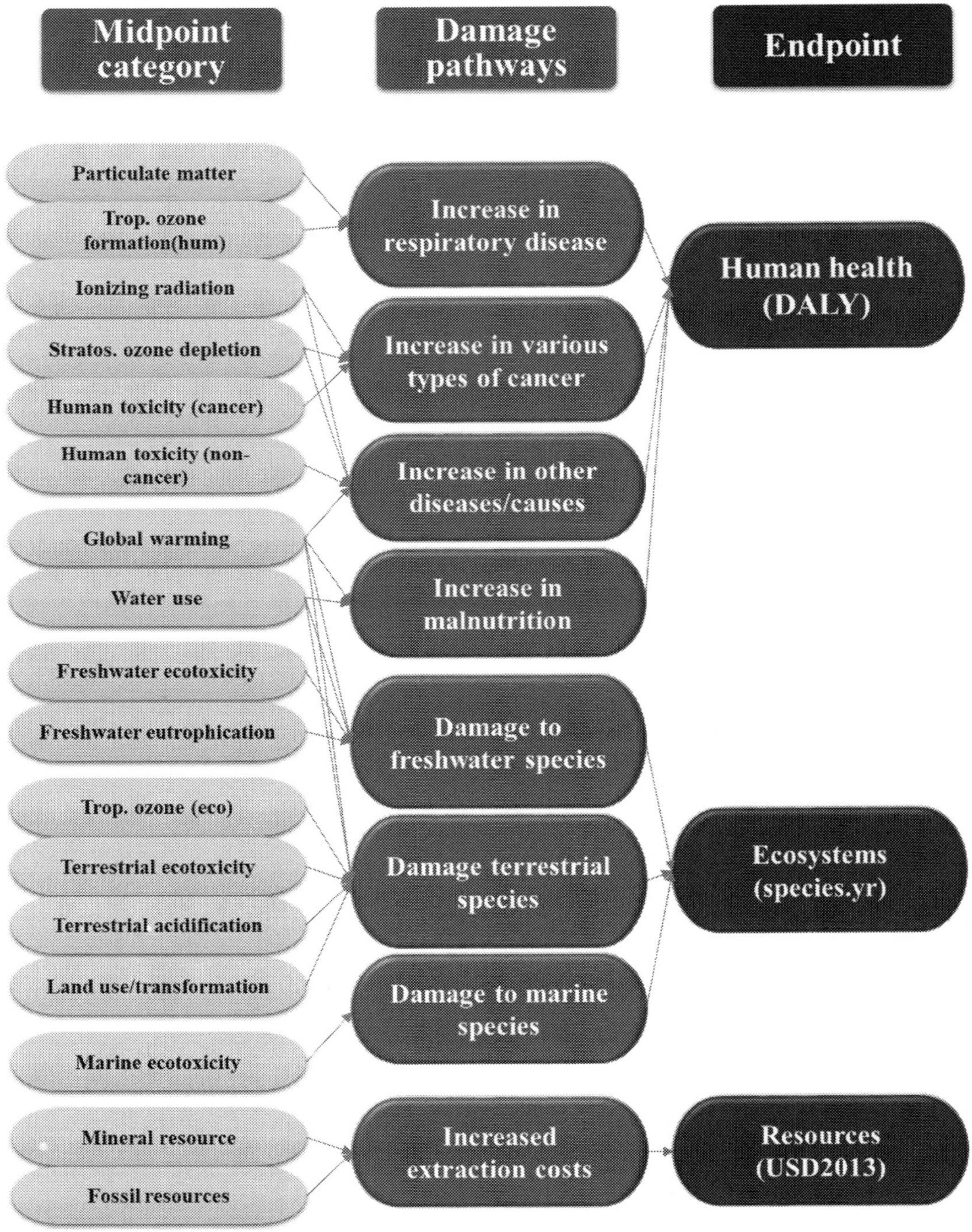

FIG. 14.8 Relationships among midpoints and endpoints in ReCiPe2016.

Linear regression is a method that has been used in agricultural studies in past years, but over the time, DEA method has been considered as an alternative. DEA have several following advantages in comparison to regression analysis (Thanassoulis, 1993):

- DEA is considered as a nonparametric technique not entailing the applicants to assume some sort of numerical ways to make the production function.
- DEA evaluates performance against efficiency instead of mean performance.
- DEA can get along much more easily with multiple outputs and inputs.
- DEA identifies the return nature to scale holding at an efficient boundary in each section.
- DEA identifies the inefficient sources in relation to exceeding usage of specific resources or/and less standards on specified outputs.
- DEA provides more precise computations of relative efficiency, as a result of being a boundary technique.
- DEA provides much more precise computation of inputs or outputs' marginal values in case it presents any variable with any marginal value.
- DEA provides efficiency instead of outputs' or inputs' mean marginal values.
- DEA considers different marginal values for various mixes of inputs–outputs.
- DEA computations of marginal values do not suffer from inaccuracy because of strong correlations or multicollinearity among explanatory variables.
- DEA provides much more targets' proper minimum or maximum of individual where inputs are not able to change independently with outputs.

The discrepancies among results by DEA and regression analysis are demonstrated in Fig. 14.9. Inputs and outputs are depicted on the vertical and horizontal axes, respectively (Fig. 14.9). Eight DMU cases comprising one output as well as one input with various ratios of output–input are identified as P1–P8 points. The linear regression model is denoted as a dotted line representing the trends. All DMUs located on or above this line are considered advantageous in this technique (P2, P3, and P4). Moreover, an envelope with a piecewise line connecting the boundary points above the dataset is

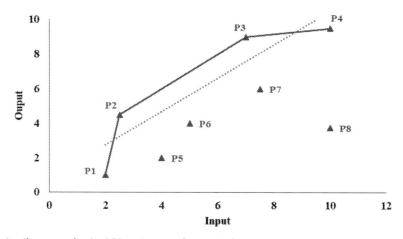

FIG. 14.9 Results discrepancies by DEA and regression analysis.

employed in the nonparametric method. Furthermore, some frontier points P1, P2, P3, and P4 are depicted in Fig. 14.9. In this regard, solid lines are employed to connect these points providing the dataset envelope. While DMUs not on the boundary are inefficient, DMUs 1, 2, 3, and 4 on the boundary are efficient (Mousavi-Avval et al., 2011b).

In DEA, two forms exist to treat returns to scale (RTS) (Banker et al., 1984). Charnes, Cooper, and Rhodes (CCR) presented CCR model with computed technical efficiency together with a trend of Constant Returns to Scale (CRS) (Charnes et al., 1978). Banker et al. (1984) introduced Banker, Charnes and Cooper (BCC) model for measuring technical efficiency and ensuring the metric had the same scale with the computed unit. The efficiency was benchmarked with the model of CCR. Efficient DMUs were the ones with the highest output and lowest input values. Besides BCC, it also considered Variable Returns to Scale (VRS). BCC and CCR models are divided into two forms, that is, output-oriented models concentrated on enhancing outputs with similar input levels, and input-oriented models aiming at minimizing inputs under the same output levels (Malana and Malano, 2006).

In DEA technique, efficiency is categorized into three various groups, involving technical efficiency, scale efficiency, and pure technical efficiency (Sabzevari et al., 2015).

Technical efficiency is considered existing if proof indicates that it is able to make better some of the inputs or outputs disregarding deteriorating some of the other inputs or outputs (Charnes et al., 1978). Furthermore, technical efficiency based on VRS or pure technical efficiency is associated with managers' ability to apply given sources of firms, and, finally, scale efficiency alludes to exploit economies of scale when CRS is represented by the frontier of production (Banker et al., 1984). If there appears to be a discrepancy for a specific DMU score between the technical efficiency and pure technical efficiency, scale inefficiency occurs (Rosman et al., 2014).

As mentioned so far, the technical efficiency can be described as the summed weighted outputs divided by the summed weighted inputs, as presented in the following formula (Cooper et al., 2006):

$$\text{TE}_j = \frac{u_{1j}y_{1j} + u_{2j}y_{2j} + \cdots + u_{sj}y_{sj}}{v_{1j}x_{1j} + v_{2j}x_{2j} + \cdots + v_{mj}x_{mj}} = \frac{\sum_{r=1}^{s} u_{rj}y_{rj}}{\sum_{i=1}^{m} v_{ij}x_{ij}} \tag{14.6}$$

In Eq. (14.6), s denotes the outputs' number, m shows the inputs' number, n presents the DMUs number, TE_j ($j = 1, 2, \ldots, n$) represents the DMU_j. technical efficiency, u_{rj} ($r = 1, 2, \ldots, s$) is the output y_r. weighting amount in the comparison, v_{ij} ($i = 1, 2, \ldots, m$) is the input x_i. weighting amount y_{rj} denotes the service output r amount produced by DMU j throughout the period of observation, and x_{ij} presents the resource input i amount applied by DMU j throughout the interval.

Each DMU defines one set of output and input weights for the valuation of efficiency in Eq. (14.6). Hence, n DMUs consist of n sets of output and input weights.

Let DMU_o ($o = 1, 2, \ldots, n$) be assessed DMU_j in any run. To estimate a DMU_o relative efficiency based on n DMUs, it is tailored as a fractional programming problem with the following formulas (Cooper et al., 2006):

$$\text{Maximize} \quad TE_o = \frac{\sum_{r=1}^{s} u_{ro} y_{ro}}{\sum_{i=1}^{m} v_{io} x_{io}}$$

$$\text{Subject to:} \quad \frac{\sum_{r=1}^{s} u_{ro} y_{rj}}{\sum_{i=1}^{m} v_{io} x_{ij}} \leq 1, j = 1,2,3,\ldots,n$$

$$u_{ro} \geq 0, \quad v_{io} \geq 0 \tag{14.7}$$

where y_{ro} denotes the output r amount produced by DMU_o throughout the interval, x_{io} presents the resource input i amount which is applied by DMU_o throughout the interval, u_{ro} shows the weight allocated to output r calculated by DEA model, and v_{io} represents the weight allocated to input i calculated by DEA model. In this regard, Eq. (14.7) is designed as the following linear programming problem (Cooper et al., 2006):

$$\text{Maximize} \quad TE_o = \sum_{r=1}^{s} u_{ro} y_{ro}$$

$$\text{Subject to:} \quad \sum_{r=1}^{s} u_{ro} y_{rj} - \sum_{i=1}^{m} v_{io} x_{ij} \leq 0, j = 1,2,3,\ldots,n$$

$$\sum_{i=1}^{m} v_{io} x_{io} = 1$$

$$u_{ro} \geq 0, \quad v_{io} \geq 0 \tag{14.8}$$

In fact, Eq. (14.8) is more difficult to solve than the dual linear programming problem, as a result of more limitation. It is mathematically designed under a vector–matrix format (Cooper et al., 2006):

$$\text{Minimum} \quad TE_o$$

$$\text{Subject to:} \sum_{j=1}^{n} Y_j \lambda_j \geq y_o$$

$$\sum_{j=1}^{n} X_j \lambda_j - TEx_o \leq 0$$

$$\lambda_j \geq 0 \tag{14.9}$$

where y_o denotes the $s \times 1$ vector of the amount produced of the initial outputs and x_o shows the $m \times 1$ vector of the amount used of initial inputs by the oth DMU. Y presents the $s \times n$ output matrix and X represents the $m \times n$ input matrix of all n involving units. λ shows an $n \times 1$ vector of weights and TE_o denotes a scalar with limits between zero and one determining the DMU_o's technical efficiency score. Eq. (14.9) is considered input-oriented.

Eq. (14.9) has a feasible solution $TE = 1$, $\lambda_0 = 1$, $\lambda_j = 0$, $j = 1, 2, \ldots, n$ and $j \neq 0$. Then, the optimal technical efficiency, designed by TE^*, is not larger than one. As all data are assumed nonzero and the constraint $\lambda j \geq 0$, λ is nonzero as $y_0 > 0$ and $y_0 \neq 0$. So, from $\sum_{j=1}^{n} X_j \lambda_j - TEx_o \leq 0$, technical efficiency must be more than zero. Thus, it leads to $0 < TE^* \leq 1$ (Cooper et al., 2006). As Eq. (14.8) would be a multiplier form of Eq. (14.9) (envelopment form), Eq. (14.8) has a feasible solution.

Besides CRS model, Banker et al. (1984) suggested another model by applying the DEA concept, which is termed BCC model, or pure technical efficiency. This technique computes the DMUs technical efficiency at different return to scale conditions and is able to distinguish scale and technical efficiencies. Its basic benefit is that the scale ineffective fields are solely compared with effective fields at an identical size (Mobtaker et al., 2012). It is provided by getting a constraint on λ ($\lambda = 1$) in Eq. (14.9) regarding the permissible RTS. Besides, VRS is assumed under this model, showing that a disproportionate change in outputs will result from a change in inputs (Pahlavan et al., 2012). The performance frontier line, in this condition, is not limited to passing through the origin, and the input enhancement may not lead to a proportionate enhancement of outputs in this case (Cooper et al., 2006). As a result, inefficient points will be projected on a convex hull constituted by the efficient DMUs. Due to envelop of VRS, the data in CRS are looser compared with this. It is more flexible, pure technical efficiency is at least equal to CRS or the overall score of technical efficiency. This relationship is applied to compute the scale efficiency (Omid et al., 2011). Based on the above, scale efficiency is identified as (Qasemi-Kordkheili and Nabavi-Pelesaraei, 2014)

$$Scale\ efficiency = \frac{Technical\ efficiency}{Pure\ technical\ efficiency} \tag{14.10}$$

This decomposition presents the inefficiency sources. Scale efficiency represents the size of impact unit on system efficiency, while pure technical efficiency denotes the energy consumption efficiency. It can be said that the inefficiency part in the consumption of energy is due to the wrong choice of unit size. If an optimal size of DMU is selected, it can attain the same overall efficiency (technical) under the same input level. In agricultural studies, for the calculation of technical efficiency and pure technical efficiency, Frontier Analyst 4 software can be used to retrieve unit's scores. Furthermore, the Kruskal–Wallis test is applied to describe remarkable discrepancies among farm groups accompanied by agro-climatic zones (Nassiri and Singh, 2009).

Cross-efficiency evaluation was largely used to rank DMUs or to determine the most effective DMU applying DEA model. Some other techniques regarding this focus on how to compute the input and output weights' uniqueness, while overlooking the accumulation of this process and just dealing with them based upon their resemblances disregarding their relative significance (Wang et al., 2013). In agricultural studies, efficiency scores are summed in the matrix. The score of efficiency for the jth farmer calculated with optimal weights of the ith farmer calculated by CCR technique is denoted by the element in the ith row and jth column. The rank of efficient farmers is based on their average score of cross-efficiency obtained through the column average in the matrix (Raheli et al., 2017).

The assumption credibility is a main issue in the studies of efficiency in which the entire production process will obtain the best practice of production frontier (Chien and Hu, 2007). In agricultural studies, it is supposed that all DMUs can access the best practice, in terms of estimating the efficiency of energy. In fact, the "best practice" production is the set on the frontier among the involved DMUs. The inefficient DMUs can decrease inputs by the identified amount (Boyd and Pang, 2000). The inefficient production process and out-of-date technology level produce an energy use redundant portion requiring further adjustment. The total required adjustment, comprising the radial and slack adjustments, is estimated by DEA model (Hu and Wang, 2006). The summation of radial and slack adjustments is the target's total reduced value disregarding the output level reduction. The above summation, regarding energy inputs, is defined as the energy saving target (EST) (Hu and Kao, 2007). In total, the efficiency is considered the best practice divided by the real operation. Therefore, the energy efficiency indicator denotes EST divided by actual energy input (AEI). In fact, the EST ratio (ESTR) is applied to identify the energy usage's inefficiency of the considered DMUs. It is represented as follows (Hu and Kao, 2007):

$$ESTR = \frac{EST}{AEI} \tag{14.11}$$

In Eq. (14.11), *AEI* presents actual energy input and *EST* denotes the EST. In fact, EST shows the energy inputs' total reduction amount disregarding output reduction. Eq. (14.11) represents the standard efficiency identified in terms of the best-practice divided by the real operation. Actually, the percentage of ESTR ranges from 0 to 100. There can be no saving at all and a larger ESTR percentage presents a larger inefficiency of energy consumption, and then a larger value of energy saving (Hu and Kao, 2007).

14.6 Integration of LCA and DEA

Environmental impact evaluation is one of the most essential sections of the sustainability investigation. LCA, on the one hand, can evaluate the environmental effects of the production ecosystems and, on the other hand, does not create any other options for improvement. DEA prepares the inputs' appropriate amount (such as more or less application of inputs) pushing an inefficient agricultural producer toward an efficient production system, while the environmental effects would be overlooked. A favored DEA and LCA integration has been one of the most controversial matters during past decades. Such integration lets the verification that decreased levels of input consumption would lead to mitigation of environmental effects. Furthermore, this method aims to help farmers and policy makers to assess both environmental and operational performance of production systems (Pishgar-Komleh et al., 2020).

The LCA + DEA technique can evaluate the material and energy flows in agricultural units to enhance their economic and environmental efficiencies. Note that LCA and DEA input/outputs items are not similar and exact (Mohammadi et al., 2015). LCA and DEA inputs consist of raw data requirements for various steps of agricultural units' production, such as fertilization, spraying, irrigation, and harvesting. Crops or products are outputs in DEA. However, outputs of LCA are direct emissions' unfavorable generation in relation to water, energy, chemicals, and fertilizers accompanied by factors associated with the production system of agriculture. Besides the assessment under common application, in this technique, an evaluation for an optimal consumption takes place such that the optimized data by DEA for LCI are employed in step 2 of LCA. Moreover, LCI developed with the optimized data is employed in stage 3 involving LCA performance for each DMU. In the last step, comparison is made between potential environmental effects for the target and current DMUs to identify environmental risk assessment in terms of inefficiency in energy consumption for agricultural production (Hosseinzadeh-Bandbafha et al., 2018). This process is shown in Fig. 14.10.

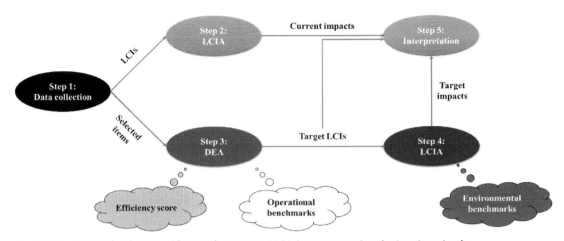

FIG. 14.10 Schematic diagram of integrating DEA and LCA for systems of agricultural production.

14.7 Result analysis
14.7.1 Energy use pattern

As mentioned above, various studies have been carried out in the field of energy consumption and production in agricultural systems in the last decade. In most of them, the results are usually expressed in the following order:

1. *Providing the physical amount of consumed inputs along with the total input and output energies*: For a better understanding, Table 14.5 presents the physical amount of consumed inputs and input and output energy to produce a sample crop.
2. *Using a pie chart to show each input's share in the consumption of total energy*: Fig. 14.11 represents the share of input consumption in total input energy for production of a sample crop.
3. *Evaluating energy indices and energy forms*: For a better understanding, as an instance, Table 14.6 shows the amount of energy indices and energy forms to produce a sample crop. The most important part of energy assessment is the analysis of energy indices, due to the physical amount of input and output energies or even the percentage of each input analysis in total energy consumption, which cannot lead to effective discussion and ultimately the adoption of appropriate policies. The most important of these indices, which in a way includes the rest of the indices, is energy use efficiency. When this index is less than one, it indicates that the amount of input energy is more than the output energy and the product is inefficient and unjustifiable from energetic prospective. On the other hand, the higher the number, the more energy efficient the product is.

In fact, all studies in the field of improving energy efficiency are aimed at enhancing this index. Of course, it should be noted that the contributions of different energy forms, especially in the category of renewable and nonrenewable have been considered in most studies. Obviously, the lower the nonrenewable energy consumption and the renewable energy replacement, the better the result from an energy viewpoint. In studies that do not have optimization, results and discussion in this section are usually

14.7 Result analysis

Table 14.5 Input-output energy analysis of a sample crop.

Item (unit)	Unit per hectare	Energy consumption (MJ/ha)	Standard deviation (MJ/ha)
A. Inputs			
1. Human labor (h)	523.56	1026.18	246.28
2. Agricultural machinery (kg)	3.60	513.72	77.06
3. Diesel fuel (l)	121.74	6855.18	1233.93
4. Gasoline (l)	272.13	12,599.62	3149.90
5. Chemical fertilizers (kg)			
(a). Nitrogen	64.96	4296.45	902.26
(b). Phosphate	75.46	938.72	215.91
(c). Potassium	132.70	1479.61	251.53
6. Farmyard manure (kg)	1755.08	526.52	110.57
7. Biocides (kg)	4.63	555.60	105.56
8. Electricity (kWh)	288.42	3440.85	997.85
9. Seed (kg)	11.34	283.50	62.37
Total input energy (MJ)	-	32,515.95	7478.67
B. Output			
1. Yield (kg)	4040.14	101,003.50	18,180.63

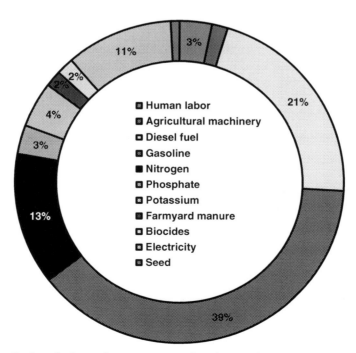

FIG. 14.11 Each input's share in the total energy consumption of a sample crop.

Table 14.6 Energy indices and energy forms of a sample crop production.

Items	Unit	Amount
A. Energy indices		
1. Energy use efficiency	-	3.11
2. Energy productivity	Kg/MJ	0.12
3. Specific energy	MJ/kg	8.05
4. Net energy	MJ/ha	68,487.55
B. Energy forms		
1. Direct energy[a]	MJ/ha	24,205.33 (74.44%)
2. Indirect energy[b]	MJ/ha	8310.63 (25.56%)
3. Renewable energy[c]	MJ/ha	1836.20 (5.65%)
4. Nonrenewable energy[d]	MJ/ha	30,679.75 (94.35%)

[a] Involves diesel fuel, human labor, electricity, gasoline, and seed.
[b] Involves chemical fertilizers, agricultural machinery, and farmyard manure.
[c] Involves farmyard manure, human labor, and seed.
[d] Involves agricultural machinery, gasoline, diesel fuel, biocides, chemical fertilizers, and electricity.

finished and finally, researchers only offer general suggestions for improving the energy of the system under study. Some of the suggestions are as follows:

- Using standard machinery and decommissioning old machinery.
- Applying renewable energy such as solar energy instead of fossil fuels.
- Application of biofertilizers rather than chemical fertilizers.
- Biological control of diseases and pests instead of using biocides.

14.7.2 Environmental life cycle analysis

In studies in which LCA is also evaluated, the results are expressed and analyzed as follows:

1. *Expression of the physical value of environmental impacts based on the method used*: For example, Table 14.7 shows environmental impacts related to the production of one ton of a sample agricultural product that is evaluated by using the ReCiPe2016 method.
2. *Representation of each input's share in a percentage of emission of each environmental impact*: Fig. 14.12, as an instance, represents each input's share in each environmental impact related to the production of 1 ton of a sample crop that is examined by using the ReCiPe2016 method.
3. *Drawing a weighting chart for dimensional display of environmental impacts*: For a better understanding, Fig. 14.13 demonstrates the share of each input in each environmental impact related to the production of 1 ton of a sample crop investigated by using the ReCiPe2016 method.

Table 14.7 Environmental impact results of a sample crop production based on ReCiPe2016 method.

Endpoint	Unit	Amount
Human health	DALY	2.22×10^{-03}
Ecosystems	species.year	6.10×10^{-06}
Resources	$	63.37

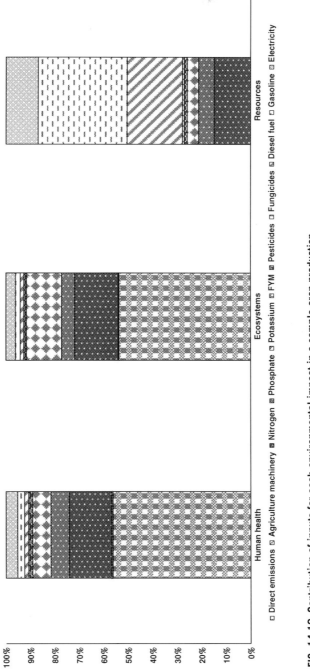

FIG. 14.12 Contribution of inputs for each environmental impact in a sample crop production.

254 Chapter 14 Coupled life cycle assessment and data envelopment analysis

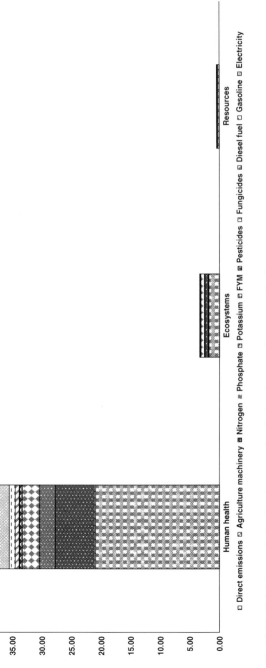

FIG. 14.13 Weighting analysis of environmental endpoints in a sample crop production.

In studies that lack optimization and optimal presentation pattern, usually after the above steps, only solutions to reduce high-consumption inputs (especially in environmental impacts, which have the highest destructive effect based on the weighting chart) are addressed. Some of the suggestions are as follows:

- To reduce the use of water pumps and consequently electricity and diesel fuel, water storage pools can be constructed in suitable positions.
- To explain the benefits of an appropriate input consumption, some educational programs can be offered to producers.
- Organic fertilizers can be applied to decrease chemical fertilizer application, enhance soil structure, and mitigate environmental pollutants.
- By determining the efficient orchardists, they can be introduced as a pattern for display to the other units.
- To enhance the efficiency, some proper schedules for maintaining agricultural machinery can be identified.
- Mechanization operations should be improved, particularly in terms of fertilizing level.

14.7.3 Energy optimization by DEA

In research works in which DEA optimization is conducted, the results are usually stated as follows:

1. *Determining the efficiencies types involving technical efficiency, pure technical efficiency, and scale efficiency*: The values of these efficiencies are between zero and one, in which the closer this number is to one, the higher is the efficiency. As an instance, the efficiency types for producing a sample crop are presented in Table 14.8.
2. *Identifying the cross-efficiency index for optimal units*: In some research works, ranking is carried out among the benchmarks by applying the cross-efficiency method and using Frontier Analyst software, and the real and ideal optimal is also determined. Also, this index is ranked between zero and one. The most efficient units have cross-efficiency closer to one. Table 14.9 represents an example of cross-efficiency results for producing a sample agricultural crop.
3. *Drawing the optimization table and presenting the optimal model*: This is the most important part of result analysis in DEA. The optimal energy level for each input is determined in this unit, and a new optimal pattern is presented by applying benchmarks. The energy amount stored for each of input and, consequently, the total saved energy, are also mentioned in this section. Table 14.10 presents an example of DEA energy optimization results for a sample crop production.
4. *Representing the contribution of each input in the optimization by application of a pie chart*: In this section, results of each input contribution in storing the total required energy are also shown.

Table 14.8 Results of different efficiencies for a sample crop production based on DEA.

Item	Technical efficiency	Pure technical efficiency	Scale efficiency
Average	0.738	0.833	0.886
Standard deviation	0.173	0.090	0.156
Minimum	0.403	0.588	0.686
Maximum	1.000	1.000	1.000

Table 14.9 Average cross-efficiency scores for 10 most efficient units based upon CCR model.

Unit No.	Average cross-efficiency	Standard deviation
7	0.790	0.11
93	0.778	0.25
191	0.776	0.14
183	0.764	0.18
195	0.736	0.12
41	0.686	0.07
36	0.685	0.29
26	0.667	0.07
69	0.658	0.34
45	0.578	0.24

Table 14.10 Requirements of optimum energy and saving energy of a sample crop production according to DEA.

Inputs	Optimum energy requirement (MJ/ha)	Saving energy (MJ/ha)	Saving energy (%)
1. Human labor	929.11	97.06	9.46
2. Machinery	436.65	77.07	15.00
3. Diesel fuel	5457.85	1397.33	20.38
4. Gasoline	9668.72	2930.90	23.26
5. Nitrogen	2933.84	1362.61	31.71
6. Phosphate	641.01	297.71	31.71
7. Potassium	1009.56	470.05	31.77
8. Farmyard manure	353.17	173.35	32.92
9. Biocides	455.61	99.99	18.00
10. Electricity	2784.40	656.45	19.08
11. Seed	258.69	24.81	8.75
Total energy input	24,928.62	7587.33	23.33

Furthermore, this section is very important in result interpretation. It is because contrary to the interpretations of energy analysis that are generally provided for all inputs, strategic analyses and solutions can be shifted to inputs that have the largest potential for total energy savings. In fact, DEA method renders it possible to interpret the results in a more practical way. For a better understanding, Fig. 14.14 provides an example of the contribution of each input in optimizing energy consumption for a sample crop production.

5. *Expression of improving energy indices and changing energy forms after optimization*: This section is stated in the last part of the analysis. For a better understanding, Table 14.11 shows the improvement of energy indices and changes in energy forms for a sample crop production after energy optimization by DEA.

14.7 Result analysis

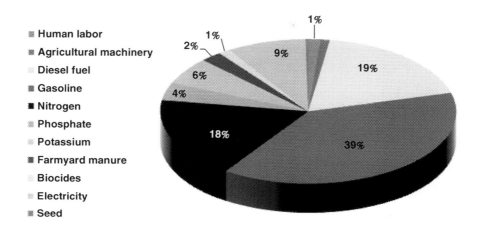

FIG. 14.14 Potential energy saving by DEA for each input in a sample crop production.

Table 14.11 Improvement of energy indices and energy forms of a sample crop production after energy optimization by DEA.

Item	Unit	Optimum quantity	Difference (%) [a]
A. Energy indices			
1. Energy use efficiency	-	4.05	30.44
2. Energy productivity	Kg/MJ	0.16	30.44
3. Specific energy	MJ/kg	6.17	-23.33
4. Net energy	MJ/ha	76,074.88	11.08
B. Energy forms			
1. Direct energy	MJ/ha	19,098.78 (76.61%)	21.10
2. Indirect energy	MJ/ha	5829.84 (23.39%)	29.85
3. Renewable energy	MJ/ha	1540.97 (6.18%)	16.08
4. Nonrenewable energy	MJ/ha	23,387.65 (93.82%)	23.77

[a] Difference (%) = [(Optimum quantity - Present quantity)/ Present quantity] × 100

14.7.4 Mitigation of environmental impacts by DEA + LCA

After having provided the optimal patterns of energy inputs by DEA, environmental impacts are computed again and presented in this section. In DEA+LCA analysis, the results are usually presented in the following order:

1. *Presentation of improved environment impacts computed after DEA optimization*: As mentioned above, new values of environmental impacts are identified and the rate of improvement of each item is expressed as a percentage in this section. An example of environmental emission results after DEA optimization for a sample crop production is represented in Table 14.12.

Table 14.12 Optimum environmental impacts after energy optimization by DEA.

Damage category	Unit	Amount after optimization by DEA	Improvement (%)
Human health	DALY	1.60×10^{-03}	27.93
Ecosystems	species.year	4.36×10^{-06}	28.52
Resources	$	48.11	24.08

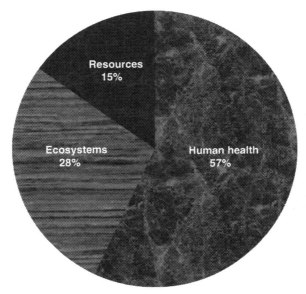

FIG. 14.15 Share of each category to mitigate total weighted environmental impacts for a sample crop production.

2. *Demonstration of each category's contribution in reducing the total environmental impacts using a pie chart*: The contribution of each category in reducing total emissions after optimization with DEA is presented after reweighting all impacts. As an instance, Fig. 14.15 displays the potential of each category to mitigate total weighted impacts for a sample crop production.

14.8 Conclusions

In previous years, the goal of enhancing agricultural systems was usually simply to increase yields, which in turn has led to a growing energy input consumption and a lot of irreparable environmental damages to the world. However, in recent years, the goals of sustainable production in agriculture have not only led to increasing yield per unit area, but also optimal amount of energy inputs and, consequently, mitigation of environmental emissions. In this regard, the use of new optimization methods has always been addressed. One of the main methods is DEA, which has been applied more in the last decade as a way to improve energy consumption in agro-ecosystems. Combining this optimization method with LCA allows researchers to study and measure the improvement of environmental impacts in addition to energy optimization. The combination of these techniques can achieve an optimal energy-environmental friendly model in agricultural systems by saving cost and time and preventing unnecessary trial and errors.

References

Abeliotis, K., Detsis, V., Pappia, C., 2013. Life cycle assessment of bean production in the Prespa National Park. Greece. J. Clean. Prod. 41, 89–96.

Ali, Q., Raza, A., Narjis, S., Saeed, S., Khan, M.T.I., 2020. Potential of renewable energy, agriculture, and financial sector for the economic growth: evidence from politically free, partly free and not free countries. Renew. Energy 162, 934–947.

Alluvione, F., Moretti, B., Sacco, D., Grignani, C., 2011. EUE (energy use efficiency) of cropping systems for a sustainable agriculture. Energy 36, 4468–4481.

Amid, S., Mesri Gundoshmian, T., Shahgoli, G., Rafiee, S., 2016. Energy use pattern and optimization of energy required for broiler production using data envelopment analysis. Inf. Process. Agric. 3, 83–91.

Angulo-Meza, L., González-Araya, M., Iriarte, A., Rebolledo-Leiva, R., Soares de Mello, J.C., 2019. A multiobjective DEA model to assess the eco-efficiency of agricultural practices within the CF + DEA method. Comput. Electron. Agric. 161, 151–161.

Atilgan, A., Hayati, K., 2006. Cultural energy analysis on broilers reared in different capacity poultry houses. Ital. J. Anim. Sci. 5, 393–400.

Aydın, B., Aktürk, D., 2018. Energy use efficiency and economic analysis of peach and cherry production regarding good agricultural practices in Turkey: a case study in çanakkale province. Energy 158, 967–974.

Aydoğan, B., Vardar, G., 2020. Evaluating the role of renewable energy, economic growth and agriculture on CO_2 emission in E7 countries. Int. J. Sustain. Energy 39, 335–348.

Bakhtiari, A.A., Hematian, A., Sharifi, A., 2015. Energy analyses and greenhouse gas emissions assessment for saffron production cycle. Environ. Sci. Pollut. Res. 22, 16184–16201.

Banaeian, N., Omid, M., Ahmadi, H., 2011. Energy and economic analysis of greenhouse strawberry production in Tehran province of Iran. Energy Convers. Manag. 52, 1020–1025.

Banker, R.D., Charnes, A., Cooper, W.W., 1984. Some models for estimating technical and scale inefficiencies in data envelopment analysis. Manage. Sci. 30, 1078–1092.

Beheshti Tabar, I., Keyhani, A., Rafiee, S., 2010. Energy balance in Iran's agronomy (1990–2006). Renew. Sustain. Energy Rev. 14, 849–855.

Bianco, I., Blengini, G.A., 2019. Life Cycle Inventory of techniques for stone quarrying, cutting and finishing: contribution to fill data gaps. J. Clean. Prod. 225, 684–696.

Boyd, G.A., Pang, J.X., 2000. Estimating the linkage between energy efficiency and productivity. Energy Policy 28, 289–296.

Cetin, B., Vardar, A., 2008. An economic analysis of energy requirements and input costs for tomato production in Turkey. Renew. Energy 33, 428–433.

Charnes, A., Cooper, W.W., Rhodes, E., 1978. Measuring the efficiency of decision making units. Eur. J. Oper. Res. 2, 429–444.

Chen, C., Habert, G., Bouzidi, Y., Jullien, A., Ventura, A., 2010. LCA allocation procedure used as an incitative method for waste recycling: an application to mineral additions in concrete. Resour. Conserv. Recycl. 54, 1231–1240.

Chien, T., Hu, J.-L.L., 2007. Renewable energy and macroeconomic efficiency of OECD and non-OECD economies. Energy Policy 35, 3606–3615.

Cochran, W.G., 1977. The estimation of sample size. Sampl. Tech. 3, 72–90.

Cooper, W.W., Seiford, L.M., Tone, K., 2006. Introduction to Data Envelopment Analysis and its Uses: With DEA-Solver Software and References. Springer Science & Business Media, Berlin.

Dreyer, L.C., Niemann, A.L., Hauschild, M.Z., 2003. Comparison of three different LCIA methods: EDIP97, CML2001 and Eco-indicator 99. Int. J. Life Cycle Assess. 8, 191–200.

EcoInvent, 2019. EcoInvent. (n.d.). System Models in Ecoinvent 3.6. EcoInvent. https://www.ecoinvent.org/database/ecoinvent-36/ecoinvent-36.html. Accessed 12 September 2019.

Elahi, E., Weijun, C., Jha, S.K., Zhang, H., 2019. Estimation of realistic renewable and non-renewable energy use targets for livestock production systems utilising an artificial neural network method: a step towards livestock sustainability. Energy 183, 191–204.

Erdal, G., Esengün, K., Erdal, H., Gündüz, O., 2007. Energy use and economical analysis of sugar beet production in Tokat province of Turkey. Energy 32, 35–41.

Fantin, V., Buscaroli, A., Dijkman, T., Zamagni, A., Garavini, G., Bonoli, A., Righi, S., 2019. PestLCI 2.0 sensitivity to soil variations for the evaluation of pesticide distribution in Life Cycle Assessment studies. Sci. Total Environ. 656, 1021–1031.

Fathollahi, H., Mousavi-Avval, S.H., Akram, A., Rafiee, S., 2018. Comparative energy, economic and environmental analyses of forage production systems for dairy farming. J. Clean. Prod. 182, 852–862.

Finnveden, G., Hauschild, M.Z., Ekvall, T., Guinée, J., Heijungs, R., Hellweg, S., Koehler, A., Pennington, D., Suh, S., 2009. Recent developments in life cycle assessment. J. Environ. Manage. 91, 1–21.

Ghasemi-Mobtaker, H., Mostashari-Rad, F., Saber, Z., Chau, K.W., Nabavi-Pelesaraei, A., 2020. Application of photovoltaic system to modify energy use, environmental damages and cumulative exergy demand of two irrigation systems—a case study: barley production of Iran. Renew. Energy 160, 1316–1334.

Hatirli, S.A., Ozkan, B., Fert, C., 2006. Energy inputs and crop yield relationship in greenhouse tomato production. Renew. Energy 31, 427–438.

Hatirli, S.A., Ozkan, B., Fert, C., 2005. An econometric analysis of energy input–output in Turkish agriculture. Renew. Sustain. Energy Rev. 9, 608–623.

Hauschild, M.Z., Potting, J., 2004. Spatial differentiation in life cycle impact assessment - the EDIP-2003 methodology. Guidelines from the Danish EPA.

Heidari, M.D., Omid, M., Akram, A., 2011. Energy efficiency and econometric analysis of broiler production farms. Energy 36, 6536–6541.

Hélias, A., Esnouf, A., Finkbeiner, M., 2020. Consistent normalization approach for life cycle assessment based on inventory databases. Sci. Total Environ. 703, 134583.

Hemmati, A., Tabatabaeefar, A., Rajabipour, A., 2013. Comparison of energy flow and economic performance between flat land and sloping land olive orchards. Energy 61, 472–478.

Hosseini-Fashami, F., Motevali, A., Nabavi-Pelesaraei, A., Hashemi, S.J., Chau, K.W., 2019. Energy-Life cycle assessment on applying solar technologies for greenhouse strawberry production. Renew. Sustain. Energy Rev. 116, 109411.

Hosseinzadeh-Bandbafha, H., Nabavi-Pelesaraei, A., Shamshirband, S., 2017. Investigations of energy consumption and greenhouse gas emissions of fattening farms using artificial intelligence methods. Environ. Prog. Sustain. Energy 36, 1546–1559.

Hosseinzadeh-Bandbafha, H., Nabavi-Pelesaraei, A., Khanali, M., Ghahderijani, M., Chau, K.W., 2018. Application of data envelopment analysis approach for optimization of energy use and reduction of greenhouse gas emission in peanut production of Iran. J. Clean. Prod. 172, 1327–1335.

Hu, J.-L., Wang, S.-C., 2006. Total-factor energy efficiency of regions in China. Energy Policy 34, 3206–3217.

Hu, J.L., Kao, C.H., 2007. Efficient energy-saving targets for APEC economies. Energy Policy 35, 373–382.

Huijbregts, M.A.J., Steinmann, Z.J.N., Elshout, P.M.F., Stam, G., Verones, F., Vieira, M., Zijp, M., Hollander, A., van Zelm, R., 2017. ReCiPe2016: a harmonised life cycle impact assessment method at midpoint and endpoint level. Int. J. Life Cycle Assess. 22, 138–147.

IPCC, 2006. IPCC Guidelines for National Greenhouse Gas Inventories. Institute for Global Environmental Strategies, Hayama, Japan.

ISO, 2006. 14040 International Standard. Environmental Management–Life Cycle Assessment–Principles and Framework. International Organisation for Standardization, Geneva, Switzerland.

Jat, H.S., Jat, R.D., Nanwal, R.K., Lohan, S.K., Yadav, A.K., Poonia, T., Sharma, P.C., Jat, M.L., 2020. Energy use efficiency of crop residue management for sustainable energy and agriculture conservation in NW India. Renew. Energy 155, 1372–1382.

Jekayinfa, S.O., Adebayo, A.O., Afolayan, S.O., Daramola, E., 2013. On-farm energetics of mango production in Nigeria. Renew. Energy 51, 60–63.

Jolliet, O., Margni, M., Charles, R., Humbert, S., Payet, J., Rebitzer, G., Rosenbaum, R., 2003. IMPACT 2002+: a new life cycle impact assessment methodology. Int. J. life cycle Assess. 8, 324.

Kaab, A., Sharifi, M., Mobli, H., Nabavi-Pelesaraei, A., Chau, K.W., 2019a. Combined life cycle assessment and artificial intelligence for prediction of output energy and environmental impacts of sugarcane production. Sci. Total Environ. 664, 1005–1019.

Kaab, A., Sharifi, M., Mobli, H., Nabavi-Pelesaraei, A., Chau, K.W., 2019b. Use of optimization techniques for energy use efficiency and environmental life cycle assessment modification in sugarcane production. Energy 181, 1298–1320.

Kaltsas, A.M., Mamolos, A.P., Tsatsarelis, C.A., Nanos, G.D., Kalburtji, K.L., 2007. Energy budget in organic and conventional olive groves. Agric. Ecosyst. Environ. 122, 243–251.

Kaveh, M., Amiri Chayjan, R., Taghinezhad, E., Rasooli Sharabiani, V., Motevali, A., 2020. Evaluation of specific energy consumption and GHG emissions for different drying methods (Case study: *Pistacia Atlantica*). J. Clean. Prod. 259, 120963.

Kazemi, H., Kamkar, B., Lakzaei, S., Badsar, M., Shahbyki, M., 2015. Energy flow analysis for rice production in different geographical regions of Iran. Energy 84, 390–396.

Kitani, O, 1999. CIGR Handbook of Agricultural Engineering, Energy and Biomass Engineering. ASABE, USA.

Kizilaslan, H., 2009. Input–output energy analysis of cherries production in Tokat Province of Turkey. Appl. Energy 86, 1354–1358.

Kosemani, B.S., Bamgboye, A.I., 2020. Energy input-output analysis of rice production in Nigeria. Energy 207, 118258.

Kouchaki-Penchah, H., Nabavi-Pelesaraei, A., O'Dwyer, J., Sharifi, M., 2017. Environmental management of tea production using joint of life cycle assessment and data envelopment analysis approaches. Environ. Prog. Sustain. Energy 36, 1116–1122.

Lasvaux, S., Achim, F., Garat, P., Peuportier, B., Chevalier, J., Habert, G., 2016. Correlations in Life Cycle Impact Assessment methods (LCIA) and indicators for construction materials: what matters? Ecol. Indic. 67, 174–182.

Li, N., Jiang, Y., Mu, H., Yu, Z., 2018. Efficiency evaluation and improvement potential for the Chinese agricultural sector at the provincial level based on data envelopment analysis (DEA). Energy 164, 1145–1160.

Malana, N.M., Malano, H.M., 2006. Benchmarking productive efficiency of selected wheat areas in Pakistan and India using data envelopment analysis. Irrig. Drain. J. Int. Comm. Irrig. Drain. 55, 383–394.

Mardani, A., Zavadskas, E.K., Streimikiene, D., Jusoh, A., Khoshnoudi, M., 2017. A comprehensive review of data envelopment analysis (DEA) approach in energy efficiency. Renew. Sustain. Energy Rev. 70, 1298–1322.

Meier, M.S., Stoessel, F., Jungbluth, N., Juraske, R., Schader, C., Stolze, M., 2015. Environmental impacts of organic and conventional agricultural products—are the differences captured by life cycle assessment? J. Environ. Manage. 149, 193–208.

Mellenbergh, G.J.(with contributions by D.J. Hand), 2008. Tests and questionnaires: construction and administration. In: Adèr, H.J., Mellenbergh, G.J. (Eds.), Advising on Research Methods: A Consultant's Companion. Johannes van Kessel Publishing. The Netherlands, pp. 235–268.

Mobtaker, H.G., Akram, A., Keyhani, A., Mohammadi, A., 2012. Optimization of energy required for alfalfa production using data envelopment analysis approach. Energy Sustain. Dev. 16, 242–248.

Mobtaker, H.G., Keyhani, A., Mohammadi, A., Rafiee, S., Akram, A., 2010. Sensitivity analysis of energy inputs for barley production in Hamedan Province of Iran. Agric. Ecosyst. Environ. 137, 367–372.

Mohammadi, A., Omid, M., 2010. Economical analysis and relation between energy inputs and yield of greenhouse cucumber production in Iran. Appl. Energy 87, 191–196.

Mohammadi, A., Rafiee, S., Jafari, A., Dalgaard, T., Knudsen, M.T., Keyhani, A., Mousavi-Avval, S.H., Hermansen, J.E., 2013. Potential greenhouse gas emission reductions in soybean farming: a combined use of life cycle assessment and data envelopment analysis. J. Clean. Prod. 54, 89–100.

Mohammadi, A., Rafiee, S., Jafari, A., Keyhani, A., Dalgaard, T., Knudsen, M.T., Nguyen, T.L.T., Borek, R., Hermansen, J.E., 2015. Joint life cycle assessment and data envelopment analysis for the benchmarking of environmental impacts in rice paddy production. J. Clean. Prod. 106, 521–532.

Mostashari-Rad, F., Ghasemi-Mobtaker, H., Taki, M., Ghahderijani, M., Kaab, A., Chau, K.W., Nabavi-Pelesaraei, A., 2021. Exergoenvironmental damages assessment of horticultural crops using ReCiPe2016 and cumulative exergy demand frameworks. J. Clean. Prod. 278, 123788.

Mostashari-Rad, F., Ghasemi-Mobtaker, H., Taki, M., Ghahderijani, M., Saber, Z., Chau, K.W., Nabavi-Pelesaraei, A., 2020. Data supporting midpoint-weighting life cycle assessment and energy forms of cumulative exergy demand for horticultural crops. Data Br 33, 106490.

Mostashari-Rad, F., Nabavi-Pelesaraei, A., Soheilifard, F., Hosseini-Fashami, F., Chau, K.W., 2019. Energy optimization and greenhouse gas emissions mitigation for agricultural and horticultural systems in Northern Iran. Energy 186, 115845.

Mousavi-Avval, S.H., Rafiee, S., Jafari, A., Mohammadi, A., 2011a. Energy flow modeling and sensitivity analysis of inputs for canola production in Iran. J. Clean. Prod. 19, 1464–1470.

Mousavi-Avval, S.H., Rafiee, S., Jafari, A., Mohammadi, A., 2011b. Improving energy use efficiency of canola production using data envelopment analysis (DEA) approach. Energy 36, 2765–2772.

Mousavi-Avval, S.H., Rafiee, S., Sharifi, M., Hosseinpour, S., Notarnicola, B., Tassielli, G., Renzulli, P.A., 2017a. Application of multi-objective genetic algorithms for optimization of energy, economics and environmental life cycle assessment in oilseed production. J. Clean. Prod. 140, 804–815.

Mousavi-Avval, S.H., Rafiee, S., Sharifi, M., Hosseinpour, S., Shah, A., 2017b. Combined application of life cycle assessment and adaptive neuro-fuzzy inference system for modeling energy and environmental emissions of oilseed production. Renew. Sustain. Energy Rev. 78, 807–820.

Musafiri, C.M., Macharia, J.M., Ng'etich, O.K., Kiboi, M.N., Okeyo, J., Shisanya, C.A., Okwuosa, E.A., Mugendi, D.N., Ngetich, F.K., 2020. Farming systems' typologies analysis to inform agricultural greenhouse gas emissions potential from smallholder rain-fed farms in Kenya. Sci. Afr. 8, e00458.

Nabavi-Pelesaraei, A., Abdi, R., Rafiee, S., Mobtaker, H.G., 2014. Optimization of energy required and greenhouse gas emissions analysis for orange producers using data envelopment analysis approach. J. Clean. Prod. 65, 311–317.

Nabavi-Pelesaraei, A., Amid, S., 2014. Reduction of greenhouse gas emissions of eggplant production by energy optimization using DEA approach. Elixir Energy Environ 69, 23696–23701.

Nabavi-Pelesaraei, A., Bayat, R., Hosseinzadeh-Bandbafha, H., Afrasyabi, H., Berrada, A., 2017a. Prognostication of energy use and environmental impacts for recycle system of municipal solid waste management. J. Clean. Prod. 154, 602–613.

Nabavi-Pelesaraei, A., Bayat, R., Hosseinzadeh-Bandbafha, H., Afrasyabi, H., Chau, K.W., 2017b. Modeling of energy consumption and environmental life cycle assessment for incineration and landfill systems of municipal solid waste management —a case study in Tehran Metropolis of Iran. J. Clean. Prod. 148, 427–440.

Nabavi-Pelesaraei, A., Kaab, A., Hosseini-Fashami, F., Mostashari-Rad, F., Chau, K.-W., 2019. Life cycle assessment (LCA) approach to evaluate different waste management opportunities,. In: Singh, R.P., Prasad, V., Vaish, B. (Eds.), Advances in Waste-to-Energy Technologies. CRC Press, FL, pp. 195–216.

Nabavi-Pelesaraei, A., Rafiee, S., Hosseini-Fashami, F., Chau, K.W., 2021. Artificial neural networks and adaptive neuro-fuzzy inference system in energy modeling of agricultural products,. In: Deo, R, Samui, P., Roy, S.S. (Eds.), Predictive Modelling for Energy Management and Power Systems Engineering. Elsevier, Amsterdam, pp. 299–334.

Nabavi-Pelesaraei, A., Rafiee, S., Mohtasebi, S.S., Hosseinzadeh-Bandbafha, H., Chau, K.W., 2018. Integration of artificial intelligence methods and life cycle assessment to predict energy output and environmental impacts of paddy production. Sci. Total Environ. 631–632, 1279–1294.

Nabavi-Pelesaraei, A., Rafiee, S., Mohtasebi, S.S., Hosseinzadeh-Bandbafha, H., Chau, K.W., 2017c. Energy consumption enhancement and environmental life cycle assessment in paddy production using optimization techniques. J. Clean. Prod. 162, 571–586.

Naderi, S., Ghasemi Nejad Raini, M., Taki, M., 2020. Measuring the energy and environmental indices for apple (production and storage) by life cycle assessment (case study: Semirom county, Isfahan, Iran). Environ. Sustain. Indic. 6, 100034.

Nassiri, S.M., Singh, S., 2009. Study on energy use efficiency for paddy crop using data envelopment analysis (DEA) technique. Appl. Energy 86, 1320–1325.

Omid, M., Ghojabeige, F., Delshad, M., Ahmadi, H., 2011. Energy use pattern and benchmarking of selected greenhouses in Iran using data envelopment analysis. Energy Convers. Manag. 52, 153–162.

Owsiniak, M., Laurent, A., Bjørn, A., Hauschild, M.Z., 2014. IMPACT 2002+, ReCiPe 2008 and ILCD's recommended practice for characterization modelling in life cycle impact assessment: a case study-based comparison. Int. J. Life Cycle Assess. 19, 1007–1021.

Ozkan, B., Ceylan, R.F., Kizilay, H., 2011. Energy inputs and crop yield relationships in greenhouse winter crop tomato production. Renew. Energy 36, 3217–3221.

Ozkan, B., Kurklu, A., Akcaoz, H., 2004. An input-output energy analysis in greenhouse vegetable production: a case study for Antalya region of Turkey. Biomass and Bioenerg. 26, 89–95.

Pahlavan, R., Omid, M., Rafiee, S., Mousavi-Avval, S.H., 2012. Optimization of energy consumption for rose production in Iran. Energy Sustain. Dev. 16, 236–241.

Pelletier, N., Audsley, E., Brodt, S., Garnett, T., Henriksson, P., Kendall, A., Kramer, K.J., Murphy, D., Nemecek, T., Troell, M., 2011. Energy intensity of agriculture and food systems. Annu. Rev. Environ. Resour. 36, 233–246.

Pishgar-Komleh, S.H., Ghahderijani, M., Sefeedpari, P., 2012. Energy consumption and CO_2 emissions analysis of potato production based on different farm size levels in Iran. J. Clean. Prod. 33, 183–191.

Pishgar-Komleh, S.H., Zylowski, T., Rozakis, S., Kozyra, J., 2020. Efficiency under different methods for incorporating undesirable outputs in an LCA+DEA framework: a case study of winter wheat production in Poland. J. Environ. Manage. 260, 110138.

Pizzol, M., Laurent, A., Sala, S., Weidema, B., Verones, F., Koffler, C., 2017. Normalisation and weighting in life cycle assessment: quo vadis? Int. J. Life Cycle Assess. 22, 853–866.

Qasemi-Kordkheili, P., Nabavi-Pelesaraei, A., 2014. Optimization of energy required and potential of greenhouse gas emissions reductions for nectarine production using data envelopment analysis approach. Int. J. Energy Environ. 5, 207–218.

Raheli, H., Rezaei, R.M., Jadidi, M.R., Mobtaker, H.G., 2017. A two-stage DEA model to evaluate sustainability and energy efficiency of tomato production. Inf. Process. Agric. 4, 342–350.

Rattanatum, T., Frauzem, R., Malakul, P., Gani, R., 2018. LCSoft as a tool for LCA: new LCIA methodologies and Interpretation. Comput. Aided Chem. Eng. 43, 13–18.

Rosman, R., Wahab, N.A., Zainol, Z., 2014. Efficiency of Islamic banks during the financial crisis: an analysis of Middle Eastern and Asian countries. Pacific-Basin Financ. J. 28, 76–90.

Roy, P., Nei, D., Orikasa, T., Xu, Q., Okadome, H., Nakamura, N., Shiina, T., 2009. A review of life cycle assessment (LCA) on some food products. J. Food Eng. 90, 1–10.

Saber, Z., Esmaeili, M., Pirdashti, H., Motevali, A., Nabavi-Pelesaraei, A., 2020. Exergoenvironmental-Life cycle cost analysis for conventional, low external input and organic systems of rice paddy production. J. Clean. Prod. 263, 121529.

Saber, Z., van Zelm, R., Pirdashti, H., Schipper, A.M., Esmaeili, M., Motevali, A., Nabavi-Pelesaraei, A., Huijbregts, M.A., 2021. Understanding farm-level differences in environmental impact and eco-efficiency: The case of rice production in Iran. Sustain. Prod. Consum. 27, 1021–1029.

Sabzevari, A., Yousefinejad-Ostadkelayeh, M., Nabavi-Pelesaraei, A., 2015. Assessment of technical efficiency for garlic production in Guilan province of Iran. Elixir Agric 81, 31994–31998.

Samavatean, N., Rafiee, S., Mobli, H., Mohammadi, A., 2011. An analysis of energy use and relation between energy inputs and yield, costs and income of garlic production in Iran. Renew. Energy 36, 1808–1813.

Soni, P., Sinha, R., Perret, S.R., 2018. Energy use and efficiency in selected rice-based cropping systems of the middle-Indo Gangetic Plains in India. Energy Rep. 4, 554–564.

Taki, M., Soheili-Fard, F., Rohani, A., Chen, G., Yildizhan, H., 2018. Life cycle assessment to compare the environmental impacts of different wheat production systems. J. Clean. Prod. 197, 195–207.

Thanassoulis, E., 1993. A comparison of regression analysis and data envelopment analysis as alternative methods for performance assessments. J. Oper. Res. Soc. 44, 1129–1144.

Unakıtan, G., Aydın, B., 2018. A comparison of energy use efficiency and economic analysis of wheat and sunflower production in Turkey: a case study in Thrace region. Energy 149, 279–285.

Valdivieso Pérez, I.A., Toral, J.N., Piñeiro Vázquez, Á.T., Hernández, F.G., Ferrer, G.J., Cano, D.G., 2019. Potential for organic conversion and energy efficiency of conventional livestock production in a humid tropical region of Mexico. J. Clean. Prod. 241, 118354.

Wang, Q., Zhao, Z., Zhou, P., Zhou, D., 2013. Energy efficiency and production technology heterogeneity in China: a meta-frontier DEA approach. Econ. Model. 35, 283–289.

Yamane, T., 1967. Elementary Sampling Theory. Prentice-Hall Englewood Cliffs, NJ, USA.

Yilmaz, H., Aydin, B., 2020. Comparative input-output energy analysis of citrus production in Turkey: case of Adana Province. Erwerbs-Obstbau 62, 29–36.

Zanghelini, G.M., Cherubini, E., Soares, S.R., 2018. How Multi-Criteria Decision Analysis (MCDA) is aiding Life Cycle Assessment (LCA) in results interpretation. J. Clean. Prod. 172, 609–622.

Zentner, R.P., Lafond, G.P., Derksen, D.A., Nagy, C.N., Wall, D.D., May, W.E., 2004. Effects of tillage method and crop rotation on non-renewable energy use efficiency for a thin Black Chernozem in the Canadian Prairies. Soil Tillage Res 77, 125–136.

Zhu, J., 2014. Quantitative Models for Performance Evaluation and Benchmarking: Data Envelopment Analysis With Spreadsheets. Springer, Cham.

CHAPTER 15

Lean integrated management system for sustainability improvement:

An integrated system of tools and metrics for sustainability management

João Paulo Estevam de Souza

National Institute of Space Research / Instituto Nacional de Pesquisas Espaciais (INPE), São José dos Campos, SP, Brazil

15.1 Introduction

After more than three decades since the emergence of the concept of "sustainable development" (World Commission on Environment and Development, 1987), development is still measured mainly by economic indicators, such as increased production or exploitation of natural resources. Actually, development policies are based on the belief of inexhaustible natural resources and the possibility of perpetual GDP growth, although according to Georgescu–Roegen hypotheses, economic growth based on a steady increase in the gross domestic product (GDP) is not sustainable by the global ecosystem (Georgescu-Roegen, 1971; Herrmann-Pillath, 2011; Levallois, 2010). Such problems persist more than four decades after the first warnings about the unsustainability of the production and consumption model and predictions of an energy and resources imbalance and total collapse of the planet Earth (Levallois, 2010; Meadows et al., 1978). According to Rockström et al. (2009), several planetary boundaries of climate change have already been crossed; if the current pace is maintained, others will be crossed as well.

Stakeholders and the society have been increasingly pressuring companies to take responsibility about environmental and social sustainability (Garcia et al., 2016); organizations are increasingly striving for corporate sustainability (CS) as it is now considered a provider of competitive advantage (RobecoSAM, 2014). There is an unprecedented demand for sustainability models that takes into account economic, social, and environmental perspectives (Fonseca et al., 2016; Fonseca, 2015a), and the implementation of management systems (MSs) and lean manufacturing systems (LMSs) is trending; however, organizations are facing difficulties in managing the complexity of integration of multiobjective MSs and high amount of human and financial resources (Nunhes et al., 2017) and the necessary tools, to integrate such systems (Souza et al., 2014). Some studies show evidence that application of the integrated MSs (IMS) can lead to sustained success (Holm et al., 2015; Kurdve et al., 2014; Rebelo et al., 2016, 2015; Siva et al., 2016; Witjes et al., 2017); however, companies are facing difficulties in operating multiple MSs simultaneously (Nunhes et al., 2016). There is a lack of tools for implementing strategies and managing CS (Burritt and Schaltegger, 2010; Garcia et al., 2016; Hansen and Schaltegger, 2014); the IMS can provide a foundation for corporate social responsibility (CSR), but the current literature on the integration of CSR into business processes remains limited and further research is required about its practical implementation (Asif et al., 2013). Although companies in the growth stage of CSR have more MS certifications, the MSs itself does not ensure CS integration (Witjes et al., 2017).

Although MSs can help organizations to improve the performance of its specific objectives and provide the basis for sustainable development initiatives (International Organization for Standardization, 2015a), they are insufficient for competitive-advantage establishment in the view of stakeholders and the society. As an example, ISO 9001 does indeed incorporate many of the principles of a Business Excellence Model it can be considered as a step toward that direction (Fonseca, 2015b). Similarly, organizations have been implementing LMSs to fill the gaps in MSs (Souza et al., 2014). Since 2009, sustainability and LMSs have been attracting ever-increasing interest. However, the literature reveals significant gaps in research on social sustainability and LMSs (Martínez-Jurado and Moyano-Fuentes, 2014); in research based on empirical evidence on IMSs, sustainability, and methods and guidelines for IMSs; and in research on IMSs and social responsibility (SR) (Nunhes et al., 2017).

Considering the increasing demands of the scientific community and practitioners regarding sustainability management and the literature gaps in research based on empirical evidence, this chapter presents the LIMSSI model. The LIMSSI model is a management model for the improvement of CS, which is based on empirical evidence of integration of IMS (quality, environmental, occupational health and safety (OHS), and SR) with the LMS as an option for the demands of the scientific community and practitioners.

15.2 Literature overview and background
15.2.1 Sustainability

According to the report entitled Our Common Future, sustainable development is the development that meets current needs without compromising the ability of future generations to meet their own needs. The new global Sustainable Development Goals (United Nations, 2015) demand that technology and social organization should be managed and improved to usher in a new era of economic growth (ISO, 2010; World Commission on Environment and Development, 1987). To achieve this demand, it is crucial to focus on three dimensions of sustainable development—environmental protection, economic growth, and social equity (Fisher and Bonn, 2011)— on a global, long-term scale (Gutowski, 2011). By nature, capitalism-based business do not inherently ensure that all humans have adequate access to environmental services for their life support systems (Milne and Gray, 2013). This creates the need for a management model that integrates economic, environmental, and social successes to enable sustainable development.

CS is defined as "meeting the needs of a firm's direct and indirect stakeholders [...], without compromising its ability to meet the needs of future stakeholders as well" (Dyllick et al., 2002). This task is complex, and it necessitates new CS management approaches (Schaltegger et al., 2013).

Sustainable manufacturing is an innovative set of practices and technologies for transforming materials into products by consumption of smaller amounts of energy and fewer nonrenewable or toxic materials and generation of fewer emissions and waste (Bhanot et al., 2016). Several organizations undertake CS initiatives at the operational level (Fisher and Bonn, 2011), instead of integrating CS at all business levels. To create a dynamic and efficient system, there is a need for synergy between the different areas and the integration at all business levels. It creates a sound basis for working toward a more sustainable MS (Jørgensen, 2008).

Sustainability is perceived as the new paradigm for the business of the twenty-first century; its concept was translated to the corporate context via the triple bottom line (TBL) framework developed by Elkington (1998). Nevertheless, TBL remains an abstract concept that is difficult to put into practice (Lozano, 2012).

One of the practices that have been increasingly used among organizations is Sustainability reporting, and the Global Reporting Initiative (GRI) is the most widely used global guideline for CS reports. However, few tools are available for implementing the required strategies for managing CS (Burritt and Schaltegger, 2010; Garcia et al., 2016; Hansen and Schaltegger, 2014).

Although companies in the growth stage of CSR have more MS certifications, the MS itself is not used to ensure CS integration (Witjes et al., 2017); therefore, the IMS has a great potential as a basis for CSR. However, a holistic systems approach that brings integration and synergy to the use of methods and tools for sustainability improvement to the table is needed (Souza and Dekkers, 2019).

15.2.2 Management systems

The business environment is often intense in competition, continual technological progress, new consumer requirements, and scarce natural resources (de Oliveira, 2013). Additionally, there has been an increase in consumer demands, which presently relate to not only quality but also the environment, OHS, and sustainability (Domingues et al., 2016).

As a response to stakeholders demands, organizations are implementing standardized MSs (Rebelo et al., 2015). In 2019 the overview of certifications issued worldwide was 883,521 certifications under the ISO 9001, 312,580 under the ISO 14001, and 38,654 under ISO 45001 (ISO, 2020). Implementation of a certifiable MS has become a common practice in various organizations seeking higher competitiveness (Nunhes et al., 2017) because of its benefits such as improved internal organization, higher customer satisfaction, and greater personnel motivation (Bernardo, 2014).

Although organizations recognize the benefits of MSs, many of them still struggle with MS implementation activities, especially regarding the integration of multiple MSs (Nunhes et al., 2016; Souza et al., 2014). Difficulties arise primarily from the complexity of internal management, low efficiency, and increased management-structure costs (Zeng et al., 2007).

When an organization has to deal with multiple MSs, the best practice is to merge different MSs into a single and more effective IMS, because doing so would improve organization performance (Chatzoglou et al., 2015); eliminate conflicts between individual MSs; and optimize human, technological, and financial resources. Additionally, fewer internal and external audits, and more added value creation for the business through elimination of organizational waste (Rebelo et al., 2016), will improve efficiency via reduced costs, improved internal organization, and exploitation of the synergies between the IMS(Bernardo, 2014; de Oliveira, 2013).

Another point to be considered is that MSs apply to and are useful for CS management at the operational and strategic levels, regardless of the organization strategy type (Baumgartner, 2014). However, the need for "development of sustainability solutions that are multilevel, systematically integrated (including inputs, processes, outputs, and feedback), and multistakeholder oriented" remains (Starik and Kanashiro, 2013).

15.2.2.1 Quality management system (QMS)

Organizations face stakeholders pressure to improve their performance continually; this has led to the emergence of quality management (QM) (Mokhtar et al., 2013). QM is vital for the economic development of organizations because it provides strategic benefits and a strong quality–productivity correlation (Movahedi et al., 2013).

The ISO 9001 standard provides requirements for a QMS; it is process-oriented and based on the plan–do–check–act (PDCA) cycle. It is used worldwide for enhancing performance, ensuring customer satisfaction, and improving supplier-customer relationships (International Organization for Standardization, 2015a; Nunhes et al., 2016). A QMS promotes continuous improvements of processes and may lead to excellence, sustainability, and competitiveness achievements (Fernandes et al., 2017). ISO 9001 is considered fundamental for the sustainable economic success of an organization (Movahedi et al., 2013; Qi et al., 2013).

To integrate CS into strategic management it is crucial that the consideration of QM, because it fosters the quality improvement, has an impact beyond the immediate production level, and is correlated with stakeholder satisfaction (Engert et al., 2015).

15.2.2.2 Environmental management system (EMS)

The environment is being adversely impacted by corporate activities (Agan et al., 2013), although certain organizations disregard environmental problems. However, several organizations have stopped endorsing pollution caused by operations. Consequently, management of activities to prevent an adverse impact on the environment is crucial (Campos et al., 2015).

To manage and prevent adverse impacts on the environment, ISO 14001 is the most commonly applied EMS standard worldwide. It uses a set of interrelated elements to develop and implement an environmental policy and manage its environmental aspects to enhance environmental performance (Campos et al., 2015; International Organization for Standardization, 2015b). The EMS offers benefits that are both environment oriented and company performance oriented (Campos et al., 2015). There is a belief that environmental processes only represent costs. However, they actually can lead to gains on a larger scale and contribute to the development of competitive advantage by reducing pollution and promoting resource savings; improve internal efficiency; and enhancement of corporate image, reputation, and performance, including long-term profits (Agan et al., 2013; Nunhes et al., 2016). Advanced environmental management practices can also improve financial performance and quality, increase manufacturing competitiveness, and promote cost reductions and development of innovative products and processes (Jabbour et al., 2013).

ISO 14001 steers companies toward green production, requiring them to systematically manage their environmental responsibilities and impacts; this is reflected as the strengthening of the environmental pillar of sustainability and consequently in CSR (International Organization for Standardization, 2015b; Nunhes et al., 2016).

15.2.2.3 Social responsibility management system (SRMS)

SR is defined as the "responsibility of an organization for the impacts of its decisions and activities on society and the environment, through transparent and ethical behavior that: contributes to sustainable development, including health and welfare of society; takes into account the expectations of stakeholders; is in compliance with applicable law and consistent with international norms of behavior; and is integrated throughout the organization and practiced in its relationships" (International Organization for Standardization, 2010).

In the business context, SR is commonly known as CSR. Organizations take corporate social actions to meet stakeholders' ethical expectations via managerial activities to advance a social good beyond law requirements in the best interests of stakeholders" (Lee, 2017). Emphasis on CSR study and practice has helped to improve corporate practices; however, incidents of unethical corporate behavior (e.g., bribery, fraud, money laundering) still occur. For CSR to have a meaningful impact, it must be integrated into every level of a corporation and must be seen as an organizational imperative (Asif et al., 2013).

CSR can foster competitive advantage by improving company reputation and image, retention of exceptional personnel, aggregate value, environmental and social performances, and corporate governance and by increasing employee motivation (Aureliano et al., 2013; Filho et al., 2010). However, many companies lack a strategic CSR approach and follow unsystematic practices (e.g., without formal processes and procedures). It is important to establish strategies that address CSR issues and integrate them into company operations (Hahn, 2013). In 2010, ISO 26000 was published to "provide guidance…on ways to integrate socially responsible behavior into the organization" (International Organization for Standardization, 2010). ISO 26000 offers guidelines on several CSR subjects and issues that will assist in internal and external assessments; additionally, this can eventually improve sustainability performance (Hahn, 2013).

15.2.2.4 Occupational health and safety management system (OHSMS)

Human rights and labor practices are core subjects of CSR, which include respect for a decent work environment. To address this demand, organizations are adopting the implementation of OHS MSs (OHSMSs). Organizations should generate value to the society and not burden the social security and health system; thus, they must guarantee OHS by incorporating an effective management response with dynamic strategies into their decision-making processes.

OHSMSs aim to support and promote good practices in the field of OHS via systematic and structured MSs (Abad et al., 2013, 2014) that mitigate accident risks and protect persons exposed to OHS risks associated with occupational activities (Nunhes et al., 2016). OHSAS 18001 had become the predominant standard for evaluating safety management processes at the firm level (Abad et al., 2013, 2014; Granerud and Rocha, 2011; Lo et al., 2014) although in 2018 the International Organization for Standardization published the ISO 45001 as an international standard for OHSMS.

The ISO 45001 standard specifies requirements for an OHSMS, and gives guidance for its use to enable organizations to provide safe and healthy workplaces by preventing work-related injury and ill health, as well as by proactively improving its OHS performance. ISO 45001 is reported as applicable to any organization that aims to establish, implement, and maintain an OHSMS to improve OHS, eliminate hazards and minimize OHS risks (including system deficiencies), take advantage of OHS opportunities, and address OHSMS nonconformities associated with its activities (ISO, 2018).

Reported drawbacks of OHSMSs are certification costs, lack of employee motivation, difficulties in changing company culture, and increased bureaucracy (Abad et al., 2013; Fernández-Muñiz et al., 2012a, 2012b; Lo et al., 2014; Nunhes et al., 2016). Besides, there are reports of OHSMs correlation with a reduction in the probability of accidents and disruptions in the production process and the improvement on internal climate, image, and overall performance. In summary, OHSMSs can be used as a strategic tool for improving competitiveness to eventually achieve a favorable market position (Fernández-Muñiz et al., 2012a, 2012b).

15.2.3 Lean manufacturing system

Efficiency and competitiveness are challenges that have led several organizations to adopt new manufacturing management strategies. Hence, the most critical issue is how to deliver products quickly, at low cost, and with good quality (Rohani and Zahraee, 2015).

To address those challenges, LM seeks to eliminate the seven waste types: overproduction, waiting, transportation, defects, inappropriate processing, unnecessary inventory, and unnecessary motion. It achieves this through lean thinking, which is defined as a business approach that delivers better value for customers by removing nonvalue-adding activities (Caldera et al., 2017). The LM may promote continuous improvement of processes; focus on customer satisfaction to increase efficiency; cost reduction (Galeazzo et al., 2014; Vinodh et al., 2011; Womack and Jones, 2004); and contributions toward improving quality, profitability, and public image (Verrier et al., 2013).

Quite a few attempts have been made to analyze how lean and sustainability relate to each other (Souza and Dekkers, 2019). However, tools and techniques of LM (e.g., just-in-time (JIT), total productive maintenance (TPM), pull production, cellular manufacturing, 5S/7S, kaizen, visual control, poka-yoke, and value stream mapping (VSM)) can also develop CS (Vinodh et al., 2011). In particular, VSM is an important method for comprehending the activity sequences and information flows used in product manufacturing and service delivery (Rohani and Zahraee, 2015; Rother and Shook, 2003; Vinodh et al., 2011).

Several organizations have achieved better results and higher competitiveness after LM implementation; however, others have not, as they were unable to sustain medium- and long-term results. Considering that sustainability is a new trend toward competitive advantage, many companies are adopting LM to improve their results including a way to be seen as socially responsible. As sustainability is considered the new LM frontier; this necessitates investigation of LM impact on the three sustainability dimensions (environmental, economic, and social) (Martínez-Jurado and Moyano-Fuentes, 2014). Here, we present a comprehensive literature review to identify the interrelationships among LM and these three dimensions.

15.2.3.1 How to integrate the environmental dimension

The environmental dimension of sustainability focuses on the impacts caused by organizations' operations. Therefore, environmental management aims to prevent adverse impacts on the environment (Campos et al., 2015).

As a result, environmental management can create synergy with other management practices. Support for environmental management tends to be greater when companies adopt LM practices, which would improve their environmental performance (Jabbour et al., 2013).

The proper conduction of an EMS that conforms to ISO 14001 can be a valuable tool for identifying and reducing wasteful activities; however, ISO 14001 only stipulates that environmental performance shall be improved, and it does not offer guidelines on how this should be achieved or what methods should be used. LMSs when integrated to ISO 14001 can offer a path to specify methods and tools for their implementation. Examples of integration of LMS practices and EMS include root-cause identification, corrective action process, prevention of failures (jidoka/poka-yoke), and provision of continuous improvement based on critical analysis by top management (TM). Examples of LMS tools that can be integrated into EMS include 5S, TPM, and VSM (Jabbour et al., 2013). For example, VSM can be helpful to map raw materials, energy, and water usage by a process or product, and pollution and environmental wastes (Vinodh et al., 2011).

Additionally, the integration of EMS with LMS is effective in standardizing continuous improvement, which may lead to holistic understanding, improved organization performance (Kurdve et al., 2014), and better prevention of pollution owing to stock reductions (Jabbour et al., 2013); the last outcome will ensure prevention of waste generation instead of the management of generated waste. Consequently, reducing the implementation cost of environmental improvements may reduce the marginal cost of pollution and the dispersion of toxic substances through the use of fewer raw materials (Martínez-Jurado and Moyano-Fuentes, 2014). Usually, an EMS requires organizational changes (e.g., more training and better interpersonal contacts), which, in turn, may lead to more committed employees and higher labor productivity (Lannelongue et al., 2017).

Hence, lean principles and practices may facilitate the achievement of environmental goals and improvements in environmental results (Martínez-Jurado and Moyano-Fuentes, 2014) as it has been reported that lean organizations with environmental practices have achieved better lean results than those without these practices. To summarize, lean and environmental practices implemented together can operate with greater potency and bring greater benefits than when implemented separately. Therefore, organizations can create synergies between lean and green actions (Dues et al., 2013; Jabbour et al., 2013).

15.2.3.2 How to integrate the economic dimension

Most of us agree that productivity and cost saving are necessary for the economic survival of most organizations. However, these tasks should be conducted with sustainable orientation, by mitigating negative environmental and social impacts and contributing to a sustainable society (Baumgartner, 2014).

For instance, initiatives leading to cost savings include EMSs, waste- and energy-reduction analysis, and implementation of improvement programs for the productivity and efficiency of material use (Engert et al., 2015). Hence, MSs can be valuable for identifying and mitigating wasteful activities, particularly when used along with LM tools such as VSM (Kurdve et al., 2014). However, the achievement of sustainable results via LM can be complex because several practices need to be implemented simultaneously (Martínez-Jurado and Moyano-Fuentes, 2014). For instance, it is important to highlight the importance of ISO 45001 for the reduction of the probability of accidents and disruptions in the production process improves the internal climate, image, and overall performance of an organization.

15.2.3.3 How to integrate the social dimension

An organization is responsible for the impacts of its decisions and activities on society and the environment (International Organization for Standardization, 2010). Considering that fact, organizations need to generate greater value and contribute to social equality. To integrate the organization´s activities to improve social performance, it can use a basic principle of LM that pertains to the role of people and their treatment. The principle of the people and their treatment includes efforts for incorporating their suggestions and for respecting and recognizing them have served as a factor for the integration of LM with ISO 26000 (Martínez-Jurado and Moyano-Fuentes, 2014) and ISO 45001. Additionally, work performed according to LM principles may generate intrinsic motivation and greater autonomy of responsibility (Taubitz, 2010; Vinodh et al., 2011). Regarding OHS, there is a consensus that LM promotes more ergonomic and secure workplace design (Taubitz, 2010; Vinodh et al., 2011). Vinodh et al. (2011) have reported that for greater integration of LM with social sustainability, it is necessary to transition from the 5S tool to the 7S tool, where the latter includes aspects of OHS and sustainability.

Moreover, LM has been associated with fewer accidents. Particularly, as workers reported OHS improvements, as a result of LM practices that improved housekeeping and material handling.

Additionally, they also reported that kaizens reduced or eliminated several workstation security risks. Finally, LM implementation can improve workplace safety, productivity, and product cost through attention to safety and ergonomics (James et al., 2013).

Considering the role of MSs, OHSMSs contribute to the social dimension of sustainability (Klute-Wenig and Refflinghaus, 2015; Rebelo et al., 2015); the QMS may help organizations to develop SR, and it can create a corporate culture that promotes and encourages ethical behavior. To ensure successful QM, it is necessary to explicitly focus on moral values, because ethical and social issues in business environments assume full quality control to be able to address moral questions satisfactorily. In short, ISO 9001 significantly influenced the first steps of SR standardization; however, an approach that incorporates QM and SR management (e.g., ISO 26000) is not yet available (Tarí, 2011).

15.3 The Lean Integrated Management System for Sustainability Improvement (LIMSSI) model

Considering the demand from society and stakeholders for sustainability improvement, organizations have the challenge to drive themselves toward their objectives improving the three dimensions of sustainability. However, there is a lack of sustainability management models that describes how organizations should manage sustainability to establish competitive advantage and generate value for society and stakeholders. To deal with this challenge the LIMSSI model is an innovative option to manage sustainability at the strategic, tactical, and operational levels. To manage the economic, environmental, and, social dimensions of sustainability, the LIMSSI model integrates the QMS, EMS, OHSMS, and SRMS into the LMS to facilitate continuous improvement of the CS performance (Souza and Alves, 2018). The integration was aimed at creating synergies to extend the benefits of each system and at promoting rational use of resources and time for conducting the system activities; such rational use would also reduce costs and overlapping of tasks. Additionally, it seeks to make the organization economically profitable, ecologically correct, operationally safe, socially fair, and culturally accepted.

To promote efficient resource usage, the LIMSSI model considers the premises of the IMS—comprising the aforementioned MS and the LMS—as it is considered state of the art in operations management. Although the LIMSSI model is described as a model for the improvement of CS, it is considered to be generic, regardless of whether the organization has one or more MSs of the model; and it can be applied in organizations of any size, type, and sector, (i.e., public entities, universities, hospitals, not for profit organizations and non-governmental organizations) (Souza and Alves, 2018).

Thus, to reinforce the promotion of efficiency, lean thinking was incorporated in the LIMSSI model to improve economic, social, and environmental performance. It is expected that LM contributes through pollution cost reduction, environmental improvements, efficient use of energy and resources, and highlighting the importance of people (incorporation of suggestions, respect for people, and their recognition) to create synergies with the social dimension. It also incorporates the transition from 5S to 7S and kaizen while addressing sustainability, environmental hazards, risks, and OHS; this is done to reduce OHS incidents via improvements in housekeeping, ergonomics, and material handling.

Those considerations resulted in the framework of the Lean Integrated Management System for Sustainability Improvement, the LIMSSI model. Fig. 15.1 provides a graphical representation of the LIMSSI model structure and its systems correlations.

15.3 The Lean Integrated Management System for Sustainability Improvement

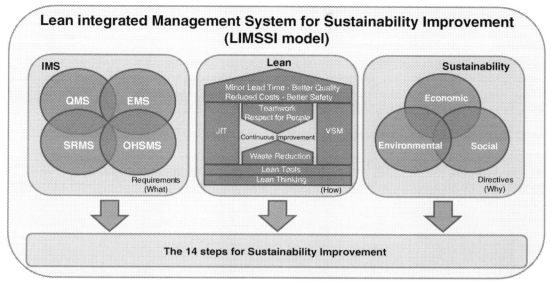

FIG. 15.1 The LIMSSI Model.
Source: Adapted from Souza, J.P.E., Alves, J.M., 2018. Lean-integrated management system: A model for sustainability improvement. J. Clean. Prod. 172, 2667–2682.

As shown in Fig. 15.1, the IMS integrates the QMS, EMS, SRMS, and OHSMS; the IMS provides the requirements for the LIMSSI model, that is, *what* shall be done. The three dimensions of sustainability: economic, environmental, and social; constitute a basis for measurement of sustainability performance; this part describes *why* the model must be implemented. The LMS brings to the LIMSSI model lean thinking; this defines *how* it can be done. Thus, the LMS acts as a liaison agent between the requirements (*what*) and the dimensions of sustainability (*why*) and also as a potentiator of synergy creation.

15.3.1 How to implement the LIMSSI model

As many organizations report difficulties in implementing sustainability MSs, one of the most reported difficulty is the lack of a description of how to conduct sustainability activities in a structured and synergic way. To supply this demand, the LIMSSI model provides a structured framework that describes 14 steps to promote the rational use of resources and energy while engaging and empowering people. It also provides an innovative description of correlations of the QMS, EMS, OHSMS, and SRMS requirements with the LMS principles and tools, including a description of how they can be integrated to create synergies. This description was structured to foster synergies by considering the difficulties organizations face in conducting sustainability improvement activities. It is expected that following the 14 steps of the LIMSSI model organizations can prevent loss of organizational efficiency due to waste, duplication, and excessively bureaucratic processes.

Next, the steps for implementation of the LIMSSI are described, which establish a continuous improvement path for sustainability performance (Souza and Alves, 2018).

15.3.1.1 Step 1: Identify the stakeholders

As the beginning of a journey toward sustainability, the organization should identify stakeholders that may be significantly affected by its activities, products, and services, or whose actions may affect the organization's ability to successfully implement strategies and achieve objectives. Subsequently, for each group of stakeholders, their expectations and interests should be identified to provide a broad view of their interrelationships. This is done to measure how these interrelationships affect and how they are affected by the environmental, social, and economic settings in which the organization operates.

15.3.1.2 Step 2: Perform a critical analysis of compliance

Compliance is a basic requirement for SR and IMS applications. Therefore, organizations must identify and meet all requirements (e.g., legal, statutory, and other requirements) to which it has subscribed. To manage this task, knowledge of stakeholders and the operational environment is crucial to support the process of anticipating and preparing for new or modified requirements and consequently to maintain compliance with applicable requirements. Therefore, the organization must maintain an up-to-date record of applicable requirements, principles, or other economic, environmental, and social initiatives that it endorses or subscribes to, in addition to a description of how to address them.

15.3.1.3 Step 3: Establish the LIMSSI policy

As a result of the knowledge of its stakeholders, applicable requirements, and operational environment, the organization has the baseline to establish policies. After the establishment of this baseline, the organization should develop the LIMSSI policy as a statement of the principal decision-maker of the organization regarding sustainability improvement. The LIMSSI policy should include a statement of the relevance of sustainability to the organization and its sustainability strategy. Hence, conflicts between MS functions should be approached based on the LIMSSI policy to find a balance point. This balance point should seek to maximize the LIMSSI efficiency in an integrated manner and minimizing its negative impact on other MSs.

15.3.1.4 Step 4: Assure TM support and involvement

Support and involvement of TM are crucial for the development of CS. Consequently, full support and commitment of TM with sustainability initiatives are requisites for the implementation of sustainable manufacturing (Amrina and Yusof, 2011; Bhanot et al., 2016).

In the same way, for successful LIMSSI implementation, the TM involvement is crucial. Hence, meetings with TM should be held to define the LIMSSI implementation approach and TM actions. During those meetings, the project scope and stages should be detailed to ensure alignment between the LIMSSI and the organization's needs and resources.

15.3.1.5 Step 5: Build awareness

The lack of awareness of sustainability is among the main barriers to the development of CS (Bhanot et al., 2016). Hence, building awareness of sustainability must be a constant effort throughout the implementation of the LIMSSI model. Therefore, meetings, lectures, and other awareness-raising activities should be promoted, and they should include TM and employees from all organization departments and levels.

To build awareness, training has tremendous potential in enhancing sustainability activities. It is necessary to create awareness among professionals about the benefits and ways of using sustainability improvement methods. For example, this can be achieved using a proper model that will mitigate any negative aspects and help TM in providing adequate support (Bhanot et al., 2016).

Providing that, these activities should address the main concepts of LM, quality, environment, SR, and OHS and clarify their main tools, advantages, and potentialities to provide more knowledge about sustainability. The activities should seek to conceptually empower those involved and break paradigms to improve the receptivity of employees to the actions that will be developed.

15.3.1.6 Step 6: Assign responsibilities

The definition of responsibilities plays an important role. Therefore, responsibilities for LIMSSI activities should be defined. The responsibilities should comprise a LIMSSI coordinator; at least one focal point for each of the areas of quality, environment, OHS, and SR; and a value stream (VS) manager. Those assigned for each responsibility should act as facilitators of change and use their knowledge to promote the integration of the dimensions of sustainability.

As an example, the VSM is not restricted to only one area or sector of the organization; therefore, a person with the ability to cross organizational boundaries should be designated as the VS manager to facilitate the process of mapping, understanding, and optimizing the VS as a whole. In the same way, the LIMSSI coordinator and the VS manager must have easy and open access to the TM, with the freedom to operate across departmental boundaries.

Moreover, the VS manager and the LIMSSI coordinator are key players in creating synergies between the IMS and the LMS. They should seek continuous improvement through kaizen activities according to the documented information process and integrate the systems to standardize actions and share best practices.

15.3.1.7 Step 7: Select a product family

The final objective should be the implementation of the LIMSSI model throughout the organization. However, the model proposes that the organization initiates implementation via a product family.

Hence, a "product–process" matrix should be developed to enable verification of product groupings with technological similarities. Thusly, product families can be formed according to such similarities. For instance, it is interesting that LIMISSI implementation starts with a product family that is meaningful and feasible for the organization.

15.3.1.8 Step 8: Draw a current-state value stream map

Knowledge of the current state of the organization's operations is crucial knowledge to support any improvement. Considering the importance of knowing the current state, after selection of a product family the organization must perform the current-state VSM. Thus, the VSM describes the VS as providing the basis for the design and implementation of a future-state VS map.

Mapping a VS is, essentially, the standardization and use of a language (Rother and Shook, 2003), which can be used as a communication and planning tool to manage the change process toward improved sustainability. Consequently, employees should be informed about the mapping objectives to start activities in a clear and aligned manner, thus reassuring them that they are not being evaluated but rather the VS.

Once finished, the current-state VS map should be integrated into the IMS, as a complement to the description of the processes and their sequence and interaction. Additionally, the VSM promotes the focus of actions toward achieving planned results and continuous improvement. For the preparation of the current-state VS map, validated data reflecting the VS operational reality should be used. Even if MS data are available, data must be collected on the shop floor. The mapping should start with a walk around the VS, "door to door" of the organization, to understand the sequence flow of processes, what activities of the IMS are conducted, and how they are conducted. Next, information on processes and operations, quality, OHS, environmental, and SR problems must be registered.

15.3.1.9 Step 9: Define targets and goals

Any initiative toward continuous improvement should begin with the definition of clear objectives. Hence, TM should establish targets and goals considering the current-state VS map, the LIMSSI policy, and legal obligations. Providing that, the objectives must be measurable and must encompass programs to meet the defined objectives; also, they must include responsibility assignment, the necessary resources and approaches, and deadlines.

15.3.1.10 Step 10: Define performance indicators

Performance indicators allow a comparison of what has happened and know tendencies of what will happen. Therefore, they provide information to support decision-making, potentially affecting the future competitive position of the organization (Amrina and Yusof, 2011).

Through the use of performance indicators, sustainability can be measured and consequently provides data to managers for evaluation and decision making. Thus, performance indicators are crucial for measuring sustainability (Rodrigues et al., 2016). By contrast, a lack of performance benchmarks hinders organizations from assessing sustainability performance and identifying their underperforming domains (Bhanot et al., 2016).

To manage sustainability using the LIMSSI model, the organization should establish, implement, and maintain a set of performance indicators that meet requirements applicable to the organization and requirements of stakeholders or agreements to which the organization has submitted. Additionally, stakeholders should take part directly or indirectly in defining performance indicators (Garcia et al., 2016).

In accord with the QMS, EMS, SRMS, OHSMS, and LMS and the GRI principles (GRI, 2016a, 2016b, 2016c), the LIMSSI model suggests at least the following indicators; however, organizations should include any additional indicators as deemed necessary.

- *Performance indicators for the economic dimension of sustainability*
    1. Quality of products and services;
    2. On-time delivery;
    3. Lead time;
    4. Overall equipment effectiveness (OEE).

- *Performance indicators for the environmental dimension of sustainability*
    1. Percentage reduction in material use;
    2. Volume of reductions in energy consumption achieved directly from improvements of conservation and efficiency; and
    3. Total volume of water recycled and reused by the organization, citing the percentage concerning the total water used.

- *Performance indicators for the social dimension of sustainability*
    1. Total number of hours spent on training in human rights and anticorruption policies or procedures related to relevant aspects of organization operations;

2. Number of implemented programs of
   * community engagement;
   * impact assessment and local development, including gender-impact assessments, based on participatory processes;
   * environmental-impact assessments and continuous monitoring;
   * public dissemination of results of environmental and social impact assessments;
   * local development based on local community needs;
   * committees and consultation processes for the local community, including vulnerable groups; and
   * work councils, OHS committees, and other representative bodies of workers to discuss impacts;
3. Formal claims and complaints processes by local communities;
4. Monetary value of fines and the total number of nonmonetary sanctions owing to noncompliance with laws and regulations;
5. Rate of occupational accidents, injuries, and diseases, stratified by type; the rates should consider the total number of employees (own and outsourced employees); and
6. The average number of hours of OHS training undergone by the organization's employees.

While defining, measuring, and analyzing its set of performance indicators, the organization should consider the principles of materiality and completeness (GRI, 2016a). It is also important that the organization uses the performance indicators with a focus on the promotion of synergy between the MSs and the LMS, thus reducing adverse effects and increasing the benefits of actions and programs toward systemic sustainability.

15.3.1.11 Step 11: Contextualize organization´s sustainability performance

The organization´s performance should be contextualized based on broader sustainability concepts and compared with the limits and demands imposed on environmental or social resources at the sectional, local, regional, and global levels. Such contextualization should clarify how the organization impacts or could impact positively or negatively the economic, environmental, and social dimensions, at local, regional, and global levels.

15.3.1.12 Step 12: Design the future state

While the current-state VS map reveals wastes and problems, it provides information that fosters the design of a more sustainable VS. Therefore, the ultimate goal is to build a more sustainable VS, wherein processes are articulated to stakeholders. Each such process would be as close as possible to producing only what customers need and when they need it; this would have the least possible impact on the environment and also be safe and socially responsible.

15.3.1.12.1 Identification of opportunities for improvement

The current-state VS map helps the identification of wastes and problems to be eliminated or reduced. From a critical analysis of the current-state VS map, the organization must identify the improvement opportunities while seeking synergy among the economic, environmental, and social dimensions for the construction of the future-state VS map, with the ultimate aim of improving sustainability. Furthermore, the VSM is an excellent precursor of kaizen events because it helps identify which VS points need these events (Rother and Shook, 2003). Hence, process improvements and kaizen events should be subordinated to the future-state VS project and the sustainability concept, instead of vague and isolated improvement efforts. Therefore, teams should work on improvements with a clear view of why they are making such improvements and how they connect to the sustainability of the entire VS.

Kaizen events usually are used for improvements, but additionally, it can lead to the fulfillment of social needs, increase self-esteem, and self-realization. These practices promote continuous improvement of processes and products; hence, they strengthen the motivation of individuals, as they incorporate the social principle of the importance of people by adopting their suggestions and respecting and recognizing them.

15.3.1.12.2 Future-state VSM

The future-state VS map must be designed aiming at the elimination of waste, and when this is not possible its reduction. Other focus areas should be the establishment of a pull system, optimization of the use of available resources, and implementation of improvement opportunities that require the use of the fewest possible resources in the easiest and fastest way possible.

15.3.1.12.3 Attaining the future state for sustainability improvement

The effectiveness of the implementation of the future-state VS map significantly depends on the planning of this implementation. A LIMSSI implementation plan should include the current- and future-state VS maps, a process-analysis report, and a description of the improvement opportunities. Finally, the organization should devise an implementation plan for the future-state VS map, where it should be treated as a strategic decision of the organization.

15.3.1.13 Step 13: Integrate the MSs and create synergy

Many organizations already have MSs; nevertheless, many still report difficulties in applying them. Despite the difficulties, there is a consensus that MSs contribute to the development of competitive advantages and can help improve the organization's products and processes. Thus, MSs establish requirements—what needs to be done; however, there is a gap on how to meet these requirements. From an analysis of MS compatibilities and incompatibilities, the LMS was chosen as an agent for such integration. Hence, to integrate the LMS with the IMS, it is important to treat them as complementary and integrated from the beginning, and not as interchangeable.

15.3.1.13.1 The Sustainability Committee

A key feature for sustainability improvement brought by the LIMSSI model is the Sustainability Committee (SC). The SC includes representatives from the economic, environmental, and social dimensions interacting to promote integration and synergy. The SC should review all proposed changes and verify if it causes problems for other dimensions. Hence, it should also determine whether there are alternatives that maximize improvements synergistically. Finally, the SC should have direct access to TM for reporting the positive and negative impacts that improvement opportunities may have on the VS and the organization's sustainability performance.

Fig. 15.2 illustrates the integration of the LMS with the IMS. For instance, the synergy column highlights its importance as a support to the LIMSSI to improve sustainability management. Synergy is created through systems integration aimed at a holistic improvement.

One of the main contributions of the LIMSSI model is provided through its description of opportunities for synergistic MSs integration. It was possible after extensive analysis of the QMS, EMS,

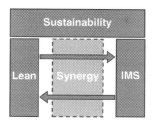

FIG. 15.2 Sustainability supported by Lean, IMS and their synergy.

Source: Adapted from Souza, J.P.E., Alves, J.M., 2018. Lean-integrated management system: A model for sustainability improvement. J. Clean. Prod. 172, 2667–2682.

OHSMS, SRMS, and LMS that allowed the identification of the opportunities for synergistic integration. Table 15.1 presents the results as an innovative description of correlations of MSs requirements with the LMS principles and tools, including a description of how they can be integrated to create synergies.

15.3.1.14 Step 14: Seek perfection

Seeking perfection can be seen as a utopia. However, the principle of continuously seek perfection can keep us moving toward improvement. By understanding and implementing the practices listed in Table 15.1, the VS manager can act to implement the future-state VS map and improve sustainability. However, the actions must be monitored using performance indicators.

Once the future state has been attained, a new cycle of continuous improvement must be pursued. Eventually, seeking perfection causes the future-state VS map to become the current-state VS map. Then, a new future-state VS map is necessary to pursue new challenges toward perfection. But, organizations should always remember that a VS should be developed with respect for people; however, respect for people should not be confused with respect for old habits (Rother and Shook, 2003).

15.4 Conclusions

Organizations still face difficulties in managing the complexity of integration of multiobjective MSs, particularly in integrating them, considering the significant gap in research and the practitioners' demand for a holistic system approach that brings integration and synergy to the use of methods and tools for sustainability improvement. The LIMSSI model was identified as an innovative approach to deal with those challenges regarding sustainability management.

The LIMSSI model is a management model for the improvement of CS based on a holistic system approach that brings synergy and integration to the IMS (quality, environmental, OHS, and SR) with the LMS. The LIMSSI can contribute to the most effective CS management. The structured framework of the LIMSSI includes 14 steps to promote the rational use of resources and energy while engaging and empowering people. Additionally, it provides an innovative description of correlations of the QMS, EMS, OHSMS, and SRMS requirements with the LMS principles and tools, including a description of how they can be integrated to create synergies. Hence, it was structured by considering the difficulties organizations face in conducting sustainability improvement activities. Therefore, its use aims to prevent loss of organizational efficiency due to waste, duplication, and excessively bureaucratic processes.

Table 15.1 Synergistic integration of IMS with LMS.

ISO 9001:2015 requirement	ISO 14001:2015 requirement	ISO 45001:2018 requirement	ISO 26000:2010 directive	LMS tool	Integration of IMS with LMS
4.1 Understanding the organization and its context	4.1 Understanding the organization and its context	4.1 Understanding the organization and its context	7.3 Understanding the social responsibility of an organization	Current-state VSM; Hoshin Kanri	The organization shall perform an analysis of the internal and external environments by considering the sustainability concept; its impact on the organization; and how the organization is affected at the local, regional, and global levels.
4.2 Understanding the needs and expectations of interested parties	4.2 Understanding the needs and expectations of interested parties	4.2 Understanding the needs and expectations of interested parties 6.1.3 Determination of legal requirements and other requirements	5.3 Stakeholder identification and engagement	Current-state VSM	The organization shall establish a procedure for identifying stakeholders and their needs, expectations, and interests by considering the quality, environment, OHS, and SR of stakeholders. The identification and value specification in VSM, including aspects of sustainability from the stakeholders' perspective, should be integrated with LIMSSI activities.
4.3 Determining the scope of the QMS	4.3 Determining the scope of the EMS	4.3 Determining the scope of the OHSMS	6.1 General	VSM	The highest level document of the LIMSSI should include the covered scope and should be mapped into a VS map by incorporating LMS practices and tools into the organization's procedures.
4.4 QMS and its processes	4.4 EMS	4.4 OH&S management system	7.3.2 Determining relevance and significance of core subjects and issues to an organization	Current-state VSM; Lean metrics	Include the current-state VS map in the documentation to comply with the process description requirement. The current-state VS map must be integrated as a complement to the description of the processes and their sequence and interaction, which would lead to actions for the accomplishment of planned results and continuous improvement. Lean metrics should be used together with performance indicators for process monitoring and measurement.
5.1 Leadership and commitment	5.1 Leadership and commitment	5.1 Leadership and commitment	7.4.1 Raising awareness and building competency for social responsibility 7.4.2 Setting the direction of an organization for social responsibility 6.2.2 Principles and considerations		TM must be committed to the LIMSSI and provide resources for its activities. Such a commitment must be included in the LIMSSI policy and must be translated into TM involvement with LIMSSI activities.

5.1.2 Customer focus	6.1.2 Environmental aspects 6.1.3 Compliance obligation	6.1.2 Hazard identification and assessment of risks and opportunities 6.1.3 Determination of legal requirements and other requirements	7.2 The relationship of an organization's characteristics to social responsibility	Current-state and future-state VS maps; takt time; pull systems; pacemaker process selection; load leveling	Identifying and specifying value from the stakeholders' viewpoint, setting the takt time, setting flow, leveling, and balancing of VS should be integrated into the LIMSSI activities. The pacemaker process operating in the rhythm of takt time carries to the value flow the rhythm of customer demand to meet client expectations and facilitates better use of resources and a pace that is adequate for workers.
5.2 Policy	5.2 Environmental policy	5.2 OHS policy	7.4.2 Setting the direction of an organization for social responsibility	Lean thinking	TM should establish an integrated policy that communicates the importance of incorporating, understanding, and applying the requirements of IMS and lean thinking to enable continuous improvement of sustainability. In addition to incorporating the principles of the MSs, the policy should include lean thinking principles (e.g., efficiency, waste reduction, and commitment to sustainability).
5.3 Organizational roles, responsibilities, and authorities	5.3 Organizational roles, responsibilities, and authorities	5.3 Organizational roles, responsibilities, and authorities	7.4.2 Setting the direction of an organization for social responsibility	Designation of VS manager and of representatives of economic, environmental, and social dimensions	Responsibilities and authorities should be defined by TM and communicated to everyone in the organization. People with responsibilities and authorities within the LIMSSI should have knowledge about the IMS and lean thinking so that they can conduct activities in an integrated way and thereby identify and implement opportunities for synergies. Designate as VS manager a person linked to IMS activities. Such designation seeks to integrate the coordination of VS improvement activities with IMS activities.
6. Planning	6. Planning	6. Planning	7.4.3 Building social responsibility into an organization's governance, systems, and procedures	Hoshin Kanri	The organization should plan its programs and activities from an integrated and sustainable perspective.

(*continued*)

Table 15.1 (Cont'd)

ISO 9001:2015 requirement	ISO 14001:2015 requirement	ISO 45001:2018 requirement	ISO 26000:2010 directive	LMS tool	Integration of IMS with LMS
6.1 Actions to address risks and opportunities	6.1 Actions to address risks and opportunities	6.1 Actions to address risks and opportunities 6.1.1 General	7.3.1 Due diligence		Describe the process and responsibilities for stakeholder consultation and describe relationship processes to identify and manage impacts, risks, and opportunities arising from economic, environmental, and social issues.
6.2 Quality objectives and plans to achieve them	6.2 Environmental objectives and planning to achieve them	6.2 OHS objectives and planning to achieve them	7.3.4 Establishing priorities for addressing issues	Hoshin Kanri	By considering lean thinking, TM should include goals and objectives to improve sustainability. Planning must include lean thinking strategies that should be deployed through Hoshin Kanri into measurable goals and objectives for relevant roles and levels. Programs should be established for conducting continuous improvement activities and improving sustainability performance.
6.3 Planning of changes		6.2 OHS objectives and planning to achieve them	7.7.3 Reviewing organization's progress and performance on social responsibility	SC	When the organization establishes a need for changes in the MS, the changes must be submitted for evaluation by the SC. The changes must comply with requirements and seek the creation of synergies in consideration of economic aspects, environment, health and safety of the workers, and SR with the aim of improving sustainability performance.
7.1 Resources	7.1 Resources	7.1 Resources	7.4.3 Building social responsibility into an organization's governance, systems, and procedures	Future-state VSM	The organization should determine and provide resources for improvement of sustainability through LIMSSI activities. The VSM within the LIMSSI aims to promote rational and optimized use of resources in consideration of sustainability aspects.
7.1.3 Infrastructure	7.1 Resources	7.1 Resources	7.4.3 Building social responsibility into an organization's governance, systems, and procedures	VSM; spaghetti diagram; TPM; OEE	The organization should design its layouts by using VSM and the spaghetti diagram to reduce waste (e.g., in the contexts of handling, transportation, optimization of energy use, consideration of ergonomics, worker health and safety, and environmental protection). TPM and OEE monitoring aid in the search for continuous improvement and in maintenance of the availability of the VS infrastructure.

7.1.4 Environment for the operation of processes		7.1 Resources	6.4.6 Labor practices issue 4. Health and safety at work	7S	The work environment should be managed to ensure compliance with requirements and improvements of VS. 7S is a tool that enables continuous improvement of the work environment to keep it safe and sustainable.
7.1.5 Monitoring and measuring resources	9.1.1 General	9.1 Monitoring, measurement, analysis, and performance evaluation	7.7.2 Monitoring activities on social responsibility	7S and TPM	Significant activities that address stakeholder concerns should be monitored to ensure that data are reliable, easy to understand, and timely and to enable supervision of sustainability performance. 7S must be used to maintain the reliability of the measuring equipment by adopting TPM principles.
7.2 Competence 7.3 Awareness	7.2 Competence 7.3 Awareness	7.2 Competence 7.3 Awareness	3.4.1 Raising awareness and building competency for social responsibility	Flexible operator and training needs	Include in the systematic of the LIMSSI the training and skills necessary to raise awareness of sustainability and lean thinking. Examples of training initiatives that collaborate for this goal are training to create flexible operators, VSM, and training kaizens defined in the VSM.
7.4 Communication	7.4 Communication	7.4. Communication 7.4.1 General 7.4.2 Internal communication 7.4.3 External communication	7.5 Communication on social responsibility	Visual control and management; Andon; VSM; continuous improvement	Visual control and management in processes and/or workstations helps in communicating the situations of processes and/or workstations, thereby highlighting problems. One such communication tool is the Andon system. VSM and continuous improvement should be performed in an integrated manner to address joint themes to highlight the interdisciplinary required for sustainability.
7.5 Documented information	7.5 Documented information	7.5 Documented information	7.3.2.1 Determining relevance / 7.4.2 Setting the direction of an organization for social responsibility / Box 15—Reporting on social responsibility	VSM, kaizen, continuous flow, leveling, 7S, etc.	It is necessary to standardize documentation and records, including the practices of continuous flow, First In-First Out (FIFO), poka-yoke, leveling, kaizen, and 7S, so as to achieve the following: to cover aspects of the economic, environmental, and social dimensions; to make better use of resources; and to accomplish greater integration of procedures and objectives.

(continued)

Table 15.1 (Cont'd)

ISO 9001:2015 requirement	ISO 14001:2015 requirement	ISO 45001:2018 requirement	ISO 26000:2010 directive	LMS tool	Integration of IMS with LMS
8.1 Operational planning and control	8.1 Operational planning and control	8.1 Operational planning and control	7.4.2 Setting the direction of an organization for social responsibility	Hoshin Kanri; current-state and future-state VS maps and planning of implementation of future-state VS map; kanban; pull systems; visual control and management	Use of Hoshin Kanri in conjunction with VSM collaborates with product realization planning with consideration of the current-state VS map and its metrics. Kanban and pull systems help in product realization planning with focus on stakeholder needs.
8.2 Requirements for products and services	6.1.2 Environmental aspects / 6.1.3 Compliance obligations	6.1.2 Hazard identification and assessment of OHS risks 6.1.3 Determination of applicable legal requirements and other requirements	7.3.2 Determining relevance and significance of core subjects and issues to an organization	Current-state and future-state VS maps; Takt time and kanban	The identification and specification of sustainable value for stakeholders in the VS map must be integrated with LMSSI activities. Kanban and takt time should be considered with focus on the economic, environmental, OHS, and SR requirements of the stakeholders.
8.3 Design and development of products and services					The organization must use a multidisciplinary approach to integrate the concept of sustainability into the design and development process of its products and services. Standardization is a way to reduce waste. Aspects of quality, environment, OHS, and SR should be considered to integrate the concept of sustainability into the design and development of products, services, and processes.

8.4 Control of externally provided processes, products, and services	8.1 Operational planning and control	8.3 Outsourcing 8.4 Procurement 8.5 Contractors	7.3.3.2 Exercising influence	Development of sustainable suppliers; continuous improvement; kanban; VSM; continuous flow	The organization must identify its suppliers in the VS map. The organization should implement a supplier development strategy based on the concepts and tools of SR and LM to achieve sustainability improvement. Waste reduction and kanban should be used to avoid unnecessary stock accumulation and order information and to know which product should be delivered and when and where it should be delivered. In a continuous flow, the supplier performance is evaluated more easily, as shortcomings of product quality and on-time delivery are evidenced by the kanban system and the low inventory levels. Any problems in the supplied product are detected more easily via the JIT method and the low inventory level.
8.5.1 Control of production and service provision	8.1 Operational planning and control	8.1.2 Hierarchy of controls 8.2 Management of change 8.3 Outsourcing 8.4 Procurement 8.5 Contractors	7.4.2 Setting the direction of an organization for social responsibility	Visual management and control; VSM; 7S; kanban; TPM; continuous flow; pull systems; OEE monitoring	Production scheduling should be sent only to the pacemaker process, which is controlled by customer requests to set the pace of previous processes and to avoid waste in other processes. The OEE of critical equipment should be monitored to promote its continuous improvement. The organization should plan and conduct production activities under controlled conditions, including the availability of work instructions for lean tools such as 7S; autonomous maintenance; visual management and control over takt time; and results of quality, environment, and OHS improvements. Production should be programmed using pull systems, kanban, and continuous flow to gradually reduce the work in process and promote efficient use of resources.

(continued)

Table 15.1 (Cont'd)

ISO 9001:2015 requirement	ISO 14001:2015 requirement	ISO 45001:2018 requirement	ISO 26000:2010 directive	LMS tool	Integration of IMS with LMS
8.5.4 Preservation				Kaizen to control perishable materials; JIT	Kaizen for improved handling and storage of products should consider product preservation requirements using JIT concepts.
8.5.6 Control of changes	8.1 Operational planning and control	8.1 Operational planning and control	7.7.3 Reviewing an organization's progress and performance on social responsibility	SC	The organization shall review and monitor changes to ensure compliance with requirements and improvement of sustainability performance by considering aspects of the environment, OHS, and SR to create synergies.
8.6 Release of products and services	9.1.1 General	9.1 Monitoring, measurement, analysis and evaluation	7.7.3 Reviewing an organization's progress and performance on social responsibility	Continuous flow	With continuous flow, any problems or defects that occur are detected immediately and the flow is interrupted; this prevents a large number of defects and creates a need for root-cause identification and eradication.
8.7 Control of nonconforming outputs	8.2 Emergency preparedness and response	8.6 Emergency preparedness and response	7.6.3 Resolving conflicts or disagreements between an organization and its stakeholders	VSM; A3	Discrepancies and problems in the VS should be addressed as nonconformities, including identification of the root cause of the problem. The A3 approach can be used to quickly identify the problem, analyze it, and propose actions and a plan to resolve it. If the problem involves additional actions to be implemented, it should be included in the VSM.
9.1.1 General	9.1.1 General	9.1.1 General	7.7.2 Monitoring activities on social responsibility 7.7.3 Reviewing an organization's progress and performance on social responsibility	Visual management; future-state VSM; lean metrics; search for perfection	The organization must implement visual management and control methods (e.g., lights, sounds, displays, pictures, and colored cards) to clearly demonstrate the objectives. With a focus on achieving improvements in the total value flow and integrating activities, indicators should be used to demonstrate sustainability performance. The process-monitoring data should incorporate the data set in the future-state VS map. Indicators should be used to monitor critical processes, where improvements to the OEE should be established and implemented.

9.1.2 Customer satisfaction			Current- and future-state VSM; VS leveling; takt time	The activities of the systems should focus on meeting customer, regulatory, and statutory requirements with the aim of adding value from the perspectives of the client and other sustainability-seeking stakeholders. VS leveling should be considered in the procedures. The organization should seek to operate within takt time to serve the customer while using resources rationally.
9.1.3 Analysis and evaluation	9.1.2 Evaluation of compliance	9.1.2 Evaluation of compliance with legal requirements and other requirements	Hoshin Kanri; visual management; search for perfection	The organization should evaluate LIMSSI effectiveness through Hoshin Kanri. Data should be analyzed at all levels. Everyone should be aware of data and visual controls.
		7.7.3 Reviewing an organization's progress and performance on social responsibility		The use of joint indicators for the LIMSSI allows the organization to evaluate the results obtained through implementations of improvement opportunities and the effectiveness of the proposed modifications to create synergy among the systems with the ultimate goal of improving the sustainability performance.
9.2 Internal audit	9.2 Internal audit	7.7.4 Enhancing the reliability of data and information collection and management	VSM and planning for implementation of future-state VS map; 7S	Integrated audits, that is, audits that consider all the MSs of the organization, can be used to optimize resources. It should be considered in the integrated audit plan whether systems are following planned arrangements and whether they are maintained and implemented effectively. The audit program should consider stakeholder, statutory, and regulatory requirements; future-state VS map; and the 7S program.

(continued)

Table 15.1 (Cont'd)

ISO 9001:2015 requirement	ISO 14001:2015 requirement	ISO 45001:2018 requirement	ISO 26000:2010 directive	LMS tool	Integration of IMS with LMS
9.3 Management review	9.3 Management review	9.3 Management review	7.7.3 Reviewing an organization's progress and performance on social responsibility	VSM and planning for implementation of future-state VS map	The management review should consider, in addition to the IMS, the LMS to enable the organization to assess compliance with the planned arrangements and also assess whether they are maintained and implemented effectively. It should also consider the results and effectiveness achieved by the implementations and the need for changes aimed at improving sustainability performance. Management-review inputs should include lean metrics and sustainability indicators. Management-review outputs should include actions, decisions, and plans focused on waste reduction and sustainability.
10. Improvement	10. Improvement	10. Improvement	7.7.5 Improving performance	kaizen; future-state VSM	Process improvements (process kaizen) should be subordinated to the VS project (kaizen of full flow), instead of vague and isolated improvement efforts; this would aid in sharing of best practices between systems and the dimensions of sustainability.
10.2 Nonconformity and corrective action	10.2 Nonconformity and corrective action	10.2 Incident, nonconformity and corrective action	7.6.3 Resolving conflicts or disagreements between an organization and its stakeholders	Actual- and future-state VSM; kaizen	Problems in the VS should be addressed with corrections and corrective actions implemented using kaizen activities.
10.3 Continual improvement	10.3 Continual improvement	10.3 Continual improvement	7.7.5 Improving performance	Hoshin Kanri; kaizen; future-state VSM; performance indicators	For continuous improvement of sustainability performance, the organization must seek a quick response to problems. Improvement opportunities should be identified and implemented by always monitoring performance indicators, seeking synergy between systems, and improving sustainability through kaizen activities.

Previous use of the LIMSSI model allowed a critical analysis by representatives of strategic, tactical, and operational levels. The strategic level representative pointed out as one of the strengths of the LIMSSI model the consideration of legal compliance, which reinforces the basis of SR and contributes to reducing the risks of negative impacts from fines, embargoes, and sanctions. For instance, the representative of the tactical level highlighted as strengths of the model: the integration between IMS and LMS, the focus on meeting legal requirements, and focus on waste reduction. However, the representative of the tactical level remarked that the needs of investment, guidelines, and corporate definitions can be potential difficulties for the implementation of the model. Hence, representatives of the operational level emphasized that the model has its implementation facilitated by allowing simple improvements to have a great impact on the organization. They also pointed out strength that the model brings to the light subjects such as ergonomics and reuse of materials and resources. Moreover, the consideration of all aspects of the organization and the use of real data instead of forecasts were emphasized. Finally, the reduction of inventories and the promotion of a holistic view of the VS were reported as the major contributions. However, the need for a cultural change of the organization and the break of paradigms of the shop floor were reported as potential difficulties.

Hence, the critical analysis allowed the conclusion that the LIMSSI model can be applied to organizations that do not have any MSs or already have them. Finally, the findings suggest that the LIMSSI model offers an opportunity to develop the organization´s potential to improve CS performance. Thereby, the LIMSSI model can render organizations more competitive and sustainable by being economically profitable, ecologically correct, operationally safe, socially fair, and culturally accepted.

References

Abad, J., Dalmau, I., Vilajosana, J., 2014. Taxonomic proposal for integration levels of management systems based on empirical evidence and derived corporate benefits. J. Clean. Prod. 78, 164–173. doi:10.1016/j.jclepro.2014.04.084.

Abad, J., Lafuente, E., Vilajosana, J., 2013. An assessment of the OHSAS 18001 certification process: objective drivers and consequences on safety performance and labour productivity. Saf. Sci. 60, 47–56. doi:10.1016/j.ssci.2013.06.011.

Agan, Y., Acar, M.F., Borodin, A., 2013. Drivers of environmental processes and their impact on performance: a study of Turkish SMEs. J. Clean. Prod. 51, 23–33. doi:10.1016/j.jclepro.2012.12.043.

Amrina, E., Yusof, S.M., Key performance indicators for sustainable manufacturing evaluation in automotive companies, 2011 IEEE International Conference on Industrial Engineering and Engineering Management, 2011, pp. 1093-1097. http://dx.doi.org/10.1109/IEEM.2011.6118084. https://ieeexplore.ieee.org/document/6118084.

Asif, M., Searcy, C., Zutshi, A., Fisscher, O.A.M., 2013. An integrated management systems approach to corporate social responsibility. J. Clean. Prod. 56, 7–17. doi:10.1016/j.jclepro.2011.10.034.

Aureliano, M., Andrade, M.De, Gosling, M., Vinícius, R., Jordão, D., 2013. A responsabilidade social de siderúrgicas mineiras e a percepção de suas comunidades de entorno. Produção 23, 793–805.

Baumgartner, R.J., 2014. Managing corporate sustainability and CSR: a conceptual framework combining values, strategies and instruments contributing to sustainable development. Corp. Soc. Responsib. Environ. Manag. 21, 258–271. doi:10.1002/csr.1336.

Bernardo, M., 2014. Integration of management systems as an innovation: a proposal for a new model. J. Clean. Prod. 82, 132–142. doi:10.1016/j.jclepro.2014.06.089.

Bhanot, N., Rao, P.V., Deshmukh, S.G., 2016. An integrated approach for analysing the enablers and barriers of sustainable manufacturing. J. Clean. Prod. 142, 4412–4439. doi:10.1016/j.jclepro.2016.11.123.

Burritt, R.L., Schaltegger, S., 2010. Sustainability accounting and reporting: fad or trend? Accounting, Audit. Account. J. 23, 829–846. doi:10.1108/09513571011080144.

Caldera, H.T.S., Desha, C., Dawes, L., 2017. Exploring the role of lean thinking in sustainable business practice: a systematic literature review. J. Clean. Prod. doi:10.1016/j.jclepro.2017.05.126.

Campos, L.M.S., De Melo Heizen, D.A., Verdinelli, M.A., Miguel, CauchickP.A., 2015. Environmental performance indicators: a study on ISO 14001 certified companies. J. Clean. Prod. 99, 286–296. doi:10.1016/j.jclepro.2015.03.019.

Chatzoglou, P., Chatzoudes, D., Kipraios, N., 2015. The impact of ISO 9000 certification on firms' financial performance. Int. J. Oper. Prod. Manag. 35, 145–174. doi:10.1108/IJOPM-07-2012-0387.

de Oliveira, O.J., 2013. Guidelines for the integration of certifiable management systems in industrial companies. J. Clean. Prod. 57, 124–133. doi:10.1016/j.jclepro.2013.06.037.

Domingues, P., Sampaio, P., Arezes, P.M., 2016. Integrated management systems assessment: a maturity model proposal. J. Clean. Prod. 124, 164–174. doi:10.1016/j.jclepro.2016.02.103.

Dues, C.M., Tan, K.H., Lim, M., 2013. Green as the new Lean: how to use Lean practices as a catalyst to greening your supply chain. J. Clean. Prod. 40, 93–100. doi:10.1016/j.jclepro.2011.12.023.

Dyllick, T., Hockerts, K., Thomas Dyllick, K.H., 2002. Beyond the business case for corporate sustainability. Bus. Strateg. Environ. 11, 130–141. doi:10.1002/bse.323.

Elkington, J., 1998. Partnerships from cannibals with forks: the triple bottom line of 21st-century business. Environ. Qual. Manag. 8, 37–51. doi:10.1002/tqem.3310080106.

Engert, S., Rauter, R., Baumgartner, R.J., 2015. Exploring the integration of corporate sustainability into strategic management: a literature review. J. Clean. Prod. 112, 2833–2850. doi:10.1016/j.jclepro.2015.08.031.

Fernandes, A.C., Sampaio, P., Sameiro, M., Truong, H.Q., 2017. Supply chain management and quality management integration. Int. J. Qual. Reliab. Manag. 34, 53–67. doi:10.1108/IJQRM-03-2015-0041.

Fernández-Muñiz, B., Montes-Peón, J.M., Vázquez-Ordás, C.J., 2012a. Occupational risk management under the OHSAS 18001 standard: analysis of perceptions and attitudes of certified firms. J. Clean. Prod. 24, 36–47. doi:10.1016/j.jclepro.2011.11.008.

Fernández-Muñiz, B., Montes-Peón, J.M., Vázquez-Ordás, C.J., 2012b. Safety climate in OHSAS 18001-certified organisations: antecedents and consequences of safety behaviour. Accid. Anal. Prev. 45, 745–758. doi:10.1016/j.aap.2011.10.002.

Filho, J.M.de S., Wanderley, L.S.O., Gómez, C.P., Farache, F., 2010. Strategic corporate social responsibility management for competitive advantage. Brazilian Adm. Rev. 7, 294–309.

Fisher, J., Bonn, I., 2011. Business sustainability and undergraduate management education: an Australian study. High. Educ. 62, 563–571. doi:10.1007/s10734-010-9405-8.

Fonseca, L., Ramos, A., Rosa, Á., Braga, A.C., Sampaio, P., 2016. Stakeholders satisfaction and sustainable success Amílcar Ramos and Álvaro Rosa Ana Cristina Braga and Paulo Sampaio. Int. J. Ind. Syst. Eng. 24, 144–157.

Fonseca, L.M., 2015a. Strategic drivers for implementing sustainability programs in Portuguese organizations—let's listen to Aristotle: from triple to quadruple bottom line. Sustainability 8, 136–142. doi:10.1089/SUS.2015.29004.

Fonseca, L.M., 2015b. From quality gurus and TQM to ISO 9001:2015: a review of several quality Paths. Int. J. Qual. Res. 9, 167–180.

Galeazzo, A., Furlan, A., Vinelli, A., 2014. Lean and green in action: interdependencies and performance of pollution prevention projects. J. Clean. Prod. 85, 191–200. doi:10.1016/j.jclepro.2013.10.015.

Garcia, S., Cintra, Y., Torres, R., de, C.S.R., Lima, F.G., 2016. Corporate sustainability management: a proposed multi-criteria model to support balanced decision-making. J. Clean. Prod. 136, 181–196. doi:10.1016/j.jclepro.2016.01.110.

Georgescu-Roegen, N., 1971. The Entropy Law and the Economic Process. Harvard University Press, Cambridge.

Granerud, R.L., Rocha, R.S., 2011. Organisational learning and continuous improvement of health and safety in certified manufacturers. Saf. Sci. 49, 1030–1039. doi:10.1016/j.ssci.2011.01.009.

GRI, 2016. GRI 101: Foundation. GRI Global, Amsterdam.

GRI, 2016. GRI 102: General Disclosures. GRI Global, Amsterdam.

GRI, 2016. GRI 103: Management Approach. GRI Global, Amsterdam.

Gutowski, T.G., 2011. Manufacturing and the Science of Sustainability In: Hesselbach J., Herrmann C. (eds) Glocalized Solutions for Sustainability in Manufacturing. Springer, Berlin, Heidelberg. https://doi.org/10.1007/978-3-642-19692-8_6.

Hahn, R., 2013. ISO 26000 and the standardization of strategic management processes for sustainability and corporate social responsibility. Bus. Strateg. Environ. 22, 442–455. doi:10.1002/bse.1751.

Hansen, E.G., Schaltegger, S., 2014. The sustainability balanced scorecard: a systematic review of architectures. J. Bus. Ethics 133, 193–221. doi:10.1007/s10551-014-2340-3.

Herrmann-Pillath, C., 2011. The evolutionary approach to entropy: reconciling Georgescu-Roegen's natural philosophy with the maximum entropy framework. Ecol. Econ. 70, 606–616. doi:10.1016/j.ecolecon.2010.11.021.

Holm, T., Vuorisalo, T., Sammalisto, K., 2015. Integrated management systems for enhancing education for sustainable development in universities: a memetic approach. J. Clean. Prod. 106, 155–163. doi:10.1016/j.jclepro.2014.03.048.

International Organization for Standardization, 2020. The ISO Survey of Management System Certifications –2019.

International Organization for Standardization, 2018. ISO 45001:2018 Occupational Health and Safety Management Systems — Requirements with Guidance for use.

International Organization for Standardization, 2015. ISO 9001:2015 Quality Management Systems - Requirements.

International Organization for Standardization, 2015. ISO 14001:2015 Environmental management systems - Requirements with guidance for use.

International Organization for Standardization, 2010. ISO 26000:2010 - Guidance on social RESPONSIBILITY.

Jabbour, C.J.C., De Sousa Jabbour, A.B.L., Govindan, K., Teixeira, A.A., De Souza Freitas, W.R., 2013. Environmental management and operational performance in automotive companies in Brazil: the role of human resource management and lean manufacturing. J. Clean. Prod. 47, 129–140. doi:10.1016/j.jclepro.2012.07.010.

James, J., Ikuma, L.H., Nahmens, I., Aghazadeh, F., 2013. The impact of Kaizen on safety in modular home manufacturing. Int. J. Adv. Manuf. Technol. 70, 725–734. doi:10.1007/s00170-013-5315-0.

Jørgensen, T.H., 2008. Towards more sustainable management systems: through life cycle management and integration. J. Clean. Prod. 16, 1071–1080. doi:10.1016/j.jclepro.2007.06.006.

Klute-Wenig, S., Refflinghaus, R., 2015. Integrating sustainability aspects into an integrated management system. TQM J 27, 303–315. doi:10.1108/TQM-12-2013-0128.

Kurdve, M., Zackrisson, M., Wiktorsson, M., Harlin, U., 2014. Lean and green integration into production system models—experiences from Swedish industry. J. Clean. Prod. 85, 180–190. doi:10.1016/j.jclepro.2014.04.013.

Lannelongue, G., Gonzalez-Benito, J., Quiroz, I., 2017. Environmental management and labour productivity: the moderating role of capital intensity. J. Environ. Manage. 190, 158–169. doi:10.1016/j.jenvman.2016.11.051.

Lee, D., 2017. Corporate social responsibility and management forecast accuracy. J. Bus. Ethics 140, 353–367. doi:10.1007/s10551-015-2713-2.

Levallois, C., 2010. Can de-growth be considered a policy option? A historical note on Nicholas Georgescu-Roegen and the club of Rome. Ecol. Econ. 69, 2271–2278. doi:10.1016/j.ecolecon.2010.06.020.

Lo, C.K.Y., Pagell, M., Fan, D., Wiengarten, F., Yeung, A.C.L., 2014. OHSAS 18001 certification and operating performance: the role of complexity and coupling. J. Oper. Manag. 32, 268–280. doi:10.1016/j.jom.2014.04.004.

Lozano, R., 2012. Towards better embedding sustainability into companies' systems: an analysis of voluntary corporate initiatives. J. Clean. Prod. 25, 14–26. doi:10.1016/j.jclepro.2011.11.060.

Martínez-Jurado, P.J., Moyano-Fuentes, J., 2014. Lean management, supply chain management and sustainability: a literature review. J. Clean. Prod. 85, 134–150. doi:10.1016/j.jclepro.2013.09.042.

Meadows, D.H., Meadows, D.L., Randers, J., Behrens, W.W., 1978. Limites do Crescimento, second ed. Perspectiva, São Paulo.

Milne, M.J., Gray, R., 2013. W(h)ither ecology? The triple bottom line, the global reporting initiative, and corporate sustainability reporting. J. Bus. Ethics 118, 13–29. doi:10.1007/s10551-012-1543-8.

Mokhtar, S.S.M., Abdullah, N.A.H., Kardi, N., Yacob, M.I., 2013. Sustaining a quality management system: process, issues and challenges. Bus. Strateg. Ser. 14, 123–130. doi:10.1108/BSS-12-2011-0032.

Movahedi, M.M., Teimourpour, M., Teimourpour, N., 2013. A study on effect of performing quality management system on organizational productivity. Manag. Sci. Lett. 3, 1063–1072. doi:10.5267/j.msl.2013.03.022.

Nunhes, T.V., César, L., Motta, F., Oliveira, O.J.De, 2016. Evolution of integrated management systems research on the Journal of Cleaner Production: identification of contributions and gaps in the literature. J. Clean. Prod. 139, 1234–1244. doi:10.1016/j.jclepro.2016.08.159.

Nunhes, T.V., Motta Barbosa, L.C.F., de Oliveira, O.J., 2017. Identification and analysis of the elements and functions integrable in integrated management systems. J. Clean. Prod. 142, 3225–3235. doi:10.1016/j.jclepro.2016.10.147.

Qi, G., Zeng, S., Yin, H., Lin, H., 2013. ISO and OHSAS certifications: how stakeholders affect corporate decisions on sustainability. Manag. Decis. 51, 1983–2005. doi:10.1108/MD-11-2011-0431.

Rebelo, M.F., Santos, G., Silva, R., 2015. Integration of management systems: towards a sustained success and development of organizations. J. Clean. Prod. 127, 96–111. doi:10.1016/j.jclepro.2016.04.011.

Rebelo, M.F., Silva, R., Santos, G., Mendes, P., 2016. Model based integration of management systems (MSs) – case study. TQM J 28, 907–932. doi:10.1108/TQM-09-2014-0079.

RobecoSAM, 2014. The Sustainability Yearbook 2014. Josefstrasse 218 8005 Zurich, Switzerland. https://www.p-plus.nl/resources/articlefiles/SustainabilityYearbook2014.pdf.

Rockström, J., Steffen, W., Noone, K., Persson, Å., Chapin, F.S.I.I.I., Lambin, E., Lenton, T.M., Scheffer, M., Folke, C., Schellnhuber, H.J., Nykvist, B., Wit, C.A.De, Hughes, T., Leeuw, S.Van Der, Rodhe, H., Sörlin, S., Snyder, P.K., Costanza, R., Svedin, U., Falkenmark, M., Karlberg, L., Corell, R.W., Fabry, V.J., Hansen, J., Walker, B., Liverman, D., Richardson, K., Crutzen, P., Foley, J., 2009. Planetary boundaries : exploring the safe operating space for humanity. Ecol. Soc. 14, 32.

Rodrigues, V.P., Pigosso, D.C.A., McAloone, T.C., 2016. Process-related key performance indicators for measuring sustainability performance of ecodesign implementation into product development. J. Clean. Prod. 139, 416–428. doi:10.1016/j.jclepro.2016.08.046.

Rohani, J.M., Zahraee, S.M., 2015. Production line analysis via value stream mapping: a lean manufacturing process of color industry. Procedia Manuf 2, 6–10. doi:10.1016/j.promfg.2015.07.002.

Rother, M., Shook, J., 2003. Learning to see: value stream mapping to create value and eliminate Muda. Lean Enterp. Inst. Brookline 102. doi:10.1109/6.490058.

Schaltegger, S., Beckmann, M., Hansen, E.G., 2013. Corporate sustainability meets transdisciplinarity. Bus. Strateg. Environ. 22, 217–218. doi:10.1002/bse.1770.

Siva, V., Gremyr, I., Bergquist, B., Garvare, R., Zobel, T., Isaksson, R., 2016. The support of quality management to sustainable development: a literature review. J. Clean. Prod. 138, 148–157. doi:10.1016/j.jclepro.2016.01.020.

Souza, J.P.E., Alves, J.M., 2018. Lean-integrated management system: a model for sustainability improvement. J. Clean. Prod. 172, 2667–2682. doi:10.1016/j.jclepro.2017.11.144.

Souza, J.P.E., Alves, J.M., Silva, M.B., 2014. Quality improvement in the aerospace industry: investigation of the main characteristics. Int. Rev. Mech. Eng. 8, 893–900.

Souza, J.P.E., Dekkers, R., 2019. Adding sustainability to lean product development. Procedia Manuf 39, 1327–1336. doi:10.1016/j.promfg.2020.01.325.

Starik, M., Kanashiro, P., 2013. Toward a theory of sustainability management: uncovering and integrating the nearly obvious. Organ. Environ. 26, 7–30. doi:10.1177/1086026612474958.

Tarí, J.J., 2011. Research into quality management and social responsibility. J. Bus. Ethics 102, 623–638. doi:10.1007/s10551-011-0833-x.

TAUBITZ, M.A., 2010. Lean, green & safe: integrating safety into the lean, green and sustainability movement. Prof. Saf. 55, 39–46.

United Nations, 2015. Transforming Our World: The 2030 Agenda for Sustainable Development https://sdgs.un.org/2030agenda.

Verrier, B., Rose, B., Caillaud, E., Remita, H., 2013. Combining organizational performance with sustainable development issues: the Lean and Green project benchmarking repository. J. Clean. Prod. doi:10.1016/j.jclepro.2013.12.023.

Vinodh, S., Arvind, K.R., Somanaathan, M., 2011. Tools and techniques for enabling sustainability through lean initiatives. Clean Technol. Environ. Policy 13, 469–479. doi:10.1007/s10098-010-0329-x.

Witjes, S., Vermeulen, W.J.V., Cramer, J.M., 2017. Exploring corporate sustainability integration into business activities. Experiences from 18 small and medium sized enterprises in the Netherlands. J. Clean. Prod. 153, 528–538. doi:10.1016/j.jclepro.2016.02.027.

Womack, J.P., Jones, D.T., 2004. A mentalidade enxuta nas empresas Lean Thinking: elimine o desperdício e crie riqueza. Campus, Rio de Janeiro.

Report of the World Commission on Environment and Development: Our Common Future, 1987. https://sustainabledevelopment.un.org/content/documents/5987our-common-future.pdf.

Zeng, S.X., Shi, J.J., Lou, G.X., 2007. A synergetic model for implementing an integrated management system: an empirical study in China. J. Clean. Prod. 15, 1760–1767. doi:10.1016/j.jclepro.2006.03.007.

CHAPTER 16

Coupled life cycle thinking and data envelopment analysis for quantitative sustainability improvement

Mario Martín-Gamboa[a], Diego Iribarren[b]

[a]*Chemical and Environmental Engineering Group, Rey Juan Carlos University, Móstoles, Spain*
[b]*Systems Analysis Unit, IMDEA Energy, Móstoles, Spain*

16.1 Introduction

The transition to resource-efficient, low-carbon, and sustainable economies worldwide is of vital importance for the prosperity of societies. In light of this need, international initiatives oriented toward sustainable development have been launched in the last years. The year 2015 marked a milestone with the adoption of the UN 2030 agenda and the sustainable development goals, and the Paris Agreement (UNFCCC, 2015; United Nations, 2015). In December 2019, the European Commission presented the European Green Deal, a set of policy initiatives within a growth strategy to make Europe climate-neutral by 2050 while achieving social welfare and economic competitiveness (European Commission, 2019). Within the frame of the European Green Deal, the adoption of the taxonomy regulation should be highlighted as a legislative tool to boost private investment in green and sustainable projects (European Union, 2020). All these initiatives have increased pressure on nations, organizations, and companies to act and report on sustainability strategies. Thus, assessing sustainability issues in terms of environmental, economic, and social indicators is required to identify positive and negative contributions of such strategies and activities. In consequence, a plethora of methodologies, tools and indicators for sustainability assessment have emerged (de Olde et al., 2017; Khan et al., 2021). However, further efforts are still required to provide well-established methodological frameworks capable of: (1) integrating reliable sustainability indicators, (2) managing decision-makers' preferences, (3) identifying life cycle sustainability hotspots, and (4) reporting in a consistent and clear way sustainability performance results.

The evaluation of sustainability should involve a holistic perspective when quantifying the short-, medium- and long-term impacts of activities and strategies. In this sense, the sustainability assessment of product systems should consider all the stages related to their supply chain, use, and disposal. This line of thinking matches the philosophy behind life cycle approaches, which—as further addressed in Section 16.1.1—are useful tools to assess sustainability issues of a product system over its lifetime (ISO, 2006a, 2006b). In the last years, many studies have proven the relevance of life cycle methods to translate the science of sustainability into valuable knowledge to support decision-making processes (Visentin et al., 2020). In a review conducted by Martín-Gamboa et al. (2017) focusing on the sustainability assessment of energy systems, environmental Life Cycle Assessment (environmental LCA) and Life Cycle Costing (LCC) were found to be common methodologies for the calculation of environmental and economic indicators, respectively. Concerning the social dimension, a reduced number of studies were found to include social indicators with a life cycle perspective, which is associated with the relatively low maturity of Social Life Cycle Assessment (S-LCA). These findings can also be extrapolated to other sectors such as agriculture (Notarnicola et al., 2017), building (Lazar and Chithra, 2020), industry (Cai and Lai, 2021), and waste management (Iqbal et al., 2020).

The transition toward sustainable societies implies the search for suitable alternatives at the level of products, technologies, policies, etc. In this regard, supporting decision-making processes under sustainability criteria is a complex task that requires the consideration of multiple indicators and preferences (Cinelli et al., 2014). Hence, robust quantitative methods are needed to jointly interpret the multiple criteria and stakeholders' views. Multicriteria decision analysis (MCDA) methods and tools have arisen in the last decades as decision-support solutions to address complex problems involving tradeoffs, uncertainties, different metrics, etc. (Munda, 2005; Wang et al., 2009). MCDA involves a process to evaluate alternatives by compiling a set of criteria and stakeholders' preferences and using them to build a preference model that allows the prioritization of alternatives. Within the MCDA tools available, and specifically within the distance-to-target methods, Data Envelopment Analysis (DEA) arises as a tradeoff solution between soundness and practicality (Cooper et al., 2007; Martín-Gamboa et al., 2017). DEA is used to compute the relative efficiencies of a set of homogenous entities (called decision-making units, DMUs) that use inputs to produce outputs. Because of its nonparametric approach, DEA can be applied to a wide range of fields (healthcare, education, banking, agriculture, energy, etc.) as long as multiple DMUs are assessed (Liu et al., 2016). Given the potential of DEA as an MCDA tool, Section 16.1.2 provides further details on this methodology.

In the past decade, the combination of life cycle approaches and DEA has proven to be a useful framework for the joint interpretation of operational efficiencies and sustainability benchmarks of multiple similar entities (Iribarren et al., 2016). The resultant combined methods are of general use and they are only conditioned by the availability of inventory data for a set of multiple homogeneous entities. As highlighted by Iribarren and Vázquez-Rowe (2013), a key feature of the LCA + DEA methodology is that it enables the assessment of environmental, economic, and social aspects in a single combined approach facilitating the joint interpretation of performance indicators. The LCA + DEA methodology was formally presented in 2010 as a combination of LCA and DEA to benchmark the operational and environmental performance of multiple resembling entities (Iribarren, 2010; Vázquez-Rowe et al., 2010). From then on, the LCA + DEA methodology has been increasingly used to assess case studies belonging to the primary (Lozano et al., 2009, 2010; Iribarren et al., 2011; Vázquez-Rowe et al., 2011, 2012; Mohammadi et al., 2013), secondary (Iribarren et al., 2015; Vázquez-Rowe and Iribarren, 2015; Martín-Gamboa et al., 2017), and tertiary (Álvarez-Rodríguez et al., 2019a, 2019b, 2020a, 2020b) sectors. However, as a growing field of research, there are still challenges linked to this combined methodology, for example its enhanced use for quantitative sustainability assessment.

Fig. 16.1 shows the roadmap of this chapter on coupled life cycle thinking and DEA for quantitative sustainability improvement. After introducing key concepts on life cycle approaches (Section 16.1.1) and DEA (Section 16.1.2), a description of the combined methodological framework is provided (Section 16.2). Afterwards, Section 16.3 presents recent progresses in sustainability-oriented LCA + DEA, while Section 16.4 proposes novel solutions according to the identified needs. Finally, conclusions and perspectives are addressed.

16.1 Introduction

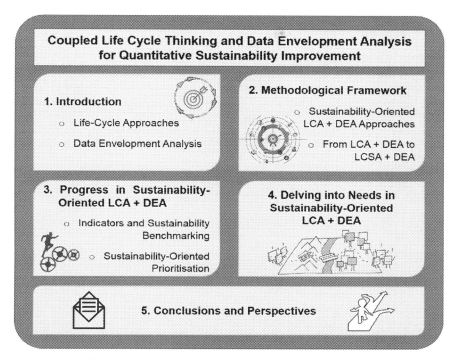

FIG. 16.1 Structure of the chapter.

16.1.1 Life cycle approaches

Due to their holistic nature, life cycle approaches allow a thorough assessment of potential impacts associated with product systems, from the acquisition of raw materials to the end of their life (ISO, 2006a, 2006b). There are specific approaches to each of the main sustainability dimensions: (environmental) LCA, LCC, and S-LCA. Furthermore, a joint evaluation of the sustainability performance of a product system could be undertaken through a Life Cycle Sustainability Assessment (LCSA) approach (UNEP/SETAC, 2011; Hu et al., 2013). Besides, there are other approaches also linked to life cycle thinking (e.g., emergy analysis).

Environmental LCA is a standardized methodology widely used to evaluate environmental aspects and potential environmental impacts linked to the life cycle of a product system (ISO, 2006a, 2006b). Four interrelated stages are distinguished in LCA studies (Fig. 16.2): goal and scope definition, life cycle inventory (LCI) analysis, life cycle impact assessment (LCIA), and interpretation. Among the numerous LCA applications, the following ones are highlighted: identification of hotspots and opportunities to improve the environmental performance of products throughout the different stages of their life cycle, source of information to support decision-making processes (e.g., strategic planning, prioritization, product/process design or re-design, implementation of an environmental labeling scheme, etc.), and marketing.

LCC summarizes all the costs associated with the life cycle of a product or service that are directly covered by one or more of the actors in the product life cycle (supplier, producer, user/consumer, etc.)

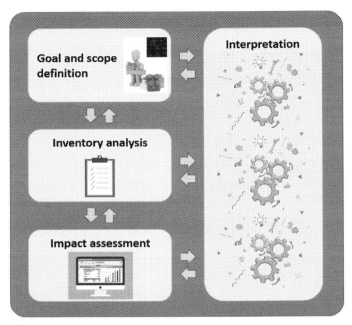

FIG. 16.2 LCA methodology.

(Swarr et al., 2011). The combined use of LCC and environmental LCA is relatively widespread for the joint environmental and economic evaluation of systems (for instance, as a standardized eco-efficiency assessment).

S-LCA addresses the social and socioeconomic impacts across the supply chain of a product (potentially including use and final disposal) and involves interrelated stages similar to those in Fig. 16.2 (UNEP/SETAC, 2009; UNEP, 2020). The social and socioeconomic aspects evaluated in S-LCA are those that can directly affect stakeholders (workers, society, etc.) positively (e.g., contribution to local economy) or negatively (child labor, forced labor, etc.) during the life cycle of a product.

Finally, LCSA brings together indicators from the previous life cycle approaches to make an "all-in-one" analysis and jointly interpret the sustainability outcomes (Kloepffer, 2008). However, despite significant methodological advancements in recent years, further efforts are required to achieve the same degree of robustness in each of the life cycle methods that separately cover the sustainability pillars (i.e., environmental LCA, LCC, and S-LCA). This is especially relevant to S-LCA as it presents open challenges that hamper a standardized approach (UNEP, 2020). Regarding the development of a cross-cutting sustainability method, significant efforts are needed to make LCSA practical (reducing its conceptual degree), communicate the results in a clear way, and improve the participation of decision-making agents in the analytical process (Costa et al., 2019).

16.1.2 Data envelopment analysis

DEA is a linear programming methodology that empirically quantifies the relative efficiency of multiple similar entities or DMUs (Cooper et al., 2007). The DMU is the homogeneous entity responsible for the

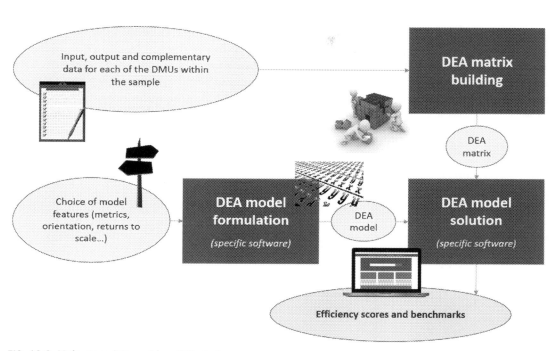

FIG. 16.3 Main steps followed in a DEA study.

conversion of inputs into outputs. As shown in Fig. 16.3, to carry out a DEA study, a matrix composed of the inputs, outputs, and complementary elements of the sample of DMUs is required. Once the DEA model has been formulated according to a set of features such as metrics and orientation, the matrix is implemented in the model to be solved, thus obtaining as main results relative efficiency scores and operational benchmarks for each DMU.

The relative efficiency scores are calculated through a nonparametric procedure based only on the observed data and basic assumptions for the resolution of an optimization model. For each DMU, an efficiency score (ϕ) is obtained. In addition, for the DMUs identified as inefficient (i.e., $\phi < 1$), a set of target values (i.e., benchmarks) that would transform these entities into efficient is computed. Hence, DEA allows discrimination between efficient and inefficient entities while promoting feasible improvements for an efficient operational performance. Once the matrix of observed data and the DEA model are ready, the mathematical procedure is based on the calculation of efficiency frontiers for the set of DMUs. The efficiency frontier is said to envelop all units, defining the production possibility set. The DMUs located on this border constitute the reference set. DEA works by projecting each DMU on the efficiency frontier according to the formulated model, calculating the maximal improvements that can be achieved on the inputs and/or outputs of the DMU (Cooper et al., 2007, 2011).

In accordance with the diversity of needs behind each study, there is a wide variety of DEA models. These models are formulated according to a set of technical features (Lozano et al., 2009): model orientation (model oriented to inputs and/or outputs), model metrics (radial or nonradial), and display of the production possibility set (e.g., constant returns to scale (CRS) or variable returns to scale (VRS)).

Regarding model orientation, three variants are distinguished: input-oriented, output-oriented, and nonoriented (Cooper et al., 2007). An input-oriented model means that an inefficient entity becomes efficient by reducing its inputs while its outputs remain at least at the same level. On the other hand, in an output-oriented model, the conversion of the DMU into efficient entity is achieved with an increase in outputs while inputs remain unchanged. A nonoriented (or mixed) model pursues both an increase in outputs and a decrease in inputs.

Depending on the model metrics, two variants are differentiated: radial and nonradial models. Radial models are represented by the CCR model developed by Charnes, Cooper, and Rhodes, and they are based on proportional changes in the levels of inputs or outputs (Charnes et al., 1978). In contrast, nonradial models such as the slacks-based measure of efficiency (SBM) model do not handle proportional changes in inputs or outputs, but specific slacks for each input or output (Tone, 2001).

Finally, the concept of returns to scale has been widely explored within the different DEA frameworks. A DMU operates at CRS if an increase in its input levels leads to a proportional increase in its output levels. If it is suspected that this proportional effect does not occur, a model considering VRS should be used (Cooper et al., 2007; Lozano et al., 2009).

16.2 Methodological framework

This section presents the LCA + DEA strategies developed so far to address the sustainability-oriented management of multiple similar entities (Section 16.2.1), as well as the progress made in the last decade to evolve LCA + DEA from an environmental and operational methodological solution to a sustainability one (Section 16.2.2).

16.2.1 Sustainability-oriented LCA + DEA approaches

Two main strategies have been proposed in the LCA + DEA field: (1) the five-step LCA + DEA approach processes information on material and energy flows and socioeconomic aspects to a sustainability outcome via the computation of consistent operational, socioeconomic, and environmental benchmarks associated with the optimized performance of the DMUs, and (2) the three-step LCA + DEA approach addresses the direct, preliminary benchmarking of performance indicators of the DMUs to prioritize the entities under assessment (Vázquez-Rowe and Iribarren, 2015). Both strategies are further detailed in Sections 16.2.1.1 and 16.2.1.2 with a focus on their sustainability scope.

16.2.1.1 Five-step sustainability-oriented LCA + DEA approach

The five-step sustainability-oriented LCA + DEA approach compiles and processes information on material and energy flows and socioeconomic aspects to an overall sustainability outcome by calculating relative efficiencies and coherent operational, socioeconomic and environmental benchmarks associated with the optimized behavior of the DMUs under assessment (Iribarren et al., 2016). Fig. 16.4 shows the five-step LCA + DEA framework first proposed by Iribarren and Vázquez-Rowe (2013). The first step refers to the collection of data concerning material and energy flows and socioeconomic aspects. This step results in the LCI of each DMU. In the second step, the LCIs are used to carry out the LCIA of each DMU, thus leading to the current environmental profile of each evaluated entity.

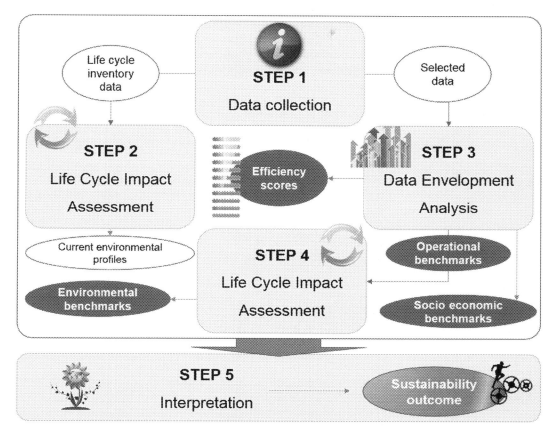

FIG. 16.4 Five-step sustainability-oriented LCA + DEA approach.

In the third stage, DEA computation is carried out for the sample of DMUs using the most relevant data from the first stage. This step provides the relative efficiency indices (ϕ) as well as the quantitative operational and socioeconomic benchmarks for the entities found inefficient. These operational benchmarks involve a modification of the LCI of each inefficient DMU. Therefore, in the fourth step, a new LCIA is conducted for each of the modified DMUs according to the new inventory data. Hence, this step leads to the computation of the environmental benchmarks for the inefficient entities.

The last step deals with the interpretation of the results from the previous stages. In this sense, the presence of two LCA blocks (steps 2 and 4) is associated with the use of the method for eco-efficiency verification, that is, to quantitatively prove that minimizing resource intensity entails reductions in environmental impacts while satisfying the same service (Iribarren et al., 2010). In addition, the operational benchmarks from the third step can be translated into economic savings associated with the minimization of resources (Iribarren and Martín-Gamboa, 2014). Regarding socioeconomic benchmarks and taking into account that working hours are often considered as a DEA element, the corresponding target values need careful interpretation (Iribarren et al., 2013). Overall, the computation of socioeconomic benchmarks along with operational and environmental benchmarks allows a joint interpretation of the results from a sustainability perspective (Iribarren et al., 2016).

16.2.1.2 Three-step sustainability-oriented LCA + DEA approach

Lozano et al. (2010) introduced an LCA + DEA approach that consists of only three stages. Initially, this approach was oriented toward the calculation of environmental impact efficiency (Martín-Gamboa et al., 2018). In the last years, the growing interest in sustainability assessment has led to the use of this method for the sustainability-oriented prioritization of DMUs. In this sense, besides life cycle environmental indicators, economic and social ones could also be used as DEA inputs with the aim of computing efficiency indices enabling the identification of best-performing entities (not only physical DMUs but also other entities such as scenarios, plans, strategies, etc.) in terms of sustainability.

As shown in Fig. 16.5, the first step of the combined approach involves input and output data collection. The second step results in the environmental characterization of the current DMUs using the LCIs developed in the first step. The third step focuses on DEA computation using sustainability-related data and indicators from the previous steps. A sustainability efficiency score of each DMU is directly determined, and target DMUs are determined without a link to operational targets.

Even though the potential of this method for sustainability-oriented prioritization of DMUs has already been proven, further efforts are required to integrate a wider range of indicators that strengthen the vision of sustainability (in particular, life cycle economic and social indicators), as well as to guarantee consistency in the calculation of sustainability benchmarks. Furthermore, the use of the three-step LCA + DEA approach for the calculation of a single sustainability score (i.e., a sustainability index) would still benefit from a larger number of case studies from different sectors.

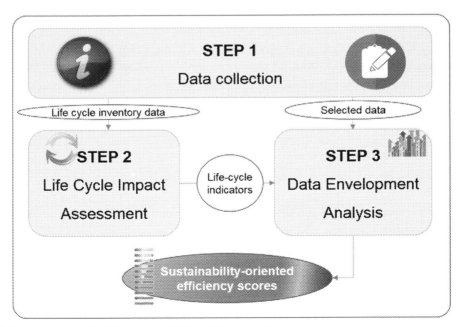

FIG. 16.5 Three-step sustainability-oriented LCA + DEA approach.

16.2.2 From LCA + DEA to LCSA + DEA

The combined LCA + DEA methodological framework was originally conceptualized as a tool oriented to the operational and environmental management of multiple similar entities (Lozano et al., 2009). Interest in concepts such as eco-efficiency motivated the development of this combined methodology as a quantitative management tool. In fact, in the last years, a considerable amount of research papers have addressed the use of LCA + DEA to assess eco-efficiency, applying different versions of this combined methodology in a wide variety of case studies (Vásquez-Ibarra et al., 2020). In parallel, the growing need for sustainability assessment tools has led to advancements in the LCA + DEA methodology in this direction.

Although the evaluation of the environmental dimension is widely consolidated in LCA + DEA, it should be noted that the number of indicators and LCIA methods used has increased significantly in the last decade (Vásquez-Ibarra et al., 2020). Regarding the economic dimension, significant progress has been made since Iribarren et al. (2010) proposed the calculation of economic reductions linked to operational benchmarks. In this regard, Tatari and Kucukvar (2012) combined environmental LCA, LCC, and DEA to assess the eco-efficiency of construction materials. In this line, Ghimire and Johnston (2017) jointly used LCA and LCC to calculate sustainability indicators and DEA to prioritize eco-efficiency measures for green infrastructure. Iribarren and Martín-Gamboa (2014) presented guidelines to enhance the economic component of LCA + DEA studies through different pathways: (1) straightforward options including the calculation of economic savings and/or LCC indicators linked to operational benchmarks, and (2) indirect pathways relying on the monetization of externalities and/or emergy benchmarks.

As regards the incorporation of the social dimension into LCA + DEA, important steps have been taken to address this issue. Iribarren and Vázquez-Rowe (2013) focused on the implementation of social parameters in LCA + DEA studies, analyzing the suitability of labor as an additional DEA item. In the wake of this study, Iribarren et al. (2016) presented a screening approach based on four criteria to test the suitability of socioeconomic indicators for LCA+DEA implementation, expanding the identification of suitable inputs beyond labor. As a consequence of this research, several LCA + DEA studies have already implemented a number of social/socioeconomic indicators. For instance, González-García et al. (2018) applied the abovementioned screening approach to identify suitable sustainability indicators for the evaluation of urban systems. Esteve-Llorens et al. (2020) integrated environmental, nutritional, and (health-related) socioeconomic indicators to evaluate sustainable dietary patterns.

Progress has also been made to adapt LCA + DEA for the evaluation of increasingly complex entities. Efforts in this direction have been based on the use of new DEA models, for example, dynamic DEA models in sustainability assessment oriented to periods (Martín-Gamboa and Iribarren, 2016; Martín-Gamboa et al., 2018, 2019; Torregrosa et al., 2018; Álvarez-Rodríguez et al., 2019b) and DEA network models to assess complex DMU structures such as supply chains (Álvarez-Rodríguez et al., 2020a). Other DEA models (e.g., super-efficiency and DEA "kaizen" models) have also been implemented providing further potentials (Iribarren et al., 2011; Mohammadi et al., 2015; Paramesh et al, 2018; Álvarez-Rodríguez et al., 2019a).

Overall, all these advancements have allowed the implementation of new indicators and the analysis of new types of DMUs, from physical entities (products, technologies, cities, etc.) to nonphysical ones (plans, strategies, diets, etc.). The main advancements in LCA + DEA are summarized in Table 16.1 and further described in Section 16.3. Nevertheless, it should be noted that the compilation of this progress toward a sustainability LCA + DEA approach is still associated with some limitations, such as the

Table 16.1 Main milestones in the progress of the LCA + DEA methodology as a sustainability-oriented tool.

Progress	Milestone	Reference
Improvement in the economic component	Methodological development	Iribarren et al. (2010); Iribarren and Martín-Gamboa (2014)
	Application to case studies	Tatari and Kucukvar (2012); Iribarren et al. (2013, 2020); Ghimire and Johnston (2017)
Improvement in the social component	Methodological development	Iribarren and Vázquez-Rowe (2013); Iribarren et al. (2016)
	Application to case studies	Iribarren et al. (2013); González-García et al. (2018); Esteve-Llorens et al. (2020)
Adaptation to complex DMUs and further potentials	Methodological development	Iribarren et al. (2010, 2014); Martín-Gamboa and Iribarren (2016); Álvarez-Rodríguez et al. (2019a, 2020a)
	Application to case studies	Iribarren et al. (2011); Mohammadi et al. (2015); Lorenzo-Toja et al. (2018); Martín-Gamboa et al. (2016, 2018, 2019); Paramesh et al. (2018); Torregrosa et al. (2018); Álvarez-Rodríguez et al. (2019b)

need to implement social indicators with a life cycle perspective and the development of an integrative LCSA + DEA method. These needs are further discussed in Section 16.4.

16.3 Progress in sustainability-oriented LCA + DEA

The environmental, economic, and social dimensions have traditionally been considered the three fundamental pillars of sustainability. Although significant efforts have been made in the last decades to combine the assessment of these three dimensions, there is still a lack of approaches that effectively integrate them. This section delves into the progress and potential of the LCA + DEA methodology to cope with economic, environmental, and social parameters when evaluating multiple similar entities, thereby providing an overview of the state of the art in sustainability-oriented LCA + DEA. Section 16.3.1 revisits LCA + DEA studies (mainly linked to the five-step approach) that explored the implementation of sustainability indicators as DEA items, with a focus on social/socioeconomic indicators. Section 16.3.2 provides illustrative examples from the scientific literature regarding the use of the three-step LCA + DEA approach for sustainability-oriented prioritization. Finally, Section 16.3.3 deals with LCA + DEA studies using novel DEA models.

16.3.1 Indicators and sustainability benchmarking

One of the milestones that has boosted the assessment of the social component in LCA + DEA studies is the inclusion of labor as a DEA item. Iribarren and Vázquez-Rowe (2013) authored the first study on the implementation of social parameters in LCA + DEA, exploring two alternative approaches to the implementation of labor in the five-step method: consideration of the number of workers or implementation of the number of working hours. Both approaches were tested using illustrative case studies within the fishing sector, evaluating vessels belonging to two separate fishing fleets. In both

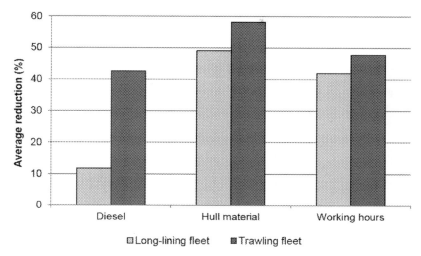

FIG. 16.6 Average reduction targets for DEA inputs for a case study of fishing fleets based on Iribarren and Vázquez-Rowe (2013).

cases, labor was implemented as a DEA input along with other operational elements of the entities under evaluation. The comparison of the two approaches for the implementation of labor led to the conclusion that the approach based on working hours is more suitable and realistic than the worker-based one as the calculation of working hours takes into account the number of working days, working hours per day, and workers. Iribarren and Vázquez-Rowe (2013) also highlighted that the inclusion of both social inputs in the same DEA matrix could lead to inconsistent results, given the relationship between both inputs. Fig. 16.6 shows the average reductions calculated for the sample of vessels in terms of both operational elements and working hours. Significant average reductions were estimated for all DEA elements. However, caution is necessary when interpreting the minimization of working hours, as discussed later.

The research conducted by Iribarren and Vázquez-Rowe (2013) sets the foundations for the implementation of labor as a feasible DEA item in LCA + DEA studies. Following this line, Iribarren et al. (2013) applied the five-step LCA + DEA approach to a set of wind farms located in Spain to benchmark their operational and environmental performance. Within the set of DEA elements implemented in the study, labor (in terms of working hours) was included to integrate a socioeconomic parameter into the analysis. Fig. 16.7 shows the average reduction calculated for each of the DEA inputs as well as the resultant average reduction in each environmental indicator. Average reductions in DEA inputs ranged from 19% (for oil) to 45% (for concrete). These reductions led to average impact reductions ranging from 18% (for eutrophication) to 29% (for global warming). Regarding the socioeconomic component, the average reduction estimated in terms of working hours should be understood as a virtual means toward the redefinition of tasks.

Following these pioneering studies, the use of working hours as a DEA element has spread throughout the LCA + DEA literature (e.g., Mohammadi et al., 2013; Álvarez-Rodríguez et al., 2019a, 2019b). The main lessons from this extended practice correspond to those already highlighted by Iribarren and Vázquez-Rowe (2013). In this respect, the main strength observed for the inclusion of labor is that it succeeds in providing a social perspective to the five-step LCA + DEA approach. Furthermore, labor

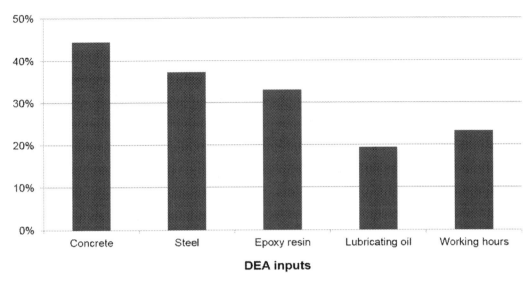
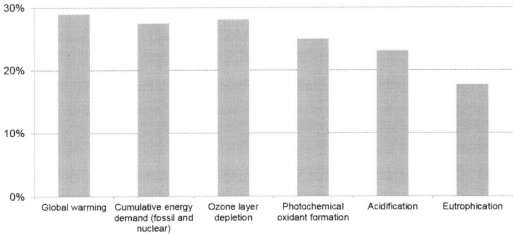

FIG. 16.7 Average reduction targets for DEA inputs and life cycle environmental impacts for a case study of wind farms based on Iribarren et al. (2013).

in terms of working hours is easy to implement because of its quantitative nature. However, it is necessary to interpret with caution the results obtained when labor is implemented in LCA + DEA. This is closely linked to the resource optimization performed through DEA, especially when an input-oriented model is selected. In this sense, when using labor as a DEA input, all actors should understand labor minimization as a virtual means toward the redefinition of tasks with socioeconomic growing purposes, but not as a tool for the identification of useless job positions.

The implementation of working hours as a DEA element awakened interest in the incorporation of further social/socioeconomic indicators in LCA + DEA. Iribarren et al. (2016) developed a screening method to identify appropriate socioeconomic indicators for LCA + DEA according to a set of four criteria: quantifiability, DMU-specificity, data availability, and data quality. The first two criteria are closely linked to the DEA concept, while the remaining two criteria refer to general requirements in analytical studies. According to the screening method, an indicator is considered to be of "straightforward" implementation when meeting the four criteria, while the failure to meet one criterion implies that the indicator should be "excluded." For some indicators, the fulfillment of certain criteria depends on the specific context of the case study involving a certain degree of uncertainty when a final decision regarding the classification of an indicator should be made. Thus, the classification system within the screening method includes two additional categories: "likely" or "unlikely" suitability for LCA + DEA implementation.

To illustrate the screening method, Iribarren et al. (2016) evaluated a set of 40 relevant socioeconomic indicators associated with two specific stakeholder categories: workers and local community. Fig. 16.8 shows the final set of 12 indicators deemed appropriate for LCA + DEA implementation and their potential level of relationship. Within the stakeholder category "workers," most of the indicators selected as suitable were related to economic aspects (e.g., capital costs and labors costs) and health and safety at work (e.g., occupational accidents and commuting accidents). In the case of "local community," the selection was limited to noise pollution and complaints due to the uncertainty regarding data availability and quality for other indicators.

The research activity carried out by Iribarren et al. (2016) led to expand the number of socioeconomic indicators suitable for LCA + DEA implementation while providing a valuable screening tool to be used in future studies, as done for example, in González-García et al. (2018) and Esteve-Llorens et al. (2020). However, even though all the studies included in this section have contributed to improving the social component of the LCA + DEA methodology, the same level of robustness as in the case of the environmental and economic dimensions has not yet been achieved. In this regard, there is a need to include social indicators with a life cycle perspective. The need for an S-LCA + DEA approach is further addressed in Section 16.4.

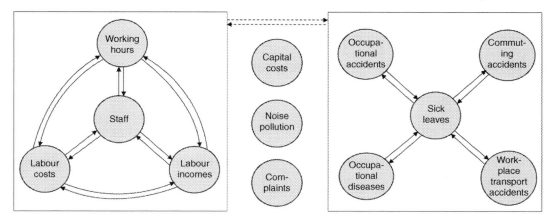

FIG. 16.8 Straightforward and likely socioeconomic indicators for **LCA + DEA** based on Iribarren et al. (2016). Solid arrows represent a hard link between indicators while dotted arrows indicate a soft link.

16.3.2 Sustainability-oriented prioritization

Decision-making support through the prioritization of a set of alternatives plays a key role in achieving sustainable products, systems, plans, etc. Over the last decade, the three-step LCA + DEA approach emerged as a suitable tool for prioritization in terms of sustainability. In particular, this approach has been thoroughly explored in the energy sector for the sustainability-oriented prioritization of energy scenarios. Martín-Gamboa et al. (2017) proposed a conceptual framework for sustainability-oriented energy planning through the combined use of three main components: energy systems modeling (ESM), LCA, and (dynamic) DEA. This methodological framework for energy planning constitutes an enlargement of the three-step LCA + DEA approach by including an ESM stage. This additional phase provides quantitative data on the evolution of energy aspects such as electricity production and installed capacity, as well as on the evolution of life cycle indicators embedded in the energy systems optimization model. These prospective results are implemented in DEA, thus obtaining efficiency scores (understood as sustainability indices) for the prioritization of energy scenarios.

Martín-Gamboa et al. (2019) applied such a methodological framework for energy planning to prioritize 15 energy scenarios concerning power generation in Spain. Following the same case study, Iribarren et al. (2020) examined the influence of the internalization of climate change external costs on the sustainability-oriented prioritization of energy scenarios. Fig. 16.9 shows the prioritization of the set of 15 energy scenarios without and with consideration of externalities. Moderate changes were found in the ranking of energy scenarios and common recommendations were drawn: avoidance of scenarios with an extension of the lifetime of coal-fired power plants, and promotion of scenarios with high penetration of renewables in the electricity production mix. Overall, this combined methodology based on ESM and the three-step LCA + DEA approach presents a high potential for sustainability-oriented energy planning.

The three-step LCA + DEA approach has also been used for the identification of sustainable dietary patterns in Spain. Esteve-Llorens et al. (2020) presented a variant of the three-stage LCA + DEA method by addressing an extended set of indicators: the carbon footprint of diets, a nutritional index, and a set of socioeconomic indicators (namely, number of deaths from tumors of the digestive system, obesity-related health expenditure, and number of people with food shortages). The efficiency scores obtained for the dietary patterns of the 17 Spanish autonomous regions led to identify seven DMUs as regions with suitable dietary habits (Fig. 16.10). All of the autonomous regions showed efficiency scores above 0.55, which indicates the presence of relatively good dietary habits in Spain. The potential of the three-step LCA + DEA method for sustainability-oriented prioritization was again stressed. Nevertheless, more case studies from different sectors would help consolidate this combined approach.

16.3.3 Other advancements

Besides the advancements explained in Sections 16.3.1 and 16.3.2, further progress has been made in the field of LCA + DEA in connection with the potentials of specific DEA models. In particular, this section focuses on three advancements: (1) gradual sustainability benchmarking for continuous improvement using the SBM-Min and SBM-Max models (Tone, 2016), (2) period-oriented sustainability benchmarking using dynamic SBM models (Tone and Tsutsui, 2010), and (3) sustainability-oriented assessment and benchmarking of DMUs with a complex structure such as supply chains by using SBM-network models (Tone and Tsutsui, 2009, 2014). Although these advancements have been separately explored through different case studies, this section focuses on a case study within the tertiary sector as it brings the three advancements together. Such a case study aims to improve the sustainability management of 30 retail stores located in the northwest of Spain (Álvarez-Rodríguez et al., 2019a, 2019b, 2020a).

16.3 Progress in sustainability-oriented LCA + DEA

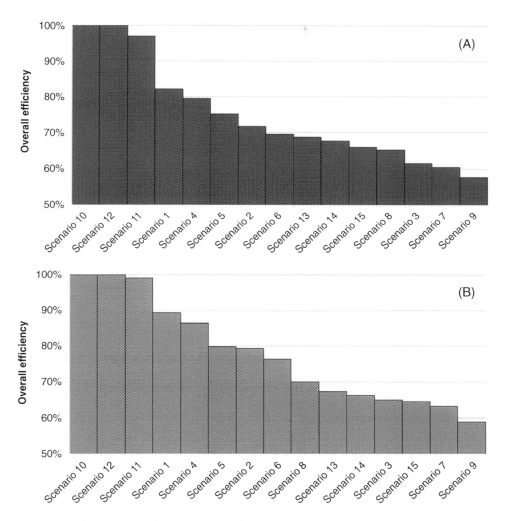

FIG. 16.9 Prioritization of prospective energy scenarios (A) without and (B) with consideration of externalities based on Iribarren et al. (2020).

The DEA models traditionally used in LCA + DEA evaluate DMU efficiency with respect to the furthest point from the efficiency frontier (Tone, 2001). This leads to projections of the DMUs (i.e., benchmarks) that may be too ambitious as target values. On the other hand, the SBM-Max model finds approximate solutions with respect to the closest reference point of the efficiency frontier. Álvarez-Rodríguez et al. (2019a) explored the combined use of both traditional (SBM-Min) and novel (SBM-Max) models as a strategy for gradual benchmarking, thereby facilitating management practices oriented toward continuous improvement via the calculation of ambitious and less ambitious benchmarks. Operational benchmark ranges involve environmental benchmark ranges thanks to the LCA steps of the overall framework.

310 Chapter 16 Coupled life cycle thinking and data envelopment analysis

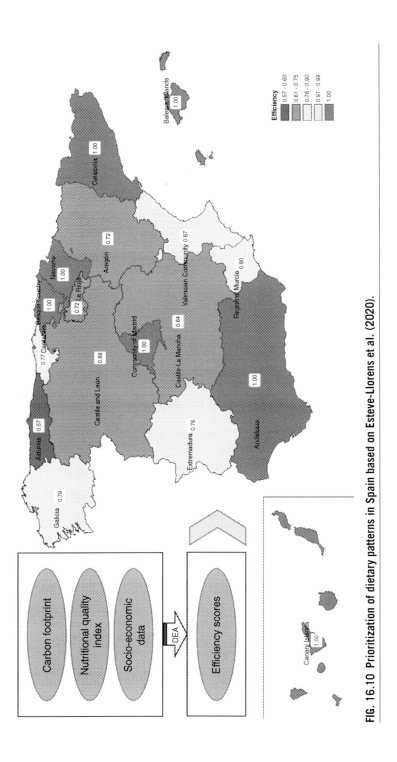

FIG. 16.10 Prioritization of dietary patterns in Spain based on Esteve-Llorens et al. (2020).

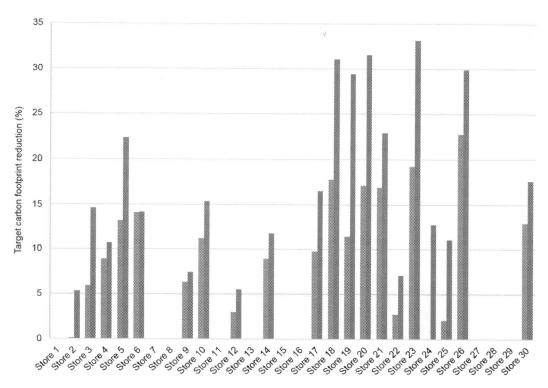

FIG. 16.11 Ranges of carbon footprint benchmarks through an LCA + DEA framework involving SBM-Min and SBM-Max models based on Álvarez-Rodríguez et al. (2019a).

For illustrative purposes, Fig. 16.11 shows the ranges of carbon footprint benchmarks for the retail stores assessed in Álvarez-Rodríguez et al. (2019a). As observed, progressive objectives for environmental improvement could be set for most of the DMUs, proving the usefulness of the proposed methodological strategy when it comes to establishing reference values for continuous improvement. The combination of the reduction objectives from both SBM models means that lower reduction values could be used as short-term benchmarks, while higher (i.e., more ambitious) values could be used as medium/long-term benchmarks. At the company level, this robust framework for the quantification of continuous improvement objectives could facilitate sustainability management (Berlin and Iribarren, 2018).

Regarding the need for five-step LCA + DEA approaches suitable for period-oriented evaluation and/or the evaluation of DMUs with a complex structure involving internal divisions (e.g., supply chains), Álvarez-Rodríguez et al. (2020a) filled this gap by using a dynamic network DEA model. They expanded the base case of retail stores by considering 30 multidivisional DMUs (i.e., 30 retail supply chains) and three time terms (years 2015, 2016, and 2017). The supply chains under assessment included not only the internal operation of each retail store but also two additional divisions involving distribution stages (namely, transport of goods from the distribution center to each store by diesel trucks, and distribution of goods purchased in stores to customers' homes using electric vans). Fig. 16.12 presents the carbon footprint benchmarks for the 30 supply chains, broken down by division and year. The average reduction in carbon footprint for the sample of retail supply chains was estimated at 22% for each

312 Chapter 16 Coupled life cycle thinking and data envelopment analysis

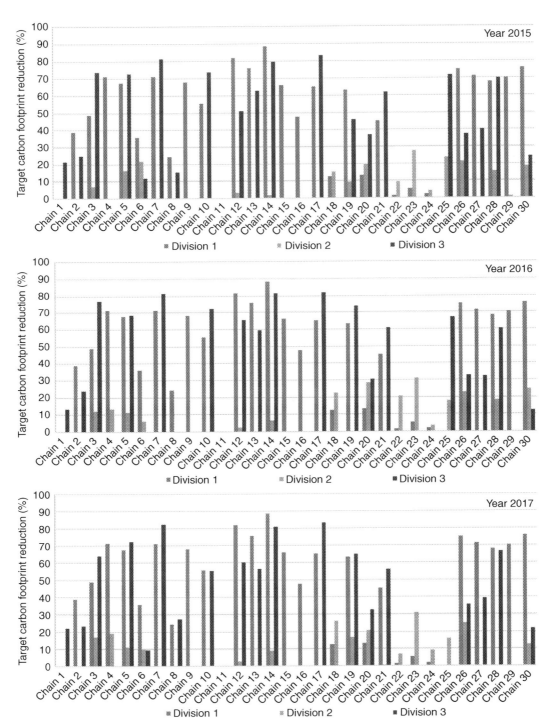

FIG. 16.12 Target carbon footprint benchmarks by division and year for a set of retail supply chains through an LCA + DEA approach involving a dynamic network SBM model based on Álvarez-Rodríguez et al. (2020a).

of the evaluated years. The minimization of both diesel demand in division 1 (central distribution) and electricity demand in division 2 (store operation) was identified as a central objective to reach these environmental reductions. In this sense, improvements related to business logistics, substitution of fossil diesel, and energy efficiency would significantly contribute to reducing the carbon footprint of the retail supply chains under assessment. Overall, the reviewed advancements help consolidate LCA + DEA as a valuable framework for the sustainability-oriented assessment and benchmarking of multiple similar entities, but still requiring further efforts to provide a complete sustainability picture (especially regarding a balanced implementation of the social component as discussed in Section 16.4).

16.4 Delving into needs in sustainability-oriented LCA + DEA

As stated in previous sections, the usefulness of combined life cycle and DEA approaches for sustainability assessment and benchmarking could be significantly enhanced by further integrating social indicators into this type of symbiotic frameworks. In particular, indicators coming from S-LCA could highly enrich both three- and five-step life cycle + DEA approaches.

Fig. 16.13 shows a novel S-LCA + DEA framework aimed at prioritizing DMUs according to social indicators coming from S-LCA. Moreover, under this three-step approach, social benchmarks would also be calculated. Beyond limitations inherent in S-LCA (Costa et al., 2019), this type of approach could suffer from the same concerns as the three-step LCA + DEA approach (regarding e.g., benchmarks' consistency) (Vázquez-Rowe and Iribarren, 2015).

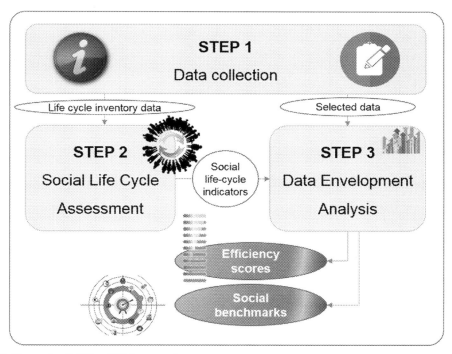

FIG. 16.13 Three-step S-LCA + DEA framework.

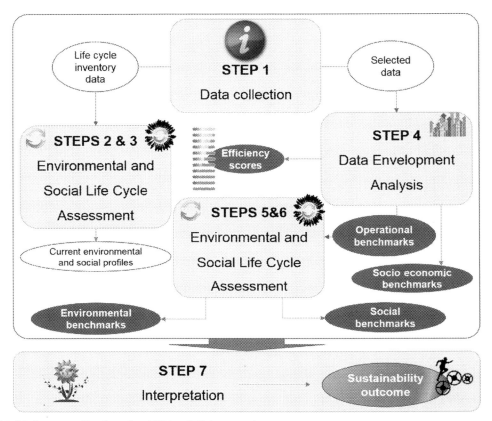

FIG. 16.14 Conceptualization of an LCSA + DEA framework.

When increasing the scope of the framework to comprehensively deal with the sustainability concept, more complex life cycle + DEA approaches would be needed. In this regard, Fig. 16.14 shows a novel life cycle + DEA framework enabling a deep and quantitative consideration of the three main dimensions of sustainability. The proposed LCSA + DEA framework involves an increased number of steps due to the enlargement of the five-step LCA + DEA framework to integrate the S-LCA of current and target DMUs. Thus, the total number of steps is seven. Further steps could be considered if LCC was also implemented. Under this scheme, the sustainability outcome would benefit from the joint interpretation of robust operational, economic, and social benchmarks. Practitioners are highly encouraged to undertake case studies using this novel methodological framework.

16.5 Conclusions and perspectives

The combination of life cycle approaches and DEA is concluded to be a synergistic strategy paving the way toward thorough sustainability assessment and benchmarking of multiple similar entities. In this sense, as long as multiple resembling units are subject to efficiency analysis from a sustainability perspective, the joint use of LC(S)A and DEA is highly recommended.

Finally, future studies in this specific field of research could deal with (1) untapped potentials of LCA + DEA, (2) the application of the novel S-LCA + DEA and LCSA + DEA approaches proposed for the first time in this book chapter, (3) use for labeling, and (4) the real-life implementation and monitoring of sustainability benchmarks derived from these methodological frameworks. Overall, the way is open for the wide use of life cycle + DEA approaches for sustainability-oriented assessment and benchmarking of multiple similar entities.

Acknowledgments

Dr. Martín-Gamboa would like to thank the Regional Government of Madrid for financial support (2019-T2/AMB-15713).

References

Álvarez-Rodríguez, C., Martín-Gamboa, M., Iribarren, D., 2019a. Combined use of Data Envelopment Analysis and Life Cycle Assessment for operational and environmental benchmarking in the service sector: a case study of grocery stores. Sci. Total Environ. 667, 799–808. doi:10.1016/j.scitotenv.2019.02.433.

Álvarez-Rodríguez, C., Martín-Gamboa, M., Iribarren, D., 2019b. Sustainability-oriented management of retail stores through the combination of life cycle assessment and dynamic data envelopment analysis. Sci. Total Environ. 15, 49–60. doi:10.1016/j.scitotenv.2019.05.225.

Álvarez-Rodríguez, C., Martín-Gamboa, M., Iribarren, D., 2020a. Sustainability-oriented efficiency of retail supply chains: a combination of Life Cycle Assessment and dynamic network Data Envelopment Analysis. Sci. Total Environ. 705, 135977. doi:10.1016/j.scitotenv.2019.135977.

Álvarez-Rodríguez, C., Martín-Gamboa, M., Iribarren, D., 2020b. Sensitivity of operational and environmental benchmarks of retail stores to decision-makers' preferences through Data Envelopment Analysis. Sci. Total Environ. 718, 137330. doi:10.1016/j.scitotenv.2020.137330.

Berlin, J., Iribarren, D., 2018. Potentials and limitations of combined life cycle approaches and multi-dimensional assessment. In: Benetto, E., Gericke, K., Guiton, M. (Eds.), Designing Sustainable Technologies, Products and Policies: From Science to Innovation. Springer, Cham, pp. 313–316. doi:10.1007/978-3-319-66981-6.

Cai, W., Lai, K, 2021. Sustainability assessment of mechanical manufacturing systems in the industrial sector. Renew. Sust. Energ. Rev. 135, 110169. doi:10.1016/j.rser.2020.110169.

Charnes, A., Cooper, W.W., Rhodes, E., 1978. Measuring the efficiency of decision making units. Eur. J. Oper. Res. 2, 429–444. doi:10.1016/0377-2217(78)90138-8.

Cinelli, M., Coles, S.R., Kirwan, K., 2014. Analysis of the potentials of multi criteria decision analysis methods to conduct sustainability assessment. Ecol. Indic. 46, 138–148. doi:10.1016/j.ecolind.2014.06.011.

Cooper, W.W., Seiford, L.M., Tone, K., 2007. Data Envelopment Analysis: A Comprehensive text with Models, Applications, References and DEA-Solver Software. Springer, New York.

Cooper, W.W., Seiford, L.M., Zhu, J., 2011. Handbook on Data Envelopment Analysis. Springer, New York.

Costa, D., Quinteiro, P., Dias, A.C., 2019. A systematic review of life cycle sustainability assessment: current state, methodological challenges, and implementation issues. Sci. Total Environ. 686, 774–787. doi:10.1016/j.scitotenv.2019.05.435.

de Olde, E.M., Bokkers, E.A.M., de Boer, I.J.M., 2017. The choice of the sustainability assessment tool matters: differences in thematic scope and assessment results. Ecol. Econ. 136, 77–85. doi:10.1016/j.ecolecon.2017.02.015 http://dx.doi.org/.

Esteve-Llorens, X., Martín-Gamboa, M., Iribarren, D., Moreira, M.T., Feijoo, G., González-García, S., 2020. Efficiency assessment of diets in the Spanish regions: a multi-criteria cross-cutting approach. J. Clean. Prod. 242, 118491. doi:10.1016/j.jclepro.2019.118491.

European Commission, 2019. Communication from the Commission to the European Parliament, the European Council, the Council, the European Economic and Social Committee and the Committee of the Regions – The European Green Deal. European Commission, Brussels.

European Union, 2020. Regulation (EU) 2020/852 of the European Parliament and of the Council of 18 June 2020 on the establishment of a framework to facilitate sustainable investment, and amending Regulation (EU) 2019/2088. Official Journal of the European Union L 198, 13–43.

Ghimire, S.R., Johnston, J.M., 2017. A modified eco-efficiency framework and methodology for advancing the state of practice of sustainability analysis as applied to green infrastructure. Integr. Environ. Assess. Manag. 13, 821–831. doi:10.1002/ieam.1928.

González-García, S., Manteiga, R., Moreira, M.T., Feijoo, G., 2018. Assessing the sustainability of Spanish cities considering environmental and socio-economic indicators. J. Clean. Prod. 178, 599–610. doi:10.1016/j.jclepro.2018.01.056.

Hu, M., Kleijn, R., Bozhilova-Kisheva, K.P., Di Maio, F., 2013. An approach to LCSA: the case of concrete recycling. Int. J. Life Cycle Assess. 18, 1793–1803. doi:10.1007/s11367-013-0599-8.

Iqbal, A., Liu, X., Chen, G.H., 2020. Municipal solid waste: review of best practices in application of life cycle assessment and sustainable management techniques. Sci. Total Environ. 729, 138622. doi:10.1016/j.scitotenv.2020.138622.

Iribarren, D., 2010. Life cycle Assessment of Mussel and Turbot Aquaculture — Application and Insights. University of Santiago de Compostela, Santiago de Compostela.

Iribarren, D., Vázquez-Rowe, I., Moreira, M.T., Feijoo, G., 2010. Further potentials in the joint implementation of life cycle assessment and data envelopment analysis. Sci. Total Environ. 408, 5265–5272. doi:10.1016/j.scitotenv.2010.07.078.

Iribarren, D., Hospido, A., Moreira, M.T., Feijoo, G., 2011. Benchmarking environmental and operational parameters through eco-efficiency criteria for dairy farms. Sci. Total Environ. 409, 1786–1798. doi:10.1016/j.scitotenv.2011.02.013.

Iribarren, D., Martín-Gamboa, M., Dufour, J., 2013. Environmental benchmarking of wind farms according to their operational performance. Energy 61, 589–597. doi:10.1016/j.energy.2013.09.005.

Iribarren, D., Vázquez-Rowe, I., 2013. Is labor a suitable input in LCA + DEA studies? Insights on the combined use of economic, environmental and social parameters. Soc. Sci. 2, 114–130. doi:10.3390/socsci2030114.

Iribarren, D., Martín-Gamboa, M., 2014. Enhancing the Economic Dimension of LCA + DEA Studies for Sustainability Assessment. MDPI, Basel. doi:10.3390/wsf-4-c005.

Iribarren, D., Vázquez-Rowe, I., Rugani, B., Benetto, E., 2014. On the feasibility of using emergy analysis as a source of benchmarking criteria through data envelopment analysis: a case study for wind energy. Energy 67, 527–537. doi:10.1016/j.energy.2014.01.109.

Iribarren, D., Marvuglia, A., Hild, P., Guiton, M., Popovici, E., Benetto, E., 2015. Life cycle assessment and data envelopment analysis approach for the selection of building components according to their environmental impact efficiency: a case study for external walls. J. Clean. Prod. 87, 707–716. doi:10.1016/j.jclepro.2014.10.073.

Iribarren, D., Martín-Gamboa, M., O'Mahony, T., Dufour, J., 2016. Screening of socio-economic indicators for sustainability assessment: a combined life cycle assessment and data envelopment analysis approach. Int. J. Life Cycle Assess. 21, 202–214. doi:10.1007/s11367-015-1002-8.

Iribarren, D., Martín-Gamboa, M., Navas-Anguita, Z., García-Gusano, D., Dufour, J., 2020. Influence of climate change externalities on the sustainability-oriented prioritisation of prospective energy scenarios. Energy 196, 117179. doi:10.1016/j.energy.2020.117179.

ISO, 2006. ISO 14040:2006 Environmental Management — Life Cycle Assessment — Principles and Framework. International Organization for Standardization, Geneva.

ISO, 2006. ISO 14044:2006 Environmental Management — Life Cycle Assessment — Requirements and Guidelines. International Organization for Standardization, Geneva.

Khan, S.A.R., Yu, Z., Golpira, H., Sharif, A., Mardani, A., 2021. A state-of-the-art review and meta-analysis on sustainable supply chain management: future research directions. J. Clean. Prod. 278, 123357. doi:10.1016/j.jclepro.2020.123357.

Kloepffer, W., 2008. Life cycle sustainability assessment of products. Int. J. Life Cycle Assess. 13, 89–95. doi:10.1065/lca2008.02.376.

Lazar, N., Chithra, K., 2020. A comprehensive literature review on development of building sustainability assessment systems. J. Build. Eng. 32, 101450. doi:10.1016/j.jobe.2020.101450.

Liu, J.S., Lu, L.Y.Y., Lu, W.M., 2016. Research fronts in data envelopment analysis. Omega 58, 33–45. doi:10.1016/j.omega.2015.04.004.

Lorenzo-Toja, Y., Vázquez-Rowe, I., Marín-Navarro, D., Crujeiras, R.M., Moreira, M.T., Feijoo, G., 2018. Dynamic environmental efficiency assessment for wastewater treatment plants. Int. J. Life Cycle Assess. 23, 357–367. doi:10.1007/s11367-017-1316-9.

Lozano, S., Iribarren, D., Moreira, M.T., Feijoo, G., 2009. The link between operational efficiency and environmental impacts: a joint application of Life Cycle Assessment and Data Envelopment Analysis. Sci. Total Environ. 407, 1744–1754. doi:10.1016/j.scitotenv.2008.10.062.

Lozano, S., Iribarren, D., Moreira, M.T., Feijoo, G., 2010. Environmental impact efficiency in mussel cultivation. Resour. Conserv. Recy. 54, 1269–1277. doi:10.1016/j.resconrec.2010.04.004.

Martín-Gamboa, M., Iribarren, D., 2016. Dynamic ecocentric assessment combining emergy and data envelopment analysis: application to wind farms. Resources 5, 8. doi:10.3390/resources5010008.

Martín-Gamboa, M., Iribarren, D., Susmozas, A., Dufour, J., 2016. Delving into sensible measures to enhance the environmental performance of biohydrogen: a quantitative approach based on process simulation, life cycle assessment and data envelopment analysis. Bioresource Technol 214, 376–385. doi:10.1016/j.biortech.2016.04.133.

Martín-Gamboa, M., Iribarren, D., García-Gusano, D., Dufour, J., 2017. A review of life-cycle approaches coupled with data envelopment analysis within multi-criteria decision analysis for sustainability assessment of energy systems. J. Clean. Prod. 150, 164–174. doi:10.1016/j.jclepro.2017.03.017.

Martín-Gamboa, M., Iribarren, D., Dufour, J., 2018. Environmental impact efficiency of natural gas combined cycle power plants in Spain: a combined life cycle assessment and dynamic data envelopment analysis approach. Sci. Total Environ. 615, 29–37. doi:10.1016/j.scitotenv.2017.09.243.

Martín-Gamboa, M., Iribarren, D., García-Gusano, D., Dufour, J., 2019. Enhanced prioritisation of prospective scenarios for power generation in Spain: how and which one? Energy 169, 369–379. doi:10.1016/j.energy.2018.12.057.

Mohammadi, A., Rafiee, S., Jafari, A., Dalgaard, T., Knudsen, M.T., Keyhani, A., Mousavi-Avval, S.H., Hermansen, J.E., 2013. Potential greenhouse gas emission reductions in soybean farming: a combined use of Life Cycle Assessment and Data Envelopment Analysis. J. Clean. Prod. 54, 89–100. doi:10.1016/j.jclepro.2013.05.019.

Mohammadi, A., Rafiee, S., Jafari, A., Keyhani, A., Dalgaard, T., Knudsen, M.T., Nguyen, T.L.T., Borek, R., Hermansen, J.E., 2015. Joint Life Cycle Assessment and Data Envelopment Analysis for the benchmarking of environmental impacts in rice paddy production. J. Clean. Prod. 106, 521–532. doi:10.1016/j.jclepro.2014.05.008.

Munda, G., 2005. Multi criteria decision analysis and sustainable development. In: Figueira, J., Greco, S., Ehrogott, M. (Eds.), Multiple Criteria Decision Analysis: State of the Art Surveys. Springer, NY, pp. 953–986. doi:10.1007/0-387-23081-5_23.

Notarnicola, B., Sala, S., Anton, A., McLaren, S.J., Saouter, E., Sonesson, U., 2017. The role of life cycle assessment in supporting sustainable agri-food systems: a review of the challenges. J. Clean. Prod. 140, 399–409. doi:10.1016/j.jclepro.2016.06.071.

Paramesh, V., Arunachalam, V., Nikkhah, A., Das, B., Ghnimi, S., 2018. Optimization of energy consumption and environmental impacts of arecanut production through coupled data envelopment analysis and life cycle assessment. J. Clean. Prod. 203, 674–684. doi:10.1016/j.jclepro.2018.08.263.

Swarr, T.E., Hunkeler, D., Klöpffer, W., Pesonen, H.L., Ciroth, A., Brent, A.C., Pagan, R., 2011. Environmental life-cycle costing: a code of practice. Int. J. Life Cycle Assess. 16, 389–391. doi:10.1007/s11367-011-0287-5.

Tatari, O., Kucukvar, M., 2012. Eco-efficiency of construction materials: data envelopment analysis. J. Constr. Eng. Manag. 138, 733–741. doi:10.1061/(ASCE)CO.1943-7862.0000484.

Tone, K., 2001. A slacks-based measure of efficiency in data envelopment analysis. Eur. J. Oper. Res. 130, 498–509. doi:10.1016/S0377-2217(99)00407-5.

Tone, K., Tsutsui, M., 2009. Network DEA: a slacks-based measure approach. Eur. J. Oper. Res. 197, 243–252. doi:10.1016/j.ejor.2008.05.027.

Tone, K., Tsutsui, M., 2010. Dynamic DEA: a slacks-based measure approach. Omega 38, 145–156. doi:10.1016/j.omega.2009.07.003.

Tone, K., Tsutsui, M., 2014. Dynamic DEA with network structure: a slacks-based measure approach. Omega 42, 124–131. doi:10.1016/j.omega.2013.04.002.

Tone, K., 2016. Data envelopment analysis as a Kaizen tool: SBM variations revisited. Bull. Math. Sci. Appl. 16, 49–61. doi:10.18052/www.scipress.com/BMSA.16.49.

Torregrossa, D., Marvuglia, A., Leopold, U., 2018. A novel methodology based on LCA+DEA to detect eco-efficiency shifts in wastewater treatment plants. Ecol. Indic. 94, 7–15. doi:10.1016/j.ecolind.2018.06.031.

UNEP, 2020. Guidelines for Social Life Cycle Assessment of Products and Organizations 2020. United Nations Environment Programme, Paris.

UNEP/SETAC, 2009. Guidelines for Social Life Cycle Assessment of Products. United Nations Environment Programme, Paris.

UNEP/SETAC, 2011. Towards a Life Cycle Sustainability Assessment. UNEP/SETAC Life Cycle Initiative, Paris.

UNFCCC, 2015. Adoption of the Paris Agreement. United Nations Framework Convention on Climate Change, Paris.

United Nations, 2015. Transforming Our World: The 2030 Agenda for Sustainable Development. United Nations, New York.

Vásquez-Ibarra, L., Rebolledo-Leiva, R., Angulo-Meza, L., González-Araya, M.C., Iriarte, A., 2020. The joint use of life cycle assessment and data envelopment analysis methodologies for eco-efficiency assessment: a critical review, taxonomy and future research. Sci. Total Environ. 738, 139538. doi:10.1016/j.scitotenv.2020.139538.

Vázquez-Rowe, I., Iribarren, D., Moreira, M.T., Feijoo, G., 2010. Combined application of life cycle assessment and data envelopment analysis as a methodological approach for the assessment of fisheries. Int. J. Life Cycle Assess. 15, 272–283. doi:10.1007/s11367-010-0154-9.

Vázquez-Rowe, I., Iribarren, D., Hospido, A., Moreira, M.T., Feijoo, G., 2011. Computation of operational and environmental benchmarks within selected Galician fishing fleets. J. Ind. Ecol. 15, 776–795. doi:10.1111/j.1530-9290.2011.00360.x.

Vázquez-Rowe, I., Villanueva-Rey, P., Iribarren, D., Moreira, M.T., Feijoo, G., 2012. Joint life cycle assessment and data envelopment analysis of grape production for vinification in the *Rías Baixas* appellation (NW Spain). J. Clean. Prod. 27, 92–102. doi:10.1016/j.jclepro.2011.12.039.

Vázquez-Rowe, I., Iribarren, D., 2015. Review of life-cycle approaches coupled with data envelopment analysis: launching the CFP + DEA method for energy policy making. Sci. World J. 2015, 813921. doi:10.1155/2015/813921.

Visentin, C., da Silva Trentin, A.W., Braun, A.B., Thomé, A., 2020. Life cycle sustainability assessment: a systematic literature review through the application perspective, indicators, and methodologies. J. Clean. Prod. 270, 122509. doi:10.1016/j.jclepro.2020.122509.

Wang, J.J., Jing, Y.Y., Zhang, C.F., Zhao, J.H., 2009. Review on multi-criteria decision analysis aid in sustainable energy decision-making. Renew. Sust. Energ. Rev. 13, 2263–2278. doi:10.1016/j.rser.2009.06.021.

CHAPTER 17

How can sensors be used for sustainability improvement?

Patryk Kot[a], Khalid S. Hashim[a,b], Magomed Muradov[a], Rafid Al-Khaddar[a]

[a]Built Environment and Sustainable Technologies (BEST), Research Institute, Liverpool John Moores University, United Kingdom
[b]Faculty of Engineering, University of Babylon, Hilla, Iraq

17.1 Introduction

In recent years, there has been a growing worldwide awareness, supported by scientific research that CO_2 emissions, carbon footprints, climate change, etc. are having a significant effect on the environment (Nejat et al., 2015, Zubaidi et al., 2019). Therefore, the United Nations has set out the Sustainable Development Goals and targets to protect the world with individual countries setting their objectives namely to address issues related to the no poverty, zero hunger, good health and wellbeing, clean water and sanitation, affordable and clean energy, industry innovation and infrastructure, sustainable cities, and communities. For example, the United Kingdom's Government is aiming to cut carbon emissions by at least 80% by 2050 (I. Development 2019).

With the consideration of sustainability in civil engineering especially in construction building materials, structural health monitoring, and environmental monitoring many types of sensors are being used to produce the data and monitor the sustainable development measures (Acharya and Lee, 2019, Shubbar et al., 2020). The applications of sensing technologies in both civil and environmental engineering fields have been developing over the years, which have initiated revolutionary phases of a better understanding and monitoring of structural and environmental health. The recent developments in the sensing technology helped engineers to have real-time accurate monitoring systems for structures' conditions, quantity, and quality of natural resources, which resulted in cost-effective and efficient management of infrastructures and the environment (Gkantou et al., 2019, Ryecroft et al., 2019, Teng et al., 2019). Application of sensing technology for online monitoring of the structural health provides advance warnings about potential serious incidents, such as hidden cracks, corrosion of reinforcing bars or chemical attacks, which can lead to high repair costs or even demolition. In environmental engineering, using advanced sensor technologies provides valuable information about the depletion of natural resources, potential natural disasters, and variation in pollutants concentrations and thereby the potential pollution incidents (Lau, 2014). Thus, it provides experts with valuable information to develop efficient management plans for potential disasters and also to manage the available natural resources.

Therefore, the aim of this chapter is to demonstrate how the sensor technologies can be used to improve sustainability in civil engineering. This work will focus on the introduction of existing industrial sensor systems and research technologies, their working principle, application, and benefits for sustainability improvement.

17.2 Sustainability in civil engineering

Civil Engineering is one of the major contributors to sustainability improvement. Civil Engineering companies are using more environmentally friendly materials, recycled materials, and artificial alternatives (to prevent extraction and transportation of the natural materials) (Kibert, 2016). The maintenance and sustainable preservation of structures and infrastructures are very essential in terms of supporting the future vitality of the economy and prosperity of a society. Their most significant contribution is sustaining wealth and social welfare in general. The structural health monitoring and proper management of these infrastructures are the key roles in keeping them safe and durable for as long as possible (Cawley, 2018, Chen and Ni, 2018). The appropriately managed structural health monitoring not only increases the life span of the structures, but it also helps to increase safety of users/occupants of these structures and infrastructures. In addition, it helps to timely plan the maintenance and supports the preservation activities as well as verifies hypotheses and decreases uncertainty on real structural behaviors and conditions. Moreover, additional knowledge on the monitored structures is obtained during the monitoring process, which can be used in the forecasting/predictive maintenance of structures. Finally, it can help to prevent the adverse impacts, which may occur due to the structural deficiency, namely social, economic, ecological, and aesthetic. This is also very critical to the emergence of sustainable and environmental engineering (Glisic, 2013).

Deterioration of infrastructures and environmental pollution are major global concerns as it determines the life span and safety of the structures and the sustainability of the natural resources (Leung et al., 2015, Qin et al., 2012, Hashim et al., 2020).

There are several ways that structural engineers can mitigate the environmental impact of structural design, of which one is the improvement of life cycle performance. Many structures are designed to reduce the initial cost rather than the whole life cost. One of the examples, in the case of bridges, the maintenance and demolition costs often exceed the initial cost of construction. While the small increase in initial costs could dramatically reduce the life cycle cost by decreasing maintenance and allowing for salvage or disposal at the end of life. The reduction of life cycle costs can contribute to a more sustainable approach than current practice (Ochsendorf, 2005).

In terms of infrastructures, after a certain period of use, structures start showing signs of deterioration due to the exposure to the harsh external environment, earthquakes, humidity, or unexpected loads. In addition, subjecting concrete of steel structures for elevated temperatures results in a sudden and sharp decrease in the strength of the structures, rapid degradation of construction of materials, and other temperature-induced deformations (Bao et al., 2019). In fact, failure of structures due to the mentioned reasons is not as infrequent of an incident as the people may believe. For example, within a period of 10 years (between 1989 and the end of 1999),134 bridges partially or completely collapsed in the United States (Wardhana and Hadipriono, 2003, Spencer Jr and Cho, 2011). In Italy, five main road bridges, namely Petrulla viaduct, Annone, Ancona, Fossano viaduct, and Polcevera bridges, collapsed during the last decade (Bazzucchi et al., 2018). While in Canada, the latest studies indicated that 26% of bridges are not in good conditions (Omar and Nehdi, 2018). Similarly, about a quarter of the bridges in the United States are threatened due to the fair structural conditions or aging (Omar and Nehdi, 2018), and about 11% of the total bridges in the United States are considered structurally deficient, which requires $76 billion to repair them (Barrias et al., 2016). Such damages and failures cost the country significant economic losses and threaten the lives of citizens, for example, 158 people were killed or injured because of the collapse of the I-35W bridge in the USA in 2007 (Koch et al., 2015). In addition, the estimated losses due to the collapse of electricity network in the United States in 2003 were between $60 to $10 billion (Luke, 2010), and about $80 billion are required to maintain the threaten bridges in the United States (Omar and Nehdi, 2018). In the United Kingdom, it has been reported that the number of "substandard" bridges has recently increased to about 3441 bridges; and £1 billion is the required cost to maintain these structures (MailOnline 2020). Although these collapses incidents occurred due to different reasons, such as harsh environment and degradation of construction materials, the literature gathered them into the lack of knowledge of the structural conditions and into the inadequate use of technologies for precisely evaluating the conditions of structures (Bazzucchi et al., 2018, Koch et al., 2015). In response to this fact, a great deal of attention has been paid for developing efficient sensing and monitoring systems, and also strong arguments have been initiated for the development of accurate real-time monitoring systems to avoid such tragic and costly accidents and also to achieve efficient management of structures in service (Myung et al., 2014, Grosse et al., 2007). All of these reasons have driven the expansion in structural monitoring by regulatory agencies, which results in the application of a variety of sensing and monitoring systems, from robotics to wireless systems.

Environmental pollution cannot be eliminated or inhibited from the eco-system of the planet as long as a single inhabitant lives on this planet, as pollution is the natural by-products of human activities (Goel, 2006, Emamjomeh et al., 2020). However, the degree of pollution that resulted from the normal activities of a single inhabitant does not represent a threat to the eco-system and apparently harmless for human health, but the combined influence of large populations, such in cities and urban areas, causes substantial pollution (Goel, 2006, Jayaswal et al., 2018, Al-Jumeily et al., 2019, Al-Saati et al., 2019). In fact, there is an irreversible relationship between the survival of human and environmental pollution as mankind needs to utilize the natural resources for their existence, which in turn results in the generation of wastes. A few centuries ago, people were living in small communities near to the freshwater bodies and disposing of their waste directly into these bodies. However, these wastes were insignificant in quantity and very simple in chemical composition (the majority were organic matter) that the receiving freshwater bodies barely showed any symptom of contamination (Goel, 2006, Varol et al., 2010, Hu and Cheng, 2013, Hashim et al., 2018). The intensity and rapid growth in the global population and the expansion in the industrialization and agricultural activities resulted in massive production of wastes that contain a broad range of harmful pollutants (Hashim et al., 2019, Omran et al., 2019, Hashim et al., 2020, Hashim et al., 2020), including gaseous wastes, such as greenhouse emissions, and liquid wastes, such as carbon dioxide, lead, mercury, nutrients, and organic matter (Duarte et al., 2015, Holt, 2000, Abdulhadi et al., 2019, Hashim et al., 2019). This huge disposal of wastes exerted severe pressure on the natural resources that caused serious deterioration of the eco-system and also caused serious health problems and outbreaks of fatal infections (Al-Saati et al., 2019, Omran et al., 2019, Vardhan et al., 2019, Singh and Singh, 2017, Hashim et al., 2018). For example, Zhang et al. (2014) reported that to meet the increasing demand for food in China, the annual consumption of pesticides reached 3,000,000 tons, and the average usage of pesticides in China is higher than the worldwide average by 2.5–5 times, which in turn substantially increased phosphate loads in the freshwater bodies. In addition, some studies demonstrated that continuous discharge of wastewaters has increased the concentrations of some heavy metals in freshwater bodies by 1000-fold of the allowable limits (Obiechina and Joel, 2018), which seriously polluted both water and soils of freshwater sources (Emamjomeh et al., 2020, Hashim et al., 2019), and resulted in spreading of serious diseases (Preker et al., 2016, Hill, 2020). Furthermore, the increasing rate in the generation of municipal solid wastes (MSW) led to serious pollution of soil and groundwater sources (Ramachandra et al., 2018, Abdulredha et al., 2017, Idowu et al., 2019). For instance, in India, the annual production of MSW from urban areas is about 48,000,000 tons, and it is currently increasing to reach about 250,000,000 tons by 2047 (Ravindra et al., 2015). Unfortunately, 90% of the MSW in India is disposed on land without proper management (Ramachandra et al., 2018), which not only pollutes the soil and groundwater but also contributes to climate change by emitting large amounts of greenhouse gases, where the global MSW is responsible for 3%–19% of the total anthropogenic methane emission (Ravindra et al., 2015). The greenhouse gases are the main reason for global warming, which in turn causes unexpected natural disasters, such as storms and floods events, that accelerate the deterioration of infrastructures. In fact, the emissions from the MSW are insignificant in comparison to the emissions from the cement industry, which is the main pile for urbanization (Shubbar et al., 2020, Cai et al., 2016). The cement industry is responsible for about 5%–7% and 26% of the global anthropogenic and industrial carbon emissions, and for 12% and 26% of the total emissions of nitrogenous oxides and particulates in China, respectively (Liu et al., 2018). Therefore, the cement industry is categorized as a major reason for global warming, which is behind the water crisis (Zubaidi et al., 2019, Zubaidi et al., 2020) and deterioration of infrastructures.

Although environmental pollution is a natural result of human activities, the pollution incidents that result from unintentional loss of containment, sudden failure of treatment units, natural disasters, and mega-events could cause substantial impacts on human health and the ecosystems (Lim et al., 2010, Otutu and Agba, 2003, Cao et al., 2018). For example, 5213 environmental incidents were recorded in China during the period between 2006 and 2015 (Cao et al., 2018). Similarly, in Scotland, United Kingdom, 179 environmental incidents were reported in 2014; of these incidents, 2 involved radiation pollution, 64 were related to microbiological pollution, and the rest were either chemical or airborne chemical pollutions (Henton and Ramsay, 2015). Mega-events cause a sudden upsurge in the generation of MSW that makes management of such waste very difficult. For instance, it was reported that more than 20 million people visit Kerbala city, Iraq during the 15-day Arba'een religious festival, which at least results in the generation of 16.6–25 million kg of MSW (according to the hotels' records) (Abdulredha et al., 2017, 2018). Due to the mentioned impact of environmental pollution on the sustainability of the infrastructures and ecosystems and also human health, monitoring of sources of pollution and quality of the environment is a concern of growing importance worldwide (Luo and Yang, 2019). Additionally, many studies demonstrated that the application of efficient sensing and monitoring systems enhances the cost effectiveness of the treatment or management process as it gives enough time to the operators or planners to take the required actions to avoid the potential disasters or impacts. For example, it has been found that accurate real-time monitoring for wastewater treatment plants could minimize the energy consumption by 40% (Qin et al., 2012).

In summary, two key facts could be concluded from these brief backgrounds; first, there is a strong relationship between the sustainability of infrastructures and environmental pollution. Second, the sustainability infrastructures and environment could be maintained and efficiently managed by applying accurate sensing systems.

17.3 Working principle of sensing technologies for sustainability improvement

Continuous structural and environmental health monitoring provides valuable data that help to have a better understanding of their current conditions, and also to predict the remaining life time (Qin et al., 2012, Myung et al., 2014, Grosse et al., 2007). The use of continuous monitoring systems positively impacts the sustainability and safety of the monitored body as it provides quantitative information that could be effectively analyzed and used to assess the current and future conditions of the measured objects (Leung et al., 2015). Therefore, engineers have started, during the 1960s, to use simple vision methods to monitor and measure the damages or deformation of structures, these vision trials utilized infrared and laser beams (Myung et al., 2014). Then, the sensing technologies have rapidly developed to utilize high-resolution cameras, microwave sensors, fiber optic, remote sensing, and satellite and drones imaging (Gkantou et al., 2019, Ghauri and Zaidi, 2011, Ryecroft et al., 2019, Lee et al., 2016).

Remote sensing usually uses cameras, scanners, or radar to collect the information in a noncontact manner from the object under investigation and passes this information to the analysis unit via a suitable type of waves, such as electromagnetic waves (Patel and Singh, 2013).

Remote sensing revolution started, during the 70s of the last century, shortly after the remarkable developments in the satellite transmission technologies, where some studies were conducted to remotely measure the moisture of soil (Malik and Shukla, 2014). Remote sensing could be conducted for short distances, using unmanned vehicles, or for very long distances, using satellites or airplanes (Ghauri and Zaidi, 2011, Myung et al., 2014). For short-distances sensing, radio waves (radio frequency signals) are usually used to transfer the collected data from the sensor to the base unit (analyzer) (Myung et al., 2014). However, recent studies utilized new types of waves to transfer data in short distances sensing. For example, continuous laser waves were used in this type of remote sensing to minimize the computational loads because of the encoding sensor measurements (Park et al., 2011, Myung et al., 2014). Long-distances remote sensing depends on the reflectance in different wavelength regions of the electromagnetic spectrum to monitor the changes in the conditions of the monitored object (Ghauri and Zaidi, 2011).

The most important advantage of the remote sensing is the nature of the technology, namely it does not require any expensive cable installation and maintenance, and its ability to monitor large areas and objects accurately (Ghauri and Zaidi, 2011). At the same time, all forms of wireless sensing still face a number of challenges, for example, these units must be designed in an accurate way to avoid high energy consumption, and also the majority of the short-distances sensing units are operated using batteries that should be replaced periodically (Myung et al., 2014).

There is an enormous number of classification methods of the currently used sensing systems in civil and environmental engineering applications, where it could be classified into stationary (such as fiber optics) and mobile (such as drones and robotic units) systems (Leung et al., 2015, Lee et al., 2016), wired and wireless systems, or according to the mechanism of detection, such as light-based, microwave-based, and vibration-based systems (Mitra and Gopalakrishnan, 2016).

This chapter reviews the sensing methods based on their working principle namely acoustics, magnetics, and electromagnetics (radio and microwaves, infrared (IR), optics and X-Ray and Gamma Ray).

17.3.1 Acoustics sensing methods

The working principle of acoustic sensing is a transmission of acoustic sound pulse towards a target, which reflects the sound back to the sensor. The acoustic system measures the time for the echo to return and computes the distance to the target using the speed of sound in the medium (air) (Latha et al., 2016). Acoustic sensing can be also used to detect physical changes of the measured material by studying and analyzing any changes to characteristics of the acoustic propagation path, see Fig. 17.1. These changes could influence the amplitude and/or velocity of the wave. Piezoelectric materials are utilized to generate the acoustic waves for acoustic sensors and devices. The piezoelectric material produces electrical charges by imposing mechanical stress. An appropriate oscillating electric field is applied to piezoelectric material to create the mechanical stress/wave, which propagates through the substrate and is converted back to an electric field for the measurement (Sun et al., 2010, Drafts, 2001). The acoustic methods could be divided into two main types; sound (ultrasonic wave) penetration and impact sound.

17.3.1.1 Sound penetration

This method depends on the ability of the sound (ultrasound) to penetrate bodies; a suitable ultrasonic wave (guided, surface, or bulk wave) is emitted from a piezoelectric transducer, which is carefully attached to one face of the element being tested, and received using an ultrasonic sensor that is usually attached to the opposite face of the element (Barrias et al., 2016). Because the velocity of the ultrasonic wave depends on the physical properties of the element, the change in the properties of the received ultrasonic wave could be used to assess some key properties of the tested element (Barrias et al., 2016). In some cases, sound waves are used instead of ultrasonic; where a sounding rod or a chain drag is used to generate the sound wave. Fig. 17.2 shows the principles of the sounding technique.

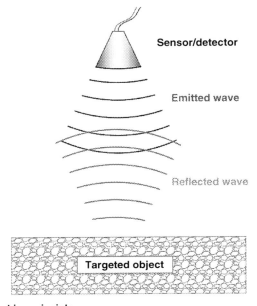

FIG. 17.1 Acoustic sensor working principle.

17.3 Working principle of sensing technologies for sustainability improvement

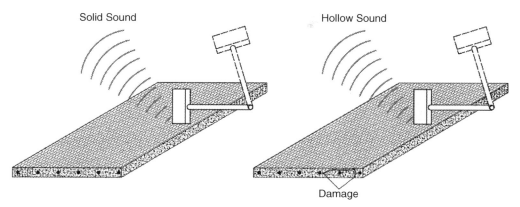

FIG. 17.2 Example of sounding technique used to assess the concrete conditions.

17.3.1.2 Impact sound

This system transforms the impact sound into electric signals, which are analyzed by specialist software to evaluate the magnitude and location of the damage in the structures. In this sensing method, a hydraulic hammer is used to impact the required location within the structure; the generated sound will be transformed into electric signals that are analyzed later to deliver the required information (Montero et al., 2015). Fig. 17.3 demonstrates the example of impact sound system used to assess the pavement conditions.

17.3.2 Magnetic sensing methods (Hall-effect sensor)

The Magnetic sensing methods (Hall-effect sensor) is constructed from a thin sheet of a conductive material. The output connections are perpendicular to the direction of the current flow. The material is then subjected to a magnetic field and responds with an output voltage proportional to the magnetic field strength (Sanfilippo, 2011).

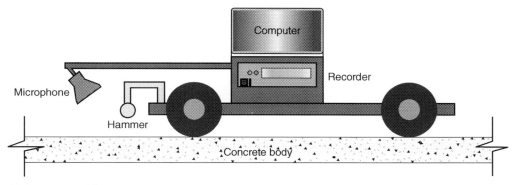

FIG. 17.3 Example of impact sound system used to assess the pavement conditions.

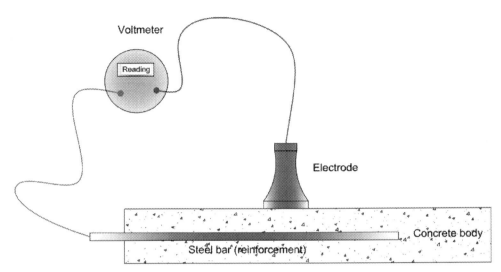

FIG. 17.4 Example of Hall-effect sensor used to assess the concrete conditions.

The Hall-effect sensor is used to detect the change in the properties around the reinforcement bars; with the development of rust on the surfaces of the steel bars, the reluctance will change because of the significant difference between the permeability of rust and steel. Therefore, the magnetic flux through the Hall-effect sensor will be changed correspondingly, which in turn is used to assess the corrosion level and corrosion rate of the measured object (Zhang et al., 2017). Although this sensing method yields reliable results and could be run quickly, its applications are limited to the detection of corrosion of reinforcement bars and the location of these bars within the concrete body. Fig. 17.4 shows an example of Hall-effect sensor used to assess the concrete conditions.

17.3.3 Electromagnetic sensing methods

The working principle of electromagnetic sensing methods is based on electromagnetic waves. The electromagnetic waves do not require a medium to propagate, that is, can travel not only through the air and solid materials but also through the vacuum of space. The electromagnetic energy can be described by frequency, wavelength, or energy. Radio and microwaves are usually described in terms of frequency (Hertz), infrared and visible light in terms of wavelength (meters), and X-rays and gamma rays in terms of energy (electron volts) (N. SCIENCE 2020). The electromagnetic spectrum is presented in Fig. 17.5.

17.3.3.1 Radio and microwave sensing methods

Radio and microwave waves are a form of electromagnetic radiation. Electromagnetic waves consist of oscillating electric and magnetic fields that propagate at the speed of light. The microwaves are characterized by various parameters namely frequency or wavelength, intensity, strength, and polarization. Frequency is the rate at which the amplitude of an electromagnetic wave changes and is related to wavelength. The intensity is a measure of the strength or amplitude of the electromagnetic radiation

17.3 Working principle of sensing technologies for sustainability improvement

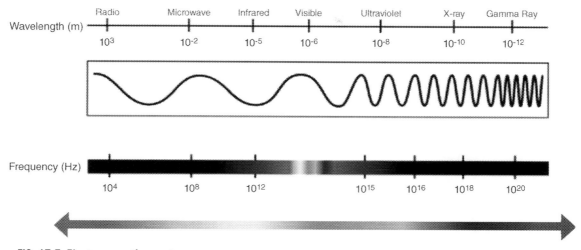

FIG. 17.5 Electromagnetic spectrum.

and is related to the amount of energy carried by the wave (Hensley and Farr, 2013). Electromagnetic waves interaction with the material can be revealed in the form of a unique signal spectrum called a reflection coefficient (S_{11}) and transmission coefficient (S_{21}). Permittivity and conductivity of measured material will vary the measurement quantities of the signal spectrum. Permittivity is a measurement of the dielectric medium response to the applied microwaves through the changing of its electric field. It depends on the material's ability to polarize in response to the applied field (Teng et al., 2019). The electromagnetic waves are currently used in various civil engineering applications namely remote sensing (Kot et al., 2020), structural health monitoring (Kot et al., 2017), and environmental monitoring (Ryecroft et al., 2018).

17.3.3.2 IR sensing methods

IR sensors emit or detect IR radiation. IR sensors are capable of measuring the heat being emitted by an object and detecting motion. This type of sensing method depends on the difference in the thermal capacities of materials to detect any changes in physical properties. Usually, a sensitive sensor is used to register the change in the properties of infrared, which is highly sensitive to the thermal radiation emitted by the body being monitored (Montero et al., 2015). The recorded changes in the properties of the infrared are analyzed to detect the abnormalities within the tested body. With the recent developments, this technique is not only limited to close sensing process, but it could be used remotely (Omar and Nehdi, 2017). The spectral radiant emittance follows Plank's relation as shown in Fig. 17.6. The infrared sensors found their use in astronomy to understand the universe, military applications for scanning the battlefield in poor visibility situations, search and rescue, meteorology, and civil engineering. There are two types of IR detectors namely, thermal and photonic. Thermal detectors function by detecting the thermal effects of the incident IR radiation, whereas in photonic detectors the incident IR radiation causes intrinsic or extrinsic electronic excitations. The IR spectrum is split into three subspectrum, namely near IR, mid IR, and far IR. The wavelength region of near, mid, and far IR are 0.75–3 μm, 3–5 μm, and above 5 μm, respectively (Karim and Andersson, 2013).

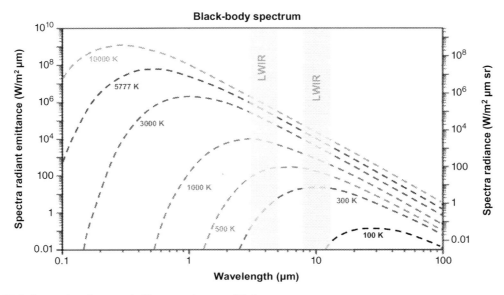

FIG. 17.6 Spectral emittance of objects at given equilibrium temperatures.

17.3.3.3 Optic sensing methods

Optical fiber technology is one of the recent technologies that witnessed rapid development during the last decades, and consequently, it has been applied in a wide range of applications, such as sensing technologies (Leung et al., 2015, Barrias et al., 2016). The principle of fiber optic sensor is the properties of the passed light through the optical fibers that are very sensitive for the external effects, such as the change in the temperature, exerted loads, and stresses, which could be calibrated to the measurement. Thus, this new technology has been widely used in structural health monitoring, where it is used to monitor the conditions of roads, buildings, geotechnical works, pipelines, and cables (Bao et al., 2019).

Optical fibers are commercially available nowadays in two types: single-fiber and multi-fiber basing on the number of waveguides in the fiber. Generally, the single-fiber type is made from three parts; the core of the fiber, cladding, and protection layers; while the multi-fiber type has almost the same structure of the single-fiber, but with a wider core (Bao et al., 2019, Barrias et al., 2016). It is noteworthy to highlight that single fiber provides high transmission speed and low transmission losses in comparison with the multi-fiber type, which makes it the most widely used type in long-distances applications (Bao et al., 2019). Fig. 17.7 shows a coaxial optical fiber sensor.

Optical fibers respond to any external effects because of a very small diameter (from 4 to 600 μm), which makes it possible to detect these responses and analyze them to assess the magnitude and type of the external effect (Krohn et al., 2014). Fiber optic sensors are divided into two types, extrinsic and intrinsic, according to the location of the produced change in the properties of the guided lights (inside or outside the optical fiber) (Bao et al., 2019). Extrinsic and intrinsic types in turn are divided into a wide range of subtypes. However, in civil engineering applications, the fiber optic sensors are divided into three types: interferometric, grating-based, and distributed sensors (Bao et al., 2019,

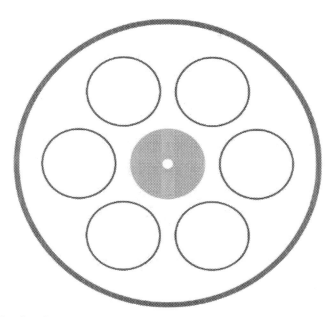

FIG. 17.7 The coaxial optical fiber sensor.

Krohn et al., 2014). Grating-based and interferometer sensors are usually for point sensing as they detect the changes where gratings or interfaces are present; however, a set of these sensors could be used together, in the form of a quasi-distributed sensor, to detect the changes in different locations simultaneously. On the other hand, the distributed fiber optic sensors possess the ability to detect the changes in several locations along the optical fiber simultaneously (Krohn et al., 2014). Distributed optical fiber sensors also have another important advantage in comparison to universal strain measurements; the huge number of sensing points in the distributed optical fiber allows the operator to develop 2D or 2D maps for strain distributions, which reflects the whole behavior of the monitored body rather than few points (Zhao et al., 2020).

17.3.3.4 X-ray and Gamma-ray sensing methods

X-ray and gamma radiations are nondestructive techniques used to analyze the internal microstructure of materials based on the radiation. The sensor is composed of an emitter to emit a ray at a given intensity, and a detector, which registers the reception intensity of the ray and the data are recorded to be analyzed later (Vicente et al., 2017), see Fig. 17.8. The irradiated materials under investigation absorb the radiation, which is related to the properties of the materials, such as thickness and density (Vicente et al., 2017, Ling et al., 2013). The captured signal data are postprocessed to produce attenuation-corrected images, which occurs with the measurement of attenuation. This means that the density of each point of the measured material can be identified and a tri-dimensional (tomographic) information is developed (Vicente et al., 2017).

There many risks related to this method, such as hazards of radiation and the required safety procedures to protect the workers and the public (Montero et al., 2015).

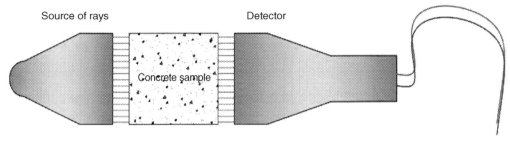

FIG. 17.8 Diagram for the gamma-rays sensing.

17.4 Installation methods of sensing technologies in structural and environmental applications

The installation of sensor technologies can be divided into three main groups namely, sensors can be embedded into the measured body, attached to the body, or be contactless.

The embedded sensors are installed inside the object. For example, microwave sensors could be installed inside the concrete structures during the construction process (Teng et al., 2019), or immersed in the solution being inspected (Ryecroft et al., 2019). This solution enables long-term continuous health monitoring of structures, real-time faults, and location identification and prediction of the structural performance to prevent expensive structural repairs (Gkantou et al., 2019).

Attached sensors are directly attached to the surfaces of the objects being monitored. For example, in the acoustic sensing method, the sensor is carefully and accurately attached to the surface of the concrete bodies to avoid the effects of any air gaps (Montero et al., 2015).

Contactless methods include sensors that do not require any physical contact with the measured object. For example, satellite imaging technology is used to monitor the changes in watercourses without any physical contact between the visual sensing unit and the surface of the water (Ghauri and Zaidi, 2011). In addition, contactless sensing systems can be attached to the robotic units, which have been broadly used in the monitoring of structural and environmental health, where robotic units provided a platform to carry a certain or a group of sensors to inspect structures, natural resources, or a source of pollution (Dunbabin and Marques, 2012, Sony et al., 2019). Generally, the robotic sensing units are classified into seven main types according to the locomotion mechanism, namely crawling, swarm, wheeled, snake-like, aerial vehicle, underwater vehicle, and complex robotic sensing units (Myung et al., 2014). Each one of these types is suitable for a certain type of inspection/sensing purposes, which are as follows (Myung et al., 2014, Nansai and Mohan, 2016, Mengüç et al., 2012, Jahanshahi et al., 2017):

1. Crawling robotic units are suitable for vertical or uneven surfaces, such as chimneys and skyscrapers. The crawling robotic units are provided with a suitable adhesive mechanism to enable them to attach themselves to the vertical surfaces; these mechanisms are suction cup adhesion, suction cup crawler adhesion, vacuum pump adhesion, magnetic adhesion, rope and/or rail gripping, bioinspired adhesion.
2. Swarm robots: these robotic units can change their shapes and functionalities automatically, which makes them very effective in sensing technology and suitable for different conditions. The

main advantages of this type of robotic sensing units are on-site reconfiguration, multifunction, ease of assembling, reusable and they have good fault-tolerance.
3. Wheeled robotic sensing systems are suitable for a horizontal surface, which makes it favorable for monitoring cracks in concrete structures, roads, bridge decks, parking slots, and inner walls.
4. Snake-like robotic units: This type of robotic units is used for confined places that cannot be accessed by humans or by mobile sensing units, such as pipes and air ducts.
5. Aerial vehicles robots (e.g., drones, helicopters, or airplane): This type of sensing unit can inspect large areas by fly through. A few decades ago, this type was very expensive and requires complex technology, but with the recent significant developments in batteries, aeronautical designs, and digital cameras the aerial vehicles robots became inexpensive, dependable, and easy to operate.
6. Underwater vehicles: This type of robotic sensing units is submersible, which enables the operator to monitor and manage the marine structures, liquid storage tanks, and pipelines efficiently.
7. Complex robotic sensing units: These types of robotic sensing units mix more than one of the mentioned locomotion mechanisms to work under complicated conditions or to work under certain conditions. For example, MOGRER (MOGRER refers to the Japanese word "Mogura", which means mole, an animal predominantly skilful at navigating in cavities (Okada and Sanemori, 1987)) is one of the complex robotic sensing units that consist of a driving wheel, spherical bearings, a leveling spring, and has a shape of a pair of scissors. This complex design allows the MOGRER to self-adjust its height during the inspection of pipelines with varying diameter.

Additionally, the robotic sensing methods are classified according to the sensing mechanism, such as the following examples discussed by (Montero et al., 2015). Recently, advanced generations of robotic sensing units have been developed that utilize the wireless technologies in both inspections and transform of data. However, all robotic sensing methods are focused on instantaneous inspections rather than continuous monitoring (Myung et al., 2014).

17.5 Applications of sensing technology in civil and environmental engineering

17.5.1 Civil engineering applications

Sensing technology has been extensively used in the field of civil engineering to monitor the structural health of tunnels, bridges, high buildings, marine structures, and cultural heritage objects. Additionally, it is used to monitor the soil conditions in the areas subjected to earthquakes, pipelines, energy lines, and in the monitoring of the vibration level. Nowadays, sensor technologies are directly or indirectly utilized to improve sustainability.

For example, cracking of concrete pavement, especially the cracks in the bottom layer of the road, is a serious global problem. Thus, Alshandah et al. (2020) used discrete strain sensors, attached to the bottom layer of the roads, to monitor the development of the cracks (bottom-up cracks). Three concrete samples, with dimensions of 15.24 cm × 15.24 cm × 5.08 cm, were prepared according to the ASTM C39 and subjected to three-point loading test. Four sensors were attached to each one of the concrete samples to monitor the propagation of cracks. The results evidenced the ability of the strain sensors to track the cracking of the bottom layers of roads. Additionally, the results showed a high agreement between the results of the sensors and the actual (observed) results, with an average accuracy of 82.4%.

The Magnetic sensors (Hall-effect sensor) were employed by Zhang et al. (2017) to monitor the corrosion of steel reinforcement. In this investigation, a concrete beam (2 m in length, 0.18 m in width, and 0.12 m in thickness) was reinforced by three steel bars (Ø 16 mm). Corrosion of steel bars, in the tensile zone, was continuously monitored using the magnetic sensors. For comparison purposes, acoustic sensors were also used to monitor the corrosion of the steel bars. To accelerate the corrosion process, the chloride solution was applied to the prepared reinforced beam. The findings of this study showed a high agreement between the measured corrosion by the magnetic sensors and the acoustic sensors. Additionally, it was found that the magnetic sensors have the ability to quantify the corrosion rate.

The electromagnetic sensor has been utilized by researchers. Teng et al. (2019) developed a cost-effective electromagnetic sensor to monitor the moisture content of concrete structures. The developed sensors were embedded in two different concrete samples, which were prepared according to the Europe EN-206-1-2013 and Malaysia MS-26-1-8: 2010 standards (Kot et al., 2016). The concrete samples were 30 cm in length, 30 cm in width, and with a thickness of 15 cm, and all samples were not reinforced.

The obtained results proved that the electromagnetic sensors, at a frequency of 2.45 GHz, were very efficient in the detection of changes in the moisture content in both European and Malaysian concrete samples. Additionally, the outcomes of this study indicated that the electromagnetic sensors were able to identify the differences in the mix ratios of utilized concrete samples.

In addition, electromagnetic sensors were embedded in concrete samples to monitor the cracking of reinforced concrete specimens. The study conducted by Gkantou et al. (2019) involved the design, manufacturing, and application of a microwave sensor. Microstrip patch antennas were installed in four reinforced concrete beams, the latter were monitored under three-point bending. In parallel to the microwave sensors, transducers and crack width gauges were used (for comparison purposes). The microwave sensors were installed along the neutral axis at the mid-span of the beams. The obtained results indicated a very good agreement between readings of the crack gauges and the microwave sensors. Additionally, a linear relationship was noticed between the cracks and the electromagnetic signals, which confirms the ability of the microwave sensors to detect the development of the cracks in concrete structures. Thus, the authors indicated that microwave sensors could be a reliable and cost-effective alternative for structural health monitoring.

Moreover, electromagnetic sensors found their application in cultural heritage structural health monitoring owing to their contactless application. In this study, Tobiasz et al. (2019) and Markiewicz et al. (2020) reviewed the methods for documentation, management, and sustainability of Cultural Heritage based on the case study in the Museum of King Jan III's Palace at Wilanow. In this study, various existing technologies for accurate architectural and archaeological documentation for management and sustainability of cultural heritage were investigated, namely, terrestrial laser scanning, photogrammetry, remote sensing, ground penetration radar, near-IR techniques, and fiber optics.

Optical fibers were also used to monitor the structural health of an old bridge, the Gotaalv Bridge in Sweden, which is 950 m in length and made from steel girders and supported by concrete bridge decks. The optical fibers were used to monitor the cracks in the bridge (wider than 0.5 mm) because this bridge shows some fatigue signs because of the heavy traffic loads and aging (built in 1939). In this study, Enckell et al. (2011) installed optical fibers at the required locations and covered with aluminum tape

17.5 Applications of sensing technology in civil and environmental engineering

to prevent the effects of the ambient environment. The outcomes of this study confirmed the ability of the optical fibers to detect, measure, and localize the cracks within the monitored bridge.

Optical backscatter reflectometer sensors (single-mode fiber) were utilized in the study conducted by Villalba and Casas (Villalba and Casas, 2013) by aiming at the application of the optical backscatter reflectometer sensors to monitor the behavior of reinforced concrete slabs under different loads, where the loading process was carried out until failure of the concrete slab, which has dimensions of 5.6 m in length, 1.6 m in width and 28.5 cm in thickness. A 50-m spread of optical fibers (sensors) was installed on the upper and lower surfaces of the concrete slabs at four stretches. The whole length of the optical fiber was coated with a polymer to insulate it from the ambient environment. The sensing process was conducted in a continuous mode during the loading process, and compared with the visual inspection. The results obtained indicated that optical fibers are not only able to detect the occurrence of cracks that are hardly observable but also, they were working perfectly until the ultimate loads (generation of cracks with a width of 0.1 cm).

Radiation sensing was utilized to assess X-ray penetration capabilities through cement mortar and the effects of the used aggregates on the shielding properties of cement mortar. Ling et al. (2013) prepared disc mortar samples using six different types of aggregates, namely river sand, fine stones, beverage glass, funnel glass-untreated, funnel glass-treated, and barite. These samples were subjected to an X-ray from one side, and the penetrated ray was measured from the other side. The obtained results showed that at X-ray energy of 140 kVp, the best shielding properties were provided by funnel glass-untreated, funnel glass-treated, and barite, while other types of aggregates provided less shielding efficiency.

Various wireless communications protocols and technologies are used to improve the sustainability of structures and the environment. One of the examples is the wireless sensing system used by Chae et al. (2012) to monitor the conditions of a long-suspended bridge (550 m) located in Korea. In this study, 45 sensors of two types, ZigBee sensors (for short distances) and CDMA (Code-division multiple access) sensors, were installed on the mentioned bridge, and connected together using two different combinations (U-Node and U-Gateway). The sensing system was operated for 3 months continuously using solar panels. The obtained results indicated that the new sensing system minimized the data loss to less than 10%, and it decreased the monitoring cost from $128,000, for the currently used system, to only $12,000 because the wireless system eliminated the need of cables and their related maintenance costs.

In summary, the above short review of sensor applications in civil engineering showed that sensor technology provides early warnings about the invisible health conditions of structures that help engineers to maintain/replace the damaged element before it developed to a complete failure, which extends the life span of the structure, prevents the sudden failure events, or enables to evacuate people before the failure event takes place to avoid lives losses, that is, make structures more sustainable.

17.5.2 Environmental engineering applications

In the field of environmental engineering, sensing technology has been used in a variety of applications, such as monitoring of climate change, global warming, pollution level monitoring, the concentration of pollutants, noise pollution, and indoor vibration levels.

Tzeng and Wey (2011) investigated the possibility of online monitoring of the indoor air quality using ZigBee wireless system in combination with air quality sensors. The study involved the measurement of three main parameters, including the temperature, humidity, and carbon dioxide. Six sensors were used to achieve the goal of this study. The results obtained showed that the used sensing unit can reliably detect the concentration of carbon dioxide gas, while other gaseous pollutants were not precisely detected. Thus, the authors claimed that the suggested new sensing system is a low-cost, low-power consumption alternative for indoor detection of carbon dioxide gas.

In addition, monitoring of the urban air quality is a very expensive process due to the elevated costs of sampling and testing procedures. Thus, Brienza et al. (2015) in their study developed a sensing system that utilizes several sets of Libelium Waspmote sensors to measure air pollution indicators (carbon dioxides, nitrogen oxides, ozone, temperatures, and humidity) at three places in Lucca city for a month. This sensing system was operated using a rechargeable battery (6600 mAh) and connected to the diagnostic tool via WiFi. The collected data were compared with official data in Lucca city. The outcomes of this study showed that this simple, low-cost, eco-friendly sensing system yielded results very similar to officially recorded data.

Another application of wireless systems was presented by Amruta and Satish (2013). In their study, the authors aimed to monitor the water quality in real time. The monitoring process involved three parameters: pH level, turbidity, and dissolved oxygen concentration. The pH level was sensed using pH probe-IH20, dissolved oxygen concentration was estimated using a redox sensor, while the turbidity was measured using a commercial sensor. The whole system was operated using a solar panel to minimize carbon emissions. The authors demonstrated that the used wireless system provided a reliable assessment of water quality (basing on the studied parameters). Additionally, this sensing unit provided a rapid prediction of any sudden changes in water quality (pollution). Hence, the authors claimed that this sensing unit is a dependable monitoring unit and it has no carbon emission, with low energy consumption.

In addition, microwave spectroscopy was used to monitor the Geosmin presence in the water. Geosmin causes a wide range of problems even when present at low concentrations (4 ng/L), such as odors. Thus, quantifying the Geosmin level is an essential step in water treatment or water quality assessment. In their study, authors, Ryecroft et al. (2019) used microwave spectroscopy to detect Geosmin in water. This sensing technology was applied to water samples that contain four different concentrations of Geosmin, ranging from 5 ng/L to 1 mg/L. The sensing system was also applied for a pure water sample (for comparison samples). The sensing system was tested at different frequencies (5.4–7.5 GHz). The results of this study showed that this sensing system could detect Geosmin at microwave frequencies in the range of 6.4–6.5 GHz.

Satellite technology has been employed in the study conducted by Thompson et al. (2014) to monitor the variation in water quality in Whitsunday Islands, Australia. This study focused on the variation in the concentrations of the total suspended solids and chlorophyll-a during low and high rainfall seasons, from 2002 to 2012. The collected satellite data were very efficient in predicting the variation in water quality with the rainfall intensities. Generally, the results of the satellite sensing revealed that chlorophyll-a concentration reached high levels during the wet seasons (3 months), while the concentrations of the total suspended solids were high even in the dry season.

Biological threats are the factors that directly affect the health and even the lives of humans and/or any living thing as well as environmental sustainability overall. The significant outbreaks of biological diseases are among the highest impact risks faced by society, threatening lives and causing disruption to public services and the economy. Teng et al. (2019), utilized electromagnetic spectroscopy to develop a non-destructive sensor to detect hazardous biological materials. The microwave cavity resonator was utilized to analyze sample solutions containing various concentrations of DPA in NaOH and DMSO (Dimethyl sulfoxide). The results showed that the varied concentrations of DPA in NaOH responded to microwaves in a linear trend at 2.45 GHz and 4.58 GHz frequencies with regression of $R^2 = 0.9201$ and 0.9215, respectively. However, DPA in DMSO at 2.36 GHz and 4.65 GHz demonstrated a linear regression of $R^2 = 0.999$ and 0.828, respectively. The electromagnetic sensor showed a high potential to be used for the detection of hazardous biological materials in a non-destructive and real-time manner.

Food pollution is another serious problem that attracted significant global attention. Thus, the study conducted by Muradov et al. (2019) was to detect plastic pieces in cheese that could easily go undetected through the traditional detection systems. This method utilized electromagnetic patch antennas to detect plastic pieces (of different sizes; 0.1 × 1 cm, 0.2 × 1.5 cm, and 0.5 × 2 cm) in cheese. The tests were conducted at different frequencies; 1–6 GHz. The obtained results were very promising, where it was found that the patch antenna could detect the pieces of Polyvinyl Chloride plastic at a frequency of 4 GHz with reliability as high as 95%.

In conclusion, the above brief literature review confirms that usage of sensors in the environmental engineering helps to provide accurate real-time monitoring of environmental pollution and depletion of natural resources, which substantially enhances the ability of engineers to take the necessary actions/procedures to prevent the deterioration of environmental health and thereby ensure a sustainable environment.

17.6 Chapter summary

In this chapter, the authors presented an overview of the sensing technologies used in the monitoring of both structural and environmental health conditions for direct and/or indirect sustainability improvement. Initially, a validation for the need to use accurate sensing systems in monitoring the structural and environmental health was presented. Afterward, a general classification for the commonly used sensing methods in the civil engineering field for structural and environmental health monitoring, their working principles, installation techniques, and applications were discussed. Then, various examples of the recent applications of the sensing methods were presented, and it was complemented with the advantages of employing the sensing technologies in the monitoring of structures, natural resources, and environmental pollution (indoors and outdoors pollution). The role of sensor technologies in the improvement of sustainability was also highlighted.

The present overview highlighted key facts; first, sensing technologies have been successfully employed in the civil and environmental applications. Second, sensing technologies significantly decrease the monitoring cost as it eliminates the need for the actual observations that are time and sources consuming. Third, due to the ability of the sensing technologies to provide precise real-time data about the unseen damages in the structures, such as corrosion of steel bars, to avoid sudden failures that cause heavy losses in the economy and human lives. Similarly, the correct application of sensing methods in monitoring environmental health helps to manage natural disasters and to avoid serious pollution incidents. Furthermore, the application of wireless sensing methods enhances the cost-effectiveness of the sensing technology due to cost reduction in the installation and maintenance. Finally, one of the most attractive advantages of sensing technology is its ability to carry out real-time continental surveying for environmental and hydrological changes, which can never be achieved by other methods.

References

Abdulhadi, B.A., Kot, P., Hashim, K.S., Shaw, A., Khaddar, R.A., 2019. Influence of current density and electrodes spacing on reactive red 120 dye removal from dyed water using electrocoagulation/electroflotation (EC/EF) process, Proc. First International Conference on Civil and Environmental Engineering Technologies (ICCEET). Iraq, 584. University of Kufa, pp. 12–22.

Abdulredha, M., Al Khaddar, R., Jordan, D., 2017. Hoteliers' attitude towards solid waste source separation through mega festivals: a pilot study in Karbala, International Conference for Doctoral Research.

Abdulredha, M., Al Khaddar, R., Jordan, D., Kot, P., Abdulridha, A., Hashim, K., 2018. Estimating solid waste generation by hospitality industry during major festivals: a quantification model based on multiple regression. Waste Manage 77, 388–400.

Abdulredha, M., Rafid, A., Jordan, D., Hashim, K., 2017. The development of a waste management system in Kerbala during major pilgrimage events: determination of solid waste composition. Procedia Eng 196, 779–784.

Acharya, T., Lee, D.H., 2019. Remote sensing and geospatial technologies for sustainable development: a review of applications. Sens. Mater 31, 3931–3945.

Al-Jumeily, D., Hashim, K., Alkaddar, R., Al-Tufaily, M., Lunn, J., 2019. Sustainable and environmental friendly ancient reed houses (inspired by the past to motivate the future), Proc. 11th International Conference on Developments in eSystems Engineering (DeSE). Cambridge, UK, 214–219.

Al-Saati, N.H., Hussein, T.K., Abbas, M.H., Hashim, K., Al-Saati, Z.N., Kot, P., Sadique, M., Aljefery, M.H., Carnacina, I., 2019. Statistical modelling of turbidity removal applied to non-toxic natural coagulants in water treatment: a case study. Desalin. Water Treat 150, 406–412.

Alshandah, M., Huang, Y., Gao, Z., Lu, P., 2020. Internal crack detection in concrete pavement using discrete strain sensors. J. Civ. Struct. Health Monit., 1–12.

Amruta, M.K., Satish, M.T., 2013. Solar powered water quality monitoring system using wireless sensor network, Proc. International Multi-Conference on Automation, Computing, Communication, Control and Compressed Sensing (iMac4s). Kerala, India. IEEE, pp. 281–285.

Bao, Y., Huang, Y., Hoehler, M.S., Chen, G., 2019. Review of fiber optic sensors for structural fire engineering. Sensors 19, 877.

Barrias, A., Casas, J.R., Villalba, S., 2016. A review of distributed optical fiber sensors for civil engineering applications. Sensors 16, 748.

Bazzucchi, F., Restuccia, L., Ferro, G.A., 2018. Considerations over the Italian road bridge infrastructure safety after the Polcevera viaduct collapse: past errors and future perspectives. Frattura e Integrita Strutturale 12, 400–421.

Brienza, S., Galli, A., Anastasi, G., Bruschi, P., 2015. A low-cost sensing system for cooperative air quality monitoring in urban areas. Sensors 15, 12242–12259.

Cai, B., Wang, J., He, J., Geng, Y., 2016. Evaluating CO_2 emission performance in China's cement industry: an enterprise perspective. Appl. Energy 166, 191–200.

Cao, G., Yang, L., Liu, L., Ma, Z., Wang, J., Bi, J., 2018. Environmental incidents in China: lessons from 2006 to 2015. Sci. Total Environ. 633, 1165–1172.

Cawley, P., 2018. Structural health monitoring: closing the gap between research and industrial deployment. Struct. Health Monit 17, 1225–1244.

Chae, M., Yoo, H., Kim, J., Cho, M., 2012. Development of a wireless sensor network system for suspension bridge health monitoring. Automat. Constr 21, 237–252.

Chen, H.-P., Ni, Y.-Q., 2018. Structural Health Monitoring of Large Civil Engineering Structures. Wiley Online Library.

Drafts, B., 2001. Acoustic wave technology sensors. IEEE Trans. Microw. Theory Tech 49, 795–802.

Duarte, K., Justino, C.I., Freitas, A.C., Gomes, A.M., Duarte, A.C., Rocha-Santos, T.A., 2015. Disposable sensors for environmental monitoring of lead, cadmium and mercury. TrAC Trends Anal. Chem. 64, 183–190.

Dunbabin, M., Marques, L., 2012. Robots for environmental monitoring: significant advancements and applications. IEEE Robot. Automat. Mag. 19, 24–39.

Emamjomeh, M.M., Mousazadeh, M., Mokhtari, N., Jamali, H.A., Makkiabadi, M., Naghdali, Z., Hashim, K.S., Ghanbari, R., 2020. Simultaneous removal of phenol and linear alkylbenzene sulfonate from automotive service station wastewater: optimization of coupled electrochemical and physical processes. Sep. Sci. Technol. 55, 3184–3194.

Enckell, M., Glisic, B., Myrvoll, F., Bergstrand, B., 2011. Evaluation of a large-scale bridge strain, temperature and crack monitoring with distributed fibre optic sensors. J. Civ. Struct. Health Monit. 1, 37–46.

Ghauri, B., Zaidi, A., 2011. Application of Remote Sensing in Environmental Studies, Proc. Second International. Conference. On. Aerospace Science & Engineering. Islamabad, Pakistan, 1–8.

Gkantou, M., Muradov, M., Kamaris, G.S., Hashim, K., Atherton, W., Kot, P., 2019. Novel electromagnetic sensors embedded in reinforced concrete beams for crack detection. Sensors 19, 5175–5189.

Glisic, B., 2013. Fiber optic sensors for subsea structural health monitoring. In: Watson, J., Zielinski, O. (Eds.), Subsea Optics and Imaging. Elsevier, Amsterdam, pp. 434–470.

Goel, P., 2006. Water pollution: causes, effects and control. Delhi: New Age International.

Grosse, C.U., Gehlen, C., Glaser, S.D., 2007. Sensing methods in civil engineering for an efficient construction management. In: Grosse, C.U. (Ed.), *Advances in Construction Materials* 2007. Springer, Cham, pp. 549–561.

Hashim, K., Kot, P., Zubaid, S., Alwash, R., Al Khaddar, R., Shaw, A., Al-Jumeily, D., Aljefery, M., 2020. Energy efficient electrocoagulation using baffle-plates electrodes for efficient *Escherichia Coli* removal from wastewater. J. Water Process Eng. 33, 101079–101086.

Hashim, K.S., Ali, S.S.M., AlRifaie, J.K., Kot, P., Shaw, A., Al Khaddar, R., Idowu, I., Gkantou, M., 2020. *Escherichia coli* inactivation using a hybrid ultrasonic–electrocoagulation reactor. Chemosphere 247, 125868–125875.

Hashim, K.S., AlKhaddar, R., Shaw, A., Kot, P., Al-Jumeily, D., Alwash, R., Aljefery, M.H., 2020. Electrocoagulation as an eco-friendly river water treatment method. In: Al Khaddar, R., Singh, R.K., Dutta, S., Kumari, M. (Eds.), Advances in Water Resources Engineering and Management. Springer, Cham, pp. 219–235.

Hashim, K.S., Al-Saati, N.H., Alquzweeni, S.S., Zubaidi, S.L., Kot, P., Kraidi, L., Hussein, A.H., Alkhaddar, R., Shaw, A., Alwash, R., 2019. Decolourization of dye solutions by electrocoagulation: an investigation of the effect of operational parameters, Proc. First International Conference on Civil and Environmental Engineering Technologies (ICCEET). Iraq, 584. University of Kufa, pp. 25–32.

Hashim, K.S., Al-Saati, N.H., Hussein, A.H., Al-Saati, Z.N., 2018. An investigation into the level of heavy metals leaching from canal-dreged sediment: a case study metals leaching from dreged sediment, Proc. First International Conference on Materials Engineering & Science. Turkey. Istanbul Aydın University (IAU, pp. 12–22.

Hashim, K.S., Hussein, A.H., Zubaidi, S.L., Kot, P., Kraidi, L., Alkhaddar, R., Shaw, A., Alwash, R., 2019. Effect of initial pH value on the removal of reactive black dye from water by electrocoagulation (EC) method, Proc. 2nd International Scientific Conference. Iraq. Al-Qadisiyah University, pp. 12–22.

Hashim, K.S., Idowu, I.A., Jasim, N., Al Khaddar, R., Shaw, A., Phipps, D., Kot, P., Pedrola, M.O., Alattabi, A.W., Abdulredha, M., 2018. Removal of phosphate from river water using a new baffle plates electrochemical reactor. MethodsX 5, 1413–1418.

Hashim, K.S., Khaddar, R.A., Jasim, N., Shaw, A., Phipps, D., Kot, P., Pedrola, M.O., Alattabi, A.W., Abdulredha, M., Alawsh, R., 2019. Electrocoagulation as a green technology for phosphate removal from river water. Sep. Purif. Technol 210, 135–144.

Hensley, S., Farr, T., 2013. 3.3 Microwave remote sensing and surface characterization. Treatise Geomorphol 3, 43–79.

Henton, I., Ramsay, C., 2015. Environment and health: Scottish Environmental Incident Surveillance System (SEISS) 2014 Annual Report. HPS Weekly Rep 49, 117–125.

Hill, M.K., 2020. Understanding environmental pollution. Cambridge University Press, UK.

Holt, M., 2000. Sources of chemical contaminants and routes into the freshwater environment. Food Chem. Toxicol. 38, S21–S27.

Hu, Y., Cheng, H., 2013. Water pollution during China's industrial transition. Environ. Dev 8, 57–73.

I. Development, 2019. Implementing the Sustainable Development Goals. Cabinet Office, UK.

Idowu, I.A., Atherton, W., Hashim, K., Kot, P., Alkhaddar, R., Alo, B.I., Shaw, A., 2019. An analyses of the status of landfill classification systems in developing countries: sub Saharan Africa landfill experiences. Waste Manage 87, 761–771.

Jahanshahi, M.R., Shen, W.-M., Mondal, T.G., Abdelbarr, M., Masri, S.F., Qidwai, U.A., 2017. Reconfigurable swarm robots for structural health monitoring: a brief review. Int. J. Intell. Robot. Appl. 1, 287–305.

Jayaswal, K., Sahu, V., Gurjar, B., 2018. Water pollution, human health and remediation. In: Bhattacharya, S., Gupta, A.B., Gupta, A., Pandey, A. (Eds.), Water Remediation. Springer, Cham, pp. 11–27.

Karim, A., Andersson, J.Y., 2013. Infrared detectors: advances, challenges and new technologies, IOP Conference Series: Materials Science and Engineering, 51, 012001.

Kibert, C.J., 2016. Sustainable construction: green building design and delivery. John Wiley & Sons.

Koch, C., Georgieva, K., Kasireddy, V., Akinci, B., Fieguth, P., 2015. A review on computer vision based defect detection and condition assessment of concrete and asphalt civil infrastructure. Adv. Eng. Inform. 29, 196–210.

Kot, P., Ali, A.S., Shaw, A., Riley, M., Alias, A., 2016. The application of electromagnetic waves in monitoring water infiltration on concrete flat roof: the case of Malaysia. Constr. Build. Mater 122, 435–445.

Kot, P., Markiewicz, J., Muradov, M., Lapinski, S., Shaw, A., Zawieska, D., Tobiasz, A., Al-Shamma'a, A., 2020. Combination of the photogrammetric and microwave remote sensing for cultural heritage documentation and preservation–preliminary results. Int. Arch. Photogramm. Remote Sens. Spat. Inf. Sci. 43, 1409–1413.

Kot, P., Shaw, A., Riley, M., Ali, A., Cotgrave, A., 2017. The feasibility of using electromagnetic waves in determining membrane failure through concrete. Int. J. Civ. Eng. 15, 355–362.

Krohn, D.A., MacDougall, T., Mendez, A., 2014. Fiber Optic Sensors: Fundamentals and Applications. Spie Press, Bellingham, WA.

Latha, N.A., Murthy, B.R., Kumar, K.B., 2016. Distance sensing with ultrasonic sensor and Arduino. Int. J. Adv. Res. Ideas Innov. Technol. 2, 1–5.

Lau, K., 2014. Structural health monitoring for smart composites using embedded FBG sensor technology. Mater. Sci. Technol 30, 1642–1654.

Lee, M., Kloog, I., Chudnovsky, A., Lyapustin, A., Wang, Y., Melly, S., Coull, B., Koutrakis, P., Schwartz, J., 2016. Spatiotemporal prediction of fine particulate matter using high-resolution satellite images in the Southeastern US 2003–2011. J. Expo. Sci. Environ. Epidemiol. 26, 377–384.

Leung, C.K., Wan, K.T., Inaudi, D., Bao, X., Habel, W., Zhou, Z., Ou, J., Ghandehari, M., Wu, H.C., Imai, M., 2015. Optical fiber sensors for civil engineering applications. Mater. Struct 48, 871–906.

Lim, S.M., Wilmshurst, T., Shimeld, S., 2010. Blowing in the wind-legitimacy theory, an environmental incident and disclosure, Proc. *Asia Pacific Interdisciplinary Research in Accounting Conference (APIRA)*. Sydney, Australia, 1–33.

Ling, T.-C., Poon, C.-S., Lam, W.-S., Chan, T.-P., Fung, K.K.-L., 2013. X-ray radiation shielding properties of cement mortars prepared with different types of aggregates. Mater. Struct. 46, 1133–1141.

Liu, J., Zhang, S., Wagner, F., 2018. Exploring the driving forces of energy consumption and environmental pollution in China's cement industry at the provincial level. J. Clean. Prod 184, 274–285.

Luke, T.W., 2010. Power loss or blackout: the electricity network collapse of August 2003 in North America. Disrupted cities: When Infrastructure Fails, 55–68.

Luo, X., Yang, J., 2019. A survey on pollution monitoring using sensor networks in environment protection. J. Sensors 2019 Article ID 6271206.

MailOnline. (2020, 29/08/2020). *Thousands of UK bridges are 'sub-standard', at risk of collapse and will cost almost £1billion to repair, experts warn after the tragedy in Genoa.*

Malik, M.S., Shukla, J.P., 2014. Estimation of soil moisture by remote sensing and field methods: a review. Int. J. Remote Sens. Geosci. 3, 21–27.

Markiewicz, J., Łapiński, S., Kot, P., Tobiasz, A., Muradov, M., Nikel, J., Shaw, A., Al-Shamma'a, A., 2020. The quality assessment of different geolocalisation methods for a sensor system to monitor structural health of monumental objects. Sensors 20, 2915.

Mengüç, Y., Yang, S.Y., Kim, S., Rogers, J.A., Sitti, M., 2012. Gecko-inspired controllable adhesive structures applied to micromanipulation. Adv. Funct. Mater. 22, 1246–1254.

Mitra, M., Gopalakrishnan, S., 2016. Guided wave based structural health monitoring: a review. Smart Mater. Struct. 25, 053001.

Montero, R., Victores, J.G., Martinez, S., Jardón, A., Balaguer, C., 2015. Past, present and future of robotic tunnel inspection. Automat. Constr 59, 99–112.

Muradov, M., Kot, P., Ateeq, M., Abdullah, B., Shaw, A., Hashim, K., Al-Shamma'a, A., 2019. Real-time detection of plastic shards in cheese using microwave-sensing technique, In: Multidisciplinary Digital Publishing Institute Proceedings, 42, 54.

Myung, H., Jeon, H., Bang, Y, 2014. Sensor Technologies for Civil Infrastructures, Volume 1: Sensing Hardware and Data Collection Methods for Performance Assessment.

Myung, H., Jeon, H., Bang, Y.-S., Wang, Y., 2014. Robotic sensing for assessing and monitoring civil infrastructures. In: Wang, M., Lynch, J., Sohn, H. (Eds.), Sensor Technologies for Civil Infrastructures. Elsevier, Amsterdam, pp. 410–445.

N. SCIENCE. Anatomy of an Electromagnetic Wave, https://science.nasa.gov/ems/02_anatomy (Accessed 28 August 2020).

Nansai, S., Mohan, R.E., 2016. A survey of wall climbing robots: recent advances and challenges. Robotics 5, 14.

Nejat, P., Jomehzadeh, F., Taheri, M.M., Gohari, M., Majid, M.Z.A., 2015. A global review of energy consumption, CO_2 emissions and policy in the residential sector (with an overview of the top ten CO_2 emitting countries). Renew. Sustain. Energy Rev. 43, 843–862.

Obiechina, G., Joel, R.R., 2018. Water pollution and environmental challenges in Nigeria. Educ. Res. Int 7, 109–117.

Ochsendorf, J., 2005. Sustainable engineering: the future of structural design, Proc. *Structures Congress* 2005: *Metropolis and Beyond*, 1–9.

Okada, T., Sanemori, T., 1987. MOGRER: A vehicle study and realization for in-pipe inspection tasks. IEEE Journal on Robotics and Automation, 3, 573–582.

Omar, T., Nehdi, M.L., 2017. Remote sensing of concrete bridge decks using unmanned aerial vehicle infrared thermography. Automat Construct 83, 360–371.

Omar, T., Nehdi, M.L., 2018. Condition assessment of reinforced concrete bridges: current practice and research challenges. Infrastructures 3, 1–23.

Omran, I.I., Al-Saati, N.H., Hashim, K.S., Al-Saati, Z.N., Patryk, K., Khaddar, R.A., Al-Jumeily, D., Shaw, A., Ruddock, F., Aljefery, M., 2019. Assessment of heavy metal pollution in the Great Al-Mussaib irrigation channel. Desalin. Water Treat 168, 165–174.

Otutu, F., Agba, S., 2003. Experience with tripod beta methodology applied to serious environmental incident investigation, Proc. *Middle East Oil Show*. Bahrain. Society of Petroleum Engineers.

Park, H.J., Sohn, H., Yun, C.B., Chung, J., Lee, M., 2011. Application of a laser-based wireless active sensing system to structural health monitoring, Proc. 1st International Conference on Smart Structures and Systems. Seoul, South Korea, 1–10.

Patel, A., Singh, S., 2013. Remote Sensing: Principles and Applications. Scientific Publishers.

Preker, A.S., Adeyi, O.O., Lapetra, M.G., Simon, D.-C., Keuffel, E., 2016. Health care expenditures associated with pollution: exploratory methods and findings. Ann. Glob. Health 82, 711–721.

Qin, X., Gao, F., Chen, G., 2012. Wastewater quality monitoring system using sensor fusion and machine learning techniques. Water Res 46, 1133–1144.

Ramachandra, T., Bharath, H., Kulkarni, G., Han, S.S., 2018. Municipal solid waste: generation, composition and GHG emissions in Bangalore, India. Renew. Sustain. Energy Rev 82, 1122–1136.

Ravindra, K., Kaur, K., Mor, S., 2015. System analysis of municipal solid waste management in Chandigarh and minimization practices for cleaner emissions. J. Clean. Prod 89, 251–256.

Ryecroft, S., Shaw, A., Fergus, P., Kot, P., Hashim, K., Moody, A., Conway, L., 2019. A first implementation of underwater communications in raw water using the 433 MHz frequency combined with a bowtie antenna. Sensors 19, 1813–1823.

Ryecroft, S.P., shaw, A., Fergus, P., Kot, P., Hashim, K., Conway, L., 2019. A novel gesomin detection method based on microwave spectroscopy, Proc. 12th International Conference on Developments in eSystems Engineering (DeSE). Kazan, Russia, 429–433.

Ryecroft, S.P., Shaw, A., Fergus, P., Kot, P., Muradov, M., Moody, A., Conroy, L., 2018. Requirements of an underwater sensor-networking platform for environmental monitoring, 2018, Proc. 11th International Conference on Developments in ESystems Engineering (DeSE), 95–99.

Sanfilippo, S., 2011. Hall probes: physics and application to magnetometry, arXiv preprint arXiv:1103.1271.

Shubbar, A.A., Sadique, M., Shanbara, H.K., Hashim, K., 2020. The development of a new low carbon binder for construction as an alternative to cement. In: Shukla, S.K, Barai, S.K, Mehta, A. (Eds.), Advances in Sustainable Construction Materials and Geotechnical Engineeringfirst ed. Springer, Berlin, pp. 205–213.

Singh, R.L., Singh, P.K., 2017. Global environmental problems. In: Singh, R.L. (Ed.), Principles and Applications of Environmental Biotechnology for a Sustainable Future. Springer, Cham, pp. 13–41.

Sony, S., Laventure, S., Sadhu, A., 2019. A literature review of next-generation smart sensing technology in structural health monitoring. Struct. Cntrl. Health Monit. 26, e2321.

Spencer Jr, B.F., Cho, S., 2011. Wireless smart sensor technology for monitoring civil infrastructure: technological developments and full-scale applications, Proc. World Congress on Advances in Structural Engineering and Mechanics (ASEM'11). Seoul, South Korea, 1–28.

Sun, M., Staszewski, W., Swamy, R., 2010. Smart sensing technologies for structural health monitoring of civil engineering structures. Adv. Civ. Eng. 2010.

Teng, K.H., Kot, P., Muradov, M., Shaw, A., Hashim, K., Gkantou, M., Al-Shamma'a, A., 2019. Embedded smart antenna for non-destructive testing and evaluation (NDT&E) of moisture content and deterioration in concrete. Sensors 19, 547–559.

Thompson, A., Schroeder, T., Brando, V.E., Schaffelke, B., 2014. Coral community responses to declining water quality: Whitsunday Islands, Great Barrier Reef, Australia. Coral Reefs 33, 923–938.

Tobiasz, A., Markiewicz, J., Łapiński, S., Nikel, J., Kot, P., Muradov, M., 2019. Review of methods for documentation, management, and sustainability of cultural heritage. case study: Museum of King Jan III's Palace at Wilanów. Sustainability 11 (24), 7046.

Tzeng, C.-B., Wey, T.-S., 2011. Design and implement a cost effective and ubiquitous air quality monitoring system based on ZigBee wireless sensor network, Proc. Second International Conference on Innovations in Bio-inspired Computing and Applications. Tzeng, 245–248.

Vardhan, K.H., Kumar, P.S., Panda, R.C., 2019. A review on heavy metal pollution, toxicity and remedial measures: current trends and future perspectives. J. Mol. Liq. 290, 111197.

Varol, M., Gökot, B., Bekleyen, A., 2010. Assessment of water pollution in the Tigris River in Diyarbakır, Turkey. Water Pract. Technol. 5.

Vicente, M., Mínguez, J., González, D.C., 2017. The use of computed tomography to explore the microstructure of materials in civil engineering: from rocks to concrete. In: Halefoglu, A.M. (Ed.), Computed Tomography: Advanced Applications. InTechopen, UK.

Villalba, S., Casas, J.R., 2013. Application of optical fiber distributed sensing to health monitoring of concrete structures. Mech. Syst. Signal Process 39, 441–451.

Wardhana, K., Hadipriono, F.C., 2003. Analysis of recent bridge failures in the United States. J. Perform. Constr. Facil 17, 144–150.

Zhang, J., Liu, C., Sun, M., Li, Z., 2017. An innovative corrosion evaluation technique for reinforced concrete structures using magnetic sensors. Constr. Build. Mater 135, 68–75.

Zhang, Y., Han, S., Liang, D., Shi, X., Wang, F., Liu, W., Zhang, L., Chen, L., Gu, Y., Tian, Y., 2014. Prenatal exposure to organophosphate pesticides and neurobehavioral development of neonates: a birth cohort study in Shenyang, China. Egypt. J. Pet 27, 1275–1290.

Zhao, L., Zhou, Y., Chen, K.-L., Rau, S.-H., Lee, W.-J., 2020. High-speed arcing fault detection: using the light spectrum. IEEE Ind. Appl. Mag 26, 29–36.

Zubaidi, S., Ortega-Martorell, S., Al-Bugharbee, H., Olier, I., Hashim, K.S., Gharghan, S.K., Kot, P., Al-Khaddar, R., 2020. Urban water demand prediction for a city that suffers from climate change and population growth: Gauteng Province case study. Water 12, 1–18.

Zubaidi, S.L., Al-Bugharbee, H., Muhsen, Y.R., Hashim, K., Alkhaddar, R.M., Al-Jumeily, D., Aljaaf, A.J., 2019. The prediction of municipal water demand in Iraq: a case study of Baghdad Governorate, Proc. 12th International Conference on Developments in eSystems Engineering (DeSE). Kazan, Russia, 274–277.

Zubaidi, S.L., Kot, P., Hashim, K., Alkhaddar, R., Abdellatif, M., Muhsin, Y.R., 2019. Using LARS –WG model for prediction of temperature in Columbia City, USA, Proc. First International Conference on Civil and Environmental Engineering Technologies (ICCEET). Iraq, 584. University of Kufa, pp. 31–38.

CHAPTER 18

Sustainable design based on LCA and operations management methods:
SWOT, PESTEL, and 7S

Manel Sansa[a], Ahmed Badreddine[b], Taieb Ben Romdhane[a]

[a]*LISI, Institut National des sciences Appliquées et de Technologie, Université de Carthage, Centre Urbain Nord BP, Tunisie.*
[b]*LARODEC, Institut Supérieur de Gestion de Tunis, Avenue de la liberté, LeBardo, Tunisie*

18.1 Introduction

The concept of sustainable development is becoming a necessity on a global scale. Indeed, current changes and transformations in our planet in terms of advanced technologies and demographical growth are having significant impacts on the wellbeing of current and future generations. This fast growth is threating the environment, the economy, and the society. In addition, the continuous increase of the consumer needs has pushed toward development of energy-intensive products and systems in large quantities to ensure the satisfaction of the consumers without considering drawbacks that affect directly sustainability pillars. For example, we are witnessing overexploitation of natural resources, the extinction of species, and increase of diseases (Dreher, 2004). Economically, we are going through price inflation that resulted an imbalance in economic rates, changes in politics, and increase of social movements (Bowman, et al., 2017). Socially, there is a significant increase in poverty, unemployment, and migration (West, 2015; Danaher, 2017). In this context, the United Nations has developed the 2030 Agenda (United Nations, 2015) that offers a plan of actions with the aim to eradicate poverty to ensure better life conditions for the human mankind.

This plan offers 17 sustainable development goals and 169 targets to ensure the balance between the sustainability pillars. This global awareness has resulted in the emergence of new standards that deal with products, systems, and their impacts (i.e., ISO14000 series (ISO, 2002; ISO, 2004; ISO, 2006; (ISO, 2006); ISO, 2015; ISO, 2011; ISO, 2000; ISO, 2018; ISO, 2016), Social responsibility (ISO, 2010), and Sustainable development of cities (ISO, 2016).

Although these standards propose requirements to deal with the aforementioned challenges but they do not provide methodologies and frameworks to deal with such problems from a sustainability perspective. To this end, in-depth scientific research has been performed to propose a variety of models and approaches to address challenges related to sustainability especially in design stages. We can mention in particular recent researches (Sansa, Badreddine, and Romdhane, 2017; Sansa, Badreddine, and Romdhane, 2019; Yang, Min, Ghibaudo, and Zhao, 2019; Watz and Hallstedt, 2020; Rafiei and Ricardez-Sandoval, 2020).

However, most of proposed solutions are generic and some of them are restricted to specific field in the industry. Hence, it is highly recommended to consider the company's internal and external issues within decisions related to the development of products and systems. This recommendation has been pointed out in the latest version of the quality, environmental, health and safety management systems standards (ISO, 2015a; ISO, 2015b; (ISO, 2018)).

In fact, the design and development of a product cannot be limited to technological options, the chosen scenario must be adapted to the context of the company. The objective is to reduce the probability of selecting an incoherent scenario. For instance, if financial constraints are not considered. The company might face challenges when implementing the chosen solution. In this context, recent studies have been conducted to propose solutions related to the integration of the company's strategies. We have reviewed most relevant works that have been published since 2019. For instance, Bai and Sarkis (2019) have developed a conceptual application including business, environmental, and social factors within the context of reverse logistics management decisions to help organizations to dynamically refine their decision-making quality for consistency and continuous improvement. Malenkov et al. (2019) have proposed a digital model based on cause-effect relationships models to take into account present and future changes in the internal and external environment of enterprises. In addition, Christodoulou and Cullinane (2019) have used a SWOT/PESTEL analysis to identify factors having positive or negative effects on the implementation of a port energy management system. The authors have focused on challenges derived from these factors to achieve successful implementation. On the other hand, Gao et al. (2020) considered external factors as risks and classified them using some of the PESTEL categories (i.e., Economic risks, Technical risks, and environmental risks) to measure the risk level for decision making in projects' investment. Rocha and Caldeira-Pires (2019) have analyzed Brazil's situation regarding life cycle assessment (LCA) and Environmental Product Declaration (EPD) development taking into account its academic, business, public policy, and consumers contexts. These contexts have been considered within SWOT analysis to adopt the principle of conjunction between strengths and opportunities, compensation for weaknesses, and neutralization of threats. Moreover, in the agricultural sector; Tsangas et al. (2020) have combined strategic planning, stakeholders' requirements along with LCA results to improve the ecological performance of products.

On the basis of the related works review and research opportunities, we have focused on the conjunction of different issues that may affect positively and negatively the organization. Our work targets the combination between the system approach and product approach at an early stage of the design process. Our general framework is proposed in Sansa et al. (2019). This chapter handles the strategic level of the proposed model for sustainable design scenarios selection.

Hence, this chapter is organized as follows. Section 18.2 recalls different management tools with a brief justification and of methods' choices and their description. Section 18.3 defines and details the adopted methodology for strategic scenarios generation. Section 18.4 validates our methodology through an illustrative example in a Tunisian company that manufactures photovoltaic panels. Section 18.5 summarizes the results with a brief discussion. The last section is dedicated to conclusions and future directions.

18.2 Methodology

The definition of the company's context means the identification of internal and external factors that impact the company directly or indirectly. These factors affect its ability to provide the best products and services for its customers. On one hand, internal factors are issues that arise from the company's structure, governance, culture, etc. They might have a positive or a negative effect on the company's performance to satisfy its customers. On the other hand, external factors include the whole environment surrounding the organization such as social, political, economic, legal environments. For instance, the current global situation due to COVID19 pandemic has affected industries and lead to economic crisis resulted from business interruptions and shutdowns (McKibbin and Fernando, 2020). This latter study also proved that the poverty rate will increase approximately 10% in the upcoming years. On the other hand, lockdowns due to the same pandemic have caused significant global air pollutions declines (Venter, Aunan, Chowdhury, and Lelieveld, 2020).

Various tools are used to identify these internal and external factors. In Table 18.1, we have selected the most relevant ones to highlight their advantages and drawbacks; a detailed description of these tools can be found in Downey (2007).

On the basis of the advantages and drawbacks of each strategic tool highlighted in Table 18.1, it is obvious that the SWOT analysis is more adapted to our approach as it gives a clear identification of different issues by classifying internal ones into Strengths, Weaknesses, and external ones into Opportunities and Threats. In addition, among the listed methods in Table 18.1, 7S and PESTEL analysis addresses a variety of significant factors that cover the internal and external environment of the company. Hence, we propose to integrate the 7S and PESTEL within the SWOT analysis to target and analyze more in depth the internal and external issues.

In the following subsections, we will present and detail each tool and demonstrate their applicability when combined.

18.2.1 SWOT analysis

SWOT analysis is the acronym of Strengths, Weaknesses, Opportunities, and Threats. It is considered as the most used tool by stakeholders to define their strategy to achieve their goals. It is a very well-known technique that helps stakeholders understand their internal and external issues by classifying them into four main categories as shown in Fig. 18.1. According to a survey performed 4 years ago by a South African enterprise, the SWOT analysis is the most used business tool having 87% of votes(Du Toit, 2016). The SWOT analysis has its origins through the work of business policy academics at Harvard University and other American business schools since the 60s (Hill and Westbrook, 1997). These works state that a successful strategy is based on ensuring the perfect fit between the company's internal advantages and throwbacks (Strengths and weaknesses) and its external situation (Threats and opportunities).

Following these interpretations, this tool is considered powerful because it uncovers the possible opportunities that the company can exploit through the understanding of its strengths. Also, by analyzing the weaknesses that prevent the company from moving forward and developing its resources, stakeholders will be able to overcome the threats that might prevent them from achieving their objectives. The matrix of the SWOT analysis is presented in Fig. 18.1.

18.2.1.1 Strengths

The strengths are the internal positive issues that lead to a successful management of the enterprise. They consist of all human and material resources that promote the image of the company and generate profits. For example, highly qualified and motivated staff are the key to the company's success. In fact,

Table 18.1 Advantages and drawbacks of most known strategic tools.

Strategic tools	Advantages	Drawbacks
SWOT (Brad and Brad, 2015)	It is a widely used simple tool that can be applied to all sectors and it is adapted to all types of organizations or individuals. It guides the users to classify the issues of the company into four categories: Strengths, weaknesses, opportunities, and threats. This classification provides a clearer vision to define the strategy. It is based on both quantitative and qualitative data integrated, which enhances the decision making. From a cost perspective, conducting a SWOT analysis does not require specific technical skills or financial resources	There is no weighting process. In fact, the SWOT provides only a list of issues. However, there is no mechanism to rank the significance of an issue with regard to the other. The identification of the issues' significance is purely subjective. Some ambiguity can be detected in the identification of issues. An opportunity might be simultaneously a threat. For instance, the opportunity to access a foreign market through export might lead to significant costs that can impact significantly the generated profits.
PESTEL (i.e., Political, Economic, Social, Technological, Environmental, Legal) (Kauskale and Geipele, 2017)	It is a simple framework that categorizes the analysis of issues. It is low cost as this tool only requires time to collect the necessary data. It provides an in-depth understanding of external issues.	It only handles the external factors. The factors of the PESTEL analysis are continuously evolving and changing with regard to the current situation. To this end, it must be constantly updated to provide an accurate and valid identification of issues. The data are not always available especially for external issues and data collection is very time-consuming.
Porter's five forces (Supplier power, Buyer power, Competitive rivalry, Threat of substitution, and Threat of new entry) (Porter, 2008)	This framework guides the company to position itself with regard to the competitive market. It deals with customers and suppliers on the basis of the competition aspect.	This framework focuses only on the strengths and competitive factors. This approach cannot be applied to all industries as in some companies, there are more important factors than what this tool offers such as legal and environmental factors.
Value chain analysis	This framework aims at boosting the profit. It can be adapted to any industry. It enables a comparison between the company business model and its competitors.	This tool involves only the economic factors. In terms of application, few experts master the use of this method and it is not popular among all organizations.
7S analysis (i.e., Systems, Strategy, Structure, Share values, Staff, Style, Skills) (Ravanfar, 2015)	This framework considers seven important factors including soft and hard elements. It guides the company to locate its strengths and weaknesses among internal factors.	This tool involves only internal factors. Similar to the SWOT and PESTEL analysis, there is no weighting process to rank the significance of the identified internal factors.
Early warning systems (Reinhardt, 1984).	This system detects or predicts significant events with regard to competitors.	The application of this tool is mainly based on assumptions and predictions. There is a risk of obtaining ineffective analysis.

18.2 Methodology

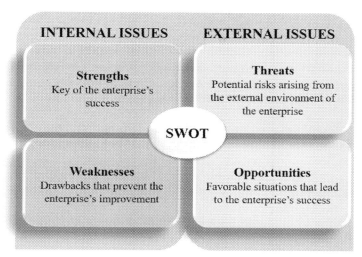

FIG. 18.1 Matrix of the SWOT analysis.

in addition to delivering a product that satisfies the requirements of customers, the quality of provided services (handling of customers' claims, continuous improvement of product's characteristics) is one of the most important elements that enables the enterprise stand within its competitors. The identification of strengths is done through answering questions about advantages and the assets of the company as well as the identification of its successful processes and what distinguish the company from the other competitors.

18.2.1.2 Weaknesses

The weaknesses are critical constraints that prevent the enterprise from achieving its goals.

To this end, the stakeholders must include these factors into their strategy and develop the adequate action plans to convert these weaknesses into strengths and overcome the constraints.

Similar to strengths, weaknesses are internal factors that include the resources of the company.

We can mention in particular a high rate of staff's absence, high import costs, and the limited space in the warehouse. These weaknesses lead to the risk of decrease in the productivity rate and delays in delivery to the customers. The identification of weaknesses is performed by answering questions about the possible lack of staff or equipment and identification of gaps limiting the company's performance in the perspective of achieving desired goals, processes that need improvements, and critical financial constraints facing the company?

18.2.1.3 Threats

These issues are considered as potential risks that might prevent the company from improving its performance. These external factors might cause a significant threat to the company as they affect its resources. These factors are the legal, political and sociocultural constraints. For instance, the implementation of new taxes and strict regulations might result unexpected fines and penalties, which affects the economic state of the company over time. In addition, the current political situation is critical and threatens the stability of workers and increase strikes' rate, and natural disasters might also destroy goods and generate potential environmental impacts (i.e., chemical spills caused by floods). To this end, stakeholders must anticipate these threats and consider them in the definition of their strategy. The identification of threats is done by answering questions about potential competitors, the availability of resources, the risks of future technological advances and innovations on the business performance, and the effects of customers' behaviors on the profits of the company.

18.2.1.4 Opportunities

The opportunities are situations that promote improvement actions and the development of the company. They are related to the external environment and have a positive impact that contributes to the business's success. For instance, the set of new laws encourages sustainability projects and provide grants to implement such projects. Moreover, the increase of the exchange rate affects positively incomes from export activities. It is important to note that this same opportunity can convert into threat in case the company imports raw material. Hence, the list of internal and external factors is not exclusive and it is dynamic according to the enterprise's situation. Usually, most of stakeholders update this analysis annually to track changes and to update the company's strategy. The identification of opportunities is done by answering questions about upcoming events that might be beneficial to the company such as seminars in the same field of activity, the political state and its influence on the economic situation, the location of the company and its influence on the activity, and the effects of new regulations on the progress of the company's activity.

Answering all questions, stakeholders will be able to define a list of the enterprise's strengths, weaknesses, opportunities, and threats to develop into a strategy of improvement.

Since the context analysis has become a requirement of international standards, most competitive industries are investing into their context analysis.

18.2.2 The 7S analysis

Also called the Mckinsey framework, the 7S is a successful tool useful for the company for better understanding of its own internal factors and to plan the adequate actions to overcome these factors. It was designed by an American consulting firm and was then applied in various organizations worldwide. The 7S refers to seven elements that share the same first letter "S" as illustrated in Fig. 18.2.

These elements are considered as a guide to the company to better understand and analyze its internal issues. The identified internal issues are classified according to these elements. The 7S are divided into two categories; hard and soft elements.

In fact, the main difference between both categories is that the soft elements (i.e., Style, Shared values, Skills, and Staff) are included in the organization's culture, whereas hard elements (i.e., Strategy, Structure, and Systems) are factors over which the organization has a direct influence.

18.2.2.1 Soft elements

Style: It includes leadership values and management styles of a successful business.
    Shared values: It includes ethics and standard values that the enterprise vision is based on.
    Skills: Not only employees' skills are concerned but also those related to the enterprise.
    Staff: It concerns the skills of the employees and their roles and responsibilities within the company.

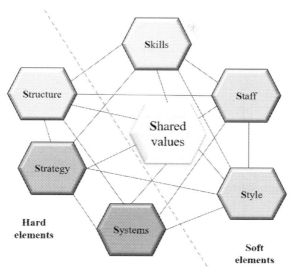

FIG. 18.2 The 7S framework.

18.2.2.2 Hard elements

Strategy: The strategy is the vision and mission of the organization to achieve its desired objectives.

Structure: The structure of the organization covers all the hierarchy and the flowchart of its different departments. Stakeholders must clearly define the hierarchy and distribute roles and authorities accordingly.

Systems: The systems are all methods and tools related to processes' operations, procedures, and communication at the operational level.

Staff: It concerns the skills of the employees and their roles and responsibilities within the company.

18.2.3 The PESTEL analysis

The PESTEL analysis is also an analytic tool to analyze the macroenvironmental factors that have a significant impact on the performance of the company. It is an acronym for Political, Economic, Social, Technological, environmental, and legal factors (See Fig. 18.3).

It is often used to classify external factors (i.e., Opportunities and threats) to better illustrate the situation of the company. In the following, we will define each factor of the PESTEL analysis.

Political factors: Each activity related to the government and has a certain impact on the industry is classified as a political issue such as the government policy, the political state whether stable or unstable, new tax policy, new laws related to labor or the environment, trade restrictions, etc. Organizations must be adapted to the current political state and must anticipate potential future legislation to adjust their strategy accordingly.

Economic factors: The economic situation of the country affects the performance of the organization. These factors include exchange rates, economic growth, interest rates, unemployment rates, and consumers' income. These factors affect the enterprise in the long term as they influence consumers'

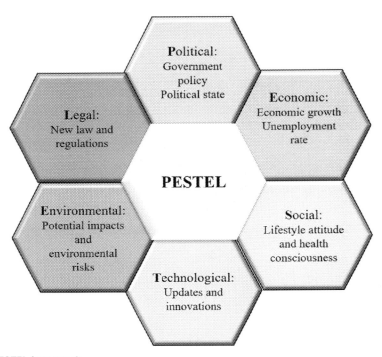

FIG. 18.3 The PESTEL framework.

behavior and may decrease their purchasing rate. Consequently, the price of the product or the service might be affected.

Social factors: The social factors include all issues related to the population. They represent the demographic characteristics, values of the population within which the company operates. We can mention in particular the growth rate of the population, safety emphasis, lifestyle attitude and health consciousness, and cultural barriers. These factors may also include the workforce and its devotion and willingness to work under specific conditions.

Technological factors: These issues are related to technological updates and innovations that may impact the operations of the industry such as technological change, research and development activities, innovations, and technological awareness. These factors have an influence on decisions to launch new products or invest in new equipment. Taking into account the technological development, stakeholders will be able to limit the costs of implementing a new technology that risks to become obsolete due to innovative technological changes worldwide.

Environmental factors: Over recent years, environmental issues have become the most important ones among other factors due to the increase of environmental impacts and significant risks facing mankind caused by the increase in the pollution rate and the decrease in the availability of natural resources. An article published in the website of Diesel service and supply (service and supply, n.d.) stated that 320 billion kilowatt-hours of energy is consumed every day and most of this quantity is obtained by burning fossil fuels, nonrenewable sources of energy. This massive consumption is leading to the depletion of these resources. To this end, environmental consciousness and policy must be initiated on the personal and industrial level. For example, climate change impact significantly industries in fields such as agriculture, tourism, and farming. Moreover, considering the potential impacts that might be generated from the company's activities leads this latter to get involved in sustainability practices.

Legal factors: These factors align with the political ones. However, they are more specific to laws and legislation that cover all interested parties. For example, these laws can be related to employment, protection of the customer, workers' health and safety, and copyrights. Companies must ensure their absolute conformity to the available laws to run their activities successfully and ethically. It is important to note that if the company is operating on the international level, stakeholders must also include the rules and laws of each country involved in the business as each country has its own set of laws and regulations. In addition, stakeholders must perform a regulatory watch periodically to be aware of any potential change in the applicable laws or the publication of new ones.

Our idea is based on integrating these tools to identify relevant and specific issues for the contextualization of design. The following subsection will detail the integration of these tools.

18.2.4 Integration of the SWOT, 7S, and PESTEL

The arguments for our methodological choices are detailed in Sansa et al. (2019). Hence, we recommend improving the deployment of SWOT with a more specific and rigorous classification of issues within a contextualized design framework for sustainable product design proposal sustainability. In fact, Ravanfar (2015) has pointed out that organizations were able to evaluate their effectiveness with regard to the seven success criteria. The author highlighted that this evaluation leads to an efficient management of organizational changes, and to a contextualized identification of new strategies and area for development.

Our methodology is deployed as follows: First, we identify internal and external factors through a deep brainstorming using the SWOT analysis. Each identified issue is affected to an ID (i.e., ST_i for strengths, WE_i, for weaknesses, OP_i for opportunities, and TH_i for threats). Second, ST_i and WE_i are categorized using 7S technique. Simultaneously, OP_i and TH_i are categorized using PESTEL analysis. The main objective is to exploit strengths and opportunities in a way that the company will be able to overcome its weakness and threats.

18.3 Creation of strategic scenarios
18.3.1 Developed algorithms

A confrontation matrix is created to generate four main types of strategic scenarios (Sansa, Badreddine, and Romdhane, 2019) as illustrated in Fig. 18.4.

This mechanism is mainly focusing on environmental, social, and economic thinking, two systems are combined in a synergy around the design process. The central one regroups all the phases of the life cycle of the product (i.e., Extraction of raw materials, Manufacturing, Distribution, Use, End of life), whereas the second represents the context of the organization (i.e., Strategies derived from the SWOT analysis integrating the 7S and the PESTEL technique). Following the process, two types of scenarios are generated on the basis of opportunities (i.e., ASS and AJSS) and the two are formed on the basis of threats (i.e., DSS and SSS) as detailed respectively in the following algorithm.

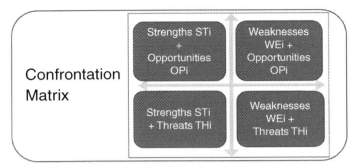

FIG. 18.4 Confrontation matrix.

18.3.2 Web application

On the basis of these algorithms, we have developed our proposed methodology as a web application called DIGITAL BRAIN to guide and assist stakeholders to generate strategic scenarios adapted to their context in a smart and effective way. DIGITAL BRAIN is user friendly and is composed of three main tabs. In this chapter we will focus on the first and last one; Manage issues and Manage scenarios as shown in Fig. 18.5.

First, the user adds the issues by selecting the type of the identified issues (i.e., internal or external), according to the selection, a new field appears to request the classification of the identified issue. If it is internal (external), the user must precise if it is a weakness or a strength (opportunity or a threat). On the basis of the user selection, a third field appears to classify the identified issue using 7S (resp. PESTEL). Once validated, this latter will be automatically inserted into the SWOT matrix. A screenshot to illustrate an example of a SWOT matrix with four added issues is provided in Fig. 18.6.

Once all issues are identified and classified into strengths, weakness, opportunities, and threats, the second tab is dedicated to the generation of strategic scenarios. It consists of creating groups of issues' combinations following Algorithm 18.1. This algorithm has been translated within the app as the Manage scenarios tab. First, the user creates a group referred to as the type of the strategic scenario. Then, runs the algorithm to check strengths and weaknesses for each opportunity and each threat according to the chosen type of scenarios. Fig. 18.7 illustrates an example of a generated strategic scenario.

As an output, a set of strategic scenarios is formed. On the basis of each one, tactics and design criteria are resulted and classified per life cycle phase to identify design alternatives to compute environmental, economic, and social impacts. The generation of alternatives and operational scenarios can be found in Sansa et al. (2019).

18.4 Illustrative example

To illustrate the proposed methodology, we have decided to apply it to a case study of a Tunisian company that manufactures photovoltaic panels and wishes to optimize the design of its panels with the aim to reduce environmental, economic, and social impacts by generating contextualized scenarios adapted to the strengths, weaknesses, threats, and opportunities of the company.

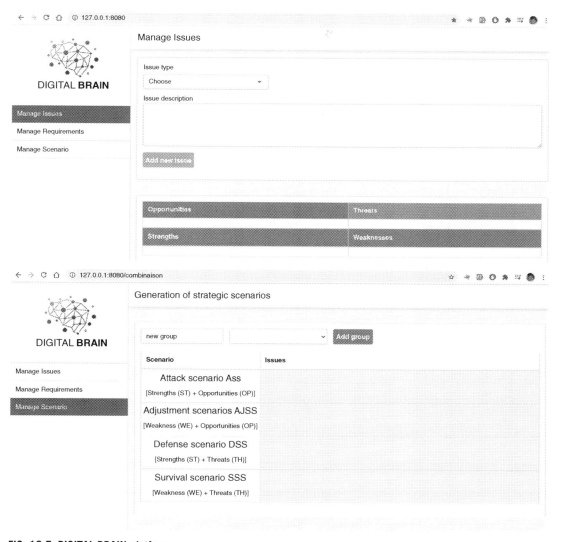

FIG. 18.5 DIGITAL BRAIN platform.

(A) Manage Issues Tab. (B) Manage Scenarios Tab.

The life cycle of a photovoltaic panel is illustrated in Fig. 18.8.

The basic life cycle of a photovoltaic panel is composed of the following five main steps:

LCP_1: *Production of raw materials.* The company imports the panel components from China and Italy. The main components are: solar cells, flux, interconnect ribbon, tempered glass, encapsulant, frame, labels, back sheet, junction box, adhesive, and silicone. The import is handled by shipment and land transportation to the factory site

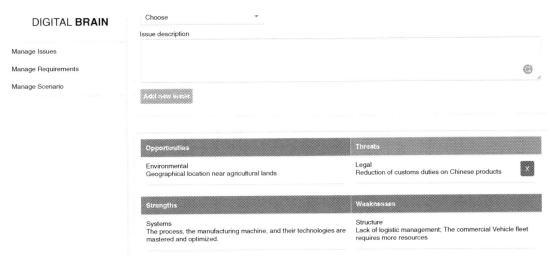

FIG. 18.6 Illustrative example of added issues.

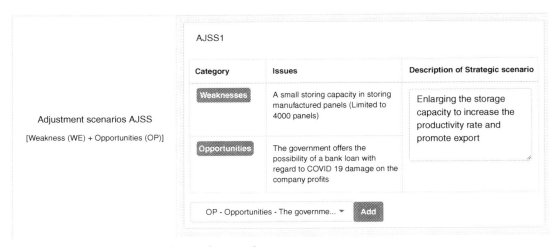

FIG. 18.7 Example of a generated strategic scenario.

LCP_2: *Manufacturing*. The assembly process is composed of five steps. First, the cells are chained and connected using the ribbon and flux to form cell chains. This operation is called stringing. Second, the module layout is layering the module using cell chains, the glass, the Ethylene Vinyl Acetate (EVA), and the back sheet to form module. Third, these layers are laminated and labeled. The fourth step is to frame each module and mount the junction box. Lastly, the performance of each module is tested through sun simulating.

LCP_3: *Distribution and installation*: The company distributes the modules locally to all regions of the country. The company also ensures the installation and offers the installation furniture (i.e., Inverter, AC and DC outlets, and installation accessories).

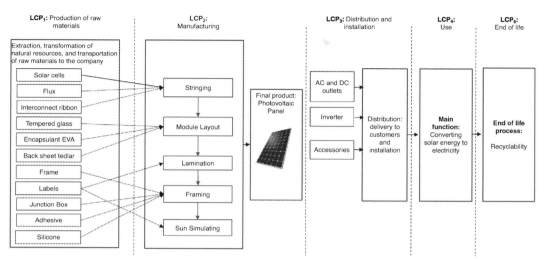

FIG. 18.8 Life cycle phases of a photovoltaic panel.

LCP_4: *The use and maintenance phase*: The utility life of the panel is estimated to be around 25 years of efficiency. Maintenance is required to ensure the performance of the installation especially the panels and the inverter.

LCP_5: *The end of life*: Panels are recyclable, they are sent to recycling centers. Also, the components of the inverter are reusable.

Based on the life cycle of the panel and the manufacturing process of the company, a deep brainstorming is done between the company's stakeholders and managers, and a set of strengths, weaknesses, threats, and opportunities has been detected, identified, and stored in Manage Issues tab. Table 18.2 summarizes the SWOT Matrix (i.e., 12 strengths, 9 weaknesses, 6 opportunities, and 9 threats).

It is important to note that issues are dynamic and might change over time, the current list is not exclusive and it is necessary to update and adjust the list of internal and external issues. For instance, the pandemic of COVID 19 has affected all internal and external factors. In fact, 2 months of lockdown has caused deterioration in profit rate and delay in salary payment which resulted in employees resignment. This external factor has been created in early 2020 and had an influence almost on all internal and external factors.

Once identified and stored in the Manage issues tab, the next step consists on combining these issues using Algorithm 18.1 to create and generate strategic scenarios. The algorithm has been developed and coded within the app and we were able to run it through the Manage Scenarios tab. At this stage, the platform combines the set of strengths, opportunities, weaknesses, and threats and the decision maker is able at this point to define the strategic scenario. To better illustrate the case study, we have simulated eight strategic scenarios (i.e., 3 Attack Scenarios ASS, 1 Adjustment scenario AJSS, 1 Defense scenario DSS, and 3 Survival Scenarios SSS) as follows:

ASS_1: Increasing the budget to enhance Sale department to target customers in the north region by promoting advertisement and highlighting the benefits of the program.

ASS_2: Planning the transformation of the manufacturing process toward the concept of Industry 4.0.

Table 18.2 Strengths, weaknesses, opportunities, and threats.

Strengths	7S	Weaknesses	Opportunities	PESTEL	Threats
ST1: Regular quality check on the product	Systems	WE1: The company is not certified ISO14001	OP1: Implication of the governance through the Tunisian social plan and PROSOL ELEC program	Political	TH1: Governance instability due to COVID19
ST2: The process, the manufacturing machine, and their technologies are mastered and optimized		WE2: A small storing capacity for manufactured panels (limited to 4000 panels)			TH2: 2 Months lockdown resulted in salaries 'payment delay
		WE3: Quality of the panel is at risk due to a defect in the layout machine and a bottleneck in the laminator	OP2: Opportunity in increasing the sales due to the significant increase in electricity costs	Economic	TH3: Critical competition based on quality and price range
		WE4: Lack of adequate means to ensure after sale services. High rate in inverters shutdown	OP3: The government offers the possibility of a bank loan with regard to COVID 19 damage on the company profits		TH4: Constant instability in raw material price range
ST3: Qualified managers and installers	Skills	WE5: Language barrier in the negotiation of raw material purchase prices			TH5: Delay in grant and customers payment
		WE6: Unqualified operators working on new technologies	OP4: The electricity company offers a bank loan for customers wishing to use renewable energy	Social	TH6: Unavailability of high-degree candidates in the site area.
ST4: High efficiency of managers with excellent qualifications	Staff	WE7: A high rate of turnover (>60%)	OP5: Availability of research in field of renewable energy	Technological	TH7: Emergence of new technologies and automated machines
ST5: Young and motivated staff with an age range between 25 and 30					
ST6: The site is located in the center of Tunisia and accessible to all cities	Structure	WE8: Lack of logistic management, the commercial vehicle fleet requires more resources	OP6: Global warming increases the performance of panels	Environmental	TH8: Geographical location near agricultural lands

			Legal	TH8: Reduction of customs duties on Chinese products TH9: Legal constraints related to material import
ST7: Manufacturing costs are optimized ST8: The site is located in a regional development area with benefits of 10 years exoneration from tax ST9: Well-known company in the sector of photovoltaic energy with a strong sale department ST10: Encouraging consumers to adopt renewable energy solutions ST11: Environmental consciousness of stakeholders ST12: A positive collaboration of the team and involvement of top management	Strategy Shared values Style	WE9: Some employees need to spend 1 h 30 min to arrive to work with additional transportation fees		

Algorithm 18.1 Combination of internal and external issues.

Data: ST, OP, WE, TH,
Result: ASS, AJSS, DSS, SSS

1	Begin
2	For each OP do
3	Find the ST leading to OP
4	Find the WE preventing OP
5	Define the sets ASS= [ST+OP]
6	Define the sets AJSS = [WE+OP]
7	For each TH do
8	Find the ST reducing TH
9	Find the WE increasing TH
10	Define the sets DSS= [ST+TH]
11	Define the sets SSS = [WE+TH]
12	End

ASS_3: Update of business plan and investments in improving the panel material.
$AJSS_1$: Enlarging the storage capacity to increase the productivity rate and promote export.
DSS_1: Creation of a head office near the city to reduce additional costs and offer opportunities for high-degree candidates.
SSS_1: Reassignment of tasks distribution, reinforce internal communication, and promoting remote work.
SSS_2: Programming weekly trainings to improve qualifications of operators.
SSS_3: Main focus on improving the quality of the panel and allocation of a budget to ISO14001 certification.

The combination of each scenario is detailed in the Appendix.

18.5 Results and discussion

The main takeaway from the generated scenarios is the possibility of identifying design criteria and classified by life cycle phase derived from the combination of issues. Hence, we can identify design criteria affected to life cycle phases as follows:

Extraction of raw material LCP_1

From SSS_3, it is mandatory to focus on improving the raw material and compare different types of material from an environmental perspective. C_{11}: *Material type*, C_{12}: *Material supply quantity*. In addition, currently, the panels are delivered without packaging. Hence, decision makers have decided to include a criterion to determine the best packaging solution. C_{13}: *Packaging type*, C_{14}: *Packaging supply quantity*.

Manufacturing phase LCP_2

From ASS_2, decision makers can mainly focus on the efficiency of machines in case of changes in the assembly process. C_{21}: *Efficiency of machines*. Also, if the company is aiming toward Industry 4.0. The contribution of operators must be measured. C_{22}: *Operators contribution*. In addition, it is mandatory to estimate the energy consumption in the current situation and using the alternative of adopting the Industry 4.0 concept. C_{23}: *Energy consumption*.

From $AJSS_1$, it is convenient to set a criterion to measure the rate of productivity, in fact, by enlarging the storage capacity. It is possible to increase the target of number of manufactured panels per day. C_{24}: *Productivity rate per functional unit*.

Distribution and installation phase LCP_3

From ASS_1, this scenario encourages the reinforcement of sales to cover all cities. To this end, decision makers propose three criteria to take into account; C_{31}: *Transport distance*, C_{32}: *Transport mode*, C_{33}: *Quantity of delivered panels per functional unit*.

Use and maintenance phase LCP_4

From ASS_3: To check the quality of the panel material. The utility life of the panel must be measured to test the efficiency of decision with regard to the chosen material. C_{41}: *Panel utility life*. Also, it is important to note that the inverter has a mandatory role in ensuring the performance and the functioning of the photovoltaic installation. C_{42}: *Inverter utility life*.

End of life phase LCP_5

From ASS_3, it is possible to improve the panel from environmental perspective by using recycled and reused material. To this end, decision has set two criteria. C_{51}: *Recycling unit*, C_{52}: *Reuse of components*. Also, it is recommended to check which recycling possibility has the best result. C_{53}: *Local recycling unit*, C_{54}: *Recycling center*.

It is important to note that the aforementioned set of identified criteria is continuously changed and updated for more accurate adoption to the context of the company through the generated scenarios. The application of our methodology has helped the company in the identification of specific criteria that are mastered. Moreover, the company was able to eliminate the risk of adopting noncoherent strategy or wrong decisions, which may result in additional costs and time loss.

18.6 Conclusion

The application of our methodology has generated four types of strategic design scenarios that take into account the strengths, weaknesses, threats, and opportunities of the company. They are oriented toward the use of existing resources for supply chain optimization, appropriate technological options, distribution, and use policies, taking into account parallel markets and prices.

The advantages of our model are its ability to create contextualized value within a sustainable development framework. In this context, it can be used as an argument for positive differentiation between marketed products. The final environmental, social, and economic balance sheet of the design scenario can be used as a relevant sales argument for consumers and informed populations.

We have developed our methodology as an online application with the aim to digitalize brainstorming and thinking toward an intelligent platform. This platform is currently under development and improvement targeting a generic application that is able to deal with more complex cases

For future directions, we aim to include the requirements of the stakeholders within the generation of scenarios on the platform and extend it to link all steps of the proposed approach in Sansa et al. (2019) to offer an intelligent tool to organization that wishes to include sustainable practices within their activities.

Acknowledgments

The authors would like to thank the company's managers for accepting to apply the proposed methodology and to share their Data for the project's support.

References

Bai, C., Sarkis, J., 2019. Integrating and extending data and decision tools for sustainable third-party reverse logistics provider selection. Comput. Oper. Res. *110*, 188–207.
Bowman, D.M., Dijkstra, A.F., Guivant, J.S., Konrad, K., Egan, C.S., Woll, S., 2017. The Politics and Situatedness of Emerging Technologies. Ios Press, Amsterdam.
Brad, S., Brad, E., 2015. Enhancing SWOT analysis with triz-based tools to integrate systematic. Procedia Eng. *131*, 616–625.
Christodoulou, A., Cullinane, K., 2019. Identifying the main opportunities and challenges from the implementation of a port energy management system: a SWOT/PESTLE Analysis. Sustainability *11* (21), 46–60.
Danaher, J., 2017. Will life be worth living in a world without work? technological unemployment. Sci. Eng. Ethics *23* (1), 41–64.
Downey, J., 2007. Strategic Analysis Tools: Topic Gateway Series. The Chartered Institute of Management Accounting, London, United Kingdom.
Dreher, K.L., 2004. Health and environmental impact of nanotechnology: toxicological. Toxicol. Sci. *77* (1), 3–5.
Du Toit, A.S., 2016. Using environmental scanning to collect strategic information: a South African survey. Int. J. Inform. Manag. *36* (1), 16–24.
Gao, J., Guo, F., Li, X., Huang, X., Men, H., 2020. Risk assessment of offshore photovoltaic projects under probabilistic linguistic environment. Renew. Energy *163*, 172–187.
Hill, T., Westbrook, R., 1997. SWOT analysis: it's time for a product recall. Long Range Plann. *30* (1), 46–52.
ISO, 2000. Environmental Labeling: General Principles. Geneva.
ISO, 2002. *ISO/TR 14062: Environmental Management - Integrating Environmental Aspects into Product Design and Development*. Geneva.
ISO, 2004. ISO 14001: Environmental Management Systems-Requirements with Guidance for Use. Geneva.
ISO, 2006. ISO 14040: Environmental Management–Life Cycle Assessment-Principles and Framework. Geneva.
ISO, 2006. ISO 14044: Environmental Management–Life Cycle Assessment, Requirements and Guidelines. Geneva.
ISO, 2010. ISO 26000: Guidance on Social Responsibility. Geneva.
ISO, 2011. ISO 14006: Environmental Management–Guidelines for Incorporating Ecodesign. Geneva.
ISO, 2015. ISO 14001: Environmental Management Systems-Requirements with Guidance for Use. Geneva.
ISO, 2015. ISO 9001: Quality Management Systems – Requirements. Geneva.
ISO, 2016. Environmental Labels and Declarations: Self-Declaration Environmental Claims, Terms and Definitions. Geneva.
ISO, 2016. ISO 37101 Sustainable Development in Communities- Management System for Sustainable Development-Requirements and Guidelines for Use. Geneva.
ISO, 2018. Environmental Labels and Declarations: Environmental Labeling Type I, Guiding Principles and Procedures. Geneva.
ISO, 2018. ISO 45001: Occupational Health and Safety Management Systems—Requirements with Guidance for Use. Geneva.
Kauskale, L., Geipele, I., 2017. Integrated approach of real estate market analysis in sustainable development context for decision making. Procedia Eng., 505–512.

Malenkov, Y., Kapustina, I., Kudryavtseva, G., Shishkin, V.I., 2019. Digital modeling of strategic sustainability assessments: new approach, recommendations, prospects, Proc. of the 2019 International SPBPU Scientific Conference on Innovations in Digital Economy. Saint Petersburg, Russia, pp.1–8.

McKibbin, W., Fernando, R., 2020. The Global Macroeconomic Impacts of COVID-19. Seven Scenarios. doi:10.2139/ssrn.3547729.

Porter, M.E., 2008. The five competitive forces that shape strategy. Harv. Bus. Rev. 86 (1), 25–40.

Rafiei, M., Ricardez-Sandoval, L.A., 2020. New frontiers, challenges, and opportunities in integration of design and control for enterprise-wide sustainability. Comput. Chem. Eng. 132, 106610.

Ravanfar, M.M., 2015. Analyzing organizational structure based on 7s model of Mckinsey. Global Journal of Management and Business Research 15, 7–12.

Reinhardt, W., 1984. An early warning system for strategic planning. Long Range Plann. 17 (5), 25–34.

Rocha, M.S., Caldeira-Pires, A., 2019. Environmental product declaration promotion in Brazil: SWOT analysis and strategies. J. Clean. Prod. 235, 1061–1072.

Sansa, M., Badreddine, A., Romdhane, T.B., 2017. A multi-leveled ANP-LCA model for the selection of sustainable design options, IFIP International Conference on Product Lifecycle Management. Sevilla, Spain. Springer, pp. 473–486.

Sansa, M., Badreddine, A., Romdhane, T.B., 2019. A new approach for sustainable design scenarios selection: a case study in a Tunisian company. J.Clean. Prod. 232, 587–607.

Tsangas, M., Gavriel, I., Doula, M., Xeni, F., Zorpas, A.A., 2020. Life cycle analysis in the framework of agricultural strategic development planning in the Balkan Region. Sustainability 12 (5), 1813.

United Nations, 2015. Transforming Our World: The 2030 Agenda for Sustainable Development. United Nations. *[Computer software manual].* https://sustainabledevelopment.un.org/: https://sustainabledevelopment.un.org/content/documents/21252030%20Agenda%20for%20Sustainable%20Development%20web.pdf. (Accessed 29 June 2020).

Venter, Z.S., Aunan, K., Chowdhury, S., Lelieveld, J., 2020. COVID-19 lockdowns cause global air pollution declines. Proc. Natl. Acad. Sci. USA 117, 18984–18990.

Watz, M., Hallstedt, S.I., 2020. Profile model for management of sustainability integration in engineering design requirements. J.Clean. Prod. 247, 119–155.

West, D.M., 2015. What Happens If Robots Take the Jobs? The Impact of Emerging Technologies. Centre for Technology Innovation, Washington DC.

Yang, S., Min, W., Ghibaudo, J., Zhao, Y.F., 2019. Understanding the sustainability potential of part consolidation design supported by additive manufacturing. J.Clean. Prod. 232, 722–738.

CHAPTER 19

The importance of integrating lean thinking with digital solutions adoption for value-oriented high productivity of sustainable building delivery

Moshood Olawale Fadeyi

Sustainable Infrastructure Engineering (Building Services) Programme, Engineering Cluster, Singapore Institute of Technology, Singapore, Singapore

19.1 Introduction

The architecture, engineering, and construction (AEC) industry is experiencing low productivity than many industries (Abdel-Wahab and Vogl, 2011; Tookey, 2011; Vogl and Abdel-Wahab 2015). Why is low productivity in the AEC industry concern for governments? An understanding of economic growth and the relationship with productivity will provide insight into the question. Economic growth is one of the major priorities of any progressive conscious government (Tan and Bhaskaran, 2015). Economic growth aids economic development, which is always a priority because of the need to create more jobs (Basnett and Sen, 2013; Melamed et al. 2011), diversify the economy into several economic friendly industries (Mishrif, and Al Balushi, 2015), retain and expand businesses (Bartik 2003; Phillips, 1996; Zhang and Warner, 2017). It also helps to prevent economic downturn (Ciccarone and Saltari, 2015) and increase tax revenue (Friedman 2003; Gurdal et al., 2021), Quality of lives of citizens will potentially improve with progressive economic development (Deller et al., 2001).

The productivity level of sectors contributing to a country's revenue affects economic growth in the country (Eifert 2009; Warr 2006). Any government that ignores the sectors' productivity levels is bound to experience a continuous economic downturn. Economic growth occurs when there is an increase in the amount of produced goods and services of value per person in the population. More revenue can be generated if more goods and services are generated with prudent use of invested resources. More revenue means additional resources available to be invested to generate more of the goods and services. The digitalization of the AEC industry could improve productivity (Woodhead et al., 2018).

Policy makers are encouraging the adoption of digital solutions in the AEC industry to increase productivity. The Building and Construction Authority (BCA) of Singapore is championing the adoption of digital solutions to streamline work processes and integrate all stakeholders involved in the building delivery process (Hwang et al., 2020). The effective utilization of digital solutions depends on people in charge of planning, monitoring, managing, and overseeing the building delivery process. Project managers of the building delivery process are tasked with these responsibilities. Project management involves planning, monitoring, and managing invested resources, including people, and activities in the building delivery process. Project managers' responsibilities start from project conception, before the construction phase begins, and sometimes continues after project completion and handover (Netscher, 2017).

The planning, monitoring, and management aim to ensure that the usefulness of sustainable building performance and other goods and services produced during the delivery process are maximized from invested resources. In the context of a sustainable building performance delivery, the usefulness is healthy indoor environments and delivery processes that respect natural systems and cycle of life, environmental resources, and people. Respect for people includes consideration for human physiology, sociology, the social and economic limit of acceptability (Fadeyi 2017). The usefulness is for building owners, building occupants and users, the environment, and other stakeholders involved during the building delivery process. The maximization of the usefulness and number of goods and services produced from invested resources is referred to as value-oriented high productivity.

The nonmaximization of the usefulness of goods and services produced from invested resources or production of nonuseful goods and services from invested resources is referred to as waste and low productivity. Eqs. (19.1) and (19.2) show a relationship between invested resources and value and wastes. Project management for value-oriented high productivity requires problem-solving philosophy. At the heart of the problem-solving process is the identification of the root cause of a problem. Digital solutions, as the name suggests, are solutions that can enhance the problem-solving process if used appropriately. However, if digital solutions are used as solutions to a problem instead of the root-cause, value delivery will be compromised. An illustration explaining the need to provide a solution to the root cause of a problem is shown in Fig. 19.1. Unfortunately, the adoption of digital solutions as solutions to a problem instead of a problem's root cause is a common practice in the industry. So, what kind of problem-solving philosophy, knowledge, skills, and approach is needed by project managers?

$$\text{Value} = \frac{\text{Goods and services' performance level that is of usefulness to a consumer}}{\text{Invested resources}} \qquad (19.1)$$

Eq. (19.1) shows the relationship between invested resources and value delivery. In Eqs. (19.1) and (19.2), invested resources may include, money, time, manpower, energy or effort, digital solutions, materials, equipment, etc.

$$\text{Waste} = \frac{\text{Goods and services' performance level that is } not \text{ of usefulness to a consumer}}{\text{Invested resources}} \qquad (19.2)$$

Eq. (19.2) shows relationship between invested resources and waste delivery.

Lean thinking provides the philosophy, knowledge, skills, and approach that can help identify the root cause of a problem and how identified solutions, including digital solutions, could be used to eliminate or reduce the effects of the root causes. This chapter aims to discuss the importance of integrating digital solutions with lean thinking for value-oriented high productivity of sustainable building delivery. The importance of eliminating the root cause of a problem using lean thinking in the building delivery process before investing in digital solutions will be emphasized. The challenges that could hinder the adoption of digital solutions and lean thinking are identified and discussed. The knowledge this chapter aims to pass across to the readers is relevant to the effective digital transformation of the

19.1 Introduction

When attempting to solve a problem, it is essential to understand and appreciate the pain caused by or associated with the problem to motivate the need for a solution. After that make efforts to identify the cause, preferably the root cause, of the problem. Your solution should be targeted toward eradicating the cause and not the problem. Unfortunately, most people focus on providing a solution to a problem instead of the cause of the problem. The existence of a problem and its intensity will persist if the cause of the problem is still active.

FIG. 19.1 Strategy for solving a problem.

(Source: Fadeyi, 2019)

AEC industry for value-oriented high productivity achievement. Practical application of the knowledge could make the AEC industry a source of significant economic contribution to a country.

19.2 Digital solutions and lean thinking adoption

This section provides information about the purpose of digital solutions and lean thinking.

19.2.1 Digital solutions adoption

The typical fragmentation among stakeholders at different stages of the building delivery leads to nonvalued added activities in the building delivery process. Digital solutions can help eliminate this fragmentation and streamline the work process as a medium of integration of people and work processes (Hwang et al., 2020). Sawhney et al. (2020) provide topical issues, opportunities, and current trends relating to digital solutions adoption for productive building delivery. Digital solutions adopted at the design stage can help optimize and coordinate designs that meet the client's regulatory and downstream requirements (Hooper et al. 2010; Sandberg et al. 2019). They can be adopted to translate the design to standardized components for automating off-site production (Li et al. 2019a; Hou et al. 2020; Vernikos et al. 2014).

Digital solutions can be adopted at the construction stage to provide just-in-time delivery, installation, monitoring of on-site activities to maximize productivity and minimize rework (Cheung et al. 2018; Dallasega, 2018; Kiani et al. 2014; Sacks et al. 2003; Wang et al. 2007). There are also opportunities to adopt digital solutions toward efforts to improve productivity at the facility management stage (Rich and Davis, 2010; Lin et al. 2014). Digital solutions can provide real-time monitoring for operations and maintenance to enhance asset values (Lee and Akin, 2011; Shalabi and Turkan, 2017). According to Wyman (2018), the benefits offer by digital solutions can be broadly categorized into three parts, namely: (1) interactive work processes; (2) connected machines, equipment, and workers; and (3) industrialized models.

Interactive work processes include digital solutions for visualization and simulation and dematerialization and reactive workflows purposes. Building information modeling (BIM) authoring software, virtual, augmented and mixed reality, and simulation software are examples of digital solutions used in the industry to provide visualization and simulation benefits. Dematerialization and reactive workflow benefits are related to optimization of document management, project schedule, resources management, bidding and contracting management, and budgeting.

Connected machines, equipment, and workers benefits provided by digital solutions can be subdivided into accelerated data collection and analysis, connected workforce and tools, smart energy management, machine performance optimization, and automation. Accelerated data collection and analysis benefits provided by digital solutions include ease in surveying, scanning, and mapping. It is also useful for monitoring work progress and performance, inventory, and any work incidents. Examples of the connected workforce and tools benefits provided by digital solutions include integrating all stakeholders and tools used for inventory management and tracking. Smart energy management benefits of digital solutions include energy consumption management, and optimization. Machine performance optimization benefits of digital solutions aid the effective usage of machines with compromising safety. The automation benefit reduces reliance on humans without compromising performance and safety.

Industrialized models benefit, which is the industrialization of processes and parts, allowed the design for manufacturing, and assembly (D*f*MA) of building systems. With the automation and visual benefits of digital solutions, building systems can be designed, and its manufacturing can be automated off-site for on-site assembly. Examples of D*f*MA are prefabricated building system component, advanced prefabricated systems, integrated subassemblies, and fully integrated integrated assemblies. Examples of structural, architectural, and MEP (mechanical, electrical, and plumbing) prefabricated building system components are precast, on-site dry applied finishes, and flexible water pipe or sprinkler dropper, respectively. Examples of structural, architectural, and MEP advanced prefabricated systems include structural steel or advanced precast or hybrid, prefinished surfaces, prefabricated ceiling modules, or prefabricated plants. Examples of structural, architectural, and MEP integrated subassemblies include mass engineered timber, PBUs, prefabricated modules with platform or catwalk, respectively. Example for all structural, architectural, and MEP fully integrated integrated assemblies is prefabricated prefinished volumetric construction (PPVC).

Unfortunately, all the stated benefits of digital solutions will not be maximized to increase productivity if the root cause of low productivity problems in the work or building delivery process is not identified and eliminated or reduced significantly. Problem-solving skills and knowledge needed to build a strong foundation for high productivity can be found in lean thinking, not digital solutions. Digital solutions can only be used to enhance the problem-solving process; they are effective at doing that. Lean thinking provides the purpose and process of, and tools for problem solving. Lean thinking provides avenues for the effective use of digital solutions. To simply put it, lean thinking provides the direction for effective digital solutions adoption. How?

19.2.2 Lean thinking adoption

Lean thinking defines a problem as the gap between the current performance and future or target performance. A problem can be "caused" or "created." A caused problem is a negative problem because there is a drop in the required performance. A created problem is referred to as a positive problem because there is an intent to raise current performance to a higher level. The approach to solving a problem is provided by Sobek II and Smalley (2011). An insight is provided in this section.

Lean thinking helps put a system, based on risk assessment, in place to reduce waste. The risk assessment involves understanding hazards that could cause harm, i.e., prevent the achievement of value-oriented productivity in the project. A hazard could be due to nature, humans, or both. Examples of hazards due to nature are poor or violent weather and environmental conditions. An example of a hazard due to humans is having a system that cannot prevent the occurrence of a problem or sustainably correct the occurrence of a problem. Harmful things (e.g., materials, equipment, etc.) created by humans are another example of hazards due to humans.

There are several areas of vulnerability that could increase the risk of low value-oriented productivity occurring. Lean thinking would necessitate the need to assess the vulnerability in the project and seek measures to address the cause of the vulnerability before and after it occurs. Exposure of construction project to a hazard due to human or nature is one vulnerability, i.e., vulnerability due to exposure. If there is no exposure to hazard, there is no risk.

Vulnerability due to poor project bodily function, i.e., project physiology, is another example of vulnerability. Technical ability and the right amount of manpower, equipment, materials, and physiological health status of people involved in the project are examples of what could cause vulnerability due to project physiology if not appropriately managed. The mental health status of people in the project and assumptions made by them, if not appropriately managed, will cause a construction project to be vulnerable due to psychology. Interactions among all stakeholders involved in the project that compromise the achievement of high value-oriented productivity will increase social vulnerability. The lack of or inadequate funds for the project delivery increases economic vulnerability.

Lean thinking increases the chances of doing an effective risk assessment, management, and avoidance when doing project schedule, resource management, and budgeting from the beginning to the end of the project. The adoption of lean thinking would also help reduce risk by increasing the effectiveness of preventing or correcting problems during the design stage, preconstruction, construction, and project closeout, handover, and completeness. It would also help to do an effective commissioning process that should start from the beginning to the end of the project.

Fundamentally, lean thinking strives to create a system where work standard is stabilized so that the occurrence or potential occurrence of problems and causes of problems can be made visible easily and solve to meet customer's expectations when and where they need it to be solved completely. It will also reduce the number of decisions that need to be made without compromising high-value-oriented productivity. The adoption of lean thinking would lead to the minimization of overload (Muri) and imbalance or variation (Mura) of tasks or work done in the construction project. The minimization or mitigation would reduce the occurrence of wastefulness (Muda) in a construction project – see Section 3.

Lean thinking requires effort to be made to understand the business reasons for embarking on solving a particular problem in a work or building delivery process. This understanding will guide the direction of risk assessment, management, and avoidance. To define a problem, understand the current situation, a situation where things stand at the moment, and the required performance level. Lean thinking requires effort to be made to know and document the actual symptoms associated with the identified problem. The appreciation of the problem's actual symptom will motivate the extent of resources to be committed. Before embarking on solving a problem, the consensus among people in charge of what success means or look like, that is, the determinants of success achievement, is essential.

Thereafter, efforts should be made to identify the root cause(s) of the identified problem. The root cause is the underlying reason why a problem and their associated symptoms are being experienced. The root cause is the constraint preventing the achievement of the required value-oriented performance of goods and services delivered. Lean thinking provides tools for identifying and strategies for eliminating the root cause(s) of a problem (Dennis, 2017). Examples of such tools are five whys questions, fishbone diagram, Six Sigma, and quality management tools (Antony et al. 2012; Furterer and Elshennawy, 2005; Sarkar et al. 2013; Sweeney 2015).

Lean thinking necessitates professionals to explore and documents countermeasures (solutions) needed to eradicate or lessen the root cause(s) (Aziz and Hafiz, 2013). The criteria for determining the effectiveness of, and choosing among the countermeasure options will be identified, documented, and agreed upon. The team involved in the problem solving or improvement efforts will also document and agreed on who will do what, when, and how things will be done, monitored, and measured. Plans will also be made for a follow-up to ensure continuous improvement. There will be plans to identify, monitor, measure, and resolve any failures and consequences of the chosen countermeasure(s). The problem-solving strategy's overarching principle is to plan, do, check, and adjust (PDCA).

Digital solutions cannot provide the problem-solving philosophy and strategies lean thinking offers. However, the philosophy and strategy are essential to avoiding waste that leads to problems that

19.3 Typical wastes in the management of the construction process

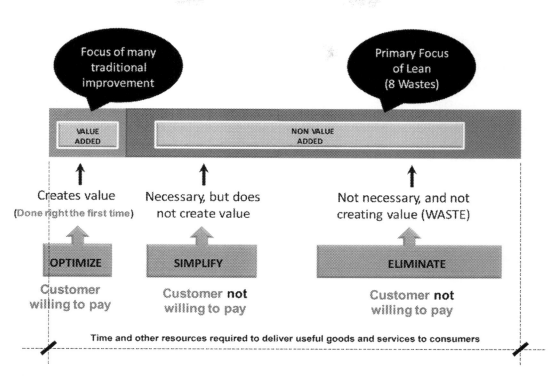

FIG. 19.2 Importance of eliminating wastes from a work process before adding value.

cause low productivity. In a broader term, waste is the first layer of problem occurrence. Waste should be eliminated before investing in resources, such as digital solutions, that can potentially add value to consumers in a process. Fig. 19.2 shows the importance of eliminating wastes from a work process before adding value. Further efforts should be made to identify why the waste is occurring.

As suggested by lean thinking, the different types of wastes are provided in Section 19.3 of this chapter. It suffices to say, digital solutions need lean thinking, and lean thinking needs digital solutions to speed the process of effectively preventing or correcting problems in the construction project. Thus, digital solutions must be grounded in lean thinking to maximise benefits inherent in digital solutions.

19.3 Typical wastes in the management of the construction process

There are typically eight wastes (Muda) according to lean thinking. These wastes are defective production, over-processing, waiting, unused talent, transportation, inventory, motion, and excessive production (Sweeney, 2015). These wastes can be found in the building delivery process. The similarity with all the eight wastes is that they consume resources without producing goods or services useful to the consumer. Examples are given for each of the wastes for better understanding. Readers are urged to reflect on more examples based on the definition of each of the wastes given.

19.3.1 Defective production

Defective production waste occurs when invested resources produced goods and services that fail to meet consumer expectations. The callback rate of contractors by clients for reworks can increase and put pressure on stakeholders' resources in the building delivery process. Rework increases the amount of jobs entering the construction work process unnecessarily. Design and coordination of errors and omissions, poor professional knowledge and skills, and lack of integrity are examples of the causes of rework on construction sites. Designers will also need to spend additional resources to get the design right. Most times, building defects are only realized after a building is delivered to the client. This occurrence may increase the defect liability period. The increase in the liability period may cause financial strain on contractors. Clients and building users will also be deprived of the on-time building value delivery.

There are also cases in which tenants call the building operator and maintenance officers and technicians several times to rectify the building system faults. The rework can occur several times if the focus is only on providing solutions to the defects and not the root cause. Lean conscious project managers would want to understand the root-cause of defects, especially if they are of significant implications. They also put a system in place, which prevent occurrence or reoccurrence of the root-cause. Errors uncorrected at the design or construction stage become a defect when the building, building system, or completed work is delivered to the consumers in the building delivery process. It does not matter if digital solutions and practices are introduced into the work process; productivity will still be compromised if the defect production's root-cause is not identified and eliminated.

Doing quality check at the source, the lean-thinking way, will help identify mistakes early enough and narrow the possible causes of the identified problem. The narrow pool of possible causes would make it easier to get to the root cause of the problem and provide a sustainable and innovative solution. It would also make it easier to redesign the work process on time. However, identifying errors after they have been passed along in the process or after a substantial amount of work has been completed would make possible causes of the error (problem) to be wider. It may also be difficult to get to the root cause of the problem. Doing the same work several times, due to the poor quality delivered, invested resources for the same work increases. Reworks contribute to low productivity in the AEC industry (Ekambaram, 2006).

19.3.2 Over-processing

Over-processing waste occurs when more resources are invested than necessary to produce the same usefulness required from goods and services that relatively lesser resources could have produced. To ensure high-quality work is done and avoid any errors leading to work delays, liability, and financial implication, many professionals at each building delivery stage, would introduce many layers of error finding or double-checking mechanisms. Sometimes, the same work with the same intended outcome is performed by different people in the work process. Such practice usually leads to additional and unnecessary information being requested, and repeated works occurrence.

Over-processing waste can occur, for example, when project managers requested information serving the same purpose from designers or requested contractors to report work progress in an unnecessary number of times to ensure high quality and a safe building is delivered. It does not matter if digital solutions are adopted in the work process; productivity will still be compromised because of the unnecessary use of resources to do the same work several times to produce the same outcome. Lean conscious project managers would design the work process to avoid over-processing waste as much as possible. They would be interested in understanding the root cause of the problem the over-processing practices are meant to prevent.

In as much as it is important to have quality check in place, not understanding the root cause of the envisaged problem to be prevented or eliminated will not prevent the problem from occurring or reoccurring. The occurrence or reoccurrence of a problem despite investing resources to avoid it will further intensify the over-processing practices. The illusion of investing in digital solutions by many project managers to optimize over-processing work activities will not necessarily improve productivity. The need to prevent over-processing waste while ensuring defects are not passed over to consumers will require an innovative solution informed by lean thinking problem-solving philosophy and approach.

19.3.3 Waiting

Waiting waste occurs when resources are put on hold for the next process or work stage. Thus, invested resources are not producing any usefulness to the consumers when they are supposed to. Waiting occurs when there is a bottleneck in a particular work process or at a stage of work. A stage of work or a particular work process will become a bottleneck if the rate at which new or reworks jobs enter it is faster than the time taken (processing time) to complete a task. The fragmentation between architects and various design engineers causes many delays at the design stage. Delays often occur when designers of a discipline wait for designers of another discipline to complete their design work or finalize revisions or corrections in a design. Waiting may occur because of the delay in getting the required information, permits, or inspection to occur. Another instance is when a subcontractor firm waits for another subcontractor firm to finish their work before starting their work.

A delay in procurement or receipt of materials and equipment on construction sites, nonsite readiness for work to be carried out, and safety problems are other examples of situations that can prolong the time taken to complete a construction project. The goal would be to ensure that the rate at which jobs is entering a stage in the construction process is equal to the rate at which job is leaving the stage. This is an ideal that must be worked toward. The callback rate for reworks will unnecessarily increase the amount of jobs entering the work process. Additionally, identifying and eliminating the root cause of delay done by project managers at any stage of the construction process will help avoid the unnecessary increase in time taken to finish work at a stage in the construction process. They will make use of digital tools to increase the productivity of identifying and eliminating the root cause.

Lean conscious project managers will analyze potential areas for work delay during the design stage, project schedule, resource management, and budgeting phase to understand and prevent or mitigate their potential root cause. The lean practice will also continue during the preconstruction, construction site work, and project handover and closeout phases. A holistic view of the impact of delay at one phase will have over another or others will also be analyzed for understanding, prevention, and mitigation. It will be easier to do the holistic-view analysis and explore many possible solutions that can potentially eliminate or reduce the intensity of the root cause with digitalization practice. Only digitalization practice grounded in lean thinking will reduce the waiting waste in the building delivery process.

19.3.4 Unused talent

Unused talent waste occurs when resources, in the form of people's talents, skills, and knowledge, are underutilized. The opportunity to produce the required usefulness from goods and services is missed. Only senior management, directors, head of departments, managers, and those holding degrees or professional qualifications are often invited to work process improvement meetings. This group of people is believed to have technical knowledge, understanding, training, and experience to improve or solve problems in the building delivery process. The laborers who are more familiar with the work process on the ground due to their daily involvement in carrying out the work are usually left out. However, the laborers may have a better understanding of the root cause of the highlighted problem.

Strategic improvement decisions are often made based on facts presented, best practices in the industry, discussion, and ideas generated. Many times these professionals do not embark on Gemba walk. That is, they do not go or go enough to the ground where actual works are being done or interact with people doing the work to decide on improvement efforts. Misalignment between the solution provided to a problem by the authority and the root cause of the problem laborers are experiencing every day causes the laborers or staff on the ground to not always implement directives from the management meetings. It is not uncommon that even when these directives are implemented on the ground, they do not yield any improvement or sometimes counterproductive.

Lean conscious project managers would want to make effective changes to reality—what is happening on the ground—and not what is believed to be happening. Ideas missed by not engaging or interacting with people doing the work could lead to wastage of invested resources. Quality and safety may also be compromised. Digital solutions can be used to enhance the quality of the Gembak walk. However, human interaction would still be necessary to appreciate human behavioral psychology contributions to an identified problem. Lean thinking mindset is needed to determine how digital solutions can be effectively utilized.

19.3.5 Transportation

Transportation waste occurs when invested resources are utilized in the unnecessary movements of products and materials. The movement will be regarded as unnecessary when it does not add additional usefulness to the consumers in the value chain. For example, contractors often spend a significant amount of time and resources to move construction supplies, materials, and equipment around different locations on site, from one site to another, or from production facility or warehouse to a site. The unresolved root cause of unnecessary transportation could increase the vulnerability of damaging the transported item. The concern is not only about wastage of resources, but safety that may be compromised because of the damaged quality of equipment and materials used for construction.

Lean conscious project managers would examine how the movement of supplies, materials, and equipment can be reduced without compromising value delivered in the construction process. They will aim to eliminate the root cause of the unnecessary movement that is occurring or could occur. If such an effort is successful, supplies, equipment, and materials will be delivered when and where they are needed with little or no unnecessary movement. Digital solutions can be very useful in generating information and knowledge needed to define the problem relating to the unnecessary movement, identify the root cause of the problem, and strategize and monitor possible solutions for eliminating the root cause. Digital solutions can also be effective in monitoring the solution's progress to the root cause and continuous improvement efforts. The possible benefits inherent in applying digital solutions cannot be actualized if it is not grounded in lean thinking.

19.3.6 Inventory

Inventory waste occurs when resources, in the form of unprocessed materials or products, and equipment, are acquired but not utilized to provide useful output to a consumer. Resources are often acquired to cover flaws in the work process. For example, the fear of lack of availability when needed often result in supplies, materials, and equipment being acquired in batches. Sometimes stocks are purchased just in case they are needed, not because there is an actual need for them. Such practice will lead to creating more storage facilities to cater to the increasing amount of supplies, materials, and equipment waiting to be processed. Wastage can also occur when environmental conditions damage the quality of supplies, materials, and equipment, rendering them damaged or ineffective or unsafe for usage.

Stocks can also be stolen from where they are stored in countries where robbery is paramount. Sometimes resources are invested in security systems or personnel to prevent a robbery in countries where they are paramount. Such invested resources are a waste as it is an unnecessary activity that could have been prevented and not needed to create the consumers' required value in the construction process. The stocks can also cause overcrowding of the construction site, thereby increasing the risk of safety problems occurring.

Lean conscious project managers will question the need for batch purchase. They would want to understand why stocks cannot be available when needed. They would want to understand why an actual consumption cannot trigger the replenishment of stocks. They would want to understand the root cause of batch purchase and eliminate or lessen the effect of the root cause. They would not favor forecasting and guessing stocks needed, an unproductive practice common in the construction industry. The lean practice will drive the kind of digital solutions required to optimize inventory replenishment triggered by actual consumption. Investing in digital solutions without making a conscious effort to eliminate or reduce the root cause's effect will further the prevalence of low productivity in the construction work process. Low productivity during the construction stage will compromise the quality and usefulness of the delivered building.

19.3.7 Motion

Motion waste occurs when invested resources, in the form of time, cost, and effort, are utilized for unnecessary movements of people without delivering any additional usefulness to the consumers in the value chain. Consumers of delivered goods and services output will not be willing to pay for people's unnecessary movement. Consumers will only be willing to pay for activities that provide the required usefulness. The call back of architects, engineers, contractors, and laborers to rectify defects creates motion waste that consumers will not be willing to pay for. The more the call back occurs, the more invested resources will be utilized. All motions that did not lead to the required value delivery will be regarded as unnecessary and a waste.

Laborers coming to the ground or lower level to take or refill supplies, materials, or equipment when working at a high rise level creates unnecessary movements that do not provide any usefulness. Another example of motion waste is laborers going to the storeroom several times to take or refill supplies, materials, or equipment every time they are needed at a particular location on site. The significant amount of waste caused by using resources to move people will depend on the size, topography, and nature of work being carried out on the site. To improve productivity, lean conscious project managers will not adopt digital solutions to solve specific unnecessary movement problems. They will ensure to make a conscious decision to define the extent of the problem, identify the root cause of the problem, and monitor possible solutions for eliminating the root cause.

19.3.8 Excessive production

Excessive production waste occurs when resources are invested in producing goods and services for consumers. Excessive production is regarded as the king of all wastes because its occurrence leads to other wastes. For example, precast systems are produced offsite, in a place known as integrated construction and prefabrication hub, and delivered to the construction site. Such practices have the potential of saving resources need for construction on site. However, if precast systems are regularly delivered to the site in quantities more than required because of uncertainty in delivering them when needed, several wastes will occur.

Contractors or clients will be forced to create spaces for storing precast systems. Invested resources, that is, precast systems, will be laying where they are stored, waiting to be used without producing any usefulness to the consumers. Depending on where, when, how, and how long the precast systems are stored, their quality may be compromised for effective usage. This is defective production waste. The precast systems will be moved regularly as part of storage practice or searching for a particular stored item. The unnecessary movement of the precast systems is a transportation waste.

The time, energy, and cost involved in getting laborers to move extra quantities of the stored precast systems currently not needed from one place to another is motion waste. The avoidable resources invested in rectifying any damages to the precast systems stored to bring them back to the required quality are over-processing waste. The opportunity missed in getting the right personnel to analyze the root cause of the excessive production before and after they occur is unused talent waste. The effectiveness of using digital solutions to prevent excessive production waste and contribute to productivity enhancement will be limited if the lean thinking concept is not adopted in problem solving.

19.4 Case study

The delivered and the process of delivering sustainable buildings must adhere to the three of spheres of sustainability and intersection between them. The three spheres are: environmental, social, and economic. The intersections include social-environmental, environmental-economic, and economic-social. A building and the process of achieving it must respect sustainability. To respect means to honor, to protect, and be a steward. Conservation of natural resources and prevention of air, water, and land pollution are examples of how a building and its delivery process can respect the environmental trait of sustainability. A building and its delivery process that is kind, considerate, and respects people's dignity respect sustainability's social trait. Profit, cost savings, and economic growth delivery are examples of how a building and its delivery process can respect the economic trait of sustainability.

Ensuring energy and water-efficiency delivery are examples of how a building and its delivery process can respect sustainability's environmental-economic trait. Conservation of natural resources and prevention of air, water, and land pollution are examples of how a building and its delivery process can respect sustainability's environmental trait. Ensuring pollution or waste generated during the delivery and operations of a building does not become a burden of other people is an example of how the social-environmental trait can be respected. The respect for workers' rights and business ethics of building delivery process are examples of how sustainability's economic-social trait can be respected.

The question that comes to mind is, how can the adoption of digital solutions and lean thinking aid the productivity of sustainable building delivery? The London 2012 Games construction project is used as a case study, as sustainability delivery was the project's primary focus (Carmichael, 2012). Using this project to answer the question is not to look at how digital solutions and lean thinking were used in the project. Instead, the direction is to let readers of this chapter appreciate how digital solutions and lean thinking could increase the productivity of delivering the sustainability of a project of that scale.

19.4.1 Digital solutions for sustainable building delivery

The complexities of the London 2012 Games project necessitate the participation of several stakeholders. Examples of buildings developed for the Games are the Aquatics Centre building, Velodrome, Olympic Stadium (now known as London Stadium), Olympic Village, Basketball Arena Building, and London 2012 Media Centre (Carmichael, 2012). The majority of these buildings have complex or irregular shapes and are large. Complex shaped and large buildings lead to complexity in integrating building systems to deliver nonconflicting building performance mandates. See Fig. 19.3 for details on design consideration for the delivery of high-performance buildings.

The need to integrate building systems to deliver the required building performance necessitates the need for building professionals of different disciplines and expertise to have the capability to work together at the same time. For example, one design professionals' changes could have several implications on the design of one or more design professionals. The adoption of digital solutions will improve design delivery productivity and allow for easy engagement of design professionals and other stakeholders to optimize and coordinate the complex buildings' design to meet the United Kingdom's sustainability regulatory and downstream requirements.

The adoption of digital solutions to visualize the construction activities' potential complexity, process and management, and the impact on contractors and other stakeholders' performance and interests at the construction stage can be simulated at the design stage. The visualization benefits provided by digital solutions will allow for corrections and preventive measures to be taken to allow for the high productivity of delivering the sustainability traits. Digital solutions can also be used to simulate how the developed design would support facility management and operations effectiveness. Fundamentally, digital solutions allow for on-time risk assessment and management. The higher the buildings' complexity, the higher the risk of poor construction, safety problems, and infringement on sustainability traits. Complex buildings increase the vulnerability of experiencing poor construction, safety problems, and infringing on sustainability traits.

The adoption of PPVC systems helps improve construction productivity (Hwang et al., 2020). The adoption of digital solutions aids fabrications of the PPVC systems effectively because it allows for easy translation of design to standardized components for automating offsite production. Complex buildings such as those used for the London 2012 Games require building systems to be fabricated offsite to allow sustainable construction productivity. During construction, productivity and safety can be maximized, and rework can be minimized or prevented by adopting digital solutions for just-in-time delivery of needed resources, installing and monitoring on-site construction activities. The benefits inherent in digital construction will be significant in complex and large buildings such as those developed for the London 2012 Games. Maintenance and operations of assets in complex buildings require real-time data generation, monitoring storage, and analysis to make informed decisions for improving asset delivery and management productivity.

The probability of designing what you expect, inspecting what you expect, and achieving what you expect regarding the mentioned sustainability traits will be much lower if digital solutions are not adopted for complex buildings. The probability will be much higher if digital solutions are adopted, especially in the advent of fast networks such as 5G networks that allows fast mobile internet connectivity. The lower the latency, the faster the speed, and the greater the reliability, capacity, and flexibility of mobile internet connectivity networks adopted with digital solutions, the higher the probability of digital solutions aiding productivity.

The adoption of digital solutions for digital design, digital fabrication, digital construction, and digital asset delivery and management, as described, for productivity achievement of delivering sustainable building

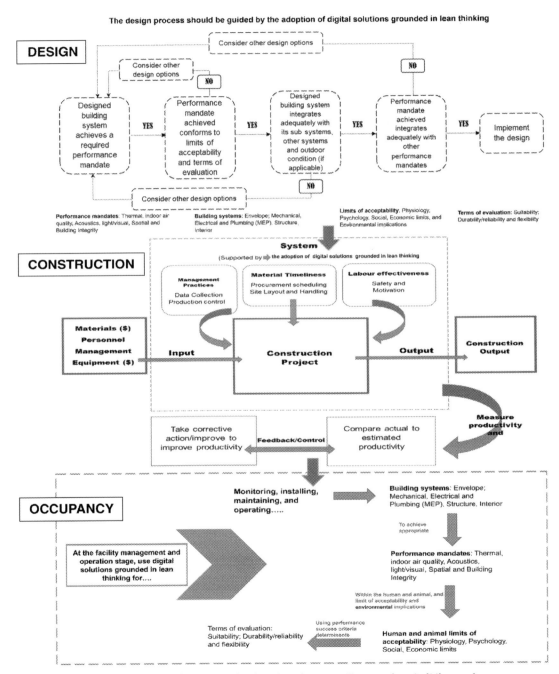

FIG. 19.3 Conceptual framework of human, animal, and environmentally conscious building performance delivery at the design, construction, and occupancy stage (Adapted from Fadeyi 2017).

and its process is informed by the initiative of the BCA of Singapore (Hwang et al. 2020). BCA encourages and leads the adoption of integrated digital delivery to streamlining work processes to improve productivity. As mentioned earlier, this intention may not be actualized effectively, if achieved at all if the root cause of wastes in the process of delivering sustainable buildings is not identified and minimized or eliminated.

19.4.2 Lean thinking for sustainable building delivery

Lean thinking is the "spectacle" that digital solutions use to enhance productivity. If the spectacle is not used, the "vision" of digital solutions to see sustainable building delivery problems will be "blurry" and performance will be compromised. What are the potential problems in delivering sustainability traits for complex buildings such as those developed for the London 2012 Games? How can lean thinking adoption help identify and reduce or eliminate the problems for the effective usage of digital solutions for high productivity achievement?

If the purpose of developing a complex or large building is not well defined, the risk of not achieving the sustainability traits will be higher. This is because if there is no purpose and determinants of success for a purpose, there will be no guidance on how to successfully achieve the purpose, its process, and metrics for measuring it. The vulnerability of experiencing failure will increase with increasing complexity and volume of a building. Adopting digital solutions in a failed system with no clear direction of monitoring and measuring success means investing in resources that would lead to relatively low performance. Even if sustainable building and process are achieved, investment in resources used for the achievement would have been very high. This is a definition of low productivity.

The adoption of lean thinking in sustainable buildings, such as those developed for the London 2012 Games, would increase productivity. Lean defines success as a value, and it does it from the perspective of a client or consumers in the process. Anyone who understands productivity will look at it from the standpoint of value delivery. A failed definition of purpose would lead to a failed process. Like those of the London 2012 Games, a highly complex and voluminous building would necessitate a well-defined and planned process, as investments of high resources will be required. According to Womack and Jones (1997), there are three characteristics of a failed process. A failed process does not have a system to identify and visualize the value stream and eliminate waste. It does not allow for a continuous flow of value. It has a push-based system instead of a pull-based system.

A pull-based system is needed for productivity because it creates an avenue for the replenishment of inventory to be triggered by actual consumption need, not by forecasting or guessing. A pull-based system helps to level the quantity of activities across the entire work process, reduce inventory, and expose hidden problems. Thus, a process that does work on a pull replenishment based system will have several wastes. A failed process's cost implications can be significant in a sustainable building delivery project on the same complexity and volume as those used for the London 2012 Games. Complex and voluminous projects necessitate people involved to be skilled and knowledgeable in increasing productivity in delivering a sustainable building, as the cost implications can be significant if they are not. People involved should be teachable, learn from experience, and strive for continuous improvement. A project with lean thinking DNA will make the development of people to their full capability a priority.

According to Martichenko (2012), lean thinking can help project managers to prevent, expose, and eliminate problems that could hinder value-oriented high productivity of sustainable building delivery by ensuring a broad view of the work process for easy planning and management. Lean thinking can provide project managers with the tools needed to simplify the complex process. The ability to streamline work processes is of significant importance in complex and voluminous buildings because it reduces human errors and aids easy identification and fixing of errors. A simple process is also easier to teach, implement, and monitor for high productivity delivery.

A lean conscious project manager would appreciate the importance of ensuring visibility in the work process. Ensuring visibility in a project at the scale of those buildings in the London 2012 Games is essential because of several stakeholders involved, and their work performances will affect one another. Visibility allows for easy identification of the source of a problem and eliminates its root cause to reduce significant cost implications. Project managers and clients should create a culture that facilitates continuous improvement among all parties involved in the work process. A work process that is simple and visible will make continuous improvement efforts easier. All parties involved should also be flexible to learn, unlearn, and relearn.

The adoption of digital solutions in a sustainable building project that is complex and voluminous will have a high probability of delivering value-oriented productivity if grounded in lean thinking. Fig. 19.4 shows the conceptual framework for integrating lean thinking with digital solutions adoption for value-oriented high productivity of goods and services—sustainable building delivery in this context. Strong leadership will be needed from project managers to achieve success suggested by the framework.

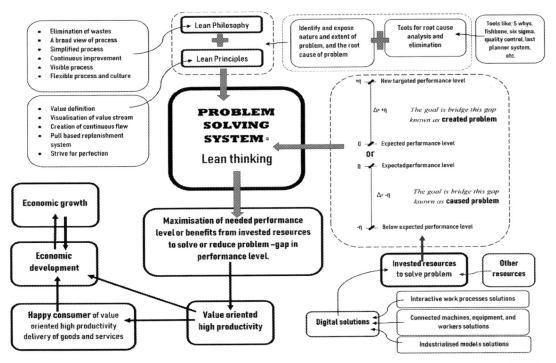

FIG. 19.4 Conceptual framework for integrating lean thinking with digital solutions adoption for value-oriented high productivity of goods and services—sustainable building delivery in this context.

19.5 Challenges of adopting digital solutions grounded in lean thinking in the industry

Despite the benefits inherent in adopting digital solutions grounded in lean thinking, several challenges must be overcome for effective adoption (Delgado et al. 2019; Netland 2016).

19.5.1 Challenges facing adoption of digital solutions

Table 19.1 shows the key challenges facing adoption of digital solutions. Effective usage of digital solutions for enhancing the process of delivering valuable goods and services depends on people using them. First, people must be conscious of the benefits inherent in the digital solutions available to them. After that, they must know how to use the available digital solutions. With the rapid emergence of several digital solutions in the AEC industry, people's willingness and ability to learn fast is essential. The learning, unlearning, and relearning required for digital solutions should cut across different age groups. The reality in practice is that many industry professionals across different stages of the building delivery process lack the required knowledge and skills to effectively adopt several fast-emerging digital solutions (Raj et al. 2020; Lavikka et al. 2018), and this often leads to pushing back in the adoption (Azhar 2011; Chen and Granitz, 2012).

The push back seems to be more prevalent among digital immigrants than digital naïve professionals (Bayne and Ross, 2007). It takes many digital immigrants' efforts to adopt digital solutions (Barnard et al., 2013). Digital immigrants are born, brought up, or trained before the widespread usage of digital solutions in the AEC industry. Digital naïve are more prone to adopt digital solutions. The integration of applied learning digital solutions training in the institute of higher learning curriculum will provide digital naïve population the digital expertise, knowledge, and experience required to effectively adopt digital solutions in the industry (Abdirad and Dossick, 2016; Chen et al. 2020; Memon et al. 2014).

The adoption of digital solutions requires a holistic perspective. The impact of adopting digital solutions to one aspect of the work process to another work process should be given due consideration

Table 19.1 Key challenges facing adoption of digital solutions.

Key challenges	References
Lack of digital solution expertise and awareness, and pushback from employees	Raj et al. (2020); Hannan and Turner (2004); Archer et al. (2008); Rahim (2007); Van Beveren and Thomson (2002); Chau and Hui (2001); Delgado et al. (2019)
Poor organization management, capability, and strategy for digital solutions adoption	Rahim (2007); Teo et al. (2004); Love et al. (2001); Jones et al. (2003)
Inability of business partner to afford or adopt digital solutions, and compatibility of digital solution with business partners	Raj et al. (2020); Porwal and Hewage (2013)
Budget constraint concern for digital solutions adoption	Hannan and Turner (2004); Rahim (2007); Teo et al. (2004); Iacovou and Benbasat (1995); Delgado et al. (2019)
Security and legal concerns	Gupta et al. (2013); Wu (2011); Esposito et al. (2016); Silverio et al. (2017); Gu and London (2010); Manderson et al. (2015); Huda (2019); Li et al. (2019); Azhar (2011)

before adoption. The digital capability should be across the board to avoid problems such as missing information or repetition of works. It could also lead to waiting waste because the nondigitalized work process could be a bottleneck in the work processes. More importantly, there should be clear policies and strategies for integrating and ensuring the effectiveness of several digital solutions available in the work process. The policies and strategies for managing information and knowledge generated should also be made clear to fulfill organization's vision. Thus, leadership is essential for implementing digital solutions across an organization's work process and structure (Cortellazzo et al. 2019; Hinings et al. 2018; Vial, 2019). Good leadership will transform digital culture (Oberer and Erkollar 2018). However, poor leadership is a major obstacle to digital transformation in many work processes, including that of the AEC industry (Larjovuori et al. 2018; Cortellazzo et al. 2019).

Digital transformation cannot occur if different digital solutions adopted in a work process cannot communicate. During the construction stage, for example, communication between a contractor and many stakeholders, for example, designers, the building owner, other contractors, etc., involved in the construction processes is expected. However, if the digital solutions that could enhance the communications are not compatible, several wastes, as suggested by lean thinking, could occur. Such a challenge would compel the contractors and other stakeholders involved in the construction process to go back to old manual practices (Dallasega et al. 2018; Watson, 2011).

There are many small and medium-sized enterprises (SMEs) involved in the construction process. Many subcontractors are SMEs. As a result, the budget for digital solution acquisition and adoption is always a major concern (Kabanda et al. 2018). Their inability to invest funds, organization structure, or resources to digital solutions acquisition and adoption usually inhibit the whole construction process's digitalization (Chau and Hui 2001; Huda, 2019; Love et al. 2001). This is why the manual operation is still favored in the construction process and other stages of the building delivery process.

Generation and sharing of data, information, and knowledge with financial or competitive business advantage implications among all stakeholders are needed at all stages of the building delivery process. Thus, data and information security are important to prevent them from being stolen or corrupted by another party. The infancy status of digital security in the AEC industry is inhibiting digital solutions adoption (Esposito et al. 2016; Gupta et al. 2013; Maskuriy et al. 2019). The sharing of any data and information is based on trust that they will not be mismanaged or copyright will be protected. However, the usual lack of trust in this regard among stakeholders involved inhibits the purpose of using the digital solution for generating and sharing of data, information, and knowledge to solve a problem (Lau and Rowlinson, 2010; Akintan and Morledge, 2013; Oliver, 2019; Tai et al., 2016).

The legal framework needed to protect the generator of data, information, and knowledge is still at the infancy stage (Alaloul et al., 2020). The potential legal vulnerabilities associated with emerging digital solutions adoption in the AEC industry are still being discovered (Gu and London, 2010; Silverio et al. 2017). The lack of a legal framework, coupled with prevalent mistrust, prevents the use of digital solutions to deliver value-oriented productivity (Barbosa et al. 2017).

19.5.2 Challenges facing adoption of lean thinking

Table 19.2 shows key challenges facing adoption of lean thinking. The unfamiliarity mostly associated with novelty usually makes people feel uncomfortable adopting it (Dawson et al. 1999; Mueller

Table 19.2 Key challenges facing adoption of lean thinking.

Key challenges	References
Poor buy-in from the top management	Ahmad (2017); Scorsone (2008); Nordin et al. (2010); Puvanasvaran et al. (2008); Malik and Abdallah (2020).
Poor organization lean culture and infrastructure	Ahmad (2017); Yadav et al. (2019); Panizzolo et al. (2012); Timans et al. (2012); Hardcopf and Shahemployee (2014); Chaple et al. (2018); Puvanasvaran et al. (2008); Malik and Abdallah (2020); Ramakrishnan and Testani (2010).
Poor employee perception and attitude toward lean adoption	Ahmad (2017); Losonci et al. (2011); Bhasin (2012); Shadur et al. (1995); Nordin et al. (2012); Nordin et al. (2010); Puvanasvaran et al. (2008); Nahm et al. (2012)
Lack of employee competency and experience on lean and its adoption	Ahmad (2017); Antony et al. (2012); Sim and Chiang (2012); Netland (2016); Nahm et al. (2012)

et al. 2012). The idea of learning, unlearning, and relearning does not come naturally to people. Top management of organizations often fears a new idea because it challenges the norm and brings many uncertainties (Nordin et al. 2010; Nordin et al. 2012). They fear it may have adverse implications on an organization's business, revenue, and competitive advantage in the market, and challenge an organization's status quo culture (Scorsone, 2008; Mueller et al. 2012). The fear of novelty is a human problem (Campbell, 2015).

Lean thinking provides a new way of thinking and doing things. Creativity and innovation are inherent in lean thinking. Even when senior management encourages employees to be creative and adopt a new way of doing things to improve organization's productivity, they often reject it in practice (van Assen, 2018). If there is a lack of lean thinking buy-in from top management of an organization responsible for managing building delivery projects, employees tasked to do the job may not see the need to adopt lean thinking in the planning, monitoring, managing, and overseeing of the project (Bashir et al. 2015; Garcia-Sabater and Marin-Garcia, 2011; Losonci et al. 2011; Shadur et al. 1995).

Assumptions drive thinking, thinking drives behavior, and behavior or practice drives culture. To create lean project management organizational culture, assumptions, thinking, and behavior or practice guiding lean adoption must be well understood and respected (Antony et al. 2012). When an organization or industry culture does not make lean adoption a priority, they will not appreciate employees' lean competency and experience (Hardcopf and Shahemployee, 2014; Malik and Abdallah, 2020; Yadav et al. 2019). Such organization will not employ people with relevant knowledge and skills or invest in their employees to embrace lean adoption (Sim and Chiang, 2012). Lean adoption will not be a reality if employees do not have knowledge about lean and its adoption (Hu et al. 2015; Nahm et al. 2012; Netland, 2016). It will not invest in resources needed for successful lean adoption. Such a nonlean culture will make creativity, innovation, continuous improvement, and productivity achievement difficult (Losonci et al. 2011).

19.6 Conclusion and future directions

Productivity status in the AEC industry can significantly influence the economic development of a country. Thus, the effective productivity management of delivering a sustainable building project that respects natural systems and the cycle of life, environmental resources, and people is essential. Project management professionals are tasked with planning and managing all activities involved in the sustainable building delivery process. Many project managers and professionals often see the building delivery process's digitalization as a solution to productivity problems.

This chapter's message is that the benefits inherent in digital solutions for productivity achievement and improvement cannot be achieved or maximized if their adoption is not grounded in lean thinking. The process of delivering sustainable buildings is filled with many problems that can potentially derail the achievement of productivity. Problem-solving philosophy, tools, and approaches that provide a solution to the root cause of the problem hindering high productivity achievement are needed. Digital tools can only enhance the problem-solving process.

Adopting digital solutions without identifying the root cause of the waste, causing productivity problems, will increase the work process's costs with little productivity achievement. Wastes in the building process can be categorized under eight lean wastes. The lean wastes are defective production, over-processing, waiting, unused talent, transportation, inventory, motion, and excessive production. The adoption of digital solutions and lean thinking in the delivery process of a sustainable building comes with challenges that need to be worked on before realizing the inherent benefits. This chapter highlights these challenges. The extent of the challenges will depend on the complexities involved in a particular work process or building project to be delivered and the hazards the project is exposed to.

Project management organizations should create a culture that adopts digital solutions grounded in lean thinking for a productive building delivery process. The digital solutions can enhance the process of doing a risk assessment, management, and avoidance guided by lean thinking to increase value-oriented productivity. High value-oriented productivity is essential for a sustainable construction project and world. This culture should also be wired into how contracts are designed and administered. The adoption of this culture provides an avenue for future research direction. The support of government in this direction will also be essential. There is currently a knowledge gap on how the adoption of digital solutions grounded in lean thinking would influence the achievement of a sustainable building process's value-oriented productivity. The attempt to bridge the knowledge gap should be made for different building types and building scale, and variation in the challenges faced should be documented for continuous improvement efforts. It is important to note that the purpose of this chapter is not to provide training on digital solutions and lean thinking adoption. Rather, the purpose is to make an argument that digital solutions without conscious lean thinking adoption will compromise high value-oriented productivity achievement and miximisation of benefits inherent in digital solutions.

Biography

About the Author: Dr. Moshood Olawale Fadeyi is an Associate Professor of Building Services Engineering at the Singapore Institute of Technology. Dr. Fadeyi holds four university degrees, including a Ph.D. in Indoor Environment and Energy from the National University of Singapore and Technical University of Denmark. He is an architect, indoor air quality expert, chartered engineer (UK)—building services, chartered construction manager (UK), and specialist adult educator. Dr. Fadeyi is a member of the Chartered Institution of Building Services Engineers, Chartered Institute of Building, and American Society of Heating, Refrigerating, and Air-conditioning Engineers. Dr. Fadeyi research aims to enhance applied learning pedagogy to make students job-ready upon graduation and improve industry and community practices. His applied learning pedagogy enhancement is informed by applied research, artistic research, and case study research grounded in design, critical, reflective, scientific, and lean thinking and digital solutions adoption. Dr. Fadeyi research and teaching domain expertise are in value-oriented productivity in healthy and energy-efficient indoor air delivery and value-oriented productivity in design, construction, and facility management practices. His expertise and leadership are highly sought after by companies and government agencies.

References

Abdel-Wahab, M., Vogl, B., 2011. Trends of productivity growth in the construction industry across Europe, US and Japan. Constr. Manag. Econ. 29 (6), 635–644.

Abdirad, H., Dossick, C.S., 2016. BIM curriculum design in architecture, engineering, and construction education: a systematic review. J. Inf. Technol. Constr. 21 (17), 250–271.

Ahmad, S., 2017. The challenges of lean implementation: a multiple case study in Malaysian aerospace companies. J. Asian Vocat. Educ. Train. 10, 103–120.

Akintan, O.A., Morledge, R., 2013. Improving the collaboration between main contractors and subcontractors within traditional construction procurement. J. Constr. Eng. 2013. 1–11.

Alaloul, W.S., Liew, M.S., Zawawi, N.A.W.A., Kennedy, I.B, 2020. Industrial revolution 4.0 in the construction industry: challenges and opportunities for stakeholders. Ain Shams Eng. J. 11 (1), 225–230.

Antony, J., Hilton, R.J., Sohal, A., 2012. A conceptual model for the successful deployment of Lean Six Sigma. Int. J. Qual. Reliab. Manag. 29 (1), 54–70.

Archer, N., Wang, S., Kang, C., 2008. Barriers to the adoption of online supply chain solutions in small and medium enterprises. Supply Chain Manag. 13 (1), 73–82.

Azhar, S., 2011. Building information modeling (BIM): trends, benefits, risks, and challenges for the AEC industry. Leader. Manag. Eng. 11 (3), 241–252.

Aziz, R.F., Hafez, S.M., 2013. Applying lean thinking in construction and performance improvement. Alex. Eng. J. 52 (4), 679–695.

Barbosa, F., Woetzel, J., Mischke, J., 2017. Reinventing Construction: A Route of Higher Productivity. McKinsey Global Institute, USA.

Bartik, T.J., 2003. "Local Economic Development Policies." Upjohn Institute Working Paper No. 03-91. Kalamazoo, MI: W.E. Upjohn Institute for Employment Research. https://doi.org/10.17848/wp03-91.

Barnard, Y., Bradley, M.D., Hodgson, F., Lloyd, A.D., 2013. Learning to use new technologies by older adults: perceived difficulties, experimentation behaviour and usability. Comput. Hum. Behav. 29 (4), 1715–1724.

Bashir, A.M., Suresh, S., Oloke, D.A., Proverbs, D.G., Gameson, R., 2015. Overcoming the challenges facing lean construction practice in the UK contracting organizations. Int. J. Archit. Eng. Constr. 4 (1), 10–18.

Basnett, Y., Sen, R., 2013. What do empirical studies say about economic growth and job creation in developing countries. Overseas Development Institute, 1–41. https://www.researchgate.net/publication/321937240_What_do_empirical_studies_say_about_economic_growth_and_job_creation_in_developing_countries (Accessed 28 April 2021).

Bayne, S., Ross, J., 2007. The 'digital native' and 'digital immigrant': a dangerous opposition, Annual Conference of the Society for Research into Higher Education (SRHE), Vol. 20. ac. uk/staff/sian/natives_final. pdf (Accessed 03 September 2020).

Bhasin, S., 2012. Prominent obstacles to lean. Int. J. Product. Perform. Manag. 61 (4), 403–425.

Campbell, K.D., 2015. Fear becomes the unintended consequence of creativity/innovation. J. Leadersh. Stud. 9 (3), 60–61.

Carmichael, L., 2012. Learning Legacy. Lessons learned from the London 2012. Games construction projecthttps://www.designcouncil.org.uk/sites/default/files/asset/document/London%202012%20Original.pdf (Accessed 28 April 2021).

Chaple, A.P., Narkhede, B.E., Akarte, M.M., Raut, R., 2018. Modeling the lean barriers for successful lean implementation: TISM approach. Int. J. Lean Six Sigma.

Chau, P.Y.K., Hui, K.L., 2001. Determinants of small business EDI adoption: an empirical investigation. J. Organ. Comput. Electron. Commer. 11 (4), 229–252.

Chen, S., Granitz, N., 2012. Adoption, rejection, or convergence: consumer attitudes toward book digitization. J. Bus. Res. 65 (8), 1219–1225.

Chen, K., Lu, W., Wang, J., 2020. University–industry collaboration for BIM education: lessons learned from a case study. Ind. High. Educ. 0950422220908799.

Cheung, W.F., Lin, T.H., Lin, Y.C., 2018. A real-time construction safety monitoring system for hazardous gas integrating wireless sensor network and building information modeling technologies. Sensors 18 (2), 436.

Ciccarone, G., Saltari, E., 2015. The decline of the Italian economy in the last twenty years. J. Mod. Ital. Stud. 20 (2), 228–244.

Cortellazzo, L., Bruni, E., Zampieri, R., 2019. The role of leadership in a digitalized world: a review. Front. Psychol. 10, 1938.

Dallasega, P., 2018. Industry 4.0 fostering construction supply chain management: lessons learned from engineer-to-order suppliers. IEEE Eng. Manag Rev. 46 (3), 49–55.

Dallasega, P., Rauch, E., Linder, C., 2018. Industry 4.0 as an enabler of proximity for construction supply chains: a systematic literature review. Comput. Ind. 99, 205–225.

Dawson, V.L., D'Andrea, T., Affinito, R., Westby, E.L., 1999. Predicting creative behavior: a reexamination of the divergence between traditional and teacher-defined concepts of creativity. Creat. Res. J. 12 (1), 57–66.

Delgado, J.M.D., Oyedele, L., Ajayi, A., Akanbi, L., Akinade, O., Bilal, M., Owolabi, H, 2019. Robotics and automated systems in construction: understanding industry-specific challenges for adoption. J. Build. Eng. 26, 100868.

Deller, S.C., Tsai, T.H., Marcouiller, D.W., English, D.B., 2001. The role of amenities and quality of life in rural economic growth. Am. J. Agric. Econ. 83 (2), 352–365.

Dennis, P., 2017. Lean Production simplified: A plain-language guide to the world's most powerful production system. CRC press, FL.

Eifert, B., 2009. Do regulatory reforms stimulate investment and growth? Evidence from the doing business data, 2003-07. Center for Global Development Working Paper 159.

Ekambaram, P., 2006. Reducing rework to enhance project performance levels, Proc. Seminar on Recent Developments in Project Management in Hong Kong. The Centre for Infrastructure & Construction Industry Development (CICID), Department of Civil Engineering, The University of Hong Kong.

Esposito, C., Castiglione, A., Martini, B., Choo, K.K.R, 2016. Cloud manufacturing: security, privacy, and forensic concerns. IEEE Cloud Comput. 3 (4), 16–22.

Fadeyi, M.O., 2017. The role of building information modeling (BIM) in delivering the sustainable building value. Int. J. Sustain. Built Environ. 6 (2), 711–722.

Fadeyi, MO, 2019. Cartoons for human development: 40 reflection topics. ISSUU Digital Publishing Platform. https://issuu.com/mofadeyi/docs/cartoon_for_human_development_ebook.

Friedman, J., 2003. The decline of corporate income tax revenues Center on Budget and Policy Priorities, 24. https://www.cbpp.org/research/the-decline-of-corporate-income-tax-revenues.

Furterer, S., Elshennawy, A.K., 2005. Implementation of TQM and lean Six Sigma tools in local government: a framework and a case study. Total Qual. Manag. Bus. Excell. 16 (10), 1179–1191.

Garcia-Sabater, J.J., Marin-Garcia, J.A., 2011. Can we still talk about continuous improvement? Rethinking enablers and inhibitors for successful implementation. Int. J. Technol. Manag. 55 (1/2), 28–42.

References

Gu, N., London, K., 2010. Understanding and facilitating BIM adoption in the AEC industry. Automat. Constr. 19 (8), 988–999.

Gupta, P., Seetharaman, A., Raj, J.R., 2013. The usage and adoption of cloud computing by small and medium businesses. Int. J. Inform. Manag. 33 (5), 861–874.

Gurdal, T., Aydin, M., Inal, V., 2021. The relationship between tax revenue, government expenditure, and economic growth in G7 countries: new evidence from time and frequency domain approaches. Econ. Change Restructur 55, 305–337.

Hardcopf, R., Shah, R., 2014. Lean and performance: the impact of organizational culture, Academy of Management Proceedings, Vol. 2014, No. 1. Academy of Management, NY, pp. 10747.

Hannan, M., Turner, P., 2004. The last mile: applying traditional methods for perpetrator identification in forensic computing investigations, 3rd European Conference on Information Warfare and Security. UK, p. 89.

Hinings, B., Gegenhuber, T., Greenwood, R., 2018. Digital innovation and transformation: an institutional perspective. Inform. Organ. 28 (1), 52–61.

Hooper, M., Ekholm, A., 2010. A pilot study: towards BIM integration—an analysis of design information exchange & coordination, Proc. CIB W,. Salford UK, Vol. 78, p. 2010.

Hou, L., Tan, Y., Luo, W., Xu, S., Mao, C., Moon, S., 2020. Towards a more extensive application of off-site construction: a technological review. Int. J. Constr. Manage, 1–12.

Hu, Q., Mason, R., Williams, S.J., Found, P., 2015. Lean implementation within SMEs: a literature review. J. Manuf. Technol. Manag. 26 (7), 980–1012.

Huda, M., 2019. Empowering application strategy in the technology adoption. J. Sci. Technol. Policy Manag. 10 (1), 172–192.

Hwang, B.G., Ngo, J., Her, P.W.Y, 2020. Integrated digital delivery: implementation status and project performance in the Singapore construction industry. J. Clean. Prod. 262, 121396.

Iacovou, C.L., Benbasat, I., Dexter, A.S., 1995. Electronic data interchange and small organizations: adoption and impact of technology. MIS Q. 19 (4), 465–483.

Jones, P., Beynon-Davies, P., Muir, E., 2003. EBusiness barriers to growth within the SME Sector. J. Syst. Inform. Technol. 7 (1), 1–25.

Kabanda, S., Tanner, M., Kent, C., 2018. Exploring SME cybersecurity practices in developing countries. J. Organ. Comput. Electron. Commer. 28 (3), 269–282.

Kiani, A., Salman, A., Riaz, Z., 2014. Real-time environmental monitoring, visualization, and notification system for construction H&S management. J. Inf. Technol. Constr. 19, 72–91.

Larjovuori, R.L., Bordi, L., Heikkilä-Tammi, K., 2018. Leadership in the digital business transformation, Proceedings of the 22nd International Academic Mindtrek Conference. NY, USA, 212–221.

Lau, E., Rowlinson, S., 2010. Trust relations in the construction industry. Int. J. Manag. Proj. Bus. 3 (4), 693–704.

Lavikka, R., Kallio, J., Casey, T., Airaksinen, M., 2018. Digital disruption of the AEC industry: technology-oriented scenarios for possible future development paths. Constr. Manag. Econ. 36 (11), 635–650.

Lee, S., Akin, Ö., 2011. Augmented reality-based computational fieldwork support for equipment operations and maintenance. Automat. Constr. 20 (4), 338–352.

Li, P., Zheng, S., Si, H., Xu, K., 2019 (b). Critical challenges for BIM adoption in small and medium-sized enterprises: evidence from China. Adv. Civ. Eng., 2019 Article ID 9482350.

Li, X., Shen, G.Q., Wu, P., Yue, T., 2019 (a). Integrating building information modeling and prefabrication housing production. Automat. Constr. 100, 46–60.

Lin, Y.C., Su, Y.C., Chen, Y.P., 2014. Developing mobile BIM/2D barcode-based automated facility management system. Sci. World J., 2014. Article ID 374735.

Losonci, D., Demeter, K., Jenei, I., 2011. Factors influencing employee perceptions in lean transformations. Int. J. Prod. Econ. 131 (1), 30–43.

Love, P.E.D., Irani, Z., Li, H., Cheng, E.W.L., Tse, R.Y.C., 2001. An empirical analysis of the barriers to implementing e-commerce in small-medium sized construction contractors in the State of Victoria, Australia. Constr. Innov. 1, 31–41.

Malik, M., Abdallah, S., 2020. The relationship between organizational attitude and lean practices: an organizational sense-making perspective. Ind. Manag. Data Syst. doi:10.1108/IMDS-09-2019-0460 (Accessed 03 September 2020).

Manderson, A., Jefferies, M., Brewer, G., 2015. Building information modelling and standardised construction contracts: a content analysis of the GC21 contract. Constr. Econ. Build. 15 (3), 72–84.

Martichenko, R., 2012. Everything I know about Lean I learned in first grade. Lean Enterprise Institute, USA.

Maskuriy, R., Selamat, A., Ali, K.N., Maresova, P., Krejcar, O., 2019. Industry 4.0 for the construction industry—how ready is the industry? Appl. Sci. 9 (14), 2819.

Melamed, C., Hartwig, R., Grant, U., 2011. Jobs, growth and poverty: what do we know, what don't we know, what should we know. Growth 18 (6), 10.

Memon, A.H., Rahman, I.A., Memon, I., Azman, N.I.A, 2014. BIM in Malaysian construction industry: status, advantages, barriers and strategies to enhance the implementation level. Res. J. Appl. Sci. Eng. Technol. 8 (5), 606–614.

Mishrif, A., Balushi, Al, H., Y., Kharusi, Al, S., Alabduljabbar, A, 2015. Economic diversification: challenges and opportunities in the GCC. In: Gulf Research Centre Cambridge Workshop, Vol. 9. https://gulfresearchmeeting.net/documents/1584599714108_WS9%20-%20Economic%20Diversification%20Challenges%20and%20%20%20opportunities%20in%20the%20GCC.pdf.

Mueller, J.S., Melwani, S., Goncalo, J.A., 2012. The bias against creativity: why people desire but reject creative ideas. Psychol. Sci. 23 (1), 13–17.

Nahm, A.Y., Lauver, K.J., Keyes, J.P., 2012. The role of workers' trust and perceived benefits in lean implementation success. Int. J. Bus. Excell. 5 (5), 463–484.

Netland, T.H., 2016. Critical success factors for implementing lean production: the effect of contingencies. Int. J. Prod. Res. 54 (8), 2433–2448.

Netscher, P, 2017. Construction Management: From Project Concept to Completion. Panet Publications, Australia.

Nordin, N., Deros, B.M., Abd Wahab, D., 2010. A survey on lean manufacturing implementation in Malaysian automotive industry. Int. J. Innov. Manag. Technol. 1 (4), 374.

Nordin, N., Deros, B.M., Wahab, D.A., Rahman, M.N.A., 2012. A framework for organisational change management in lean manufacturing implementation. Int. J. Serv. Oper. Manag. 12 (1), 101–117.

Oberer, B., Erkollar, A., 2018. Leadership 4.0: digital leaders in the age of industry 4.0. Int. J. Organ. Leadersh 7 (4), 404–412. https://ssrn.com/abstract=3337644.

Oliver, S., 2019. Communication and trust: rethinking the way construction industry professionals and software vendors utilise computer communication mediums. Vis. Eng. 7 (1), Article No. 1.

Phillips, P.D., 1996. Business retention and expansion: theory and an example in practice. Econ. Dev. Rev. 14 (3), 19–24.

Porwal, A., Hewage, K.N., 2013. Building information modeling (BIM) partnering framework for public construction projects. Automat. Constr. 31, 204–214.

Puvanasvaran, A.P., Muhamad, M.R., Megat, M.H.M.A., Tang, S.H., Hamouda, A.M.S., 2008. A review of problem solving capabilities in lean process management. Am. J. Appl. Sci. 5 (5), 504–511.

Rahim, M.M., 2007. Factors affecting adoption of B2E e-Business systems: a case of the Australian higher education industry, In Proc. Pacific Asia Conference on Information Systems 2007. Auckland, New Zealand, 16.

Raj, A., Dwivedi, G., Sharma, A., de Sousa Jabbour, A.B.L., Rajak, S., 2020. Barriers to the adoption of industry 4.0 technologies in the manufacturing sector: an inter-country comparative perspective. Int. J. Prod. Econ. 224, 107546.

Ramakrishnan, S., Testani, M.V., 2010. The role of Kaizen events in sustaining a lean transformation, Annual Conference of Institute of Industrial and Systems Engineers (IISE). Cancun Mexico, 1.

Rich, S., Davis, K.H., 2010. Geographic Information Systems (GIS) for Facility Management. IFMA Foundation, Houston, TX.

Sacks, R., Akinci, B., Ergen, E., 2003. 3D modeling and real-time monitoring in support of lean production of engineered-to-order precast concrete buildings, Proc. 11th Annual Conference of the International Group for Lean Construction, Virginia, USA.

Sandberg, M., Mukkavaara, J., Shadram, F., Olofsson, T., 2019. Multidisciplinary optimization of life-cycle energy and cost using a BIM-based master model. Sustainability 11 (1), 286.

Sarkar, S.A., Mukhopadhyay, A.R., Ghosh, S.K., 2013. Root cause analysis, Lean Six Sigma and test of hypothesis. Total Qual. Manage. Bus. Excell. 16 (10), 1179–1191.

Sawhney, A., Riley, M., Irizarry, J., 2020. Construction 4.0: An Innovation Platform for the Built Environment. Routledge, UK.

Scorsone, E.A., 2008. New development: what are the challenges in transferring Lean thinking to government? Public Money and Manag. 28 (1), 61–64.

Shadur, M.A., Rodwell, J.J., Bamber, G.J., 1995. Factors predicting employees' approval of lean production. Hum. Relat. 48 (12), 1403–1425.

Shalabi, F., Turkan, Y., 2017. IFC BIM-based facility management approach to optimize data collection for corrective maintenance. J. Perform. Constr. Facil. 31 (1), 04016081.

Silverio, M., Renukappa, S., Suresh, S., Donastorg, A., 2017. Mobile computing in the construction industry: main challenges and solutions. In: Benlamri, R., Sparer, M. (Eds.), Leadership, Innovation and Entrepreneurship as Driving Forces of the Global Economy. Springer, Cham, pp. 85–99.

Sim, K.L., Chiang, B., 2012. Lean production systems: resistance, success and plateauing. Rev. Bus. 33 (1), 97–110.

Sobek II, D.K., Smalley, A., 2011. Understanding A3 thinking: A critical component of Toyota's PDCA management system. CRC Press, FL.

Sweeney, B., 2015. Lean QuickStart Guide: A Simplified Beginner's Guide To Lean Paperback. ClydeBank Media LLC, NY.

Tan, K.S., Bhaskaran, M., 2015. The role of the state in Singapore: pragmatism in pursuit of growth. Singap. Econ. Rev. 60 (03), 1550030.

Tai, S., Sun, C., Zhang, S., 2016. Exploring factors affecting owners' trust of contractors in construction projects: a case of China. SpringerPlus 5 (1), 1783.

Teo, T.L., Chan, C., Parker, C., 2004. Factors affecting e-commerce adoption by SMEs: a meta-analysis, ACIS 2004. Hobart, Australia, 54.

Tookey, J.E., 2011. Labour productivity in the New Zealand construction industry: a thorough investigation. Constr. Econ. Build. 11 (1), 41–60.

Wang, L.C., Lin, Y.C., Lin, P.H., 2007. Dynamic mobile RFID-based supply chain control and management system in construction. Adv. Eng. Inform. 21 (4), 377–390.

Warr, P.G., 2006. "Productivity Growth in Thailand and Indonesia: How Agriculture Contributes to Economic Growth," Working Papers in Economics and Development Studies (WoPEDS) 200606, Department of Economics, Padjadjaran University, Indonesia.

Van Assen, M.F., 2018. The moderating effect of management behavior for lean and process improvement. Oper. Manag. Res. 11 (1-2), 1–13.

Van Beveren, J., Thomson, H., 2002. The use of electronic commerce by SMEs in Victoria, Australia,. J. Small Bus. Manag. 40 (3), 250–253.

Vernikos, V.K., Goodier, C.I., Broyd, T.W., Robery, P.C., Gibb, A.G., 2014. Building information modelling and its effect on off-site construction in UK civil engineering, 167, 152–159.

Vial, G., 2019. Understanding digital transformation: a review and a research agenda. J. Strateg. Inf. Syst. 28 (2), 118–144.

Vogl, B., Abdel-Wahab, M., 2015. Measuring the construction industry's productivity performance: critique of international productivity comparisons at industry level. J. Constr. Eng. Manag. 141 (4), 04014085.

Watson, A., 2011. Digital buildings–challenges and opportunities. Adv. Eng. Inform. 25 (4), 573–581.

Womack, J.P., Jones, D.T., 1997. Lean thinking—banish waste and create wealth in your corporation. J. Oper. Res. Soc. 48 (11), 1148.

Woodhead, R., Stephenson, P., Morrey, D., 2018. Digital construction: from point solutions to IoT ecosystem. Automat. Constr. 93, 35–46.

Wu, W.W., 2011. Mining significant factors affecting the adoption of SaaS using the rough set approach. J. Syst. Softw. 84 (3), 435–441.

Wyman, O., 2018. Digitalization of the construction industry: the revolution is underway. Marsh and Mclennan Companies. https://www.oliverwyman.com/our-expertise/insights/2018/sep/digitalization-of-the-construction-industry.html. Accessed 7 September 2020).

Yadav, V., Jain, R., Mittal, M.L., Panwar, A., Lyons, A.C., 2019. The propagation of lean thinking in SMEs. Prod. Plan. Control 30 (10-12), 854–865.

Zhang, X., Warner, M.E., 2017. Business retention and expansion and business clusters–a comprehensive approach to community development. Community Dev. 48 (2), 170–186.

CHAPTER 20

Robust optimization and control for sustainable processes

Alessandro Di Pretoro[a,b], Ludovic Montastruc[b], Xavier Joulia[b], Flavio Manenti[a]

[a]*Politecnico di Milano, Dipartimento di Chimica, Materiali e Ingegneria Chimica "Giulio Natta, Milano, Italy*
[b]*Laboratoire de Génie Chimique, Université de Toulouse, CNRS/INP/UPS, Toulouse, France*

20.1 Robust design challenges in a renewables-based landscape
20.1.1 The renewables challenge

Over the last years sustainability has become by far the topic of major concern in the chemical industry domain. The energy challenge is the key point of the environmental friendly policies adopted by all the most developed and industrialized countries worldwide. For this purpose indeed several research programs, such as Horizon 2020, have been funded by the European Union and other institutions for the EU member states. As regards this decarbonization policy, the International Energy Agency defines renewables as the center of the transition to a less carbon-intensive and more sustainable energy system. It estimates them to satisfy around 30% of energy consumption demand in transport by 2060. Their role is particularly important in sectors that are difficult to decarbonize, such as aviation, shipping, and other long-haul transport (IEA Bioenergy Annual Report, 2018).

For all those reasons new energy sources and bioprocesses have seen a considerably renewed interest during the last decades. However, to be competitive with the petrochemical industry and to replace at least part of it as soon as possible, these new technologies need to be able soon to cover a substantial part of the market demand and to be flexible with respect to the market needs and the raw material properties (IEA Renewables, 2019).

The main issues delaying this transition are related to the substantially different nature of sustainable resources with respect to those derived from crude oil processing, that is also the reason why petrochemical resources were the first to settle since the very first phase of industrial revolution and are still the most established in the twenty-first century.

One of the main advantages of petroleum is the huge storage in the earth crust resulting in the feeling of a virtually endless availability. This is due to the fact that oil fields have been accumulating crude oil for thousands of years while we started consuming it since a couple of centuries only. The debate concerning the oil sources depletion rate and the technological improvements able to detect new oil fields to be exploited is not in the interest of this research. The fact is that we are consuming crude oil faster and faster and, besides the sustainability issues, alternatives should be available on time to cope with this possibility.

On the contrary, renewables (with few exceptions such as geothermal energy) are not stored in nature and their availability as well as their properties (e.g., biomass) are not even stable in time or accurately predictable. As the two key conditions to be fulfilled to meet the industrial demand are the average availability and the maximum capacity, new strategies should be adopted to merge the new resources with the already established market.

For this reasons, the next sections are focused on the conventional process design procedure and on the methodology requirements for sustainable processes design to compare them and highlight the analogies and the aspects that should be reconsidered. An innovative approach indeed is already required from the very next years and the purpose of this research is to actively contribute to outline its main features to be able to accomplish the sustainability tasks as soon as possible.

20.1.2 Recasting the petrochemical industry process design procedure

The conventional process design procedure employed so far can be outlined in the scheme represented in Fig. 20.1. With some exceptions, this block diagram, with some loops in it, summarizes the sequencing of the key steps to be carried out during the design of any kind of process and system in general.

For a general problem to be solved, the process choice comes first. Indeed all the following steps, such as optimal design and control, will depend on the specific process/unit that has been selected at the very first step. It often occurs that the feasibility study at least is carried out for a set of possible candidates and the best option is identified afterwards. This procedure requires more calculations but does not substantially modify the remaining steps. For a given design then, the feasibility analysis is carried out according to the expected operating conditions and, in case the process is suitable for the initial purpose, the design phase will follow.

The equipment design and system configuration is then performed for the nominal operating conditions according to the model equations and the desired process specifications. Although the way this phase is carried out strictly depends on the particular operation under analysis, some optimization criteria are common to any kind of system. In particular the profitability of the operation is a necessary condition to be satisfied otherwise the process is not worthy to be carried out.

Sometimes the process environmental impact and its sustainability are evaluated as well mainly according to the process energy demand. This choice provides further criteria to be used when more than one design solution is possible or whether the economic aspects can be partially sacrificed to achieve sustainability goals or to stay within the limits imposed by the safety or emissions legislation.

When possible, an eventual process intensification is considered to further optimize the system performances and reduce the energy demand with the purpose of decreasing the operating expenses and, if possible, the investments and emissions as well.

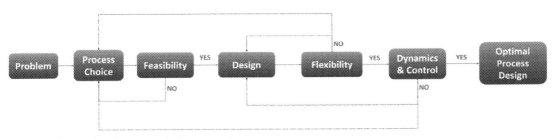

FIG. 20.1 Conventional process design procedure.

For the obtained design, an a posteriori flexibility analysis is usually performed. A more proper way to define this step would be sensitivity analysis or scenarios comparison as it is not really the flexibility of the system that is assessed. In the former case, some process variables of particular interest are perturbed and the system response as well as the feasibility boundaries are evaluated. In the latter case, different scenarios are compared for some values of the independent variables only. This phase provides as main outcomes the limit values that each variable can assume without making infeasible the desired operation and the quantitative understanding of the system response.

Once the selected design has proved to be sufficiently flexible, the final dynamic validation and control system configuration design are performed according to the several available solutions such as Proportional-Integral-Derivative (PID) feedback controllers with optimal tuning methods, dynamic optimization, Model Predictive Control (MPC), etc.

During the last century, this procedure resulted to be effective and has satisfied the requirements of the petrochemical industry in terms of methodologies and obtained results. The sequential solution of each step for the given nominal operating conditions has been possible thanks to the stable properties of the crude oil feedstock and the related long-term contracts between suppliers and companies.

On the contrary, renewable energy sources and sustainable raw materials are characterized by unstable nature as well as uncertain availability during the year's seasons and according to the geographical location. The different properties of the incoming green chemical industry forced the research and industrial sectors to reconsider the choice of processes aiming at the same final products accounting for a different feedstock. Another crucial task is the reduction of the energy consumption to replace part of the power plants processing fossil fuels with innovative solutions taking advantage of renewable energy sources such as solar, wind, geothermal, bioenergy, etc.

Given these premises, it is evident that innovative processes cannot be designed and optimized by means of the conventional methodologies as different standards are demanded. For this reason the usual design procedure needs to be revised and new optimization criteria such as flexibility, sustainability, and controllability should be integrated to cope with the sustainability related challenges.

The unified multicriteria design and control procedure is of course the final goal of Process Systems Engineering and the entire community is still working on it. However, some compromises exist and allow for a more robust design optimization and effective operations according to the incoming demand of more sustainable and flexible processes.

The new needs of an optimal design and control procedure for sustainable processes will be introduced and analyzed in detail in the following section to provide a preliminary outline of the revisions to be applied.

20.1.3 Sustainability-oriented design

The goal of this section is to list the new criteria and modifications, or at least the main ones, that should be applied to the conventional procedure to obtain robust optimization and control design of sustainable processes.

First of all, a robust design for sustainable processes should be based on high flexibility requirements. The flexibility assessment indeed should be performed according the rigorous methodologies proposed in the scientific literature throughout the entire procedure and contribute to the decision making of the optimal design. An a posteriori sensitivity analysis reliable only for an already designed system under nominal operating conditions is not enough. Moreover flexibility should be taken into account as the very first feasibility study. The purpose is indeed to evaluate if the operation is feasible or not for the entire range of expected deviation of operating conditions given by the uncertainty related to raw materials and energy supply.

Once the constraints that outline the feasible domain have been accurately defined, a combined economic and flexibility approach should be used for the design phase. Not only the optimal solution in terms of total costs needs to be considered but the decision maker should know also what is the price to pay for a higher flexibility and opt for the best compromise accounting for both the expected deviations and the affordability of the project.

The environmental impact as well needs to be evaluated over the entire perturbation range and included in the decision phase as the higher sustainability related to green feedstock and renewable energy sources could be easily undermined by a poorly adaptable system requiring a much higher energy consumption or producing more wastes than expected.

The final challenge for a robust design for sustainable processes concerns process dynamics and control. If disturbances in the operating conditions become the norm for the process, the control configuration goal is not anymore to suppress them but to cope with them and to optimize the transition between perturbed operating conditions. This ability to switch between different states affects both the system and the control configuration design. Therefore, the different design solutions and control strategies should be compared from a flexibility perspective and, eventually, some design choices could need to be reconsidered to have smoother transitions.

In conclusion, the final goal of this entire research work is to revise the conventional process design procedure to make it suitable for sustainable operations based on uncertain process variables. To accomplish this task, all the requirements discussed above will be integrated to each step of the design block diagram and the obtained outcome and eventual improvements will be analyzed in the dedicated section.

20.2 Feasibility assessment
20.2.1 Feasibility limits in biomass processing

The feasibility assessment usually represents the very first step of a design procedure whatever the system, process, or operation under study. The expected outcome of this design step is generally one of the two following alternatives:

- Yes, the selected option is a feasible and suitable candidate for the problem solution;
- No, the selected option does not meet the feasibility requirements (with an eventual detection of the inherently violated constraint).

The detailed feasibility quantification and the feasible boundaries outline are usually part of an eventual sensitivity analysis later performed in the procedure. Anyways, this sensitivity analysis evaluates the feasibility limitations after the process design phase has been already carried out, that is, it is strictly related to it and accounts for "weak" constraints, concerning for instance the equipment sizing, that could have been adjusted if an a priori analysis was performed instead.

It can be then concluded that the procedure described so far results to be effective and reliable only in case nominal operating conditions can be precisely detected and defined. Unfortunately this is not the case of sustainable processes.

As already discussed in the Introduction, the properties of green feedstock, such as biomass, are uncertain in nature during the year's seasons and according to the geographical locations. The same remark can be applied to the availability, stability, and costs of renewable energy sources. Therefore, in all these cases, an "a priori" feasible domain quantification is of critical importance to preliminarily assess the flexibility limitations of the process as well as the most critical constraints and the related process variables.

The feasible domain from a flexibility perspective is defined as the subregion of the space defined by the uncertain system parameters that is bounded both by strong and weak constraints (adjustable during the design phase).

Before going further, it is worth to better define these two categories of constraints. The expression "strong constraint" will be used hereafter when referring to constraints that do not depend on their affordability. The most significant examples of strong constraints that cannot be violated in a system design are as follows:

- the principles of mathematics and physics;
- the legislation;
- the safety restrictions, etc.

On the other hand, the so-called "weak" constraints are all those limitations that, ideally, can be relaxed during the phase of a flexibility-based design. Examples of weak constraints are the equipment sizing and the process profitability. Of course they should be considered as strong constraints in case of flexibility assessment for already existing systems except in case modifications or oversizing could still be made. The affordability of these design choices will be later discussed during the decision making phase following the process design results but will not be taken into account when performing the feasibility assessment.

In conclusion, it can be stated that, for all those systems subject to uncertain operating conditions and for which a flexibility based design is required, the operation feasibility should be quantified as the very first phase of the design procedure to know a priori what are the process variables perturbations causing the violation of strong constraints independently from the system design later carried out.

The following section presents the biorefinery case study that will be used to show how this new procedure of feasibility assessment could be carried out. However, with the proper adaptations, the main features of the analysis can be applied to any system other than chemical processes.

20.2.2 The biorefinery distillation case study

As introduced in the previous sections, biorefineries are part of those cases where the uncertainty of the operating conditions is an inherent feature of the process nature itself. The feedstock nature varies all along the year's seasons and according to the kind of biomass suitable for farming in the specific geographical location.

For all this reasons the acetone-butanol-ethanol (ABE) mixture separation case proposed by Di Pretoro et al. (2020b) will be used throughout the chapter to show the importance of a robust optimization accounting for more than the profitability criterion only for the process design.

The ABE fermentation is an anaerobic digestion process first introduced by the chemist Chaim Weizmann (1919) during the World War I with the purpose to produce acetone from carbohydrates such as starch and glucose for the British war industry. The most effective and widely used bacterial strain is the *Clostridium acetobutylicum* and its main products in the fermentation broth are acetone and butanol. Several studies aimed at the identification of the optimal bacterial strain and operating conditions to carry out this process are available in the literature (Ezeji et al. 2004; Groot et al. 1989).

Besides the feedstock uncertain nature, an additional challenge for this process comes from the fact that the ABE fermentation is strongly affected by product inhibition (Garcìa et al. 2011), that is there exists an upper limit for butanol concentration over which the produced butanol becomes poisoning for the bacteria. However, butanol is nowadays the most valuable product of the process and its production needs to be maximized. For all these reasons the effective recovery of butanol and acetone at least in the product purification section is of critical importance for the entire process effectiveness and profitability.

Moreover, it is worth to highlight that the most abundant component in the fermentation broth is water. The presence of water in alcoholic mixtures, even in case of small quantity, leads to relevant issues for the separation process. From a thermodynamic point of view indeed both ethanol and butanol show nonideal interactions with water giving rise to a homogeneous and a heterogeneous azeotropic species, respectively (Dortmund Data Bank) and thus to a liquid phase demixing in a particular subregion of the butanol-water composition space. The presence of those non-ideal behaviors makes the separation process even more challenging due to the presence of strong constraints (i.e., physical laws).

Given these premises, it is evident that a robust design for the separation section of a sustainable process strongly requires to be flexible to avoid eventual perturbations reducing the plant productivity with consequent economic losses leading to the unprofitability of the entire process.

The proposed biorefinery separation section case study will refer then to a series of distillation columns aiming at the purification of the multicomponent mixture. The process feed comes from a preliminary dewatering process performed upstream via liquid–liquid extraction. Therefore, in this case study, the flexibility assessment is not only useful to optimize the system response for unstable feedstock properties but also to discuss how malfunctioning in the preliminary operations can be withstood by the downstream processing section.

The main case study process parameters are listed in Table 20.1. The butanol and water partial flow rates have been selected as uncertain parameters as they have been detected as the most critical variables affecting the operation feasibility. The separation process aims at the effective recovery of butanol and acetone without accounting for ethanol whose recovery worthiness can be discussed in further studies.

The methodology and the results of the feasibility assessment carried out for this example are discussed in detail in the next section.

20.2.3 Feasibility assessment via residue curve maps

Several methodologies have been proposed in the literature throughout the last century for the feasibility assessment of distillation processes. One of the most effective and well-established among them is the one based on residue curve mapping. Residue curves are defined as curves in the composition space that describe the residue composition evolution during an open evaporation as well as the composition profile of a packed bed distillation column. The RCs coordinates and plot can be, respectively, calculated and traced in the composition space by means of the differential equation representing the mass balance for each generic component i as

$$\frac{dx_i}{d\xi} = x_i - y_i$$

where x and y are the liquid and vapor composition, respectively, and ξ is the dimensionless time variable.

20.2 Feasibility assessment

Table 20.1 Process variables, uncertain parameters, and separation specifications.

Variable	Component	Value	Unit	
Flow rate	Acetone	12.030	mol/s	
	n-Butanol	61.328	mol/s	uncertain
	Ethanol	3.839	mol/s	
	Water	12.428	mol/s	uncertain
Pressure		101.325	kPa	
Temperature		361.26	K	
Specifications				
Purity	Acetone	0.995	w/w	
	n-Butanol	0.99	w/w	
Recovery	Acetone	0.985	/	
	n-Butanol	0.9604	/	

This curves move from an unstable node to a stable one, representing the less and most volatile mixture compounds respectively, passing close to a saddle point that, in general, is the one with intermediate volatility. The residue curves connecting a stable or unstable node to a saddle one generate the so-called distillation bundles and are called separatrices of the composition space. As a consequence, all the residue curves belonging to the same distillation bundle are characterized by the same stable and unstable nodes.

It can be proved that, with few exceptions, the residue curves passing through the feed and products characteristic points of a distillation column should lie in the same distillation region, that is, need to have the same stable and unstable nodes. This condition will be used in this section to assess the feasibility boundaries of the separation case study. For further details about residue curve mapping, the book written by Petlyuk (2004) is nowadays the most complete reference in this domain.

The feasibility assessment proposed in this section via residue curve mapping refers to the work carried out by Di Pretoro et al. (2020a) for the ABE/W mixture. Given the feed properties listed in Table 20.1, the content of each component has been individually perturbed and the new distillate and bottom products compositions have been consequently calculated according to the process specifications and the partial mass balances. The non-ideal thermodynamic behaviour of this mixture was described by means of the Non-Random Two-Liquid (NRTL) activity coefficient model (Renon and Prausnitz, 1968). Under nominal operating conditions the stable node of the corresponding distillation region is pure acetone, while the unstable one is pure butanol as it is the bottom product. The perturbation value for which the residue curve passing through one of the characteristic points falls in the distillation bundle defined by pure acetone and pure water determines the feasibility boundary for the specific component.

Being the number of components equal to 4, the RCs can be represented in a 3D quaternary composition space with the pure compounds located at the vertex of the tetrahedron of unit edge length. Increasing the content of one of the mixture components can be graphically seen as shifting the feed characteristic point closer to the corresponding vertex along their connecting line (cf. black arrow in Fig. 20.2A).

The results in the composition space are shown in Fig.20.1A and B for the water content perturbation as it resulted to be the most critical parameter for the process feasibility with a maximum allowed deviation of +8%. In particular, it can be noticed that, for higher deviations, the characteristic point related to the water-rich stream falls in the distillation bundle characterized by pure water as a stable

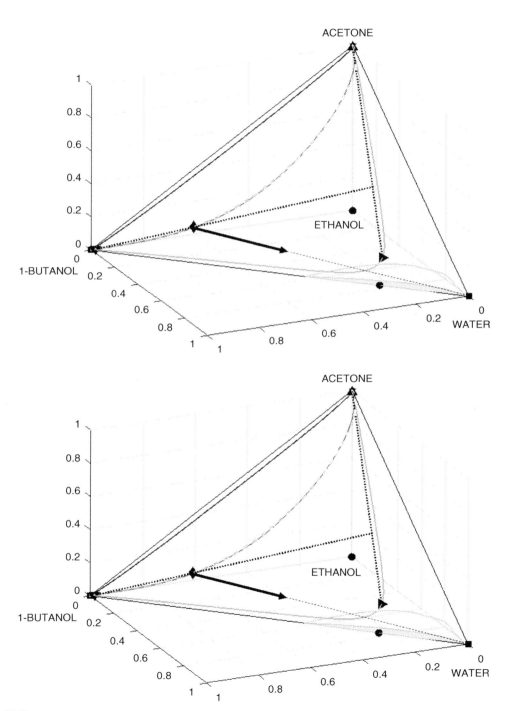

FIG. 20.2

(A) RCM diagram under nominal operating conditions. (B) Feasibility boundaries.

node. The second most critical parameter is the butanol content with a maximum negative allowed deviation of -14%. That is why they will be used as uncertain parameters throughout the entire chapter to perform the optimal flexibility–based design of the separation process.

The next step of the procedure is the "design" phase. In the light of the results obtained in this feasibility assessment, it will be discussed in detail in the following section.

20.3 Flexible, sustainable, and economic optimal process design procedure

20.3.1 Premises on the uncertainty characterization

As discussed so far, uncertain operating conditions are part of the sustainable processes nature and we should cope with them whenever a system design needs to be performed. However, before explaining the methodology for the uncertainty quantification in a flexible process design, a brief introduction about the uncertainty characterization is worth to be presented.

In recent years, uncertainty has attracted a lot of attention in the design optimization domain. The different kinds of uncertainty have been indeed formally classified by Oberkampf and Helton (2012) in the following three main categories:

- aleatory uncertainty (also referred to as stochastic uncertainty, irreducible uncertainty, inherent uncertainty, or variability);
- epistemic uncertainty (also referred to as reducible uncertainty, subjective uncertainty, model form uncertainty, or simply uncertainty);
- error.

In particular, aleatory uncertainty is defined as the inherent variation associated to the physical environment under consideration, while the epistemic uncertainty is defined as any lack of knowledge or information in any phase or activity of the modeling process. Finally, error is defined as a recognizable deficiency in any phase or activity of modeling and simulation that is not due to lack of knowledge (Oberkampf et al. 1998, 1999).

Among them, aleatory uncertainty is particularly suitable for the description of variable input data, such as process variables, and it can be characterized by using the probability theory, that is, by means of probability distribution functions (PDFs) (Agarwal et al., 2004). Therefore, it is straightforward that this kind of design under stochastic uncertainty is the one we are referring to when talking about flexibility-based design.

Given this classification background, further details about uncertainty quantification will be then presented in the following section.

20.3.2 Introduction to the flexibility indices

To have a unified convention concerning the system flexibility quantification, suitable and well-defined indices should be used. Several of them have been proposed in the literature during the last decades and their detailed analysis and comparison on a distillation column case study can be found in Di Pretoro et al. (2019).

The most widely used, both deterministic and stochastic, indices are represented in Fig. 20.3A and 20.3B and will be further detailed below. Among them, the most commonly used flexibility index in

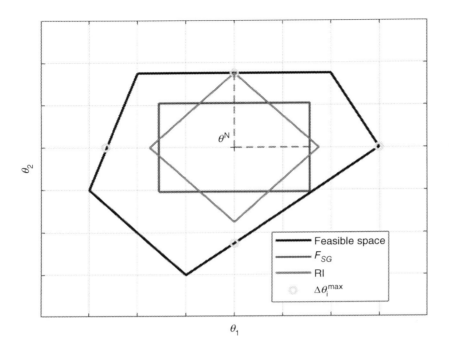

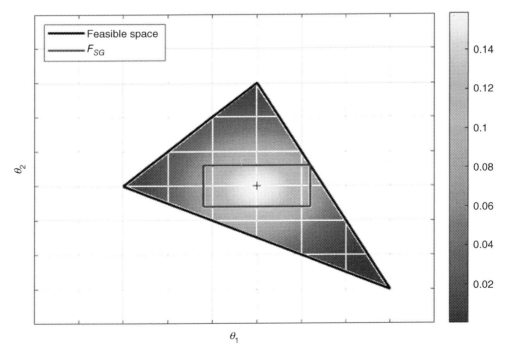

FIG. 20.3 Graphical representation of:

(A) Deterministic flexibility indices (B) Stochastic flexibility index.

process systems engineering is by far the one proposed by Swaney and Grossmann (1985), hereafter denoted as F_{SG}. It is used to assess the maximum fraction of the expected deviation of all the uncertain variables at once that can be accommodated by the system. The related flexibility index problem can be formulated as

$$F_{SG} = \max \delta$$

$$\text{s.t.} \max_{\theta \in T(\delta)} \min_z \max_{j \in J} f_j(d,z,\theta) \leq 0$$

$$T(\delta) = \{\theta \mid \theta^N - \delta \cdot \Delta\theta^- \leq \theta^N \leq \theta^N + \delta \cdot \Delta\theta^+\}$$

where θ are the uncertain parameters, d the design variables and z the control variables, respectively. Then T is the hyperrectangle representing the expected deviations, δ is the related scale factor, and f_j the set of constraints.

Another deterministic flexibility index, particularly suitable when substantial deviations of a single uncertain parameter at a time are expected, is the so-called resilience index (RI) introduced by Saboo and Morari (1985). In fact, it estimates the largest total disturbance load, independent of the direction of the disturbance, that a system is able to withstand without becoming unfeasible. This different definition makes it less conservative than the F_{SG} index. However, their different values for the same physical system are due to the fact that they assess different properties and therefore the most suitable index should be employed according to the expected deviation nature. The related flexibility index problem can be mathematically formulated as

$$RI = \min_i \Delta\theta_i^{max}$$

$$\text{s.t.} f_j(d,z,\theta) \leq 0, \forall \, \Delta\theta_i : \sum_i |\Delta\theta_i| \leq RI$$

The two indices presented so far are based then on geometrical properties of the uncertain domain, that is, they evaluate a maximum disturbance magnitude without accounting for its probability to occur. To give a different weight to all those operating conditions that are more likely to occur with respect to those whose occurrence is remotely possible, Pistikopoulos and Mazzuchi (Pistikopoulos and Mazzuchi, 1990) introduced in 1990 a so-called stochastic flexibility index (SF) that has soon become the best established indicator among those accounting for the uncertainty probability. Given a PDF of the uncertain parameters $P(\theta)$, the SF index formulation is given by

$$SF = \int_\Psi P(\theta) \cdot d\theta$$

where ψ represents the feasible space. Thus, according to this definition, while the deterministic flexibility indices represent a perturbation magnitude, the SF value assesses the probability that an incoming deviation will be withstood or, analogously, the fraction of deviations that the system is expected to withstand.

The three indicators presented above will be then included in the following step of the procedure to account for flexibility during the process design phase and to highlight how the expected costs are affected by the need of a more flexible unit according to the sustainable processes requirements.

20.3.3 Application to the biorefinery case study

This section presents the results obtained when the design step of the conventional procedure is reconsidered accounting for an a priori flexibility analysis by means of the previously presented indices. Its application to ABE/W case study is discussed below according to the research work carried out by Di Pretoro et al. (2020b).

The conventional system aimed at the purification of multicomponent mixtures into their single components by distillation is the so-called distillation train. It consists of a sequence of distillation columns arranged either in parallel or series layout. The possible alternatives are presented in Fig. 20.4 and mainly classified as follows:

- Direct configuration: the most volatile compound is purified first;
- Indirect configuration: the least volatile compound is purified first;
- Midsplit configuration: an additional column preliminarily splits the more from the less volatile compounds.

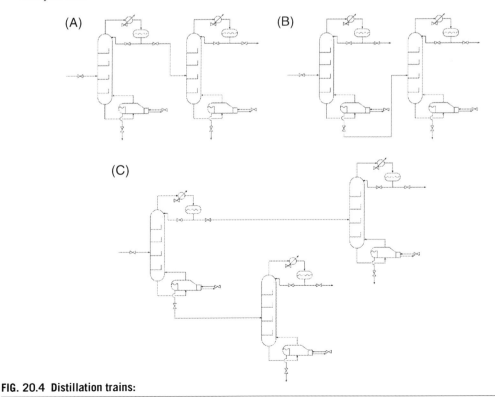

FIG. 20.4 Distillation trains:

(A) Direct, (B) indirect, (C) midsplit.

20.3 Flexible, sustainable, and economic optimal process design procedure

To perform the ABE multicomponent mixture separation with the purpose of recovering pure butanol and acetone at least, all of them have been considered and tested. Under nominal operating conditions, due to the high butanol content in the feed stream, the indirect configuration resulted cheaper than the midsplit one. On the other hand, the direct layout is barely feasible due to larger number of equilibrium stages required by the first column to purify acetone at the desired purity from a feed stream with such a low relative acetone content.

The economic assessment, presented below, was carried out by means of the CAPital EXpenses (CAPEX) and OPerating EXpenses (OPEX) correlations provided by Guthrie (1969, 1974), Ulrich (1984), and Navarrete (2001).

The indirect configuration still results to be more profitable than the midsplit one due to the convenience of the immediate removal of the most abundant component, that is butanol, in the first column of the train. The indirect configuration Total Annualized Costs (TACs), calculated for operating conditions ranging over the entire uncertain domain, was then assessed and the resulting trend is graphically represented in Fig. 20.5 as a function of the partial water and butanol flow rate deviations.

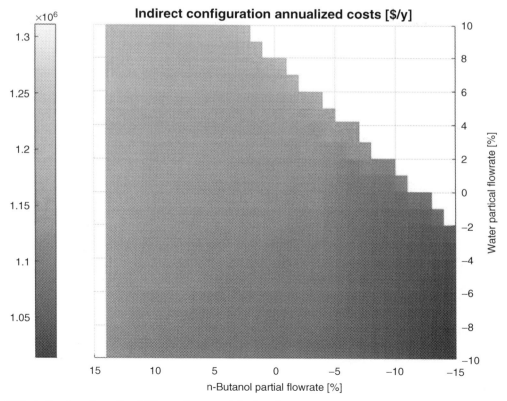

FIG. 20.5 Indirect configuration TAC over the uncertain domain.

Looking in detail at the chart, two main subregions of the uncertain domain where a TAC increase is observed can be distinguished. The first almost linearly increasing trend is observed in the direction of positive deviations both for water and butanol flow rate. This is obviously due to the higher feed flow rate to be treated and thus to the higher capacity demanded of the system. Besides the fact that it was an expected behavior, it is of poor interest for what concerns flexibility.

On the contrary, the region of higher costs that is much more interesting from a flexibility perspective is the one close to the feasibility boundary, that is higher water and lower butanol content. When approaching these operating conditions, as already highlighted during the feasibility assessment phase, the separation becomes much more challenging from a physical point of view and, therefore, to keep the specifications fulfilled, for a given number of equilibrium stages, a higher reflux ratio is required. This higher reflux ratio implies then a higher flow rate circulating along the unit and, as a consequence, both higher duties in the condenser and in the reboiler as well as higher investments in terms of required heat transfer surface areas and column diameter.

The results thus obtained were then coupled with the flexibility analysis with the final purpose to conceive a graphical tool able to resume with one single trend all the required information. Both the F_{SG} and the RI indices were used for the deterministic assessment. On the other hand, as no detailed information was available about the uncertainty likelihood, a Gaussian probability distribution function was employed for the stochastic analysis. Although this choice necessarily affects the analysis outcome both from a qualitative and quantitative perspective, the procedure keeps nevertheless remain the same whatever the system under study and the deviation PDF as well as some main features of this index that depend on its formulation only.

The trends in Fig. 20.6A show the additional costs as a function of the deterministic flexibility indices. As it can be noticed, a maximum value of about 5% can be achieved for the F_{SG} index before exceeding the feasible domain corresponding to a maximum of 6.5% additional costs. These values become about 8% and 9.5%, respectively, when using the RI instead. As expected, the RI index results are less conservative than the Swaney and Grossmann one as it accounts for the deviation of just one of the uncertain parameters at a time. Both of them show an almost linear trend for moderate perturbation with a much steeper increase when approaching the feasibility boundary as already discussed according the economic assessment results of Fig. 20.5.

On the other hand, the stochastic flexibility assessment presents some different features worth discussion. First of all, the SF value in correspondence of the nominal operating conditions is different from 0. This is due to the fact that, under nominal operating conditions, the system is already able to withstand all that part of deviations that would have required a smaller equipment sizing. Second, it shows a vertical asymptote in correspondence of the value 76.5%. The value itself is specific to the employed PDF; it represents indeed the maximum stochastic flexibility that can be achieved as the remaining part of the uncertain conditions lies outside the feasible region. Independent of the selected PDF, the vertical asymptote is the graphical consequence of the characteristic definition of any probability distribution function. Its integral over the entire space should be indeed a finite value and, in particular, equal to 1.

The cost of the higher accuracy of the SF lies behind the need of more information, that is the knowledge of the uncertainty PDF. However, this plot allows for a decisional analysis as the very first part of the curve provides higher flexibility for small additional costs, but still in the range of small SF; while the last part has higher value of SF but for a much more relevant investment increase. Therefore, the best design compromise for flexible operations can be detected in the region with the highest

20.3 Flexible, sustainable, and economic optimal process design procedure

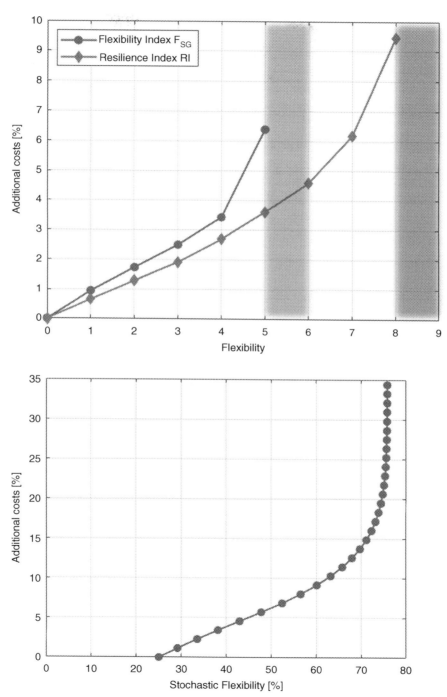

FIG. 20.6 Flexibility analysis results:
(A) deterministic flexibility, (B) stochastic flexibility.

curvature, that is SF = 65%–72%, where the flexibility value is not far from the upper boundaries and additional costs still lie in the range of 10%–15%.

In any case, besides the general remarks, the final design choice always depends on the affordability of the project and the particular requirements of the specific system. The purpose of this research is just to provide an effective and easily understandable decisional tool for sustainable flexible and profitable process design.

A further step during the process design procedure could be the consideration of more effective solutions. The robust optimization of energy-effective systems is one of the topics of major concern in the sustainable processes domain. For this purpose, the effect of process intensification on the ABE case study will be assessed and commented in detail in the next section.

20.4 Process intensification
20.4.1 General features

In recent years, a growing interest toward process intensification can be observed due to the need of more effective and profitable systems and to the general CO_2 emissions concerns related to the energy consumption (Reay et al. 2011). In process engineering, process intensification is a well-established practice in both the research and industrial domains and covers applications widely ranging from heat exchanger networks to reaction/separation processes.

- Process intensification is defined by Stankiewicz and Moulijn (2000) as the analysis of two main areas:
- Process-intensifying equipment, such as novel reactors, and intensive mixing, heat-transfer, and mass-transfer devices;
- Process-intensifying methods, such as new or hybrid separations, integration of reaction and separation, heat exchange, or phase transition (in the so-called multifunctional reactors), techniques using alternative energy sources (light, ultrasound, etc.), and new process-control methods (such as intentional unsteady-state operation).

Of course new methodologies might imply innovative types of equipment to be designed and vice versa. On the other hand, novel units make often use of new, unconventional processing methods.

The main advantages of this innovative concept for process systems is that it allows both for very relevant savings and reduced energy consumption, that is lower environmental impact (Reay et al. 2011). However, all those advantages have been tested and proved to be very effective when the intensified system works under the operating conditions considered during the design optimization. The limitations of this design solution have been poorly studied in the available literature and mainly refer to the difficulties related to the process controllability.

The purpose of the study carried out in this section is then to assess how process intensification affects the ability of a system to deal with external disturbances without becoming infeasible or less profitable than the conventional process configurations. The same analysis will be carried out in terms of energy consumption and environmental impact. Basically, the final goal is to quantify the price to pay in terms of reduced conduction and flexibility of the different units when parts of them are integrated one another.

To discuss this aspect of process integration a dedicated analysis for distillation based on the ABE case study is presented and discussed in the next section.

20.4.2 Applications to distillation

The most common application of process intensification to the distillation unit operation for a multicomponent mixture separation is by far the so-called Dividing Wall Column (DWC). It consists of several columns of a distillation train arranged in a single column shell and separated by an internal wall. It was first conceived by Petlyuk in 1965 (Petlyuk, 1965) but its detailed analysis and development was deepened as deserved during the last decades only.

An example of equivalence between the columns of a distillation train and the corresponding sections of its DWC configuration is shown in Fig. 20.7. Besides the need of a smaller surface on the plant field, this intensified solution was proved to be very effective and to allow about the 30% savings both in terms of investments and energy consumptions (Kiss 2013).

We want then to test if it is possible to design a DWC equivalent to the distillation train discussed in Section 20.3 for the ABE case study and to assess and compare their performance under perturbed operating conditions. The optimized DWC layout with respect to the nominal operating conditions was designed according to a feasible path-based procedure (Di Pretoro et al., 2021). The obtained configuration is shown in Fig. 20.8 and results in 32.9% lower investments and 29.8% lower energy consumption with respect to the equivalent indirect distillation train.

Being the obtained DWC configuration the cheapest one under nominal operating conditions, the flexibility assessment results will be referred to its value. The feasibility assessment via residue curve map for the DWC layout was already carried out by Di Pretoro et al. (2020a) according to the methodology explained in Section 20.2.

The flexibility analysis was performed for other two DWC configurations with lower and higher number of trays, namely 37 and 52, to compare the effect of more equilibrium stages in process-intensified solutions under uncertain operating conditions. The economic assessment was performed as usual by means of the Guthrie–Ulrich–Navarrete correlations, while for the environmental impact estimation the methodology and correlations proposed by Gadalla et al. (2006) for crude oil distillation units were used. In the latter research article, it is proved that the main parameters affecting the CO_2 emissions in distillation columns are the reboiler heat duty first (about 90%) and the electricity for pumping second (about 6%), while the emissions related to the material for the equipment construction are lower and distributed over the entire plant lifetime. Moreover, as raw materials and products streams are the same for all the configurations to be compared, they are not worth to be taken into account for in the sustainability assessment.

The results for both the economic and the environmental impact assessment are shown in Fig. 20.9A and B as a function of the Swaney and Grossmann flexibility index F_{SG}. As it can be noticed, the DWC configuration still results more profitable and sustainable under perturbed operating conditions only in the proximity of the nominal ones, that is for low deviation magnitude. In fact, while the conventional distillation train requires higher additional costs within the 10% range for a disturbance intensity increase, the additional costs related to the optimal DWC configuration, as well as for the nonoptimal ones, show an exponential growth. In particular, there exists a flexibility index value, that in this case is about 2.5%, after which the process-intensified unit becomes more expensive than the conventional one.

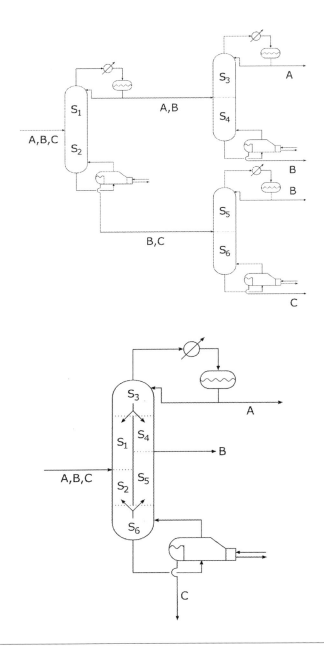

FIG. 20.7
(A) Conventional distillation train versus (B) equivalent DWC Configuration.

The exponential trend of those curves reflects the system response in terms of higher reflux ratio required to compensate the feed composition disturbance to maintain the process specifications. As the DWC is the integration of two distillation columns, a higher reflux ratio involves indeed the entire

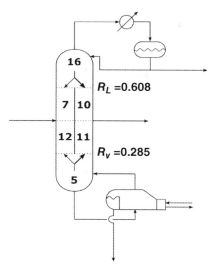

FIG. 20.8 Optimized DWC for ABE separation.

system propagating the external perturbation on both sides of the wall. On the contrary, in the conventional distillation train, just the first column is involved affected by the disturbance suppression, leaving the second column undisturbed.

This relationship between flexibility and reflux ratio is further proved when observing the trends for 37 and 52 stages, respectively. In the former case, costs and emissions are always higher than the optimal 42-stage layout as the lower number of stages always implies a higher value of the required reflux ratio. On the other hand, for a higher number of equilibrium stages two different regions can be distinguished, namely a moderately and a highly perturbed one. The presence of more trays allows indeed a better distribution of the external perturbation and helps to mitigate the reflux ratio disturbance. There exists then a flexibility index value beyond which the 52-stage layout performs better than the optimal 42-stages layout. However, both of them are way more expensive than the nonintensified distillation column train configuration.

The same remarks apply to the CO_2 emissions as they are strictly related to the reboiler heat duty, that is to the flow rate circulating along the column due to the reflux ratio exponential increase.

In the light of these results, it can be then concluded that, when dealing with oscillating operating conditions, that is the case of processes based on renewables and green feedstock, the better performance of the process-intensified solutions with respect to the conventional ones should be carefully evaluated. In fact, their lower costs and emissions need to be proved not only with respect to the nominal operating conditions but all over the expected deviation range. Inversely, it can be stated that, for a given design, the process intensification reduces the system flexibility and involves higher energy demand. For this reason, if looking for a robust optimal solution, in case of sustainable processes, flexibility should be included among the criteria used to compare intensified and conventional design solutions otherwise a substantial bias could result between the expected and the actual costs and emissions.

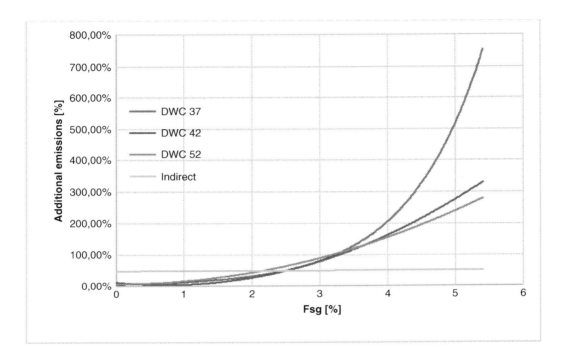

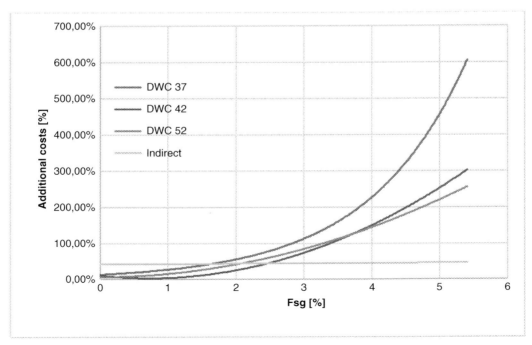

FIG. 20.9

(A) Economic assessment. (B) CO_2 emissions assessment.

20.5 Process dynamics and control
20.5.1 The concept of switchability

Although the very last step before the final design validation, the analysis of the system dynamics and the design of the most suitable control configuration according to the process needs is of the primary importance for the effective plant conduction.

Even though, by definition, the dynamics and control domain deals with process perturbations, its optimal design is conventionally conceived to perform at its best in the proximity of nominal operating conditions. The external perturbations are mainly seen as eventual disturbances that the system should be able to manage for an effective control configuration.

On the contrary, accounting for dynamics in flexible process design means that the presence of perturbations is completely normal and expected and that the system design should be performed so that it remains feasible during the entire transient and not only in correspondence of the nominal and perturbed steady states. The same idea can be used to assess the dynamic flexibility of already existing equipment. In any case, process dynamics cannot be decoupled from process control, therefore the obtained results are always function of the selected control configuration.

A robust design procedure for sustainable processes should be then a procedure optimized with respect to the continuous switching between operating conditions as well. The switchability analysis is the procedure of assessing the ability to switch between two different states. It was discussed first in the chemical engineering domain by White and Perkins in 1996 (White et al., 1996). This tool can be used for several purposes:

- During the design phase to optimize the system design;
- For existing units to assess how much "switchable" they are;
- To compare different control strategies;
- To compare different start-up or shut-down procedures, etc.

From an operational point of view, the only index proposed in the literature to study the process dynamics from a flexibility perspective is the dynamic flexibility index (DF) proposed by Dimitriadis and Pistikopoulos (1995); unfortunately it has received poor attention during the last decades. The DF definition is based on the Swaney and Grossmann steady-state flexibility index by including the semiinfinite time domain as a system variable. This means that both the operating conditions and the feasibility constraints might vary in time.

Although this index results to be really effective for the dynamic flexibility estimation, its value provides poor information about how much dynamics is more constraining for the system flexibility with respect to the steady-state analysis. Therefore, to assess the switchability property on a general system, a new index can be defined by combining the two behaviors as suggested by Di Pretoro et al. (2020c) and better described in the following section.

20.5.2 Application of the switchability assessment

According to the studies performed by Di Pretoro et al. (2020c), the switchability assessment is carried out by means of the corresponding index. The switchability index (SI) is indeed defined in the aforementioned research article by combining dynamic and steady-state flexibility indices as follows:

$$SI = \frac{F_{SG}}{DF}$$

Given this definition, being the dynamically feasible domain a subregion of the steady-state one, the proposed index can vary in the range [0, 1] where the boundary values represent the two extreme cases where process dynamics results to be either totally constraining or negligible from a flexibility point of view. In case an index other than the one proposed by Swaney and Grossmann is used, it should be employed for the dynamic flexibility assessment as well to ensure the compliance of the obtained values.

To provide a quicker understanding of the way this index works, it has been applied to the ABE/W case study to quantify the related implications and have a better idea of the outcome to be expected from a switchability assessment of a distillation separation process. As already discussed in the previous section, the column mainly affected by the external disturbances in the indirect configuration is the first one, that is, one with the purpose to purify the biobutanol. This first distillation unit not only is the most constraining parameter from a feasibility point of view, as proved in Section 20.2, but also in terms of additional costs for a flexible design as discussed in Section 20.3.

For this reason, the results presented below will be focused on the first distillation unit of the indirect sequencing. Being the SI related to the control strategies, it could be interesting to compare two possible control structures, namely the Reflux-boilup control structure (LV) and Distillate-boilup control structure (DV). They are named after the manipulated variables aimed at the process specifications control in a dual composition control system (Skogestad et al. 1990; Ryskamp 1980). In particular, the manipulated variables in the LV are the reflux ratio and the vapor boil-up, while, for the DV, the reflux ratio is replaced by the distillate flow rate.

The dynamic simulation was performed by means of AVEVA DynSim process simulator and the obtained results, obtained over a discretized deviation domain, have been regressed to have smooth trends. Both the graphics presented hereafter are traced up to a flexibility value equal to 5% as, for higher value, the infeasibility of the separation yet at steady state was proved in the previous sections. The first chart in Fig. 20.10 shows the additional costs as a function of flexibility as usual. In particular, the blue line refers to the steady-state results, while the yellow and red ones to the dynamics of the DV and LV, respectively. First of all, it should be highlighted that, for a given design, the steady-state index always overestimates the actual flexibility of the column or that, inversely, for a given flexibility requirement it always underestimates the additional costs. If we compare the two different control structures we can distinguish two main regions. In particular, the DV results to be better than LV for low deviation magnitude and vice versa. To have a more immediate understanding, let us try to use the SI instead of the flexibility ones.

One of the possible graphics that can be obtained from the switchability assessment is the SI versus Flexibility plot. In this plot, the critical point beyond which the LV configuration performs better than the DV one in terms of switchability can be straightforwardly detected for DF equal about to 1.2%. Moreover, the step at DF = 3.6% shows the effect of the column diameter behavior. For a dynamic flexibility with the DV control structure higher than this value, indeed, the minimum and maximum diameter of the column achieve the same value, that is flooding and weeping conditions cannot be both avoided with a single diameter column. Therefore a column with two diameters is required with much higher investment costs. For this reason then, the switchability drastically drops in the proximity of this point as, to maintain the SI value, much higher additional costs are required soon. In any case, besides the results that are obviously case specific, this kind of analysis clearly shows the general advantages obtained when including the SI assessment in the design phase.

In conclusion, in this section it was proved that, when designing process units for sustainable and flexible operations, the steady-state flexibility assessment only can be not sufficient as it does not account for the behavior of the system during the transition from one state to another.

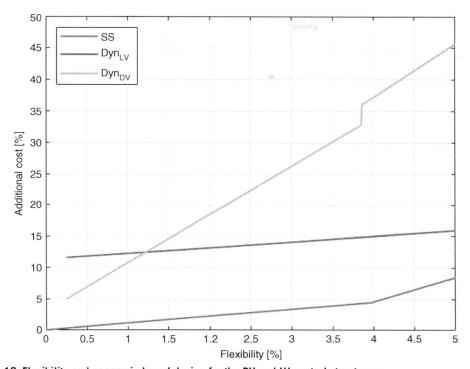

FIG. 20.10 Flexibility and economic based design for the DV and LV control structures.

For this reason, the switchability criterion needs to be integrated in the "Dynamics and control" design step to perform the very last complete check before the final validation and compare the different control strategy options according to their ability to smoothly switch between different operating conditions selecting the optimal one.

Once this task is accomplished, the obtained design could be validated as a flexible, sustainable, and operable optimal design. The revision of the conventional design procedure performed in this research work is thus complete. This domain remains nevertheless worthy to be still studied in deep as further improvements could be possible. The main outcome of each section will be then resumed and commented as deserved in the conclusion part that follows.

20.6 Conclusions

As stated since the very beginning of this chapter, the conventional process design procedure has been thoroughly analyzed and reconsidered throughout this research work. A substantial revision of the initially presented process scheme was performed and a new block diagram can be proposed as shown in Fig. 20.11.

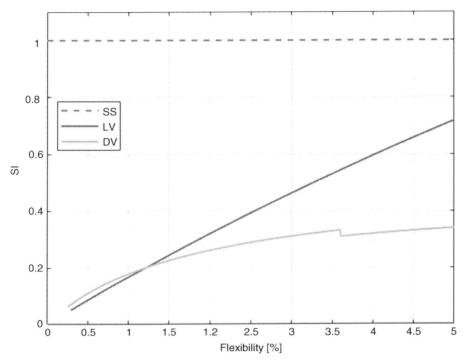

FIG. 20.11 Switchability assessment results.

This new process design procedure, that may still leave room for improvement, results to be much more suitable to the challenges related to the renewables based energy and process industry. It aims at flexible and sustainable design solutions without sacrificing the process profitability. The presence of a higher number of connections between the different design phases with respect to the conventional procedure implies an eventually higher number of decision-making iterations before achieving the optimally flexible, sustainable, and operable solution. This higher computational effort is the price to be paid for a more demanding technology and can hardly be avoided. However, in terms of computational effort, the proposed solution still represents an affordable compromise between the outdated procedure and the multicriteria unified process and control design, that is the very final goal of Process Systems Engineering, as the partial decoupling of the different steps still allows to sequentially perform them even in case several iterative loops are present. Moreover, with the exception of very complex Mixed-Integer NonLinear Programming (MINLP) case studies, the current technologies are suitable for the solution of the vast majority of the design problems in the engineering domain.

Although in this chapter a biorefinery case study with uncertain feedstock properties was used for the stepwise discussion of the design procedure, the new scheme represented in Fig. 20.12 can be nevertheless applied to whatever case study subject to aleatory uncertainty, such as variable energy sources or feedstock availability during the day/seasons or according to the geographical location, the variable costs of feedstock and energy duties as well as the fluctuating products demand or market price, etc.

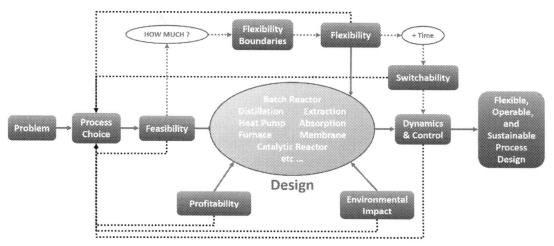

FIG. 20.12 Sustainable and flexible process design procedure.

The same remark applies to the specific process or equipment to be designed. The distillation unit operation was selected as a case study as it is the most spread separation process in chemical industry and resulted particularly suitable to highlight the implications of uncertainties over the entire domain ranging from the feasibility limitations to the control loop design challenges. However, the proposed procedure stays unchanged in case a different process, operation, or equipment, for example power plants, aerospace applications, building sector, supply chain management, etc., needs to be optimally designed. The only exception is the "Design" box itself that is the only one strictly related to the specific system under analysis and whose content is always case specific.

Given these premises, the proposed modifications for each part of the diagram can be resumed to draw some conclusions about the main outcome of this research work.

The "Process Choice" box is the one toward which almost all the other steps can redirect. In fact, increasing the amount of criteria to be satisfied for the optimality conditions, implies that the initial solution can be reconsidered several times and might result difficult, if not sometimes impossible, to achieve. Anyways, this inconvenience was already present in the conventional procedure. The very first part that was considerably revised is the feasibility assessment. In case of flexible operations the qualitative answer "yes/no" is not sufficient to determine whether an operation is feasible or not. Given the absence of uniquely defined operating conditions, the feasibility limitations need an a priori quantification according to the selected operation to provide the upper and lower boundaries for the following flexibility assessment.

As already discussed, the "Design" box is strictly case specific and related to the system under analysis. As regards the example proposed in this research work, particular attention was dedicated to optimal number of equilibrium stages in multistage operations and optimal sequence of operations in case of a distillation train. The obtained results showed that the uncertain operating conditions and the corresponding fluctuations in the energy demand could compromise the profitability of the system in case the economic criterion was the only one taken into account during the design optimization phase. For all these reasons, an original procedure has been proposed to identify the optimal number of stages accounting for flexibility, profitability, and sustainability and the same improvements have been included in the selection of the best column sequencing.

What can be concluded for this phase of the research activity is that interesting results concerning the distillation operation were obtained and that, more importantly, an original tool of general validity able to couple economic and flexibility assessment was conceived. The methodology that combines equipment sizing, operation sequencing, and flexibility along with environmental impact assessment is indeed a relevant outcome that could improve the design of several systems in the most various domains by adapting its different steps to each specific case study.

After that, an outcome of critical importance has been found when extending the analysis to process-intensified solutions. Although they are considered the "best practice" both in the research and in the industrial domains, it was found that their higher profitability and sustainability can be considerably undermined in case of uncertain operating conditions. While in the proximity of nominal operating conditions the intensified equipment, in this case the DWC, fully satisfies the 30% savings in terms of costs and emissions, the performed intensification results in an exponentially increase in the costs/emissions versus flexibility trends for substantial perturbations magnitude due to the fact that the entire system is involved. On the contrary, for the conventional design, an incoming disturbance could be eventually absorbed by the directly involved part of the process only. The process intensification domain showed then to deserve much deeper studies and more caution when drawing conclusions without accounting for eventual flexibility requirements.

As regards the final process dynamic validation, the methodologies conventionally used to assess the optimal control configuration layout need to be integrated with appropriate tests concerning the quantification of the operability boundaries for a wide range of incoming external disturbances. This may be done by means of a dedicated switchability analysis and of the corresponding SI. This indicator was obtained by coupling the results provided by the previous steady-state flexibility assessment performed during the design phase and the system dynamic behavior. As for the control loop design, the switchability assessment can be seen under two following different perspectives:

- Regulation problem: The system should be able to accommodate incoming disturbances without becoming infeasible during the transition from the old to the new manipulated variables steady-state values required to keep the specification setpoint satisfied;
- Servo problem: The required flexibility refers to the output variables, such as process productivity or market demand, and the system should be able to switch from the previous to the new operating conditions to meet the newly specified setpoint.

Once all those criteria are satisfied, the obtained design could be defined as the optimal design satisfying the imposed criteria concerning profitability, sustainability, flexibility, and operability.

Even though the final goal of the newly proposed procedure is to find an optimal design for the system under analysis that meets all the aforementioned requirements, there is no denying it considerably reduced the amount of suitable design solutions.

It could be possible indeed that no solution is able to satisfy all the optimality criteria at the same time. Either way, a design solution needs to be provided. In case such a design could not be achieved, some constraints could be relaxed, and a quasi-optimal solution can be obtained to approach the desired result.

The purpose of this work is to provide a useful procedure and related tools to the design engineer to obtain the optimal design according to his specific needs. However, a unique solution to engineering problems can be seldom attained due to the complexity of the criteria to be satisfied. Therefore, despite the considerable improvements provided by this research work, it can be concluded that it is always up to the skills of the decision maker to find the best compromise between them.

Greek letters

δ: F_{SG} index scale factor
θ : Uncertain variable
ξ: Dimensionless time
ψ: Feasible space

List of acronyms and symbols

ABE/W: Acetone-Butanol-Ethanol/Water
CAPEX: CAPital EXpenses
d: Design variable
DF: Dynamic Flexibility index
DV: Distillate-boilup control structure
DWC: Dividing Wall Column
f_j: System constraints
F_{SG}: Swaney and Grossmann flexibility Index
LV: Reflux-boilup control structure
NRTL: Non-Random Two-Liquid
OPEX: OPerating EXpenses
RCMs: Residue Curve Maps
RI: Resilience Index
SF: Stochastic Flexibility index
SI: Switchability Index
TAC: Total Annualized Costs
x: Liquid composition
y: Vapor composition
z: Control variable

References

Agarwal, H., Renaud, J.E., Preston, E.L., Padmanabhan, D., 2004. Uncertainty quantification using evidence theory in multidisciplinary design optimization. Reliab. Eng. Syst. Saf. 85, 281–294.

Dimitriadis, V.D., Pistikopoulos, E.N., 1995. Flexibility analysis of dynamic systems. Ind. Eng. Chem. Res. 34 (12), 4451–4462.

Di Pretoro, A., Montastruc, L., Manenti, F., Joulia, X., 2020a. Exploiting residue curve maps to assess thermodynamic feasibility boundaries under uncertain operating conditions. Ind. Eng. Chem. Res. 59, 16004–16016.

Di Pretoro, A., Montastruc, L., Manenti, F., Joulia, X., 2020b. Flexibility assessment of a biorefinery distillation train: optimal design under uncertain conditions. Comput. Chem. Eng. 138, 106831.

Di Pretoro, A., Ciranna, F., Fedeli, M., Joulia, X., Montastruc, L., Manenti, F., 2021. A feasible path-based approach for Dividing Wall Column design procedure. Comput. Chem. Eng. 149, 107309.

Di Pretoro, A., Montastruc, L., Joulia, X., Manenti, F., 2020c. Accounting for dynamics in flexible process design: a switchability index. Comput. Chem. Eng., 107149.

Di Pretoro, A., Montastruc, L., Manenti, F., Joulia, X., 2019. Flexibility analysis of a distillation column: indexes comparison and economic assessment. Comput. Chem. Eng. 124, 93–108.

Dortmund Data Bank, www.ddbst.de.

Ezeji, T.C., Qureshi, N., Blaschek, H.P., 1 February 2004. Acetone butanol ethanol (ABE) production from concentrated substrate: reduction in substrate inhibition by fed-batch technique and product inhibition by gas stripping. Appl. Microbiol. Biotechnol. 63 (6), 653–658.

Gadalla, M., Olujić, Ž., Jobson, M., Smith, R., 2006. Estimation and reduction of CO_2 emissions from crude oil distillation units. Energy 31, 2398–2408.

Garcia, V., Päkkilä, J., Ojamo, H., Muurinen, E., Keiski, R.L., 2011. Challenges in biobutanol production: how to improve the efficiency? Renew. Sustain. Energy Rev. 15, 964–980.

Groot, W.J., van der Lans, R.G.J.M., Luyben, K.Ch.A.M, 1989. Batch and continuous butanol fermentations with free cells: integration with product recovery by gas-stripping. Appl. Microbiol. Biotechnol 32 (3), 305–308.

Guthrie, K.M., 1969. Capital cost estimating. Chem. Eng. 76 (3), 114–142.

Guthrie, K.M., 1974. Process Plant Estimating, Evaluation, and Control. Craftsman Book Company of America, CA.

Horizon 2020 https://ec.europa.eu/programmes/horizon2020/sites/horizon2020/files/H2020_IT_KI0213413ITN.pdf, (Accessed February 1, 2021).

IEA Bioenergy Annual Report 2018 https://www.ieabioenergy.com/wp-content/uploads/2019/04/IEA-Bioenergy-Annual-Report-2018.pdf, (Accessed February 1, 2021).

IEA Renewables 2019 https://www.iea.org/reports/renewables-2019, (Accessed February 1, 2021).

Kiss, A.A., 2013. Advanced Distillation Technologies: Design, Control and Applications, first ed. Wiley, Chichester, West Sussex, United Kingdom.

Navarrete, P.F., Cole, W.C., 2001. Planning, Estimating, and Control of Chemical Construction Projects, second ed. CRC Press, FL.

Oberkampf, W., Helton, J., Sentz, K., 2012. Mathematical representation of uncertainty, 19th American Institute of Aeronautics and Astronautics Applied Aerodynamics Conference. CA, USA. American Institute of Aeronautics and Astronautics.

Oberkampf, W.L., DeLand, S.M., Rutherford, B.M., Diegert, K.V., Alvin, K.F., 1999. New methodology for the estimation of total uncertainty in computational simulation, Collection of Technical Papers - AIAA/ASME/ASCE/AHS/ASC Structures, Structural Dynamics and Materials Conference, 4, 3061–3083.

Oberkampf, W.L., Diegert, K.V., Alvin, K.F., Rutherford, B.M., 1998. Variability, uncertainty, and error in computational simulation. American Society of Mechanical Engineers 357, 259–272 Heat Transfer Division, (Publication) HTD.

Petlyuk, F.B., 1965. Thermodynamically optimal method for separating multicomponent mixtures. Int. Chem. Eng. 5, 555–561.

Petlyuk, F.B., 2004. Distillation Theory and its Application to Optimal Design of Separation Units. Cambridge University Press, Cambridge, UK.

Pistikopoulos, E.N., Mazzuchi, T.A., 1990. A novel flexibility analysis approach for processes with stochastic parameters. Comput. Chem. Eng. 14, 991–1000.

Reay, D., Ramshaw, C., Harvey, A., 2011. Process Intensification: Engineering for Efficiency, Sustainability and Flexibility. Butterworth-Heinemann, Oxford, UK.

Renon, H., Prausnitz, J.M., 1968. Local compositions in thermodynamic excess functions for liquid mixtures. AIChE J 14 (1), 135–144 S.

Ryskamp, C.J., 1980New strategy improves dual composition control59. Hydrocarbon processing, pp. 51–59 6.

Saboo, A.K., Morari, M., Woodcock, D., 1985. Design of resilient processing plants—VIII. A resilience index for heat-exchanger networks. Chem. Eng. Sci. 40, 1553–1565.

Skogestad, S., Lundström, P., Jacobsen, E.W., 1990. Selecting the best distillation control configuration. AIChE J. 36, 753–764. doi:10.1002/aic.690360512 doi:.

Stankiewicz, A.I., Moulijn, J.A., 2000. Process intensification: transforming chemical engineering. Chem. Eng. Prog. 96, 22–34.

Swaney, R.E., Grossmann, I.E., 1985. An index for operational flexibility in chemical process design. Part I: Formulation and theory. AIChE J 31, 621–630.

Ulrich, G.D., 1984. A Guide to Chemical Engineering Process Design and Economics. Wiley, New York, USA.

Charles Weizmann, GB application 191504845, Improvements in the Bacterial Fermentation of Carbohydrates and in Bacterial Cultures for the Same, published 1919-03-06, assigned to Charles Weizmann.

White, V., Perkins, J.D., Espie, D.M., 1996. Switchability analysis. Comput. Chem. Eng. 20, 469–474.

Index

Page numbers followed by "*f*" and "*t*" indicate, figures and tables respectively.

A

ABE fermentation, 395
Accelerated data collection and analysis, 350
Acoustic sensor working principle, 326*f*
Acoustics sensing methods, 326
Additive ratio assessment (ARAS), 62
Aggregate the alternative evaluations, 154–155
American Institute of Chemical Engineers Sustainability Index (AIChE SI), 1, 3
Analytic hierarchy process (AHP). *See also* Sustainability assessment; Sustainability science
  application, 6
  method, 198–199
  steps, 55, 56
Analytic network process (ANP), application, 6–7, 39, 62
Architectural design, in integrate sustainability objectives, 78–80
Architecture, engineering, and construction (AEC) industry, 347
Asian Development Bank midterm 2020 report, 19
Authoritative sustainability indexes, 1
AVEVA DynSim process simulator, 412

B

Benchmarking activities, in firm, 88
Biodiesel and ethanol assessment, 130
Biofuels, 130
Biorefinery distillation, 395
Bottom-up approach, in sustainability assessment, 51
Boustead model, 3–4
Building and Construction Authority (BCA), 347
Bus rapid transit (BRT) systems, 124

C

Carbon Disclosure Project (CDP), 82
Chemical processes, 169
Chilled water air conditioning, 217
Chlorination of methane, 182*f*
Circular economy, in sustainability assessment, 31
Civil Engineering, 322
Clean Development Mechanism (CDM), 29–30
Clostridium acetobutylicum, 395
Coaxial optical fiber sensor, 331*f*

Complex proportional assessment (COPRAS)
  application, 62
  experts weight for, 66*t*
  final score and ranking from, 65*t*
  ranks for different technologies using, 65*f*
  steps, 61. *See also* (Sustainability assessment)
Composite Sustainability Indices (CSI), 170
Confrontation matrix, 354*f*
Construction sector, LCA study, 113. *See also* Life cycle assessment (LCA), for environmental impacts
Contactless methods, 332–333
Contextual factors, in industrial sustainability PMS, 92
Contingent valuation method, steps, 36*f*, 38*f*
Conventional process design procedure, 392*f*
Core ISPMS, indicators in, 93, 97
Corporate social responsibilities (CSR) activities, in Indonesian companies, 19–20
Cost-benefit analysis (CBA) technique, 36, 36*f*
COVID-19 pandemic and rate of poverty, 20
Cradle-to-cradle integrated process, 187, 189*f*
Creating shared value (CSV), 81
Cross-efficiency evaluation, 248
CSI methodology, 170

D

Dairy farms, 208*f*
Dashboard of Sustainability (DoS), 135–136, 141
  communication strategies of, 138
Data collection, 228
Data envelopment analysis (DEA), 243, 296
  application, 6. *See also* Sustainability science
  models, 308–311
Decision makers and stakeholders, choice made, 5*f*
Decision-making trial and evaluation laboratory (DEMATEL), 6–7
Decoupling effect, in sustainability assessment, 76
Defective production waste, 351
Deterministic flexibility index, 401
Digital brain, 354
Digital brain platform, 355*f*
Digital dashboard, 141
Digital solutions, 349, 350
Distillation trains, 402*f*
Distributed optical fiber sensors, 330–331
DoS cartogram, 142

421

E

Ecological footprint (EF), 217
Ecological footprints, sustainability assessment, 31
Economic sustainability, 28
ELECTRE methods, 199–200, 207
ELECTRE TRI variant, 200
Electricity-generation technologies, for sustainability, 48–49
Electric vehicles (EVs), usage, 114
Electromagnetic sensing methods, 328
Electromagnetic sensor, 334
Electromagnetic spectrum, 329f
Electromagnetic waves, 328–329
Embodied energy, definition, 29
Endpoint method, in life cycle assessment, 3–4
Enel's materiality matrix, 79f
Energy impact index, 181–182
Energy Index, 173
Energy sector, LCA study, 114. *See also* Life cycle assessment (LCA), for environmental impacts
Energy system sustainability assessment, common indicators, 51t–52t
Environmental and sustainability rating systems (ESRS), usage, 39
Environmental issues in Indonesia, 14
Environmental management accounting (EMA), 73
Environmental management system (EMS), 268
Environmental metrics, objective, 29
Environmental pollution, 323–324
Environmental problems, in world, 27
Environmental Product Declaration (EPD), 346
Environmental Product Declaration (EPD), LCA methodology in, 111–112
Environment conservation, 29
Environment index, 173
European Commission life cycle assessment forum, 122
European Committee for Standardization (CEN), standards developed, 111–112
European green deal, 295
Excessive production waste, 354
External stakeholders
　role, 88
　usefulness, 90

F

Fiber optic sensors, 330–331
Financial Service Authority (FSA), 13
Firm performance
　adoption of PMS, 88–89
　assessment, 88
Fishbone diagram method, 6–7. *See also* Sustainability science
Flexibility index value, 409
Food and agriculture, LCA study, 114. *See also* Life cycle assessment (LCA), for environmental impacts
Food pollution, 337
Four Pillars of SDGs in Indonesia, 15f
Full ISPMS, indicators in, 93, 97

G

Gamma-rays sensing, 332f
Georgescu-Roegen hypotheses, 265
Global engagement services (GES), 76
Global reporting initiative (GRI), 267
　definition on sustainability report, 15–16
　standards, 76
Global sustainable development, 47
Global warming potential, 109–110
Goal programming model, for sustainability-oriented design and optimization, 7
"Good Health and Well Being," in Sustainable Development Goals, 22, 23
Grating-based and interferometer sensors, 330–331
Greenhouse gas (GHG), 169, 203
Greenhouse gas Protocol Corporate Standard, in company's emissions assessment, 80–81
Gross domestic product (GDP), 14, 31, 126–127, 265

H

Health and education in Indonesia, concern, 22
Hedonic pricing method (HPM), importance, 38
Hera's approach to create and measure shared value, 81b
Hydrogenation process, 187f, 188f

I

ILCD 2011 indicators, 128–129
Index development, for sustainability PMS, 91–92
Indicators
　in PMS
　　selection of, 92
　　usability and manageability, 91, 92
　　selection, what-why/how-whom framework, 53f
Indonesian companies, sustainable development goals in, 13, 23
Indonesian National Development Plan, 15
Indonesia Stock Exchange (IDX), sectors, 17
Industrial sector
　LCA purposes, 113. *See also* (Life cycle assessment (LCA), for environmental impacts)
　SDGs disclosures on, 17
Industrial sustainability performance measurement system (ISPMSs), 87
　analysis, 93
　completeness and balance on, 90
　context of application, 92
　features in development, 89, 89f

Index

reason for, 88
scalable framework for, 93, 94t–96t
selection of indicators, 92
usability and manageability, 88, 91
ways for measurement, 88
Industry process design procedure, 392
Interactive work processes, 350
Intergovernmental Panel on Climate Change, 109–110
Internal stakeholders, usefulness, 90
International Organization for Standardization (ISO), 31
  for LCA studies, 106
International regulatory reference for LCA studies, 106
International standardization organization (ISO), 119–120
International standards and documents, for LCA methodology, 112t
Inventory waste, 353
Irradiated materials, 331
IR sensors, 329

K

Kyoto Protocol, consequences, 29–30

L

LCSD implementation, 143
Lean Integrated Management System for Sustainability Improvement (LIMSSI) model, 272, 273f
  policy, 274
Lean manufacturing systems (LMSs), 265–266, 270
Lean thinking, 350, 357
Libelium Waspmote sensors, 336
Life cycle assessment (LCA), 119, 128, 153, 235
  application of, 125
  assessment of environmental sustainability, 120
  consequential, 122
  distribution of, 127
  environmental burdens, 121
  interpretation step, 123–124
  iterative process, 127
  limitations, 131
  methodologies
    characteristics of, 106
    data collection, elaboration, and application, 108
    definition, 105–106
    goal and scope, 107
    impact assessment, 109
    interpretation of results, 110
    phases of, 107f
  multicriteria methods integration, 120
  quantitative and qualitative information, 123
  sustainability approach, 128
  system boundaries of, 126f
  disadvantages, 115
  international standards, 111

Life cycle costing (LCC), 3, 4, 138, 295–296. *See also* Sustainability science
Life cycle impact assessment (LCIA), 157, 297, 302f
Life cycle impact assessment phases, 34
Life cycle inventory, 235–237
Life cycle inventory analysis (LCIA), 3–4, 33
Life cycle sustainability assessment (LCSA), 119, 135
Life cycle sustainability dashboard (LCSD), 135
  advantages and future perspectives, 145f
Life cycle sustainability triangle (LCST), 138
Life cycle thinking concept, scheme, 106f
"Life on Land", in Sustainable Development Goals, 21

M

Magnetic sensing methods (Hall-effect sensor), 327
Magnetic sensors (Hall-effect sensor), 334
Material flow analysis (MFA), 4, 5f. *See also* Sustainability science
Material intensity per service unit (MIPS) method, 39
Materiality analysis, importance, 79b
Materiality matrix, in sustainability assessment, 78
Material recovery facility (MRF), 121–122
Micro, Small, and Medium Enterprises (MSME), 215
Microstrip patch antennas, 334
Midpoint method, in life cycle assessment, 3–4
Midpoints and endpoints, 244f
Millennium development goals (MDGs), 13, 142
Ministry of National Development Planning, 13
Mixed integer nonlinear programing (MINLP) model, 7. *See also* Sustainability science
MNNIT Industrial Complex, 218
Monte Carlo approach, in LCA study, 111. *See also* Life cycle assessment (LCA), for environmental impacts
Motion waste, 354
Multiattribute decision making (MADM), 197
Multiattribute value theory (MAVT), 198–199
Multicriteria analysis (MCA) approach, 38
Multicriteria Decision Analysis (MCDA), 154, 197
  evaluation methods, 198
Multicriteria decision analysis (MCDA) method, 49, 53, 54, 54f
  analytic hierarchy process method, 55, 56
  complex proportional assessment, 61, 62, 64, 66
  multiattribute value theory, steps, 55
  preference ranking organization method for enrichment of evaluations, steps, 59, 59f
  sensitivity analysis for, 66
  technique for order preference by similarity to ideal solution, steps, 57
  VlseKriterijumska Optimizacija I Kompromisno Resenje, steps, 60
  weight assignment, 66
  weighted product method, steps, 56
  weighted sum method, 55

Multicriteria Decision Making (MCDM), 153, 197
  basic process, 155f
  methods classification, 155, 156f
Multicriteria decision-making (MCDM) models, 2, 5–6
Multifunctional processes, in LCA, 109
Multiobjective evolutionary optimization model, for sustainability-oriented design and optimization, 7
Municipal solid waste management, 125–127
Municipal wastewater management, 128

N

Nanotechnology, LCA studies, 114. *See also* Life cycle assessment (LCA), for environmental impacts
Net-zero energy (NZE), 128
Neyman allocation, 229–230
"No poverty" information, in Sustainable Development Goals, 23
Normalization, 243

O

Occupational health and safety management system (OHSMS), 269
Optical backscatter reflectometer sensors, 335
Optical fibers, 330, 334–335
Optical fiber technology, 330
Overall equipment effectiveness (OEE), 276
Over-processing waste, 351

P

Paris Climate Agreement in 2015, 48
"Peace, Justice, and Strong Institutions," in Sustainable Development Goals, 22
Performance measurement and management (PMM) literature, 76
Performance measurement systems (PMS), 73, 88, 89. *See also* Industrial sustainability performance measurement system
Perturbation analysis in LCA, purposes, 110–111. *See also* Life cycle assessment (LCA), for environmental impacts
PESTEL technique, 353
Philadelphia Energy Solutions (PES), 175–176
Photovoltaic (PV) system, 215
Piezoelectric material, 326
Poultry production systems, 123
Poverty reduction in Indonesia, efforts, 19, 20
PPVC systems, 357
Prefabricated prefinished volumetric construction (PPVC), 350
Preference Ranking Organization Method for Enrichment of Evaluations (PROMETHEE), steps, 59, 59f. *See also* Sustainability assessment
Presidential Decree no. 59/2017, 13
Process intensification, 406, 407
Process systems engineering, 170
Product category rule (PCR), 111–112

Product-process" matrix, 275
Pull-based system, 360

Q

'Quality education,' information, 21

R

Radio and microwave waves, 328–329
Rainwater harvesting system, 220
Reflux-boilup control structure (LV), 412
Remote sensing revolution, 325
Renewable resources, 47
Residue curves, 396
"Responsible Consumption and Production," in Sustainable Development Goals, 23
RETScreen 4 software, 216
Risk assessment, 123
Risk assessment index, 175–176
Robotic sensing units, 332–333
Robust multiobjective optimization model, for sustainability-oriented design and optimization, 7
Robustness analysis, 211
Rooftop rainwater harvesting system (RHS), 217
Rooftop solar photovoltaic system, 216
Rooftop solar PV system, 218

S

Satellite technology, 336
Scenario's analysis, for LCA study, 111. *See also* Life cycle assessment (LCA), for environmental impacts
Sensing technology, 333
Sensitivity analysis in LCA, importance, 110–111. *See also* Life cycle assessment (LCA), for environmental impacts
Sensor technologies, 332
Sicilian Town Master Plan, 142
SimaPro software, 241–243
Small and medium-sized enterprises (SME), 20, 73
  performance measurement, 92–93
Social impact assessment (SIA), role, 33f, 37, 38f
Social life cycle assessment (SLCA), 3, 4. *See also* Sustainability science
Social responsibility management system (SRMS), 268
Society of Environmental Toxicology and Chemistry (SETAC), 105–106
Socioeconomic issues, in world, 27, 28
Softwares and toolboxes, in LCI and LCA assessment process, 3–4. *See also* Sustainability science
Solar daylight harvesting, 217
Solar day-lighting System, 220
Solar power, 128
Sounding technique, 327f

Sound penetration, 326
Spanish agri-food system, using DEA-LCA approach, 6
Stakeholder
  map
    in importance of stakeholders assessment, 77
    in sustainability measurement process, 77
    typology, 78f
Steady-state economy, concept, 30
Stepwise weight assessment ratio analysis (SWARA), 62
Stocks, 354
Strategic environmental assessment, 142
Strategic environmental assessment (SEA), in environmental impact assessment, 35
Sustainability accounting standards board, 146
Sustainability assessment
  circular economy, 31
  comparison of methods, 39
  COPRAS method application, 62
  ecological footprints, 31
  and energy system, 48
  hedonic pricing method, 38
  indicators, 50
  multicriteria decision analysis method, 53
  need for, 29
    circular economy, 31
    ecological footprints, 31
    steady-state economy, 30
  objective, 28, 29
  pillars, 28f
  principle, 48
  steady-state economy, concept, 30
  sustainability indicators, 50
  various methods of, 32, 32f
    analytic network process, 39
    contingent valuation method, 37
    cost-benefit analysis, 36
    environmental and sustainability rating systems, 39
    life-cycle assessment, 32–33
    material intensity per service unit, 39
    multicriteria analysis, 38
    social impact assessment, 37
    socioeconomic impact assessment, 34
    strategic environmental assessment, 35
    travel cost analysis, 37
Sustainability balanced scorecard (SBSC), 78–80
Sustainability concept, 49f
Sustainability evaluation, 119
Sustainability index value ranges, 50f
Sustainability measurement
  in current business landscape, 71
  future of, 82
  literature on, 73–74
  methods and tools
    core issues and stakeholder mapping, 77
    reporting guidelines, 81
    sustainability performance measurement system, 78
  research, evolution of, 72
    assurance of sustainability reporting, 74
    carbon accounting, 74
    critical environmental accounting, 73
    determinants of sustainability disclosure, 73
    diffusion of sustainability standards, 74
    emerging clusters, 74
    lack of homogeneity, 74
    literature review on, 72, 72f, 75t
    from performance measurement literature, 76
    sustainability disclosure and performance, 73
    sustainability metrics, 73
    sustainability performance measurement system, 78
    sustainable operations and supply chain management, 73–74
Sustainability Measurement and management laboratory (SuMM Lab), 82b
Sustainability-oriented prioritization, 308
Sustainability performance measurement system, 78. *See also* Sustainability measurement
  industrial sustainability PMS, 87
    completeness and balance on, 90
    context of application, 92
    features in development, 89, 89f
    reason for measurement, 88
    scalable framework for, 93, 94t–96t
    selection of indicators, 92
    usability, 88
    usability and manageability, 91
    ways for measurement, 88
Sustainability reports
  in Indonesia, 15
  SDGs disclosures scores, 18f, 19t. *See also* (Sustainable Development Goals (SDGs))
Sustainability science
  assessment and analysis, 1, 2
    material flow analysis, 4, 5f
    sustainability analysis tools, 3
    sustainability metrics/indicators, 2
  design and optimization, 7
  enhancement and improvement, 6
  methods introduced, 8
  ranking and prioritization, 4
  studies, 1
Sustainability tools, 145
Sustainable criteria and indicators, in COPRAS method, 63t

Sustainable Development Goals (SDGs), 1, 14. *See also* Sustainability science
   global indicators, 3
   in Indonesia, business contributions to, 13
      goals, disclosures on, 18
      industrial sector, disclosures on, 17
      materials and methods, 16
      sustainability reports, 15
   targets and goals, according to WHO, 3
Sustainable system, definition, 48
Sustainable systems for the Industrial Complex, 216f
SWOT analysis, 346, 347
   matrix of, 349f
SWOT/PESTEL analysis, 346

T

Technique for Order Preference by Similarity to Ideal Solution (TOPSIS), steps, 57. *See also* Sustainability assessment
Territorial LCA analysis, 115. *See also* Life cycle assessment (LCA), for environmental impacts
Thai Eco Scarcity methodology evaluation, 130
Three-step sustainability-oriented LCA, 297
Top-down approach, in sustainability assessment, 51
Total Annualized Costs (TACs), 403
Toxic Release Inventory (TRI), 122
Transformity, definition, 29
Transportation waste, 353
Travel cost analysis (TCA) method, importance, 37
Triple bottom line of sustainability, 2f
Turbo ventilators, 217, 222

U

Uncertainty analysis, for LCA study, 111. *See also* Life cycle assessment (LCA), for environmental impacts
United Nations, sustainable goals declared, 48
United Nations Conference on Environment and Development (UNCED), 29–30
United Nations Conference on Human Environment, 1972, 29–30
United Nations Framework Convention on Climate Change (UNFCCC), 1992, 29–30
United States Environmental Protection Agency (USEPA), 1, 3
Unused talent waste, 353
Urban transport systems, 124

V

Value stream mapping (VSM), 270
VIP analysis, 159

W

Waiting waste, 352
WAR algorithm, 170, 174–175
Waste management, LCA in, 114. *See also* Life cycle assessment (LCA), for environmental impacts
Waste Reduction algorithm, 174–175
Waste stream scheme, 191f
Wastewater treatment, 127
Wastewater Treatment Plant, 144
Weighted aggregates sum product assessment (WASPAS), 62
Weighted normalized decision matrix, 64t
Weighted product method (WPM), steps, 56. *See also* Sustainability assessment
Weighted Sum Method, 198–199
Weighted sum method (WSM), steps, 55. *See also* Sustainability assessment
Weighting method, 243
World, challenges faced, 00017#0024, 27
World Bank, report on Indonesia economic growth, 14

X

X-ray and gamma radiations, 331
X-ray penetration capabilities, 335

Z

ZigBee sensors, 335
ZigBee wireless system, 335–336

Printed in the United States
by Baker & Taylor Publisher Services